# Climatology: Climate Indices, Models, Forecasting and Observations

# Climatology: Climate Indices, Models, Forecasting and Observations

Edited by **Vivian Moritz**

SYRAWOOD
PUBLISHING HOUSE
New York

Published by Syrawood Publishing House,
750 Third Avenue, 9th Floor,
New York, NY 10017, USA
www.syrawoodpublishinghouse.com

**Climatology: Climate Indices, Models, Forecasting and Observations**
Edited by Vivian Moritz



International Standard Book Number: 978-1-68286-271-1 (Hardback)



Printed in China.

# Contents


| | | |
|---|---|---|
| | **Preface** | IX |
| Chapter 1 | **Chemical weather forecasting: a new concept of integrated modelling**<br>A. Baklanov | 1 |
| Chapter 2 | **Multi-methodical realisation of Austrian climate maps for 1971–2000**<br>J. Hiebl, S. Reisenhofer, I. Auer, R. Böhm and W. Schöner | 6 |
| Chapter 3 | **DRIAS: a step toward Climate Services in France**<br>J. Lémond, Ph. Dandin, S. Planton, R. Vautard, C. Pagé, M. Déqué, L. Franchistéguy, S. Geindre, M. Kerdoncuff, L. Li, J. M. Moisselin, T. Noël and Y. M. Tourre | 14 |
| Chapter 4 | **Sensitivity of a long-range numerical weather forecast model to small changes of model parameters**<br>M. B. Gavrilov, G. R. Jovanović and Z. Janjić | 22 |
| Chapter 5 | **Comparative verification of different nowcasting systems to support optimisation of thunderstorm warnings**<br>K. Wapler, M. Goeber and S. Trepte | 28 |
| Chapter 6 | **First outcomes from the CNR-ISAC monthly forecasting system**<br>D. Mastrangelo, P. Malguzzi, C. Rendina, O. Drofa and A. Buzzi | 35 |
| Chapter 7 | **The new WMO RA VI Regional Climate Centre Node on Climate Monitoring**<br>J. Rapp, H. Nitsche and P. Bissolli | 41 |
| Chapter 8 | **Determining the accuracy of gridded climate data and how this varies with observing-network density**<br>T. P. Legg | 46 |
| Chapter 9 | **Quality control procedures at Euskalmet data center**<br>R. Hernández, M. Maruri, K. Otxoa de Alda, J. Egaña and S. Gaztelumendi | 50 |
| Chapter 10 | **On the skill of various ensemble spread estimators for probabilistic short range wind forecasting**<br>A. Kann | 56 |
| Chapter 11 | **Climate indicators for Italy: calculation and dissemination**<br>F. Desiato, G. Fioravanti, P. Fraschetti, W. Perconti and A. Toreti | 62 |
| Chapter 12 | **Processing and analysing an ensemble of climate projections for the joint research project KLIWAS**<br>F. Imbery, S. Plagemann and J. Namyslo | 66 |

Chapter 13 **A cloud-to-ground lightning climatology for north-eastern Italy** **74**
L. Feudale, A. Manzato and S. Micheletti

Chapter 14 **An empirical method for estimating probability density functions of gridded daily minimum and maximum temperature** **82**
C. Lussana

Chapter 15 **Multivariate testing of spatio-temporal consistence of daily precipitation records** **88**
H. Mächel and A. Kapala

Chapter 16 **Evaluating multi-scale precipitation forecasts using high resolution analysis** **94**
C. Wittmann, T. Haiden and A. Kann

Chapter 17 **The French contribution to the Global Climate Observing System** **104**
R. Juvanon du Vachat

Chapter 18 **The future climate characteristics of the Carpathian Basin based on a regional climate model mini-ensemble** **108**
I. Krüzselyi, J. Bartholy, A. Horányi, I. Pieczka, R. Pongrácz, P. Szabó, G. Szépszó and Cs. Torma

Chapter 19 **A data portal for regional climatic trend analysis in a Peruvian High Andes region** **113**
M. Schwarb, D. Acuña, Th. Konzelmann, M. Rohrer, N. Salzmann, B. Serpa Lopez and E. Silvestre

Chapter 20 **Improving short-term forecasting during ramp events by means of Regime-Switching Artificial Neural Networks** **121**
C. Gallego, A. Costa and A. Cuerva

Chapter 21 **A quantitative evaluation of the high resolution HARMONIE model for critical weather phenomena** **125**
E. V. van der Plas, B. Wichers Schreur and K. Kok

Chapter 22 **Statistical processing of forecasts for hydrological ensemble prediction: a comparative study of different bias correction strategies** **132**
I. Zalachori, M.-H. Ramos, R. Garçon, T. Mathevet and J. Gailhard

Chapter 23 **Observations and WRF simulations of fog events at the Spanish Northern Plateau** **139**
C. Román-Cascón, C. Yagüe, M. Sastre, G. Maqueda, F. Salamanca and S. Viana

Chapter 24 **Evaluating meteorological climate model inputs to improve coastal hydrodynamic studies** **147**
D. Bellafiore, E. Bucchignani, S. Gualdi, S. Carniel, V. Djurdjeviæ and G. Umgiesser

Chapter 25 **Statistical analysis of inter-arrival times of rainfall events for Italian Sub-Alpine and Mediterranean areas** **152**
C. Agnese, G. Baiamonte, C. Cammalleri, D. Cat Berro, S. Ferraris and L. Mercalli

Chapter 26 **The influence of the new ECMWF Ensemble Prediction System resolution on wind power forecast accuracy and uncertainty estimation** **159**
S. Alessandrini, S. Sperati and P. Pinson

Chapter 27 **Surface high-resolution temperature forecast in southern Italy** **164**
S. Federico, E. Avolio, F. Fusto, R. Niccoli, and C. Bellecci

Chapter 28 **Quality control of 10-min soil temperatures data at RMI** 171
C. Bertrand, L. González Sotelino and M. Journée

Chapter 29 **Satellite-based climate information within the WMO RA VI Regional Climate Centre on Climate Monitoring** 179
A. Obregón, H. Nitsche, M. Körber, A. Kreis, P. Bissolli, K. Friedrich and S. Rösner

Chapter 30 **Digital Climate Atlas of the Carpathian Region** 188
I. Antolović, V. Mihajlović, D. Rančić, D. Mihić and V. Djurdjević

Chapter 31 **Modelling static 3-D spatial background error covariances – the effect of vertical and horizontal transform order** 193
M. A. Wlasak and M. J. P. Cullen

Chapter 32 **WegenerNet climate station network region Feldbach, Austria: network structure, processing system, and example results** 198
T. Kabas, A. Leuprecht, C. Bichler and G. Kirchengast

Chapter 33 **Diagnostic study of the influence of lateral boundary conditions for the REMO RCM simulations over the Carpathian Basin** 204
G. Szépszó

Chapter 34 **Observations and modelling of 1/*f*-noise in weather and climate** 212
R. Blender, X. Zhu and K. Fraedrich

Chapter 35 **The historical pathway towards more accurate homogenisation** 216
P. Domonkos, V. Venema, I. Auer, O. Mestre and M. Brunetti

Chapter 36 **National Climate Observing System of Switzerland (GCOS Switzerland)** 224
G. Seiz and N. Foppa

Chapter 37 **Sunshine duration climate maps of Belgium and Luxembourg based on Meteosat and in-situ observations** 232
M. Journée, C. Demain and C. Bertrand

Chapter 38 **Quality control of 10-min air temperature data at RMI** 237
C. Bertrand, L. Gonzalez Sotelino and M. Journée

Chapter 39 **INCA-CE: a Central European initiative in nowcasting severe weather and its applications** 242
A. Kann, G. Pistotnik and B. Bica

Chapter 40 **On the predictability of outliers in ensemble forecasts** 251
S. Siegert, J. Bröcker and H. Kantz

**Permissions**

**List of Contributors**

# Preface

Climatology falls under the umbrella of atmospheric science. It is primarily used for weather forecasting apart from keeping a record of the weather dynamics. This book is a collective contribution of internationally renowned climatologists and scientists and is bound to provide significant insights into weather forecasting models, techniques and technological advances, observation tools, etc. It aims to serve as a resource guide for students and experts alike and contribute to the growth of the discipline. For all readers who are interested in pursuing research or course in climatology, the case studies included in this book will provide innovative topics which can be taken up for research.

The researches compiled throughout the book are authentic and of high quality, combining several disciplines and from very diverse regions from around the world. Drawing on the contributions of many researchers from diverse countries, the book's objective is to provide the readers with the latest achievements in the area of research. This book will surely be a source of knowledge to all interested and researching the field.

In the end, I would like to express my deep sense of gratitude to all the authors for meeting the set deadlines in completing and submitting their research chapters. I would also like to thank the publisher for the support offered to us throughout the course of the book. Finally, I extend my sincere thanks to my family for being a constant source of inspiration and encouragement.

**Editor**

# 1

# Chemical weather forecasting: a new concept of integrated modelling

**A. Baklanov**

Danish Meteorological Institute, Lyngbyvej 100, 2100 Copenhagen, Denmark

**Abstract.** During the last decade a new field of atmospheric modelling – the chemical weather forecasting (CWF) – is quickly developing and growing. However, in the most of the current studies and publications, this field is considered in a simplified concept of the off-line running chemical transport models with operational numerical weather prediction (NWP) data as a driver. A new concept and methodology considering the chemical weather as two-way interacting meteorological weather and chemical composition of the atmosphere is suggested and discussed. The on-line integration of mesometeorological models and atmospheric aerosol and chemical transport models gives a possibility to utilize all meteorological 3-D fields in the chemical transport model at each time step and to consider feedbacks of air pollution (e.g. urban aerosols) on meteorological processes/climate forcing and then on the atmospheric chemical composition. This very promising way for future atmospheric simulation systems (as a part of and a step to Earth System Modelling) will lead to a new generation of models for meteorological, environmental and chemical weather forecasting. The methodology how to realise the suggested integrated CWF concept is demonstrated on the example of the European Enviro-HIRLAM integrated system. The importance of different feedback mechanisms for CWF is also discussed in the paper.

## 1 Introduction

During the last decade a new field of atmospheric modelling – the chemical weather forecasting (CWF) – is quickly developing and growing (Lawrence et al., 2005). This was possible mostly due to the quickly growing supercomputer capability and operationally available high-resolution numerical weather prediction (NWP) data for atmospheric chemical transport models (ACTMs). However, in the most of the current systems, studies and publications this new direction is considered in a simplified concept. It includes only operational air quality forecast for the main pollutants significant for health effects and uses numerical ACTMs with operational NWP data as a driver (see e.g. the COST Action ES0602: Towards a European Network on Chemical Weather Forecasting and Information Systems, website: http://www.chemicalweather.eu/, leader Prof. J. Kukkonen).

Such an approach has limitations due to the off-line way of coupling the ACTMs with NWP or mesometeorological models (MMMs) (which are running fully independently and NWP does not get any benefits from the ACTM) without a possibility to consider any feedback mechanisms. Many experimental studies (Rosenfeld, 2000; Bellouin et al., 2005; Ramanathan and Carmichael, 2008) and numerical research simulations (Jacobson, 2002; Grell et al., 2004; Schulz et al., 2006) show that atmospheric processes (meteorological weather, including the precipitation, thunderstorms, radiation budget, cloud processes and planetary boundary-layer (PBL) structure) depend on concentrations of chemical components in the atmosphere.

Chemical species influencing weather and atmospheric processes include greenhouse gases which warm near-surface air and aerosols such as sea salt, dust, primary and secondary particles of anthropogenic and natural origin. Some aerosol particle components (black carbon, iron, aluminium, polycyclic and nitrated aromatic compounds) warm the air by absorbing solar and thermal-infrared radiation, while others (water, sulphate, nitrate, most of organic compounds) cool the air by backscattering incident short-wave radiation to space. The aerosols, beside the above mentioned direct effect (via radiation), include many other different mechanisms of aerosol-chemistry-cloud-radiation-PBL-emission interactions, in particular the semi-direct, first and second indirect effects.

Therefore ACTMs have to be run together with NWP models on the same computational grid and with the same time step using online coupling and considering two-way interaction between the meteorological processes, from one side, and chemical transformation and aerosol dynamics, from the other side.

## 2 Concept and methodology for integrated CWF modeling

Proceeding from the above mentioned limitations, a new concept and methodology considering the chemical weather as two-way interacting physical and chemical weather is suggested. The CWF should include not only health-effecting pollutants (air quality components) but also green-house gases and aerosols effecting climate, meteorological processes, etc. Such a concept of CWF requests a strategy of a new generation integrated meteorology and ACT modelling systems for predicting atmospheric composition, meteorology and climate change. The on-line integration of meteorological or NWP models and atmospheric aerosol and chemical transport models gives a possibility to utilise all meteorological 3-D fields in ACTM at each time step and to consider feedbacks of atmospheric aerosols on meteorological processes and climate forcing, and further on the chemical composition (as a chain of dependent nonlinear processes). This promising way for future atmospheric simulation systems will lead to a new generation of integrated models for meteorological, environmental and chemical weather forecasting.

The current COST728 Action "Enhancing meso-scale meteorological modelling capabilities for air pollution and dispersion applications" (http://www.cost728.org, leader Prof. R. Sokhi) addresses key issues concerning the development of meso-scale models for atmospheric environmental problems and, in particular, it encourages the advancement of science in terms of integration methodologies and strategies in Europe. The final integration strategy will not be focused around any particular model, instead it will be possible to consider an open integrated system with a fixed architecture (module interface structure) and with a possibility of incorporating different ACTMs and MMMs/NWP models. Such a strategy may only be realised through jointly agreed specifications of module structure for easy-to-use interfacing and integration.

The overall aim of COST728 working group 2 (WG2), "Integrated systems of MMM and ACTM: strategy, interfaces and module unification", is to identify the requirements for the unification of MMM and ACTM modules and to propose recommendations for a European strategy for integrated meso-scale modelling capabilities. The first report of WG2 (COST-WMO, 2007) compiles the existing state-of-the-art methodologies, approaches, models and practices for building integrated (offline and online) meso-scale systems in different, mostly European, countries. The report also includes an overview and a summary of the existing integrated models and their characteristics as they are presently used. The model contributions were compiled using COST member contributions, each focusing on national model systems.

The modern strategy for integrating MMMs and ACTMs is suggested to consider air quality modelling as a combination of (at least) the following factors: air pollution, regional/urban climate/meteorological conditions and population exposure. This combination is reasonable due to the following facts: meteorology is the main source of uncertainty in air pollution and emergency preparedness models; meteorological and pollution components have complex and combined effects on human health (e.g., the Paris deadly heat wave in July–August 2003 was caused by different combined risk factors: high temperature, ozone concentrations, etc.); pollutants, especially aerosols, influence climate forcing and meteorological events (radiation, precipitation, thunderstorms, etc.).

The integration/coupling of the NWP/MMM and ACT models can be realized by different ways using the online and offline modelling approaches. In more details the definition and specifics of the approaches, as well as the advantages and disadvantages of the online and offline modelling are described by Baklanov and Korsholm (2007) and in the COST-WMO report (2007). It can be realized using the following possible variants (see Fig. 1):

One-way integration (offline coupling):

1. MMM (or any other regional climate or NWP model) meteorological fields used to drive ACTM (this way is traditionally used by many air pollution modellers) (⟵);
2. ACTM chemical composition fields as a driver for regional climate modelling or for NWP (e.g. for aerosol forcing on meteo-processes) (-->).

Two-way integration:

1. Driver and partial feedbacks, for ACTM or for NWP (data exchange via an interface with a limited time period: offline or online access coupling) (⇐⇒);
2. ACTM built inside MMM or NWP model with full feedbacks included on each time step (online coupling) (⟺).

In this context, several levels of MMM and ACTM coupling/integration can be considered:

off-line:

- separate ACTMs driven by meteorological input data from meteo-preprocessors, measurements or diagnostic models,

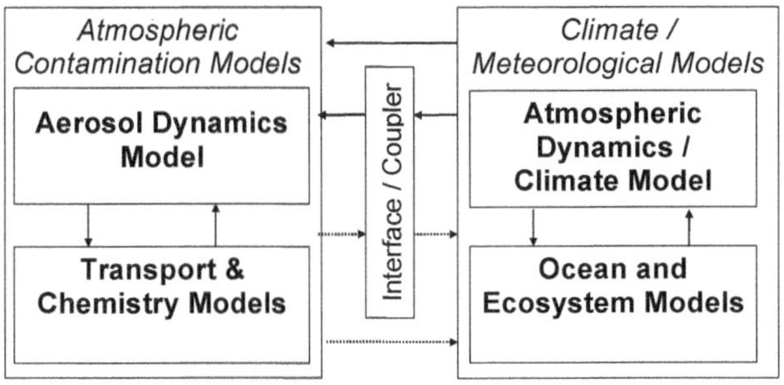


**Figure 1.** Schematic diagram of different approaches for the integration of meteorological/climate and atmospheric chemical transport models with one- or two-way interactions. Online coupling can be achieved through the use of various available coupling tools or through directly inlining the chemical and aerosol modules into the NWP models.

- separate ACTMs driven by analysed or forecasted meteorological data from NWP archives or datasets,
- separate ACTMs reading output-files from operational NWP models or specific MMMs at limited time intervals (e.g. 1, 3, 6 h).

on-line:

- on-line access models, when meteorological data are available at each time-step (possibly via a model interface as well),
- on-line integration of ACTM into MMM, where two-way feedbacks may be considered. We will use this definition for on-line coupled/integrated modelling.

The last level of the on-line integration with two-way feedbacks can consider the full chain of feedbacks and is recommended as the main strategy for integrated CWF models. These methodologies of the integrated CWF have been realised by several of the COST action partners such as the Danish Meteorological Institute (DMI), with the Enviro-HIRLAM model (Baklanov et al., 2008a; Korsholm et al., 2008; Korsholm, 2009) and the COSMO consortium with the Lokal Modell (Vogel et al., 2006; Wolke et al., 2003), as well as the American WRF-Chem (Grell et al., 2004, 2005) integrated system. These model developments for chemical weather forecasting are an activity of increasing importance which is due to be supported by the ongoing COST action ES0602.

The Enviro-HIRLAM integrated system, recently developed by DMI and other collaborators[1] (Chenevez et al., 2004; Baklanov et al., 2008a; Korsholm et al., 2008, Korsholm, 2009), is an online coupled model for research and forecasting of both meteorological and chemical weather. It includes two-way feedbacks between air-pollutants and meteorological processes. Atmospheric chemical transport equations are implemented inside the meteorological corner on each time step (Chenevez et al., 2004). To make the model suitable for CWF in urban areas, where most of the population is concentrated, the meteorological part is improved by the implementation of urban sublayer modules and parameterisations (Baklanov et al., 2008b). The aerosol module in Enviro-HIRLAM comprises a thermodynamic equilibrium model and the aerosol dynamics model CAC (Gross and Baklanov, 2004) based on the modal approach. Parameterisations of the aerosol feedback mechanisms in the Enviro-HIRLAM model are described in Korsholm et al. (2008) and Korsholm (2009). Several chemical mechanisms can be chosen depending on the specific tasks.

Model validation and sensitivity tests of the on-line versus off-line integrated versions (Grell et al., 2004, 2005; Korsholm et al., 2009) show that the online coupling improves the modelling results. Grell et al. (2005) show improved predictions of average ozone concentrations when using the online version of WRF-Chem. Korsholm et al. (2009) using the online Enviro-HIRLAM for the ETEX-1 experiment release show that the off-line coupling interval increase leads to considerable error and a false peak (not found in the observations), which almost disappears in the online version that resolves meso-scale influences during plume development.

In the case study by Zhang (2008) it was shown that including only the direct aerosol effects leads to a temperature change about 0.2 °C and the water vapor mixing ratio increases more than 3% at/near surface. The effects of urban aerosols on the urban boundary layer height, can be of the same order of magnitude as the effects of the urban heat island ($\Delta h$ is up to 100–200 m for stable boundary layer) (Baklanov et al., 2008a). Korsholm et al. (2009) show that aerosol feedbacks through the second indirect effect induce considerable changes in meteorological fields and large changes in chemical composition, in particular $NO_2$, in a case of convective cloud cover and little precipitation.

[1]At the current stage the Enviro-HIRLAM model is used as the baseline system for the HIRLAM chemical branch, and additionally to the HIRLAM community the following groups join the development team: University of Copenhagen, Tartu University (Estonia), Russian State Hydro-Meteorological University, Bilbao University (Spain), Odessa State Environmental University (Ukraine), etc.

## 3 Discussion and conclusions

The new concept and methodology considering the chemical weather as two-way interacting meteorological weather and chemical composition of the atmosphere is suggested for future chemical weather forecasting systems.

These on-line coupled model developments will lead to a new generation of integrated models not only for the chemical weather forecasting, but also for improved meteorological weather forecasting (e.g., in urban areas, severe weather events, fog and visibility, UV-radiation and solar energy, etc.), bio-meteorological forecasting, climate change modelling, air quality analysis and mitigations, long-term assessment chemical composition.

Main advantages of the on-line modelling approach include: (i) Only one grid for MMM and ACTM, no interpolation in space and time, (ii) Physical parameterizations and numerical schemes are the same, no inconsistencies; (iii) All 3-D meteorological variables are available at the right time at each time step; (iv) No restriction in variability of meteorological fields; (v) Possibility to consider two-way feedback mechanisms; (vi) No need for meteo- pre/post-processors.

However, for specific tasks the off-line approach can also be useful and includes the following advantages in separate cases, e.g. for risk assessments: (i) Possibility of independent parameterizations; (ii) More suitable for ensemble activities; (iii) Easier to use for the inverse modelling and adjoint problem; (iv) Independence of atmospheric pollution model runs on meteorological model computations; (v) More flexible grid construction and generation for ACTMs, (vi) Suitable for emission scenarios analysis and air quality management.

The COST-WMO (2007) and further COST-NetFAM (2008) overviews show a quite surprising number of on-line coupled MMM and ACTM model systems already being used in Europe. However, many of the on-line coupled models are not built for the mesometeorological scale, and they (e.g. GME, ECMWF GEMS, MESSy) are global-scale modelling systems and first of all designed for climate change modelling. Besides, at the current stage most of the on-line coupled models do not consider feedback mechanisms or include only direct effects of aerosols on meteorological processes (like COSMO LM-ART and MCCM). Only two meso-scale on-line integrated modelling systems (WRF-Chem and Enviro-HIRLAM) consider feedbacks with indirect effects of aerosols.

It is necessary to highlight that the interactions of aerosols and other chemical species with meteorological processes have many different pathways and they have to be prioritised and considered in on-line coupled modelling systems for CWF. Further sensitivity studies are needed to understand the relative importance of different feedback mechanisms for different species, scales and conditions relevant to air quality and meteorology interactions. A concerted action to mobilise and coordinate European research in this area is needed.

**Acknowledgements.** This study was supported by the COST Actions 728 and ES0602, NetFAM, EC FP7 Project MEGAPOLI and the Copenhagen Global Change Initiative (COGCI). The author is grateful to a number of COST728, FUMAPEX, Enviro-HIRLAM, MEGAPOLI and DMI colleagues, who participated in the above-mentioned projects, for productive collaboration and discussions. Especial thanks are to my former PhD student Ulrik Korsholm (DMI) who realised most of the Enviro-HIRLAM model coding, simulation runs and studies of the aerosol indirect effects.

Edited by: M. Piringer

## References

Baklanov, A. and Korsholm, U.: On-line integrated meteorological and chemical transport modelling: advantages and prospective, in: ITM 2007: 29th NATO/SPS International Technical Meeting on Air Pollution. Modelling and its Application, 24–28.09.2007, University of Aveiro, Portugal, 21–34, 2007.

Baklanov, A., Korsholm, U., Mahura, A., Petersen, C., and Gross, A.: ENVIRO-HIRLAM: on-line coupled modelling of urban meteorology and air pollution, Adv. Sci. Res., 2, 41–46, 2008a.

Baklanov, A., Mestayer, P. G., Clappier, A., Zilitinkevich, S., Joffre, S., Mahura, A., and Nielsen, N. W.: Towards improving the simulation of meteorological fields in urban areas through updated/advanced surface fluxes description, Atmos. Chem. Phys., 8, 523–543, 2008b, http://www.atmos-chem-phys.net/8/523/2008/.

Bellouin, N., Boucher, O., Haywood, J., and Reddy, M. S.: Global estimate of aerosol direct radiative forcing from satellite measurements, Nature, 438, 1138–1141, 2005.

Chenevez, J., Baklanov, A., and Sørensen, J. H.: Pollutant transport schemes integrated in a numerical weather prediction model: Model description and verification results, Meteorol. Appl., 11(3), 265–275, 2004.

COST-NetFAM: Integrated systems of meso-meteorological and chemical transport models, edited by: Baklanov, A., Mahura, A., and Sokhi, R., Materials of the COST-728/NetFAM workshop, DMI, Copenhagen, 21–23 May 2007, 183 pp., Springer (in press), available at: http://www.cost728.org, 2008.

COST-WMO: Overview of existing integrated (off-line and on-line) meso-scale systems in Europe, edited by: Baklanov, A., Fay, B., Kaminski, J., Sokhi, R., Joint Report of COST728 and GURME, May 2007, WMO-COST publication, GAW Report No. 177, WMO TD No. 1427, available at: http://www.cost728.org, 2007.

Grell, G. A., Knoche, R., Peckham, S. E., and McKeen, S.: On-line versus offline air quality modeling on cloud-resolving scales, Geophys. Res. Lett., 31, L16117, doi:10.1029/2004GL020175, 2004.

Grell, G. A., Peckham, S. E., Schmitz, R., McKeen, S. A., Frost, G., Skamarock, W. C., and Eder, B.: Fully coupled "online" chemistry within the WRF model, Atmos. Environ., 39(37), 6957–6975, 2005.

Gross, A. and Baklanov, A.: Modelling the influence of dimethyl sulphide on the aerosol production in the marine boundary layer, Int. J. Environ. Pollut., 22, 51–71, 2004.

Jacobson, M. Z.: Control of fossil-fuel particulate black carbon and organic matter, possibly the most effective method of slowing global warming, J. Geophys. Res., 107(D19), 4410, doi:10.1029/2001JD001376, 2002.

Korsholm, U.: Integrated modeling of aerosol indirect effects – development and application of a chemical weather model, PhD thesis University of Copenhagen, Niels Bohr Institute and Danish Meteorological Institute, Research department, http://www.dmi.dk/dmi/sr09-01.pdf, 2009.

Korsholm, U. S., Baklanov, A., Gross, A., Mahura, A., Sass, B. H., and Kaas, E.: Online coupled chemical weather forecasting based on HIRLAM – overview and prospective of Enviro-HIRLAM, HIRLAM Newsletter, 54, 151–168, 2008.

Korsholm, U. S., Baklanov, A., Gross, A., and Sørensen, J. H.: On the importance of the meteorological coupling interval in dispersion modeling during ETEX-1, Atmos. Environ., 43(31), 4805–4810, 2009.

Lawrence, M. G., Hov, Ø., Backmann, M., Brandt, J., Elbern, H., Eskes, H., Feichter, H., and Takigawa, M.: The Chemical Weather. Environ. Chem., 2, 6–8, 2005.

Ramanathan, V. and Carmichael, G.: Global and regional climate changes due to black carbon, Nature Geosci., 1(4), 221–227, 2008.

Rosenfeld, D.: Suppression of Rain and Snow by Urban and Industrial Air Pollution, Science, 287(5459), 1793–1796, 2000.

Schulz, M., Textor, C., Kinne, S., Balkanski, Y., Bauer, S., Berntsen, T., Berglen, T., Boucher, O., Dentener, F., Guibert, S., Isaksen, I. S. A., Iversen, T., Koch, D., Kirkevåg, A., Liu, X., Montanaro, V., Myhre, G., Penner, J. E., Pitari, G., Reddy, S., Seland,Ø., Stier, P., and Takemura, T.: Radiative forcing by aerosols as derived from the AeroCom present-day and pre-industrial simulations, Atmos. Chem. Phys., 6, 5225–5246, 2006,
http://www.atmos-chem-phys.net/6/5225/2006/.

Vogel, B., Hoose, C., Vogel, H., and Kottmeier, Ch.: A model of dust transport applied to the Dead Sea area, Meteorol. Z., 14, 611–624, 2006.

Wolke, R., Hellmuth, O., Knoth, O., Schröder, W., Heinrich, B., and Renner, E.: The chemistry-transport modeling system LM-MUSCAT: Description and CITYDELTA applications. Proceedings of the 26-th International Technical Meeting on Air Pollution and Its Application. Istanbul, May 2003, 369–379, 2003.

Zhang, Y.: Online-coupled meteorology and chemistry models: history, current status, and outlook, Atmos. Chem. Phys., 8, 2895–2932, 2008,
http://www.atmos-chem-phys.net/8/2895/2008/.

# 2

# Multi-methodical realisation of Austrian climate maps for 1971–2000

**J. Hiebl, S. Reisenhofer, I. Auer, R. Böhm, and W. Schöner**

Central Institute for Meteorology and Geodynamics, Vienna, Austria

**Abstract.** Constantly changing climate, the availability of a higher resolved digital elevation model and further development of geostatistical interpolation methods gave reason for updating the most frequently demanded climate maps out of the Austrian digital climate atlas from 1961–1990 to 1971–2000. To achieve a station density as high as possible, data from eleven national and foreign institutes were collected and gap-filled. According to the climate parameter, different geostatistical interpolation methods (including regionalised multilinear regressions, geographically weighted regressions and curve fitting to base parameter) were applied. The resultant 17 grids concern 30-year-means of air temperature, precipitation and snow parameters as well as derived indices. They are now available for a variety of scientific and planning purposes.

## 1 Introduction

The construction of DEMs (digital elevation models) together with the launch of more and more powerful GIS (geographical information system) software products prompted the rapid development of sophisticated geostatistical interpolation methods in climatology since the late 1980s (e.g. Bénichou and LeBreton, 1987; Daly et al., 1994; Hutchinson, 1995; cf. Dobesch et al., 2007; Carrega, 2010). The advantage of these methods is obvious: the resultant areal raster fields contain information for a number of points, which is of several orders of magnitude larger than the number of ingoing meteorological stations, allowing applications in a wide range of fields such as climate (impact) research, all other branches of geosciences, agriculture and forestry, engineering, water management, tourism and natural resource conservation.

Consequently, nearly all national meteorological services established digital maps of standard normal values (e.g. Cegnar, 1995; DWD, 1999; Zaninović et al., 2008). For Austria, Auer et al. (2001) at the Central Institute for Meteorology and Geodynamics (ZAMG) provided a multimedia climate atlas (ÖKLIM) including, among other things, 35 digital maps for the period 1961–1990. The 17 most frequently applied maps thereof have now been adapted and updated to the more recent period 1971–2000 (Table 1). This has been done for several reasons: (1) Ongoing climate change – only the ten-years-shift in the reference period causes a rise of mean annual temperature of 0.3–0.4 °C – made the former maps inappropriate for the analysis of current climate conditions, e.g. considering risk of frost, heating demand or snow cover duration. (2) In the meantime, a spatially higher resolved DEM for Austria has been introduced. (3) Above all, the updating of the digital maps permitted to stay abreast of the methodical developments in geostatistical interpolation during the last years.

## 2 Data

The quality of any interpolation result is limited by the quality of the underlying observational data. Therefore, two goals were set for data collection and preparation: not to confine to the national borders (unlike the first ÖKLIM project) and to achieve a spatial station density as high as possible.

Internationally arranged initiatives (e.g. Hiebl et al., 2009; Klein and Persson, 2009) have shown that the quality of national grids in border-near regions considerably depends on the availability of foreign station data. Especially in meridionally narrow Western Austria (where e.g. strong horizontal precipitation gradients are noted), knowledge of climate measurements from outside the borders is of use. In order to maintain a homogeneous level of validity across all parts of Austria, the study region for interpolation was extended to 45.5–50° N and 8.5–18° E (Fig. 1). For this domain, data were requested and kindly provided from almost

**Table 1.** The final data collection according to climate parameter, country of origin and data provider. Bold station numbers denote gap-filling in the respective subdataset.

| | Country | int. | AT | AT | DE | CZ | SK | HU | HR | SI | IT | IT | CH | |
|---|---|---|---|---|---|---|---|---|---|---|---|---|---|---|
| Para-meter | Institute | ZAMG (HISTALP) | ZAMG | HZB | DWD | CHMI | SHMU | OMSZ | DHMZ | ARSO | OSMER-ARPA | Hydr. Office, South Tyrol | Meteo Swiss | Total |
| 1 | Air temperature, year | 96 | **153** | **232** | 100 | 147 | 7 | 13 | 28 | 29 | 6 | **41** | **42** | 894 |
| 2 | Air temperature, January | 96 | **153** | **232** | 100 | 147 | 7 | 14 | 28 | 29 | 6 | **41** | **42** | 895 |
| 3 | Air temperature, July | 96 | **153** | **232** | 100 | 147 | 7 | 15 | 28 | 29 | 6 | **41** | **42** | 896 |
| 4 | Ice days, year | – | **207** | – | 100 | 148 | – | 15 | 8 | 32 | 6 | 44 | **46** | 606 |
| 5 | Frost days, year | – | **207** | – | 100 | 148 | – | 15 | 27 | 32 | 6 | 44 | **45** | 624 |
| 6 | Freeze-thaw days, year | – | **207** | – | 100 | 148 | – | 15 | 8 | 32 | 6 | 44 | **45** | 605 |
| 7 | Summer days, year | – | **207** | – | 100 | 148 | – | 15 | 27 | 32 | 6 | 44 | **45** | 624 |
| 8 | Hot days, year | – | **207** | – | 100 | 148 | – | 15 | 8 | 32 | 6 | 44 | **45** | 605 |
| 9 | Heating degree days, year | – | **200** | – | 8 | 148 | – | – | – | – | 6 | 44 | **23** | 429 |
| 10 | Precipitation sum, year | 89 | **133** | **623** | 98 | 147 | – | 11 | 25 | 28 | 11 | **53** | **181** | 1399 |
| 11 | Precipitation sum, winter half-year | 90 | **134** | **580** | 98 | 147 | – | 11 | 25 | 28 | 11 | **53** | **181** | 1358 |
| 12 | Precipitation sum, summer half-year | 91 | **135** | **581** | 98 | 147 | – | 11 | 25 | 28 | 11 | **53** | **181** | 1361 |
| 13 | Days with precipitation, year | – | **164** | 500 | 107 | 148 | – | 16 | 12 | 32 | 11 | 56 | **191** | 1237 |
| 14 | Solid precipitation, year | – | **155** | – | – | 148 | – | 16 | – | – | – | – | – | 319 |
| 15 | Sum of fresh-fallen snow | – | **141** | **483** | 107 | 148 | – | – | 8 | – | – | – | 38 | 925 |
| 16 | Snow cover duration | – | **147** | **479** | – | 148 | – | 16 | 15 | 32 | – | – | 25 | 862 |
| 17 | Maximum snow depth | – | **147** | **504** | – | 148 | – | 16 | 7 | – | – | – | 34 | 856 |

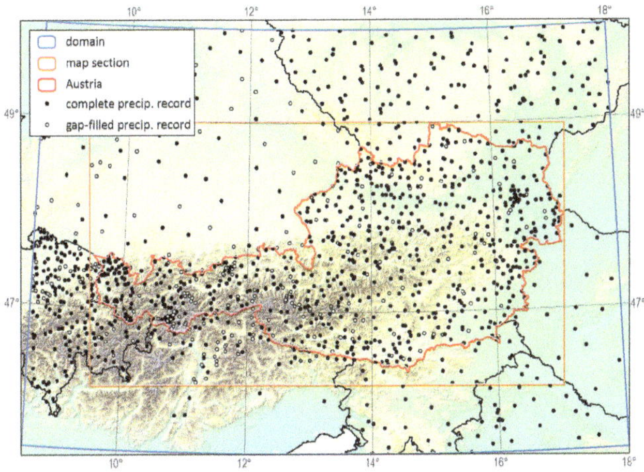


**Figure 1.** Location of the domain (for data collection and interpolation) and the map section (for printed products) in relation to the Austrian national borders (for digital products). In addition, the most extensive network of stations with means of annual precipitation sum is depicted.

all neighbouring countries and Croatia. In cases when, besides the national record version, a homogenised record version out of the HISTALP (historical instrumental climatological surface time series of the greater Alpine region) project (Auer et al., 2007) was available, the latter was preferred. For instance, a network of nearly 900 stations could be utilised for air temperature and nearly 1400 stations for precipitation (Fig. 1, Table 1).

This data collection would not have been possible without extensive gap-filling. The challenge was to find a favourable compromise between the gain of spatial information by gap-filling and growing uncertainty by too long temporal gaps. The applied method of gap-closing was strongly depending on the considered parameter: For parameters with high spatial representation like air temperature a maximum gap of ten out of 30 yr was set. For precipitation and snow parameters, however, only five missing years were allowed because of their lower spatial transferability.

Gaps in air temperature, precipitation and snow records were closed relating to highly correlated, preferably homogenised (HISTALP), reference stations. Missing values in records of parameters derived from temperature or precipitation were filled by curve fitting to the base parameter at the very same station (mostly hyperbolic tangent or tangent thereof; Auer et al., 2005). In any case, even if the annual mean value of a parameter was to be mapped, gap-filling was performed on basis of the monthly series. This, together with the conservative choice of gap length, keeps the induced uncertainties low. Even though comprehensive validation of gap-filling was beyond the scope of the project, exemplary plausibility checks argue for the applied procedures. The gained results are, in any case, to be preferred to averaging over incomplete series or series concerning different periods, as misleadingly often practiced.

In case of air temperature data, inhomogeneities due to nationally different methods of the estimation of daily means were finally corrected, relying on existing adjustment factors used in Hiebl et al. (2009).

The DEM underlying all grid fields is provided by the Federal Office of Metrology and Surveying (BEV) in an original resolution of 50×50 m with an estimated mean error between ±1 m in agricultural and settlement area and ±10 m in forested and high Alpine area (BEV, 2010). In order to increase usability and to achieve a maintainable spatial resolution, the DEM has been transferred into geographic coordinates featuring a cell size of 9.65″, corresponding to 200 m in the longitudinal and 300 m in the latitudinal direction.

## 3 Methods

The choice of the geostatistical interpolation method most appropriate for a certain climate parameter requires knowledge about the parameter's physical background and climatological-spatial behaviour as well as about the method's functionality, strengths and shortcomings (cf. Dobesch et al., 2007). In the following sections, the applied procedures are subsumed to three groups.

### 3.1 Regionalised multilinear regressions

Three grid fields, those on annual air temperature and monthly air temperature in January and July, are based on regionalised multilinear regressions. In course of this method, multilinear regressions of air temperature against macroclimatic explanatory variables are calculated in several horizontal subregions and three vertical layers (Hiebl et al., 2009). The approach has been chosen because it is appropriate for spatially highly representative parameters with strong elevation dependency. Moreover, other (not auto-correlated) large-scale variables controlling the climate parameter can be considered and modifications in subregions (like valleys and basins prone to temperature inversions) can be incorporated. Generally, the highly transparent method is easily manageable.

High station density allowed for dividing the domain into 37 horizontal subregions of temperature climate corresponding to geographic entities such as forelands, valleys and basins. Such small-scale subregions better reflect the strong regional peculiarities of temperature climate in Alpine terrain and justify problems of smoothing at the subregions' borders. In addition, the vertical dimension was divided into three layers: (1) In the lower layer (extra-Alpine <600 m to central-Alpine <1800 m), station data were analysed separately for each horizontal subregion by calculating multilinear regressions against elevation, (topographically weighted) distance from the coast, latitude and longitude in extra-Alpine forelands and basins. Simple linear regressions against elevation were calculated in inner-Alpine valleys and basins, where, accordingly to the smaller extension of the subregions, large-scale dependencies are too weak. (2) In the upper layer (mostly >1700 m to central-Alpine >2300 m), all stations were brought together to one single high-Alpine subregion to be processed in multilinear regression calculation. (3) In the intermediate layer, the temperature value for each grid point was averaged between the lower layer's upper level and the upper layer's lower level, weighted according to the grid point's elevation. This threefold layering facilitates the description of the, especially winterly, ground-level layer of reduced or even inverted temperature decrease with elevation. For the GIS-based creation of the raw temperature fields, raster grids of the regression coefficients were generated and smoothed at the subregions' borders. In a second step, mesoclimatological temperature modifications, namely

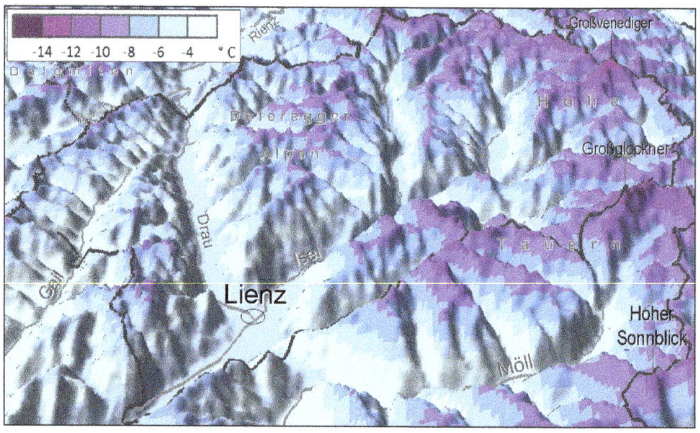


**Figure 2.** Map example 1. Mean monthly air temperature in January, bird eye's view on map section for East Tyrol (westward view, vertical exaggeration 2:1). Note the embedded effect of mesoclimate modification on slopes.

extra-Alpine inversions, lakeside effects, urban heat islands and slope effects were integrated following the tailored techniques described in detail in Hiebl et al. (2009) (Fig. 2).

### 3.2 Geographically weighted regressions

Six grid fields, those on annual, winter half-year and summer half-year precipitation sums as well as on sum of fresh-fallen snow, snow cover duration and maximum snow depth, rely on geographically weighted regressions. Such procedures have been introduced by Daly et al. (1994, 2008), applied by Frei and Schär (1998) and Schwarb (2000) in the Alpine region and adapted methodically for the present purpose. The method again strongly involves elevation dependencies but allows for a higher degree of spatial variability compared to the (regionalised) global regressions used for temperature interpolation, making it the better choice for the interpolation of precipitation sum and snow parameters in mountain regions. However, it is sensitive to sparse station coverage in high-elevation areas.

Applying geographically weighted regressions, an individual regression of the climate parameter against elevation was calculated at each grid point involving the closest stations. To control their contribution, the stations have been weighted according to their representativeness for the topographical conditions at the grid point. Checking different weights and parameterisations, the 15 nearest records were selected to be weighted by their horizontal distance and by their location north or south to the main Alpine crest. Because the method produces unrealistic gradients in regions with few high-elevation observations, several so-called control points were set. This was done by duplicating an observed value along a mountain range or by estimating an artificial value out of climatological experience. Even though this introduces a certain degree of subjectivity, it is inevitable in steep terrain with a systematic lack of sites at high altitudes.

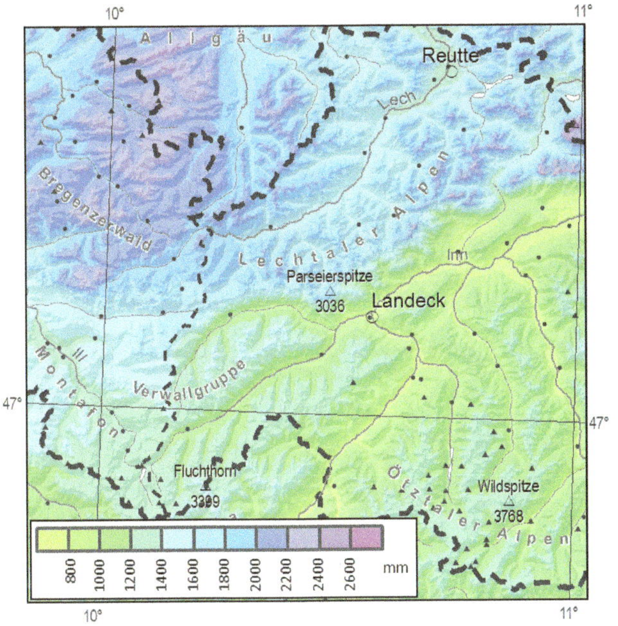


**Figure 3.** Map example 2. Mean annual precipitation sum, map section for a part of Western Austria.

Similarly, Peterson et al. (1998) emphasise the importance of the experienced climatologist's subjective judgement in addition to objectified methods in the editing of climate data (in a different context).

Like for air temperature, an effect of mesoscale climate was embedded into the precipitation fields. Böhm (1979) statistically proofed a precipitation increase triggered by higher concentration of condensation nuclei and thermal lift in the lee of Vienna. The then revealed pattern, expressed as a ready-to-use field of relative precipitation modification, was integrated into the precipitation fields. The final precipitation and snow fields realistically simulate the strong spatial variations such as abrupt transitions from *Nordstau* ranges to inner-Alpine dry valleys (e.g. from the Lechtal to the Ötztal Alps; Fig. 3).

### 3.3 Curve fitting

The remaining eight grid fields, all on secondary parameters of air temperature or precipitation climate, are based upon curve fitting to the primary parameters. On the one hand side, these indices are by definition, mostly by a threshold, directly derived from air temperature or precipitation sum. On the other hand side, the close climatic relations permit to attach the interpolation of the derived parameters to the already existing grids of the base parameters. This approach's clear advantage is due to the considerably more extensive observation networks of air temperature and precipitation compared to the derived parameters (Table 1), which allows their spatial description with higher accuracy (Fig. 4).

The different types of curve fitting used for modelling the spatial distributions of the indices are listed in Table 2. Annual air temperature gave better results explaining ice and frost days compared to January temperature, which seems reasonable for a country dominated by high-elevated areas. Because of the generally high correlations between primary and secondary parameters, further regionalisation was not necessary. An exception can be seen in terms of the remarkable dichotomy in the relation between the annual number of days with precipitation and the annual precipitation sum (Fig. 5). The suspicion of a climate-divide-effect was confirmed by data splitting along the main Alpine crest. Separate exponential fits formed the bases for interpolation in two regions, whereupon one-sided smoothing, from north to south, was performed along the region border.

Among the interpolation procedures, the grid field on freeze-thaw-days is the exceptional case, which has, accordingly to its definition, simply been constructed as the difference between the grids on frost and ice days.

To finalise the grids, the respective stations residuals were Kriging-interpolated and added whenever it explained a relevant fraction of the residuals. For example, residuals of air temperature did show distinct (even tough weak) spatial patterns of remaining variance, but residuals of precipitation sum or snow parameters did not.

## 4 Evaluation and discussion

Six common error measures have been calculated for the spatial analyses of all parameters to describe the differences (residuals) between "true", observed station data and simulated data at the station locations (Table 3). This has been done at two stages: (1) Simulated values as directly calculated by the model (regionalised multilinear regressions, geographically weighted regressions, curve fitting) were compared to measurements. This allows for estimating the applicability of the chosen interpolation approaches per se. (2) Simulated values as indicated on the grid fields were set in relation to the measured station values. By doing so, the quality of the final interpolation products is specified. This involves potentially higher errors due to deviations of the DEM from true station elevations as well as potentially lower errors due to, optionally, integrating mesoscale effects, smoothing at subregion borders or residual Kriging.

The spatial interpolations of the three mean air temperature grids are featured by negligible biases (mean errors). The regionalised multilinear regressions produce typical MAE (mean absolute error) values around 0.45 °C and RMSE (root mean squared error) values around 0.57 °C. The error values are still somewhat lower on the final climate maps and benefit, aside from other things, from the coordinate correction that has been accomplished for temperature stations only (RMSE of differences between DEM and station elevations by 11 m).

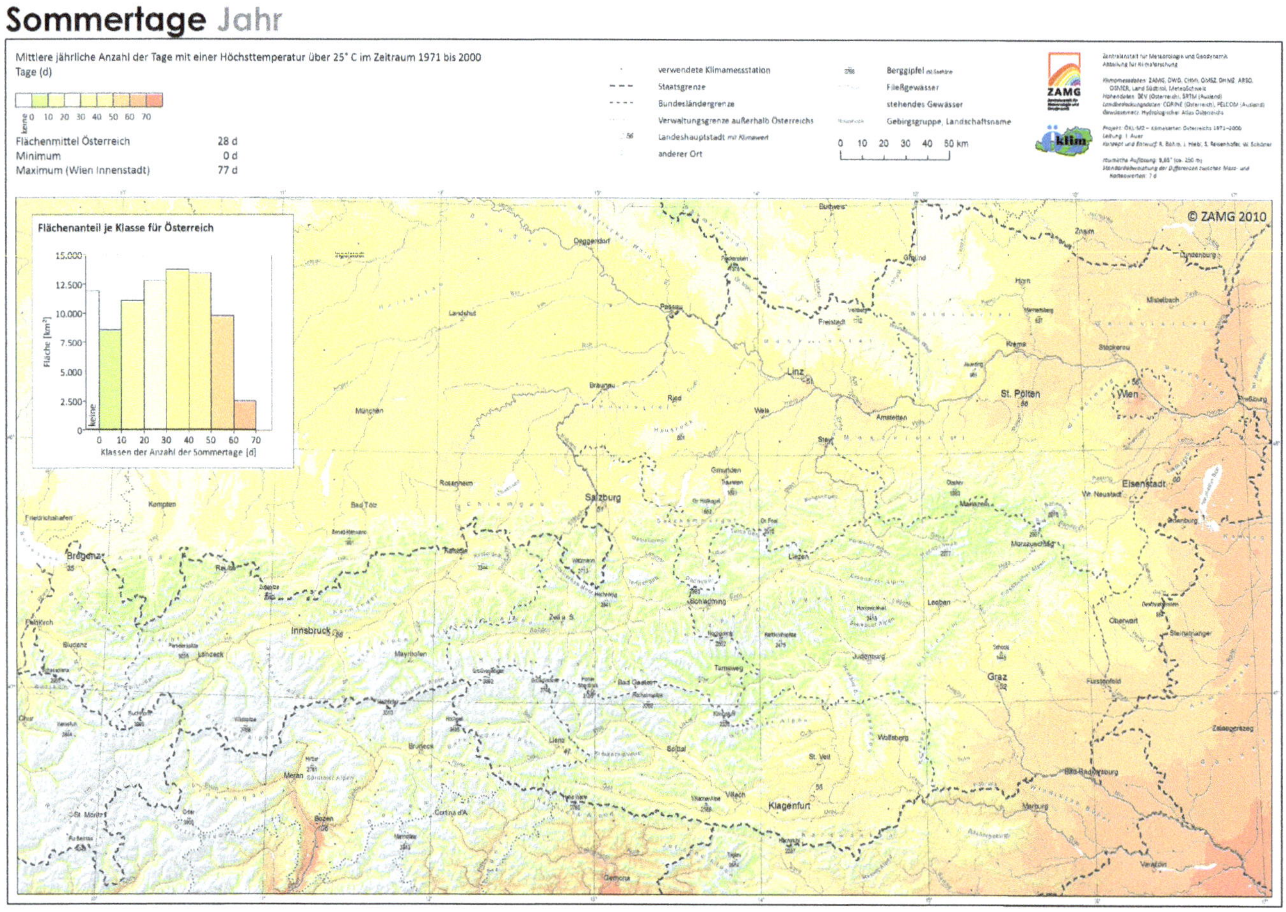


**Figure 4.** Map example 3. Mean annual number of days with a maximum temperature above 25 °C, entire map.

**Table 2.** Types of curve fitting of derived climate indices to primary parameters as basis for interpolation ($R^2$ – coefficient of determination).

| Parameter (cf. Table 1) | Independent variable | Type of function | $R^2$ |
|---|---|---|---|
| 4 | Air temperature, year | Hyperbolic tangent | 0.99 |
| 5 | Air temperature, year | Hyperbolic tangent | 0.98 |
| 6 | – | – | – |
| 7 | Air temperature, year | Linear regression | 0.76 |
| 8 | Air temperature, year | Hyperbolic tangent | 0.73 |
| 9 | Air temperature, year | Polynomial function | 0.99 |
| 13 | Precipitation sum, year | Exponential function | 0.82 |
| 14 | Air temperature, year | Hyperbolic tangent | 0.93 |

Geographically weighted regressions applied on three precipitation sum and three snow parameter datasets give MRE (mean relative error) values of 0.8–3.1%. The MARE (mean absolute relative error) and RMSRE (root mean squared relative error) values are lowest for precipitation sum of the summer half-year (5.9 and 8.1% respectively) and highest for the sum of fresh-fallen snow (13.5% and 17.8% respectively). The final grid field errors are about the same size or slightly higher compared to the pure model errors, which can mainly be explained by larger elevation inconsistencies. Cross-validation proved the applicability of the method: If successively each observational value is left out once and used as validation data, the RMSRE would increase by only about 1.9% modelling fields of precipitation totals and by about 3.7% modelling fields of snow parameters.

**Table 3.** Error measures (bias – mean error, MAE – mean absolute error, RMSE – root mean squared error, MRE – mean relative error, MARE – mean absolute relative error, RMSRE – root mean squared relative error) describing the deviations between observed values and simulated values as generated by the respective interpolation models (a) and as indicated on the final grid fields (b).

| | (a) Model | | | | | | (b) Grid field | | | | | | |
|---|---|---|---|---|---|---|---|---|---|---|---|---|---|
| Parameter (cf. Table 1) | Bias | MAE | RMSE | MRE | MARE | RMSRE | Bias | MAE | RMSE | MRE | MARE | RMSRE | RMSE (stations to DEM) [m] |
| 1 (°C) | −0.04 | 0.40 | 0.52 | – | – | – | 0.00 | 0.35 | 0.45 | – | – | – | 11 |
| 2 (°C) | −0.03 | 0.52 | 0.67 | – | – | – | 0.02 | 0.42 | 0.55 | – | – | – | 11 |
| 3 (°C) | −0.06 | 0.43 | 0.53 | – | – | – | 0.00 | 0.37 | 0.47 | – | – | – | 11 |
| 4 (d) | −0.6 | 5.2 | 6.4 | −1.7% | 17.0% | 20.4% | 0.1 | 4.0 | 6.3 | 2.4% | 12.3% | 16.0% | 52 |
| 5 (d) | 2.6 | 8.0 | 9.9 | 2.8% | 6.9% | 8.7% | 0.6 | 9.4 | 11.8 | 0.6% | 7.5% | 9.2% | 52 |
| 6 (d) | – | – | – | – | – | – | 0.4 | 9.0 | 11.5 | 1.4% | 10.2% | 13.1% | 52 |
| 7 (d) | −0.8 | 6.0 | 7.8 | – | – | – | −0.6 | 5.1 | 6.7 | – | – | – | 52 |
| 8 (d) | −0.2 | 1.6 | 2.1 | – | – | – | 0.0 | 1.2 | 1.7 | – | – | – | 52 |
| 9 (°C) | 30.3 | 61.7 | 70.9 | 0.8% | 1.6% | 1.8% | 15.7 | 124.7 | 170.9 | 0.5% | 3.0% | 3.8% | 53 |
| 10 (mm) | 2 | 81 | 121 | 1.1% | 6.7% | 9.2% | 4 | 81 | 120 | 1.3% | 6.7% | 9.1% | 71 |
| 11 (mm) | 1 | 42 | 64 | 1.9% | 9.1% | 12.4% | 3 | 43 | 63 | 2.4% | 9.2% | 12.3% | 71 |
| 12 (mm) | 1 | 44 | 65 | 0.8% | 5.9% | 8.1% | 1 | 43 | 65 | 0.8% | 5.9% | 8.0% | 71 |
| 13 (d) | −0.1 | 5.1 | 10.2 | 0.2% | 4.5% | 7.9% | −0.1 | 4.4 | 5.7 | 0.2% | 3.8% | 4.9% | 32 |
| 14 (%) | 0.0 | 2.4 | 3.0 | 1.1% | 9.9% | 12.7% | 0.2 | 1.7 | 2.2 | 1.7% | 6.9% | 8.9% | 49 |
| 15 (cm) | 1 | 30 | 51 | 3.1% | 13.5% | 17.8% | 6 | 33 | 59 | 4.8% | 14.1% | 19.2% | 43 |
| 16 (d) | 0 | 6 | 8 | 1.4% | 7.7% | 10.3% | 1 | 6 | 9 | 1.8% | 7.9% | 10.8% | 43 |
| 17 (cm) | 0 | 8 | 13 | 2.7% | 12.8% | 17.1% | 1 | 7 | 12 | 3.8% | 12.9% | 16.9% | 43 |

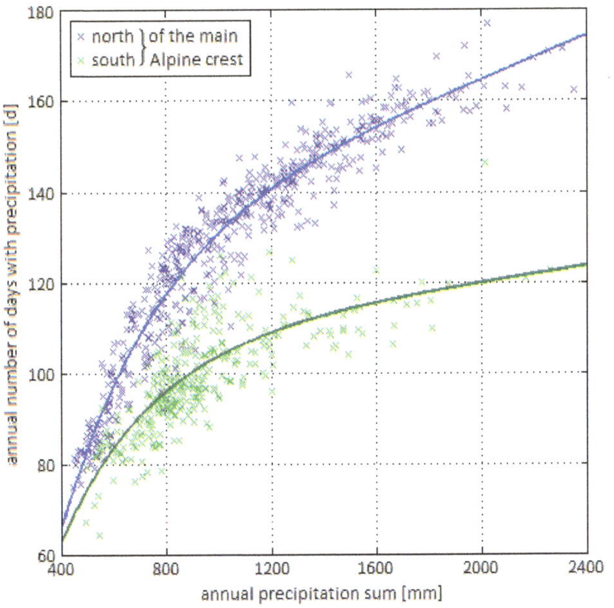


**Figure 5.** Exponential fitting of annual number of days with precipitation to annual precipitation sum over two subsets of Austrian station data, i.e. stations north and south of the main Alpine crest.

Curve fitting performed on secondary parameters of air temperature and precipitation leads to small additional systematic biases. The RMSRE values range widely between only 1.8% for the fitting of the number of heating degree days to air temperature and 20.4% for the fitting of the number of ice days to air temperature. That this measure of statistical dispersion does not necessarily go along with statistical correlation is shown by the fact that the correlation coefficient for the fitting of heating degree days is as high as 0.99 (cf. Table 2). As the annual number of summer days and hot days are parameters reaching values of zero in the study region, the calculation of relative error measures does not make sense and only the absolute errors are indicated.

The given error estimation has to be qualified by adding that station measurements as reference datasets are not holding climate reality but, to a certain degree, are biased themselves. Shifting measurement practices, station relocations, microclimatic superimpositions and all other kinds of inhomogeneities induce unwanted noise to the climate sign. Even if its fraction in the remaining error cannot be determined, seeking for a final error of zero is misleading in any case and cannot be the target of interpolation. Altogether, it can be concluded that the resulting error values are satisfactory as they are within the typical range of measurement insecurities.

A comparison to the preceding ÖKLIM maps of Auer et al. (2001) cannot be presented in terms of quantitative error reduction, as evaluation was not carried out then. However, extending the dataset by measurements from neighbouring countries has potentially improved interpolation performance in border-near regions, gap-filling incomplete time series has considerably increased spatial data coverage and applying an updated DEM has decreased elevation-caused errors. In addition, geographically weighted regressions, capable to capture higher spatial climate variability, had not been used before as well as mesoclimatic effects have been considered in more detail. For these reasons, an improvement in the quality of the new climate grid field appears very reasonable.

## 5 Results

The 17 finished interpolation results are available either as digital grid fields or as cartographically prepared products. Examples of the results are shown in Figs. 2–4. The digital grids, which have been cut out to the Austrian national borders, are featured by a cell size of 200×300 m. As the result of a collateral project (Jurković et al., 2010), similar maps for Austria and the period 1971–2000 on monthly, seasonal and annual global radiation, absolute and relative sunshine duration can be obtained additionally.

For cartographic preparation, the mapping of each climate grid has been complemented by statistical measures, a histogram chart on the areal distribution of the depicted parameter and technical specifications (Fig. 4). For further information, contact details and a preview of all maps see the project web page (http://www.zamg.ac.at/forschung/klimatologie/klimamodellierung/oeklim2).

## 6 Conclusions

The reissued collection of Austrian climate maps (ÖKLIM 1971–2000) suggests a quality improvement that has been achieved by several data and methodical adaptations: By including foreign observational data border effects towards data free regions were avoided. Spatial climate information could be increased by closing incomplete data series. Employing a new DEM helped capture the Eastern Alps' topography better. Accounting for the developments in interpolation methodology allowed for higher spatial variability and additional mesoclimatic effects were resolved by tailored techniques. The digital fields facilitate the spatial interpretation of features of temperature, precipitation and snow climate distribution over Austria. They offer extensive application for various scientific and planning purposes.

**Acknowledgements.** The project ÖKLIM 1971–2000 was funded by the Austrian Federal Ministry of Science and Research. The work considerably benefitted from data provision by the ZAMG's sister institutes in different countries. The authors expressly thank Viktor Weilguni (HZB), Gerhard Müller-Westermeier (DWD), Vít Květoň (CHMI), Zita Bihari (OMSZ), Ksenija Zaninović (DHMZ), Mojca Dolinar (ARSO), Stefano Micheletti (OSMER-ARPA), Günther Geier (Hydrographical Office, South Tyrol) and Christof Appenzeller (MeteoSwiss). Furthermore, the authors appreciate the constructive remarks of the two reviewers, which substantially improved the paper.

Edited by: M. Dolinar

## References

Auer, I., Böhm, R., Mohnl, H., Potzmann, R., Schöner, W., and Skomorowski, P.: ÖKLIM: digitaler Klimaatlas Österreichs, Zentralanstalt für Meteorologie und Geodynamik, Vienna, Austria, CD-ROM, 2001.

Auer, I., Matulla, C., Böhm, R., Ungersböck, M., Maugeri, M., Nanni, T., and Pastorelli, R.: Sensitivity of frost occurrence to temperature variability in the European Alps, Int. J. Climatol., 25, 1749–1766, 2005.

Auer, I., Böhm, R., Jurković, A., Lipa, W., Orlik, A., Potzmann, R., Schöner, W., Ungersböck, M., Matulla, C., Briffa, K., Jones, P. D., Efthymiadis, D., Brunetti, M., Nanni, T., Maugeri, M., Mercalli, L., Mestre, O., Moisselin, J. M., Begert, M., Müller-Westermeier, G., Květoň, V., Bochníček, O., Šťastný, P., Lapin, M., Szalai, S., Szentimrey, T., Cegnar, T., Dolinar, M., Gajíć-Čapka, M., Zaninović, K., Majstorović, Z., and Nieplova, E.: HISTALP: Historical instrumental climatological surface time series of the greater Alpine region 1760–2003, Int J. Climatol., 27, 17–46, 2007.

Bénichou, P. and LeBreton, O.: Prise en compte de la topographie pour la cartographie de champs pluviométriques statistiques, La Météorologie, 7/19, 23–34, 1987.

BEV – Bundesamt für Eich- und Vermessungswesen: http://www.bev.gv.at/, last access: 9 December 2010.

Böhm, R.: Untersuchung des urbanen Einflusses auf das Niederschlagsfeld in der Umgebung einer Großstadt mit Hilfe verschiedener statistischer Methoden, Arch. Meteor. Geophy. B, 27, 225–246, 1979.

Carrega, P. (Ed.): Geographical information and climatology, ISTE, London, UK, 288 pp., 2010.

Cegnar, T. (Ed.): Climate of Slovenia, Hydrometeorological Institute of Slovenia, Ljubljana, Slovenia, 70 pp., 1995.

Daly, C., Neilson, R. P., and Phillips, D. L.: A statistical-topographical model for mapping climatological precipitation over mountainous terrain, J. Appl. Meteorol., 33, 140–158, 1994.

Daly, C., Halbleib, M., Smith, J. I., Gibson, W. P., Doggett, M. K., Taylor, G. H., Curtis, J., Pasteris, P. P.: Physiographically sensitive mapping of climatological temperature and precipitation across the conterminous United States, Int. J. Climatol., 28, 2031–2064, 2008.

Dobesch, H., Dumolard, P., and Dyras, I. (Eds.): Spatial interpolation for climate data: the use of GIS in climatology and meteorology, ISTE, London, UK, 284 pp., 2007.

DWD (German Weather Service): Klimaatlas Bundesrepublik Deutschland: Teil 1, Deutscher Wetterdienst, Offenbach, Germany, CD-ROM, 1999.

Frei, C. and Schär, C.: A precipitation climatology of the Alps from high-resolution rain-gauge observations, Int. J. Climatol., 18, 873–900, 1998.

Hiebl, J., Auer, I., Böhm, R., Schöner, W., Maugeri, M., Lentini, G., Spinoni, J., Brunetti, M., Nanni, T., Perčec Tadić, M., Bihari, M., Dolinar, M., and Müller-Westermeier, G.: A high-resolution 1961–1990 monthly temperature climatology for the greater Alpine region, Meteorol. Z., 18/5, 507–530, 2009.

Hutchinson M. F.: Interpolating mean rainfall using thin plate smoothing splines, Int. J. Geogr. Inf. Syst., 9/4, 385–403, 1995.

Jurković, A., Türk, K., Orlik, A., Auer, I., Böhm, R., Potzmann, R., and Schöner, W.: SON-ALP: Interpolationen der Sonnenscheindauer sowie Globalstrahlung in Hinblick auf Klimaänderungen im Alpenraum, Zentralanstalt für Meteorologie und Geodynamik, Vienna, Austria, Internal technical report, 66 pp., 2010.

Klein, T. and Persson, C.: Showcase EUROGRID – towards a European resource for gridded climate data, products and services, Adv. Sci. Res., 3, 85–89, doi:10.5194/asr-3-85-2009, 2009.

Peterson, T. C., Easterling, D. R., Karl, T. R., Groisman, P., Nicholls, N., Plummer, N., Torok, S., Auer, I., Böhm, R., Gullett, D., Vincent, L., Heino, R., Tuomenvirta, H., Mestre, O., Szentimrey, T., Salinger, J., Førland, E. J., Hanssen-Bauer, I., Alexandersson, H., Jones, P., and Parker, D.: Homogeneity adjustments of in situ atmospheric climate data: a review, Int. J. Climatol., 18, 1493–1517, 1998.

Schwarb, M.: The Alpine precipitation climate: evaluation of a high-resolution analysis scheme using comprehensive rain-gauge data, Swiss Federal Institute of Technology, Zürich, Switzerland, Dissertation 13911, 131 pp., 2000.

Zaninović, K., Gajić-Čapka, M., and Perčec Tadić, M.: Klimatski atlas Hrvatske: climate atlas of Croatia 1961–1990, 1971–2000, Meteorological and Hydrological Service of Croatia, Zagreb, Croatia, 200 pp., 2008.

# 3

# DRIAS: a step toward Climate Services in France

J. Lémond[1,2], Ph. Dandin[1], S. Planton[2], R. Vautard[4], C. Pagé[3], M. Déqué[2], L. Franchistéguy[1], S. Geindre[1], M. Kerdoncuff[1], L. Li[5], J. M. Moisselin[1], T. Noël[4], and Y. M. Tourre[1]

[1]Direction de la Climatologie, Météo-France, 42 av. G. Coriolis, 31057 Toulouse Cedex, France
[2]CNRM-GAME, URA1357, Météo-France, CNRS, 42 av. G. Coriolis, 31057 Toulouse Cedex, France
[3]Sciences de l'Univers au CERFACS, URA1875, 42 av. G. Coriolis, 31057 Toulouse Cedex 01, France
[4]LSCE, UMR8212, IPSL, FR636, laboratoire CEA-CNRS-Université Versailles Saint Quentin, Orme des Merisiers, 91191 Gif sur Yvette, France
[5]LMD, UMR8539, IPSL, FR636, CNRS-Université Pierre et Marie Curie, 4 place Jussieu, 75252 Paris Cedex 05, France

**Abstract.** DRIAS (Providing access to Data on French Regionalized climate scenarios and Impacts on the environment and Adaptation of Societies) is a 2-yr project (2010–2012). It is funded by the GICC (Management and Impact of Climate Change) program of the French Ministry of Ecology, Sustainable Development, Transportation, and Housing (MEDDTL). DRIAS is to provide easy access to French regional climate data and products in order to facilitate mitigation and adaptation studies. The DRIAS project focuses on existing French regional climate projections obtained from national modelling groups such as: IPSL, CERFACS, and CNRM. It is more than a data server, it also delivers all kinds of climate information from numerical data to tailored climate products. Moreover, guidance is to be provided to end-users in order to promote best practices and know-how. Whilst the project is coordinated by the Department of Climatology at Météo-France, a multidisciplinary group of users and stakeholders at large concerned by climate change issues is also involved with the project. The ultimate goal will be to identify societal needs, validate the decision making processes, and thus facilitate exchanges between producers and practitioners. Key results from the DRIAS project will contribute to the implementation of French Climate Services.

## 1 Introduction

Since the 4th IPCC assessment report in 2007 (AR4), climate adaptation has become a real challenge for societies. There is an increasing need to understand climate change (CC) impacts on different scales, in order to design multi-disciplinary and effective long-term plans. A broad spectrum of users is since then involved in CC impact studies, and subsequently needs for relevant information and support have increased. Climate scientists are not the sole users of climate projections; other stakeholders are from local and/or national authorities, including policy/decision makers and industrials.

Progress made in regional climate modelling has improved results from impact studies (Christiensen et al., 2007; Nobre et al., 2010). Many scenarios and global simulations under the aegis of IPCC are used for different regional modelling and downscaling experiments. As a result, a large amount of new information becomes available but it is not always in the most suitable format.

The concept of "Climate Services" (CS) is brand-new (see GFCS, 2011). CS are thought to serve various needs related to CC issues. The gap between climate laboratories and society is still wide, both from technical (data formatting, volume, and management) and knowledge (climate scientific background, scenarios) points of view. The first action to implement is to facilitate users to access available data and products in useful formats.

The aim of the DRIAS project is to develop such a *facility*: "Providing access to Data on French Regionalized climate scenarios and Impacts on the environment and Adaptation of Societies". The project is a joint effort between the French climate modelling groups and Météo-France. DRIAS is basically to integrate already existing but widely distributed components from partners and bring together their knowledge, tools, and practices. It is also to facilitate the transfer and delivery of relevant information to stakeholders at large. The project is also to promote good practices and raise awareness regarding uncertainties attached to CC.

## 2 DRIAS rationale

### 2.1 Adaptation and Climate Services

Adaptation to CC has become a critical challenge throughout the world. It has received paramount attention at the international (United Nations' Global Framework for Climate Services, GFCS, 2011), European (European Union White Paper, EUWP, 2009), and national (National Plan of Adaptation to Climate Change in France, PNACC, 2010) levels. The adaptation concept has been introduced in a new French law (the so-called "Grenelle II", http://www.developpement-durable.gouv.fr/-L-adaptation-au-changement-.html), with regional implementation starting in 2011.

Recent estimates on investment for CC adaptation are over 400 billion dollars/year (Pary et al., 2009). In France, the law makes it compulsory to consider CC issues in the managing process of any administrative entity with more than 50 000 inhabitants.

CS are still in their definition stages. Theoretically, they are the results of crossing products and services enabling persons/groups to assess their vulnerability and take appropriate adaptative and preventive measures. In this approach, DRIAS is to basically evaluate all information available from national climate modelling laboratories, process it, and make it accessible for complementary and additional studies. In that sense, DRIAS provides the necessary step towards climate services.

### 2.2 Strong users needs, requirement, and involvement

The Project is accompanied by two supplemental activities: *governance* through its steering committee and *guidance* provided by a multi-disciplinary users committee.

DRIAS aims at serving a wide range of interests and appears to be a critical milestone for the French climate adaptation strategy. The French Ministry of Ecology, Sustainable Development, Transportation, and Housing (MEDDTL) is currently developing such national strategy. Thus, the State is a primary stakeholder in supporting DRIAS. It comes through the GICC program (Management and Impacts of Climate Change, http://www2.gip-ecofor.org/publi/page.php?id=2&rang=0&domain=38&lang=en_GB). DRIAS is governed by a steering committee, which includes representatives of the Ministry and of GICC.

Various types of users are tackling multi-faceted impacts ranging from ecosystems to socio-economy. Skilled users who have recently emerged need access to tailored information. Nonetheless the disparity of skills is quite large. It starts with those in the scientific community who do not have the mandate to deliver data and products on an operational basis. DRIAS is to fill-in this gap, answering to the ever increasing number of requests.

- A first category of DRIAS users are scientists coming from various disciplines, dealing with various impacts and/or adaptation strategies. They are environmentalists, agronomists, hydrologists, geographers, economists, sociologists, and so forth. Some of them have historically established linkages between meteorology and climatology (such as agronomists, hydrologists), and have already initiated contacts with climate groups. Others need advice and guidance.
- A second category is made of so-called intermediate users. They are, for example, consulting companies involved in environment and policy management and planning, or outfits supporting adaptation for private companies and/or administrative entities. They also need appropriate information and support. They will act as intermediate users for climate scenarios and products. In some cases, they still need to raise awareness of local impacts from global changes.
- The third category of users includes climate scientists identified as partners of the project. The added pressure on scientific teams has been increasing, and in some cases is bringing scientists away from theoretical science (as opposed to applied). There is a need to deliver climate scenarios operationally, as well as allow scientists to focus on the challenges populations will face. This is the reason why Météo-France, used to deliver meteorological information, took the lead in DRIAS.

The user committee is comprised of 20 members from various areas (public and private organizations, associations, research and decision/policy makers). Its role is to help:

1. define the effective and strategic needs;
2. evaluate prototypes by beta testers;
3. validate choices made by the project team, and
4. ensure that DRIAS will continue to meet user expectations in the forseeable future.

Users have been invited to join the project during its pilot-project phase. A seminar was held in September 2009 at the MEDDTL. The latter took this opportunity to re-affirm its commitments for success. A sub-group of motivated and representative users were thus selected to follow DRIAS progress. The latter sub-group mission is also to feed requirements and specifications into DRIAS, and test its feasibility.

### 2.3 DRIAS within on-going community efforts

DRIAS is fully integrated into the French community, displaying interests regarding CC, impacts, and adaptation. Associated to the Météo-France climatology team, DRIAS also benefits from contacts with an international network of users. It benefits from the involvement of meteorologists from the World Meteorological Organization (WMO), the current GFCS, and the Commission of Climatology from the United Nations.

DRIAS is linked with GICC ECOFOR and reports to the managing and scientific advisory committees. Agencies clarify procedures to be followed for scenarios deliveries and tailor products to the public sector (according to the European Commission directive on the re-use and updating of public sector information). Such procedures are applied through the GICC programme and are closely tracked by the project manager. Moreover, seminars are held yearly with all projects funded by GICC, enabling fruitful exchanges between multi-sectoral communities. This gives additional opportunity for DRIAS to hear about new users' requirements and/or refine specifications.

Lastly, the partners are also involved in other projects at national, European, and international levels. Many French projects provide useful inputs for DRIAS. Since they take part in the IPCC exercise, they deal with data exchange, formatting, metadata, and delivery of scenarios (e.g. GIS Climat Prodiguer, http://www.gisclimat.fr/projet/prodiguer; FP Metafor, http://metaforclimate.eu/). Another project of interest for DRIAS is IS-ENES (Infra-Structure for the European Network for the Earth System Modelling, https://is.enes.org/). It aims is to develop a commonly distributed modelling research infrastructure in Europe. This is to facilitate the development and use of global climate models and better fulfil the societal needs with regards to CC issues. DRIAS and IS-ENES have similar objectives and agenda, but different spatial scales. DRIAS will benefit from the IS-ENES experience and vice-versa.

The GICC programme, the users' committee representing a broader community, and the Ministry relay reports and information on DRIAS through their various channels of information (e.g. GICC Newsletter[1], GICC Seminars[2],...).

### 2.4 DRIAS is a facility

DRIAS basically aims at transferring on-line a tool-box currently used by scientists on demand. This requires a transfer of know-how to operational teams. A range of climate information, ranging from numerical data to tailored graphical products, will be delivered. Standard formats, easy access, quick looks, FAQ, and simple products for a wide range of users will be offered. A dedicated web portal, presented in detail in Sect. 3.4, will be set up. It will enable access to various simulations obtained from different models, scenarios, and downscaling techniques. The scope of this service is limited to French regional climate simulations.

It is recalled that DRIAS is firstly a *facility*. It will be based upon existing tools and know-how accumulated by the climate related groups and agencies, partners of the project to namely:

- Form a generic system with up-to-date internet technologies;
- Provide access to different climate data and products;
- Enrich and host new services through a friendly-user environment

Figure 1 illustrates the complexity of processes leading to climate products. Explaining these processes to users and guiding them through such complexity is a real challenge. It is of the utmost importance, however, to clearly show that climate scenarios remain only scenarios, and that they must be used – and delivered as such. The web portal will also guide users on how to get used to available materials. It will try to give expertise and guidance, especially addressing the various sources of uncertainties.

Access to several greenhouse gases (GHG) emission scenarios using different regional climate models and downscaling methods will be facilitated. Different delivery phases have been proposed by the French climate groups leaders. At the end of the project, scenarios elaborated within the Scampei (Climate Scenarios for Mountain Areas: Extreme Events, Snow Cover, and Uncertainties, http://www.cnrm.meteo.fr/scampei/index.php) project will be offered to users. Scampei is not only dealing with orographic impacts of CC, but it also promotes a better harmonization of French climate models. DRIAS, together with Scampei, are concrete endeavors highlighting the wishes of the climate community to improve the inter-operability of various models and products.

The first version of the DRIAS web portal will at best include products showing the existing diversity coming from different models (ARPEGE-Climat, IPSL-CM models, among others), different scenarios (A1B, B1, A2...), and different downscaling techniques (statistical and dynamic approaches). The wide diversity of simulations allows for taking into account uncertainties associated with climate simulations. An example of a basic product following this approach will be taken from a report to be delivered to the French government about the scenario to be used for national adaptation: it will clearly emphasize the need for a multi-model approach, multi-scenarios, and downscaling techniques (see Fig. 2). This report will also illustrate how the project benefits from various actions and features of the DRIAS web portal as the repository for information and knowledge transfer between research and operational activities.

[1] http://www.gip-ecofor.org/publi/page.php?id=502&rang=0&domain=38&lang=fr_FR, Newsletter GICC, July 2010

[2] http://www.gip-ecofor.org/publi/page.php?id=521&rang=0&domain=38&lang=fr_FR, GICC seminar in Aussois, January 2011

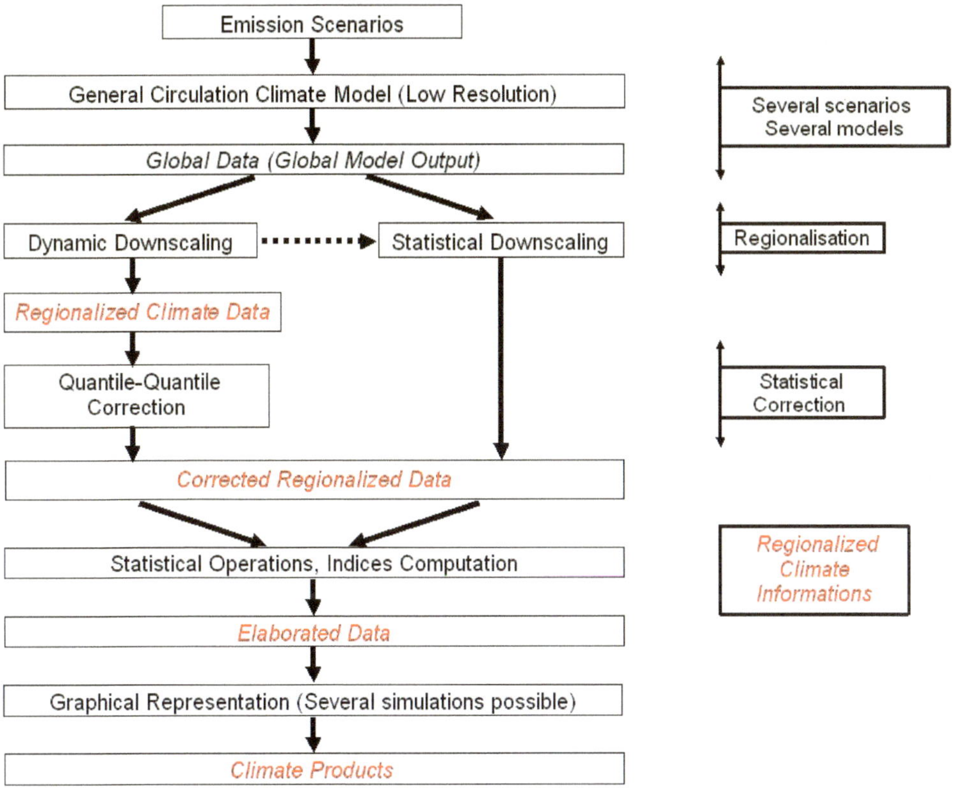


**Figure 1.** Production process modelling of regionalized climate information. Such a suite is complicated to understand for users. Moreover, each step introduces its own source of uncertainties, which add on the complexity of the delivered informations and products.

## 3 Preliminary accomplishments of the project

The project started with basic guidelines:

- Tested benchmarks and best practices for the partnership;
- Availability of existing scenarios. *No new experiment to be run by (or for) DRIAS*;
- Existing infrastructure and know-how to be used as much as possible. *Besides the existing "Météo-France Climathèque" whose back office system will be partly re-used, it was decided to build a dedicated website not only for product delivery, but also for guidance and information*;
- Users involvement. DRIAS is to take into account users' requirements. *Apart from the existing users' committee, a survey of users at large has been conducted.*

A brief summary of the first tasks and preliminary results, as well as insights for the upcoming website and foreseen products are given hereafter.

### 3.1 Benchmarking: a quick glance at other services

The study of already existing and similar and partial initiatives was the first task conducted. This benchmark covered various programmes in Europe and worldwide, pin-pointing best practices for each one. The British program UKCIP (United Kingdom Climate Impacts Program) is one the best examples: UKCIP provides a broad spectrum of climate information, from numerical data to decision-making tools; it supports and guides users whilst spreading knowledge and know-how for climate adaptation. Other relevant initiatives were noted in Germany, Spain, and various countries. Differences were due to local characteristics of different development stages. This quick survey provided a strong backbone for DRIAS.

DRIAS builds naturally on partners' knowledge and expertise. The numerous simulations, tools, and structures developed by the climate community in France during the last two decades have been extremely useful for getting users' feedback and developing appropriate tools and products. This invaluable asset constitutes a DRIAS pillar. For example, the choice was made to use the same delivery mechanisms as in "Climathèque" (http://climatheque.meteo.fr/). But it was also decided to build a new and dedicated web portal, which should become a key part of a "*French portal on*

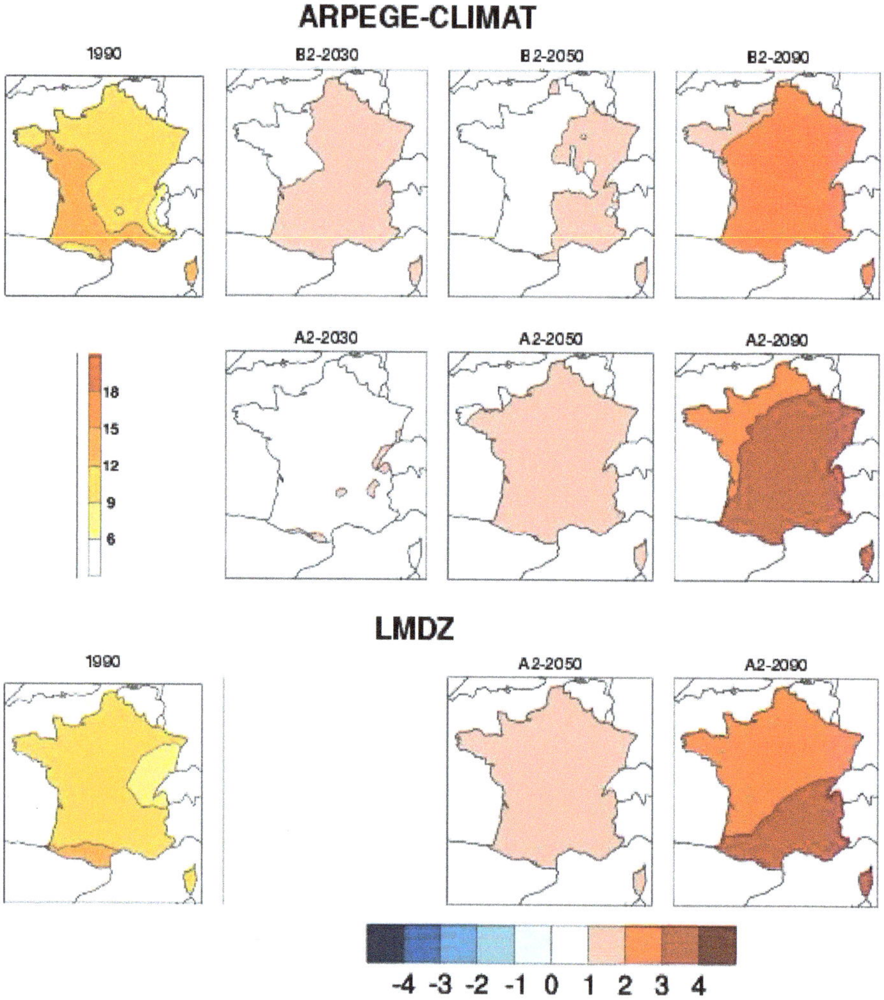


| | | | | | | | |
|---|---|---|---|---|---|---|---|
| **FRANCE** | | | | | | | |
| ARPEGE | 1990 | B2-2030 | B2-2050 | B2-2090 | A2-2030 | A2-2050 | A2-2090 |
| | 11.4 | 0.7/1.5 (-0.3/2.2) | 0.6/1.4 (-0.1/2.1) | 1.9/2.6 (1.2/3.1) | 0.5/1.2 (0.0/1.8) | 0.9/1.8 (0.0/2.6) | 2.7/3.7 (1.9/5.3) |
| LMDZ | 1990 | | | | | A2-2050 | A2-2090 |
| | 10.3 | | | | | 1.3/1.8 (0.2/3.6) | 2.6/3.0 (1.6/4.1) |
| **NORD-OUEST** | | | | | | | |
| ARPEGE | 1990 | B2-2030 | B2-2050 | B2-2090 | A2-2030 | A2-2050 | A2-2090 |
| | 11.5 | 0.6/1.4 (-0.7/2.2) | 0.6/1.3 (-0.1/2.2) | 1.7/2.5 (1.0/3.0) | 0.4/1.1 (-0.1/1.8) | 0.9/1.7 (-0.1/2.3) | 2.5/3.5 (1.8/4.9) |
| LMDZ | 1990 | | | | | A2-2050 | A2-2090 |
| | 10.3 | | | | | 1.3/1.8 (0.2/3.4) | 2.3/2.8 (1.3/3.9) |
| **NORD-EST** | | | | | | | |
| ARPEGE | 1990 | B2-2030 | B2-2050 | B2-2090 | A2-2030 | A2-2050 | A2-2090 |
| | 10.6 | 0.7/1.6 (-0.3/2.3) | 0.6/1.5 (-0.1/2.4) | 1.9/2.8 (1.0/3.4) | 0.4/1.3 (0.0/2.0) | 0.9/1.9 (-0.2/2.8) | 2.8/3.9 (1.9/5.5) |
| LMDZ | 1990 | | | | | A2-2050 | A2-2090 |
| | 8.7 | | | | | 1.4/1.9 (0.1/3.7) | 2.6/3.1 (1.4/4.2) |
| **CENTRE** | | | | | | | |
| ARPEGE | 1990 | B2-2030 | B2-2050 | B2-2090 | A2-2030 | A2-2050 | A2-2090 |
| | 11.5 | 0.6/1.4 (-0.7/2.3) | 0.6/1.4 (-0.2/2.3) | 1.8/2.6 (1.0/3.2) | 0.4/1.2 (-0.2/1.8) | 0.9/1.8 (-0.1/2.6) | 2.7/3.7 (1.8/5.2) |
| LMDZ | 1990 | | | | | A2-2050 | A2-2090 |
| | 10.3 | | | | | 1.2/1.8 (0.2/3.3) | 2.2/2.7 (1.2/3.8) |
| **SUD-OUEST** | | | | | | | |
| ARPEGE | 1990 | B2-2030 | B2-2050 | B2-2090 | A2-2030 | A2-2050 | A2-2090 |
| | 12.2 | 0.7/1.5 (-0.3/2.3) | 0.6/1.3 (-0.2/2.1) | 1.8/2.6 (1.3/3.1) | 0.4/1.2 (-0.3/1.8) | 0.9/1.8 (-0.1/2.6) | 2.7/3.8 (1.8/5.6) |
| LMDZ | 1990 | | | | | A2-2050 | A2-2090 |
| | 11.3 | | | | | 1.4/1.9 (0.2/3.9) | 2.7/3.2 (1.8/4.4) |
| **SUD-EST** | | | | | | | |
| ARPEGE | 1990 | B2-2030 | B2-2050 | B2-2090 | A2-2030 | A2-2050 | A2-2090 |
| | 11.1 | 0.8/1.6 (-0.1/2.5) | 0.6/1.4 (-0.1/2.1) | 2.0/2.7 (1.4/3.2) | 0.5/1.3 (0.0/2.0) | 1.0/1.9 (0.2/2.9) | 2.9/4.0 (1.8/5.7) |
| LMDZ | 1990 | | | | | A2-2050 | A2-2090 |
| | 10.3 | | | | | 1.4/1.9 (0.1/3.9) | 2.9/3.4 (1.9/4.6) |

**Figure 2.** Example of typical climate product that could be available on DRIAS web portal. In this case, mean daily temperature in annual average is represented (Unit: °C) for the reference period and deviations between projections and reference. Simulations from two models (ARPEGE and LMDZ) are shown. Interval of uncertainty for mean value and extreme value (in brackets) are computed for 5 major French regions. From a 2011 report given to the Ministry, written by the partners of the DRIAS project (available on the website: http://www.developpement-durable.gouv.fr/Les-scenarios-climatiques-de.html).

*CC and adaptation*" currently being constituted by the ONERC (National Observatory on the Effects of Global Warming) website[3].

### 3.2 Users survey

An important step was to conduct a survey of users at large. From the survey's preliminary results it became apparent that users have difficulties in expressing their specific needs. Exceptions, of course, were from research groups already used to CC impact studies. By and large three categories of users need to be fed data appropriately:

- "*Skilled researchers*" requesting numerical data to run impact models. DRIAS will only be a facilitator in this case (i.e. a relay for researchers who will no longer have to deal with the time consuming data delivery process);
- "*Other users*" looking for simpler data sets, indices, products, charts, that they could use for illustrating CC future, and thus design new plans within their own communities. Of course there is a continuous spectrum of users as revealed by the survey.
- "*Intermediate users*" could eventually emerge as a third category with a much large spread of requirements.

All of the above means that the work to be accomplished depends upon organizing data and products from the research realm, transferring tools (tool-box), data, and tailored products for operational purposes.

The above results have been shared with the IS-ENES community, complementing the work carried on at the European level (e.g. Copenhagen workshop, https://is.enes.org/events/is-enes-workshop-on-bridging-climate-research-data-and-the-needs-of-the-impact-community). The IS-ENES own survey gives results which are pretty close to those obtained at the national scale. It is generally agreed that the results obtained from climate modelling groups are much too sophisticated for many users. This is additional feedback with similar results from other climate "services", e.g. UKCIP (see http://www.ukcip.org.uk/ukcp09/what-users-wanted/; EumetNet, among others).

The challenge of our community is thus: *How to simplify data and products without loosing the content*? At the intermediate level DRIAS is to alleviate this challenge.

### 3.3 Available regionalized climate information for DRIAS

DRIAS is not to handle and deliver global climate models scenarios. As said before, other initiatives are already involved in these efforts: the IS-ENES European project, or the French Prodiguer project. They aim at providing easy access to CMIP and other research scenarios and work. DRIAS will basically deliver the following regional climate products for France. They were selected by the scientific partners since they were already in use nationally for many impact studies:

- A first set of data corresponds to results from the ARPEGE-Climat (Salas y Mélias et al., 2005) and LMDz (Hourdin et al., 2006) models. Both models have a "*zooming*" capability, increasing spatial resolution within a given area. The horizontal resolution is of 50 and 20 km. DRIAS will facilitate the delivery of products from the "*nested*" limited area model MM5 (5th Penn-State University community model, after Grell et al., 1995) with a configuration similar to that of Cattiaux et al. (2009).
- A second set of data corresponds to the DSCLIM downscaling method based on weather types and analogues, and developed at CERFACS (Boé et al., 2006; Pagé et al., 2009). The downscaling method includes a learning process from a reference period with known climate characteristics in order to build a statistical model (or transfer function). Technique can then be applied for future climate simulations. The spatial resolution is of 8 km, which is used by Météo-France (Quintana-Segui et al., 2008; Vidal et al., 2010).
- A third set of data is from the "Scampei project" and other "Scampei-like" scenarios if and when available. One set is currently being prepared by the IPSL group using MM5. The spatial resolution is of 8 km.

These various data sets correspond to different stages of climate modelling, scenario delivery, and impact studies in France. While old scenarios are kept because some users require stability, newer scenarios are introduced since they benefit from recent progress made in climate modelling. The integration of the next generation Scampei scenarios to be used in France will be of paramount importance for DRIAS. They will also include and illustrate the uncertainties from climate projections. Biases in dynamical downscaling outputs can be corrected using the quantile-quantile correction method (Déqué, 2007), associating each percentile of the model to the corresponding percentiles from observations.

The integrated effort was proposed not only by DRIAS. Indeed, the scientific community has been working hard for a long time to become more inter-operable. However, a facility such as DRIAS invites scientists to work effectively toward such harmonization. They are linked with operational partners in charge of integrating and disseminating scenarios. For example, if "biodiversity" is important for science, a relative homogenization is vital for operational processing and inclusion into the end-to-end channel. This is the push/pull model between producers and users.

[3] http://www.developpement-durable.gouv.fr/-Impacts-et-adaptation-ONERC-.html

After putting together "Regionalized Climate Data", "Elaborated Data" is obtained by computing quality-controlled (QC'ed) statistical parameters and indices. Finally, graphical representations of improved products will be called "Climate Products". This chain production is synthesized in Fig. 1. All outputs and products will be delivered through the DRIAS web portal.

### 3.4 The upcoming DRIAS website

The DRIAS web portal will comprise three main areas: *Discovery*, *Support*, and *Delivery*:

- The *Discovery* area will contain general information for a wide range of people, thus allowing a first grasp on climate scenarios. It is not only thought of as a place where a wide public can visualize climate scenarios outputs (e.g. Météo-France Climate Simulator, see http://climat.meteofrance.com/chgt_climat/simulateur/), it is also positioning DRIAS to fulfil the INSPIRE European directive. Users will be able to "*discover*" data and "*unveil*" new products: quick looks, bar charts, and maps.

- In the *Support* area, a panoply of documents will be offered in order to assist users on how to make the best use of available climate information (glossary, description of methods and climate models, FAQ). The user will find information on scenarios, on CC, and on past experiences. This part of the site will link toward already existing educational and information sites. It will be more specific about products to be delivered, explaining how they were created, and trying to guide users' choices. A hot-line mechanism involving experts is to be implemented at this level.

- The *Delivery* area is more classical. The user shall be able to order, in digitized format, data and products which have first been identified from the Discovery area. The partners will re-use as much as possible the production engine at Météo-France for delivering meteorological products through the so-called "Climathèque". With this user-friendly system delivered products may take various forms. Figure 2 illustrates a possible future DRIAS product. It basically concatenate on the same diagram various maps for a single given parameter, with different time periods being considered and different GHG scenarios issued from various models. There is indeed no "one reference scenario", or even worse "the reference scenario", but an ensemble of equi-probable scenarios that need to be handled properly. Added-values will include uncertainties and error-bars – so that the user remains constantly aware that CC information is not just a deterministic short range weather forecast!

## 4 Summary and discussion

An initial integration phase has been conducted by the teams from the Directorate of Climatology at Météo-France, the leader of the DRIAS project. The implementation of the web portal will take place this year, 2011.

DRIAS climate services have been described and preliminary outputs from DRIAS should appear at the end of 2011. A "Climate Service" is not only a data base, but also a facility to ease the transfer of climate knowledge from data producers to stakeholders at large. It has to provide different kinds of information with relevant support to assist in the diffusion of good practices and users training and education. Linkages between producers and users are improved through communication and two-way dialogues. This is true for any service, but especially with CS since climate modelling and uncertainties are rather complex to deal with. DRIAS is to explore the most efficient ways to provide this dialogue, at the metropolitan French scale.

In this context, DRIAS is a main ingredient of the upcoming "French Climate Services". It proposes to make available, in different forms (from numerical data to sophisticated and tailored products), climate information from regionalized simulations developed at IPSL, CNRM, and CERFACS. These climate simulations are based on several assumptions of GHG emissions and different downscaling methodologies (all taking uncertainties into account). Climate data products will be associated with appropriate support methods, user guides, and documentation. It should facilitate the transfer of information to the ever-increasing users' demands, from research teams to dedicated teams, enabling modellers to focus more on research activities, as well as spreading climate information on a broader scale. Furthermore, DRIAS will help users get a first assessment of future possible CC with uncertainties.

The project takes advantage of existing tools, know-how, and expertize available in various laboratories, the national meteorological service, and other initiatives worldwide such as UKCIP and the new promising Climate Service Centre in Germany.

While DRIAS facilitates the delivery of the CC scenarios, it contributes also to the developing climate modelling activities of French groups by showing how harmonization becomes essential in order to ease products delivery and access.

## Acronyms

| | |
|---|---|
| CEA | Commissariat à l'Energie Atomique (Atomic Energy Commission) |
| CERFACS | Centre Européen de Recherche et Formation Avancées en Calcul Scientifique (The European Centre for Research and Advanced Training in Scientific Computation) |

| | |
|---|---|
| CNRS | Centre National de la Recherche Scientifique (National Center for Scientific Research) |
| CNRM | Centre National de la Recherche Météorologique (National Center of Meteorological Research) |
| GAME | Groupe d'étude de l'Atmosphère Météorologique (Group Study of Meteorological Atmosphere) |
| IPSL | Institut Pierre Simon Laplace (Pierre Simon Laplace Institut) |
| LMD | Laboratoire de Météorologie Dynamique (Laboratory of Dynamic Meteorology) |

**Acknowledgements.** This project is funded by the Management and Impact of Climate Change program of the French Ministry of Ecology, Sustainable Development, Transportation, and Housing. Special thanks are due to the GICC scientific committee, its president, Claude Millier, CGDD, Isabelle Bénézeth, Daniel Martin, and ONERC, Nicolas Bériot, Michel Galliot, colleagues who strongly support, guide and challenge the project's team and leader. A particular appreciation is due to the DRIAS users committee and its Présidente, Sylvie Parey.

Edited by: J. Prior

The publication of this article is also financed by CNRS-INSU.

## References

Boé, J., Terray, L., Habets, F., and Martin, E.: A simple statistical-dynamical downscaling scheme based on weather types and conditional resampling, J. Geophys. Res., 111, D21106, doi:10.1029/2005JD006889, 2006.

Cattiaux, J., Vautard, R., and Yiou, P.: Origins of the extremely warm European fall of 2006, Geophys. Res. Lett., 36, L06713, doi:10.1029/2009GL037339, 2009.

Christensen, J. H., Hewitson, B., Busuioc, A., Chen, A., Gao, X., Held, I., Jones, R., Kolli, R. K., Kwon, W.-T., Laprise, R., Magaña Rueda, V., Mearns, L., Menéndez, C. G., Räisänen, J., Rinke, A., Sarr, A., and Whetton, P.: Regional Climate Projections, in: Climate Change 2007: The Physical Science Basis. Contribution of Working Group I to the Fourth Assessment Report of the Intergovernmental Panel on Climate Change, edited by: Solomon, S., Qin, D., Manning, M., Chen, Z., Marquis, M., Averyt, K. B., Tignor, M., and Miller, H. L., Cambridge University Press, Cambridge, United Kingdom and New York, NY, USA, 2007.

Déqué, M.: Frequency of precipitation and temperature extremes over France in an anthropogenic scenario: model results and statistical correction according to observed values, Global Planet. Change, 57, 16–26, 2007.

EU White Paper: Adapting to climate change: Towards a European framework for action, 17 pp., 2009.

GFCS: Knowledge for action: a global framework for climate services – empowering the most vulnerable, WMO Publications, ISBN 978-92-63-11065-7, 248 pp., 2011.

Grell, A. G., Dudhia, J., and Stauffer, D. R.: A Description of the Fifth-Generation Penn State/NCAR Mesoscale Model (MM5), NCAR/TN-398 + STR NCAR TECHNICAL NOTE, 122 pp., 1995.

Hourdin, F., Musat, I., Bony, S., Braconnot, P., Codron, F., Dufresne, J.-L., Fairhead, L., Filiberti, M.-A., Friedlingstein, P., Grandpeix, J.-Y., Krinner, G., Levan, P., Li, Z.-X., and Lott, F.: The LMDZ4 general circulation model : climate performance and sensitivity to parametrized physics with emphasis on tropical convection, Clim. Dynam., 27, 787–813, 2006.

IPCC: Climate Change, 2007: The Physical Science Basis. Contribution of Working Group I to the Fourth Assessment Report of the Intergovernmental Panel on Climate Change, edited by: Solomon, S., Qin, D., Manning, M., Chen, Z., Marquis, M., Averyt, K. B., Tignor, M., and Miller, H. L., Cambridge University Press, Cambridge, United Kingdom and New York, NY, USA, 996 pp., 2007.

Nobre, C., Brasseur, G. P., Shapiro, M. A., Lahsen, M., Brunet, G., Busalacchi, A., Hibart, K., Noone, K., and Ometto, J.: Addressing the complexity of the Earth system, B. Am. Meteorol. Soc., 91, 1389–1396, doi:10.1175/2010BAMS3012.1, 2010.

PNACC: Plan National d'Adaptation. Rapport des groupes de travail de la concertation nationale, 151 pp., 2010.

Pagé, C., Terray, L., and Boé, J.: dsclim: A software package to downscale climate scenarios at regional scale using a weather-typing based statistical methodology, Cerfacs Technical Report TR/CMGC/09/21, SUC au CERFACS, URA CERFACS/CNRS No. 1875, Toulouse, France, 2009.

Parry, M., Arnell, N., Berry, P., Dodman, D., Fankhauser, S., Hope, C., Kovats, S., Nicholls, R., Satterthwaite, D., Tiffin, R., and Wheeler, T.: Assessing the cost of adaptation to climate change: a review of the UNFCCC and other recent estimates", IIED and Grantham Institute for Climate Change, London, 2009.

Quintana-Segui, P., Moigne, P. L., Durand, Y., Martin, E., Habets, F., Baillon, M., Canellas, C., Franchisteguy, L., and Morel, S.: Analysis of Near Surface Atmospheric Variables: Validation of the SAFRAN analysis over France, J. App. Meteo. Clim., 47, 92–107, 2008.

Salas y Mélia, D., Chauvin, F., Déqué, M., Douville, H., Guérémy, J. F., Marquet, P., Planton, S., Royer, J. F., and Tyteca, S.: Description and validation of CNRM-CM3 global coupled climate model, Note de centre GMGEC, CNRM, 103, 2005.

Vidal, J.-P., Martin, E., Franchistéguy, L., Habets, F., Soubeyroux, J.-M., Blanchard, M., and Baillon, M.: Multilevel and multiscale drought reanalysis over France with the Safran-Isba-Modcou hydrometeorological suite, Hydrol. Earth Syst. Sci., 14, 459–478, doi:10.5194/hess-14-459-2010, 2010.

# 4

# Sensitivity of a long-range numerical weather forecast model to small changes of model parameters

**M. B. Gavrilov**[1], **G. R. Jovanović**[2], **and Z. Janjić**[3]

[1]Chair of Physical Geography, Faculty of Sciences, University of Novi Sad, Trg D. Obradovića 3, 21000 Novi Sad, Serbia
[2]Hydrometeorological Service of Serbia, Kneza Višeslava 66, 11000 Belgrade, Serbia
[3]National Centres for Environmental Predictions, Washington, USA

**Abstract.** Sensitivity of extended-range numerical weather forecasts to small changes of model parameters is studied for two cases. In the first case the Earth radius was perturbed. In the other case changes of the gravity were introduced. The results for the 500 hPa geopotential fields are presented on hemispheric maps and intercompared visually and using RMS differences of the perturbed and reference forecasts. During about the first 10 days of integration the results indicate modest sensitivity of the forecasts to the parameter variation. After this period the forecasts diverge rapidly and start to differ significantly. Repeated integrations on the same computer using the same model setup and the same initial conditions yield identical results.

## 1 Introduction

Predictability of the atmosphere and accuracy of atmospheric models are still among the central issues of meteorology and atmospheric modeling. Thompson (1957) was among the first to investigate the limits of predictability of the barotropic and baroclinic atmospheres. Lorenz (1969a, b, 1973) continued and expanded this research. These investigations showed that the limit of predictability for an atmospheric model was about 12 days. The sensitivity tests of numerical integrations to small changes in initial conditions in the works of Lorenz (1963, 1965) were not only important for testing the accuracy and predictability of atmospheric models, but also led to the discovery of a new phenomenon in nonlinear systems known as deterministic chaos. The problem of sensitivity of numerical integrations to small disturbances of initial conditions was addressed in more recent papers, e.g., by Zhao-Xia et al. (1997) and Yoden (2007).

In the present paper, we keep the initial conditions unchanged, and instead vary within the respective limits of their natural variability two model parameters that are commonly assumed to be constant. The two parameters chosen for testing are the Earth radius and gravity. By doing so we wanted to demonstrate that the growth of differences between the solutions can be influenced not only by uncertainties in the initial conditions, but also by other small uncertainties present in the system. The specific mechanisms of the growth of the differences may be similar in both cases, though.

## 2 The model

The new NCEP's Nonhydrostatic Multiscale Model on the B grid (NMMB) (Janjić, 2005; Janjić and Black, 2007; Janjić et al., 2011) was used in the tests. The NMMB was designed for a broad range of spatial and temporal scales and follows the general modeling philosophy of the NCEP's WRF NMM grid-point dynamical core (Janjić et al., 2001; Janjić, 2003,; Janjić et al., 2010). However, in contrast to the WRF NMM that was defined on the Arakawa E grid, the NMMB was reformulated for the Arakawa B grid. The NMMB uses the regular latitude-longitude grid for the global domain. Finite-volume horizontal differencing is employed that conserves a variety of basic and derived dynamical and quadratic quantities and preserves some important properties of differential operators. The conservation of energy and enstrophy (Janjić, 1984) improves the accuracy of the nonlinear dynamics of the model on all scales, and renders the model suitable for extended integrations. Conservative polar boundary conditions are used in global integrations. The polar filter selectively reduces tendencies and thus decelerates the wave components that would otherwise propagate faster in the zonal direction than the fastest wave propagating in the meridional direction. In the vertical, the hybrid pressure-sigma coordinate is used (Simmons and Burridge, 1981; Eckerman, 2009). Explicit, no split time differencing is applied in the model dynamics (Janjić, 2003). An affordable conservative, positive definite and monotone scheme was developed for model tracers (Janjić et al., 2009). The model physics is an evolved version of the NCEP's standard WRF NMM physical package. It includes land surface processes (Vuković et al., 2010), turbulence (Janjić, 1990, 2001), microphysics (Ferrier et al., 2002), moist convection (Janjić, 1994, 2000) and radiation (Harshvardhan et al., 1987).

## 3 Experiment setup

Two sensitivity tests of extended-range numerical simulations to small changes of Earth radius and gravity were made. The experiments were performed using reduced reference value of the Earth radius $a = 6\,376\,000$ m by 0.000002%, which resulted in $a = 6\,375\,999$ m, and by reducing the reference value by 0.5%, which yielded $a = 6\,344\,000$ m (Gavrilov, 2006). In the spherical approximation used in the atmospheric equations the Earth radius is assumed to be a constant (e.g. $a = 6\,376\,000$ m), although the Earth's shape is a geoid with the radius at the equator being $a = 6\,378\,000$ m and at the poles being $a = 6\,357\,000$ m. Moreover, the shallow approximation ignores the radius change with height. Instead of varying between the bottom and the top of the atmosphere, that makes a distance of about 30 000 m in atmospheric models, with this approximation the radius is taken to be constant and equal to its value at the sea level (Phillips, 1966). It is therefore evident that the variation of the Earth radius used in the experiments remains within the range of natural variability of the Earth radius. Thus, the perturbations used here have a realistic scale.

In the second set of experiments, the reference value of the gravity $g = 9.8060226$ ms$^{-2}$ was increased by 0.0001% of the reference value to $g = 9.8060326$ ms$^{-2}$, and reduced by 0.1% of the reference value to $g = 9.7960226$ ms$^{-2}$ (Gavrilov et al., 2007). In the atmospheric models, the gravity is mostly taken to be constant (e.g. $g = 9.8060226$ ms$^{-2}$) although it is well known that it is a function of latitude, altitude and inhomogeneity of the Earth's interior. For instance, because of changes of centrifugal force along the latitude and flattened shape of the Earth at the poles, the gravity is 0.5% higher at the poles than at the equator. Also, as the altitude is increasing, the gravity is decreasing, so that at the altitude of 1000 m above the sea level it is less than at the sea level by 0.00389 ms$^{-2}$. Beside these theoretically predictable changes of gravity, there are gravitational anomalies occurring as the effect of inhomogeneity of the inner layers of the Earth. For instance, in North Pacific close to Hawaii, the gravity is for 0.002 ms$^{-2}$ greater than the reference value. Thus, it is evident that the variation of gravity in our esperiments and the natural variability of gravity are of the same order of magnitude.

In the tests the model integrations were performed up to 40 days. The horizontal resolution was about 0.67 degrees and there were 32 levels in the vertical. The time step was 160 s.

## 4 Results

The results shown were obtained from numerical integrations starting from the NCEP's GFS analysis for 4 February 2006, at 00:00 UTC. In Fig. 1 the 500 hPa geopotential fields at days 10 and 14 are presented, respectively. The results of the reference run and the two runs with perturbud values of the Earth radius are shown in the top to bottom panels in the figures. The results with perturbed gravity are presented in Fig. 2 using the same arrangement. The RMS differences of the geopotential fields at 500 hPa between the perturbed and reference runs are computed at days 5, 10, 14, 20, 30 and 40 and shown in Fig. 3 for the perturbations of the Earth's radius and gravity, respectively.

As can be seen in Figs. 1 and 2 (left columns), during the first 10 days of integration the forecasts remain generally similar. At the 5th day the RMS differences were less than 2 gpm for both tested parameters, while at the 10th day the RMS difference was about 12 gpm for the gravity perturbation case, and 18.5–20.7 gpm for the Earth radius perturbation case. There was no indication of regularity in the appearance of the differences and they did not show clear dependence on which parameter is perturbed and for how much.

Between the 11th and 13th day of integration a rapid degradation of previously existing similarity occurred in all cases. As shown in Figs. 1 and 2 (right columns), at day 14 the differences between the reference fields and the fields obtained using pertubed parameters are no longer small and the fields do not resemble each other any more. The differences are such that one cannot detect any indication that the integrations started from the same initial conditions. The similarity between the forecasts has never been regained in further integrations up to 40 days. As can be seen from Fig. 3, the increasing visual differences in forecast fields are confirmed by significant increase of RMS differences at days 14, 20, 30 and 40. These RMS differences take on values from 45 gpm to 95 gpm, which are comparable to typical RMS errors for forecasts of 5–10 days (Shukla, 2005).

Analogous results were obtained in integrations starting from NCEP's GFS analyses for the whole period 4–14 February 2006 at 00:00 and 12:00 UTC. Comparable results were obtained also for other values of the Earth radius and gravity within the range of the changes discussed here. However, repeated integrations on the same computer and using the same values of model parameters gave identical final results.

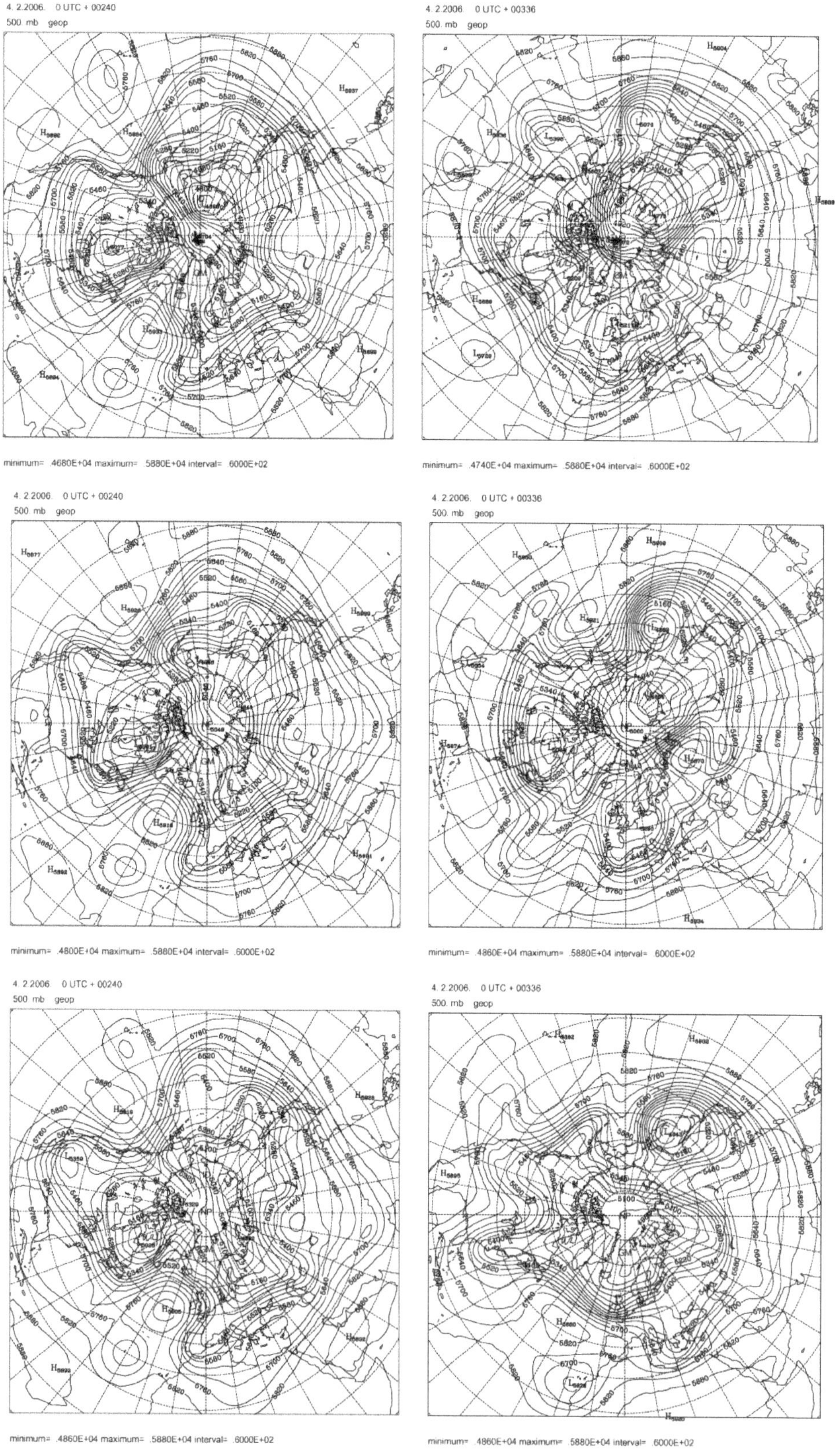


**Figure 1.** 10-days forecasts (left) and 14-days forecasts (right) of geopotential at 500 hPa for $a = 6\,376\,000$ m (top), $a = 6\,375\,999$ m (middle) and $a = 6\,344\,000$ m (bottom).

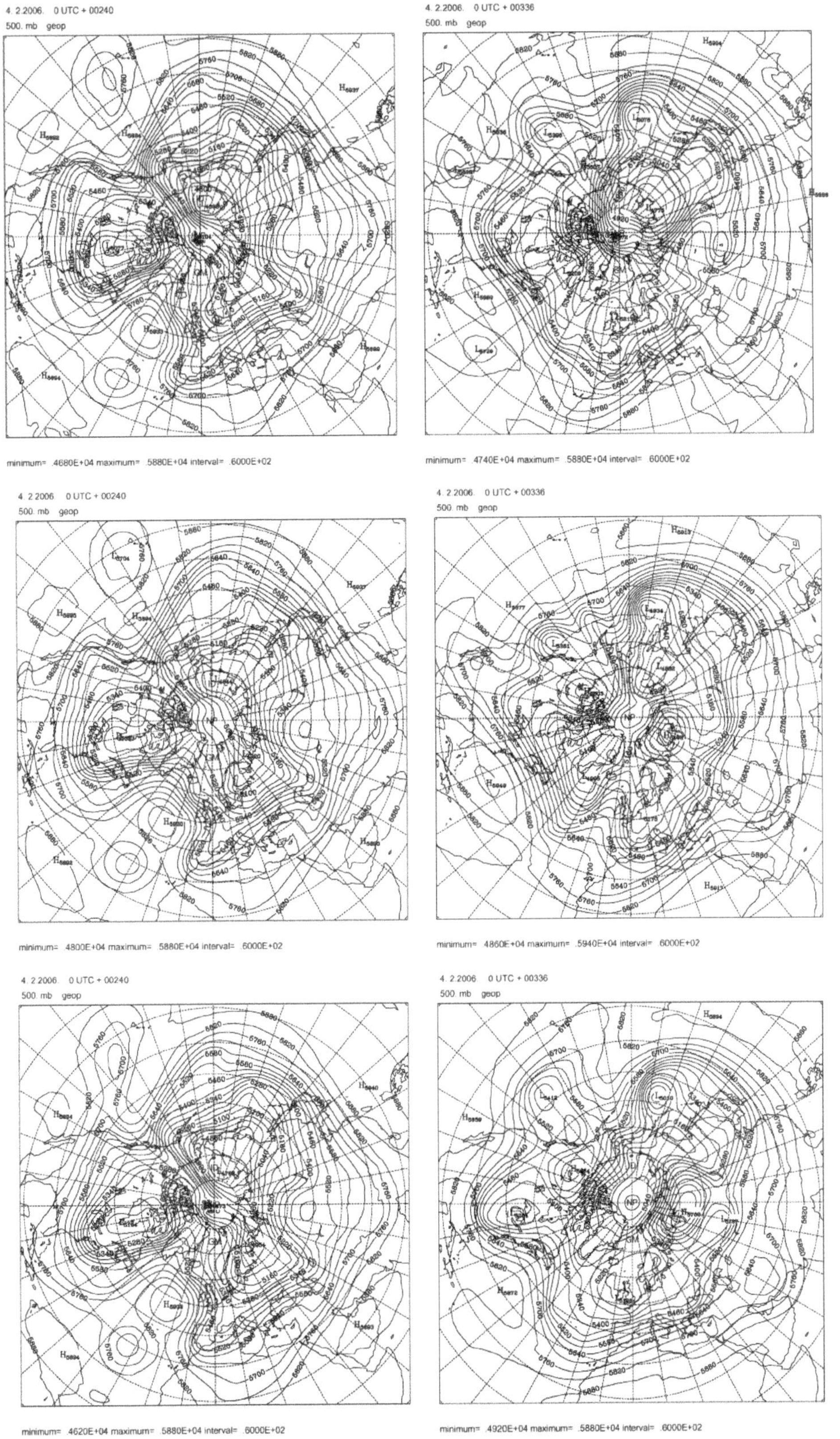


**Figure 2.** 10-day forecasts (left) and 14-days forecasts (right) of 500 hPa geopotential for $g = 9.8060226\,\mathrm{ms}^{-2}$ (top), $g = 9.8060326\,\mathrm{ms}^{-2}$ (middle) and $g = 9.7960226\,\mathrm{ms}^{-2}$ (bottom).

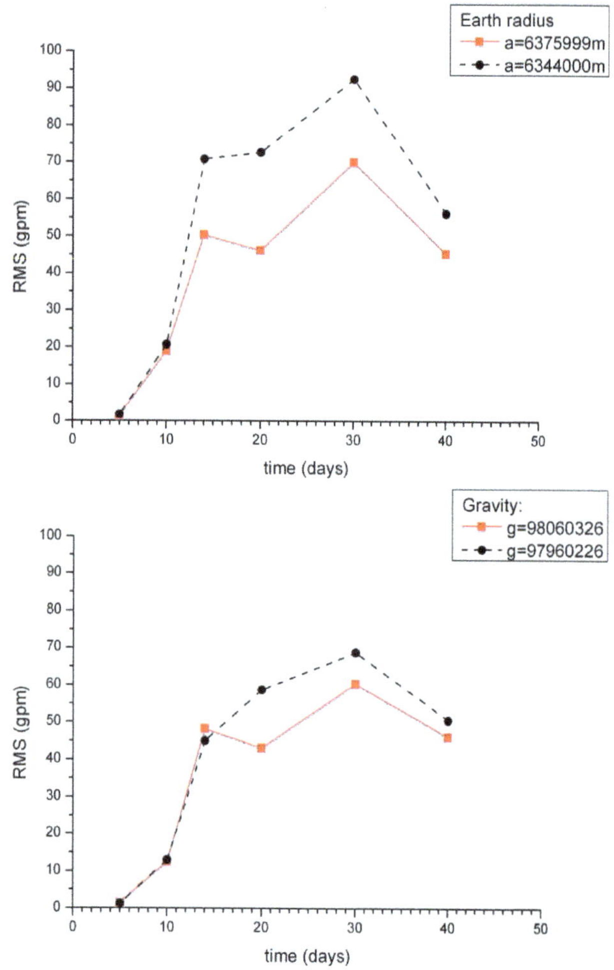


**Figure 3.** RMS differences for small changes of Earth radius (top) and gravity (bottom) as indicated in the legend.

## 5 Summary and conclusions

The sensitivity has been investigated of extended-range numerical simulations to small perturbations of Earth radius and gravity within the limits of their natural variabilities. Regardless of the value used for the Earth radius and gravity, the weather patterns have kept mutual visual similarity during the first 10 days. High similarity of the fields is confirmed by low values of their RMS differences. Within this period, the forecasts still gave an impression that they all started from the same initial conditions.

After this period of similarity, the results started to differ. Between the 11th and the 13th day, the differences became significant even on large scales, and the similarity never reappeared again in further integrations. High diversity of the fields is confirmed by large values of their RMS differences. In these later stages the predicted flow patterns give an impression that the forecasts were made starting from different initial conditions.

Since the results of numerical weather prediction have not shown considerable sensitivity to the small changes of model parameters in the first 10 days, these forecasts may be considered to fall in the predictable range and may have practical value. Contrary to that, great sensitivity of the deterministic weather forecasts to the small changes beyond day 10 makes the possibility of practical use of such forecasts highly questionable.

The obtained results appear to be in accord with the theoretical assessment that the deterministic predictability of the atmosphere is limited to about 12 days (Lorenz, 1973). However, the results of this study show that the growth of differences between the solutions can be influenced not only by uncertainties in the initial conditions, but also by other small uncertainties present in the system, although the specific mechanisms of the growth of the differences may be similar in both cases. It would be interesting to further investigate the influence of small changes of other model parameters (e.g., the angular velocity of the Earth and various physical and numerical constants) on the results of deterministic medium-range numerical weather forecasts.

**Acknowledgements.** This work was supported by Project #176020 of the Serbian Ministry of Science and Technological Development. The authors sincerely appreciate the efforts of Igor Pisnichenko and another two anonymous reviewers in improving this manuscript.

Edited by: B. Lalic

## References

Eckermann, S.: Hybrid $\sigma$-p coordinate choices for a global model, Mon. Wea. Rev., 137, 224–245, 2009.

Ferrier, S. B., Jin, Y., Lin, Y., Black, T., Rogers, E., and DiMego, G.: Implementation of a new grid-scale cloud and precipitation scheme in the NCEP Eta model, 19th Conf. on Weather Analysis and Forecasting/15th Conf. on Numerical Weather Prediction, San Antonio, TX, Amer. Meteor. Soc., 280–283, 2002.

Gavrilov, M. B.: Sensitivity of Long Range Numerical Weather Prediction on Small Change of Initial Condition, Physica Macedonica – J. Exp. Theo. Phys., 56, 23–28, 2006.

Gavrilov, M., Janjić, Z., and Jovanovic, G.: The sensitivity of a long numerical weather forecasting model, 8th European Conf. on Applications of Meteorology (ECAM), San Lorenco de El Escorial, Spain, 1–5 October 2007, Abstr. EMS2007-A-00586, 2007.

Harshvardhan, R. D., Randall, D. A., and Corsetti, T. G.: A fast radiation parameterization for general circulation models, J. Geophys. Res., 92, 1009–1016, 1987.

Janjić, Z. I.: Non-linear advection schemes and energy cascade on semi-staggered grids, Mon. Wea. Rev., 112, 1234–1245, 1984.
Janjić, Z. I.: The step-mountain coordinate: physical package, Mon. Wea. Rev., 118, 1429–1443, 1990.
Janjić, Z. I.: The step-mountain eta coordinate model: further developments of the convection, viscous sublayer and turbulence closure schemes, Mon. Wea. Rev., 122, 927–945, 1994.
Janjić, Z. I.: Comments on "Development and Evaluation of a Convection Scheme for Use in Climate Models", J. Atmos. Sci., 57, p. 3686, 2000.
Janjić, Z. I.: Nonsingular Implementation of the Mellor-Yamada Level 2.5 Scheme in the NCEP Meso model, NOAA/NWS/NCEP Office Note #437, p. 61, 2001.
Janjić, Z. I.: A Nonhydrostatic Model Based on a New Approach, Meteor. Atmos. Phys., 82, 271–285, 2003.
Janjić, Z. I., Gerrity, J. P., and Nickovic, S.: An Alternative Approach to Nonhydrostatic Modeling, Mon. Wea. Rev., 129, 1164–1178, 2001.
Janjić, Z. I.: A unified model approach from meso to global scales, Geophys. Res. Abstr., 7, 05582, 2005, European Geosciences Union, General Assembly, Vienna, Austria, 24–29 April 2005.
Janjić, Z. I. and Black, T.: An ESMF unified model for a broad range of spatial and temporal scales, Geophys. Res. Abstr., 9, 05025, 2007, European Geosciences Union, General Assembly, Vienna, Austria, 15–20 April 2007.
Janjić, Z., Huang, H., and Lu, S.: A unified atmospheric model suitable for studying transport of mineral aerosols from meso to global scales, IOP/Electronic J., Conf. Ser.: Earth and Envir. Sci. (collection of papers from WMO/GEO Expert Meet. on an Inter. Sand and Dust Storm Warning System Meeting, Barcelona, Spain), 9, 6, 7–9 November 2009.
Janjić, Z., Gall, R., and Pyle M. E.: Scientific Documentation for the NMM Solver, NCAR Tech. N. NCAR/TN-477+STR, 54, 2010.
Janjić, Z., Janjić, T., and Vasic, R.: A class of conservative fourth order advection schemes and impact of enhanced formal accuracy on extended range forecasts, Mon. Wea. Rev., doi:10.1175/2010MWR3448.1, in press, 2011.
Lorenz, E. N.: Deterministic nonperiodic flow, J. Atmos. Sci., 20, 130–141, 1963.
Lorenz, E. N.: A study of the predictability of a 28 variable atmospheric model, Tellus, 17, 321–333, 1965.
Lorenz, E. N.: Atmospeheric Predictability as Revealed by Naturally Occurring Analogues, J. Atmos. Sci., 26, 636–646, 1969a.
Lorenz, E. N.: The predictability of a flow which proccesse many scales of motions, Tellus, 21, 289–307, 1969b.
Lorenz, E. N.: On the Existence of Extended Range Predictability, J. Appl. Meteor., 12, 543–546, 1973.
NCEP: www.ncep.noaa.gov, ftp://ftpprd.ncep.noaa.gov/pub/data/nccf/com/gfs/prod/, 2006.
Phillips, N. A.: The equations of motions for a shallow rotating atmosphere and the traditional approximation, J. Atmos. Sci., 23, p. 626, 1966.
Shukla, J.: Have We Reached the Limit of Weather Predictability? Lorenz Symposium, Amer. Meteor. Soc. Ann. Meet., San Diego, CA, 13 January 2005.
Simmons, A. J. and Burridge, D. M.: An energy and angular-momentum conserving vertical finite-difference scheme and hybrid vertical coordinates, Mon. Wea. Rev., 109, 758–766, 1981.
Thompson, P. D.: Uncertainty of Initial State as a Factor in the Predictability of Large Scale Atmospheric Flow Patterns, Tellus, 9, 275–295, 1957.
Vuković, A., Rajković, B., and Janjić, Z.: Land Ice Sea Surface Model: Short Description and Verification, Proceedings, paper S.02.08, 5th Inter. Con. on Environmental Modelling and Software, Ottawa, Ontario, Canada, 5–8 July 2010.
Yoden, S.: Atmospheric Predictability, J. Meteor. Soc., Japan, 85B, 77–102, 2007.
Zhao-Xia, P., Kalnay, E., Sela, J., and Szunyogh, I.: Sensitivity of forecast errors to initial conditions witha quasi-inverse linear method, Mon. Wea. Rev., 125, 2479–2503, 1997.

5

# Comparative verification of different nowcasting systems to support optimisation of thunderstorm warnings

**K. Wapler[1,2], M. Goeber[1,3], and S. Trepte[1]**

[1]Deutscher Wetterdienst, Offenbach, Germany

[2]Hans-Ertel-Centre for Weather Research, Atmospheric Dynamics and Predictability Branch, Germany

[3]Hans-Ertel-Centre for Weather Research, Optimal Application of Weather Forecasts Branch, Germany

*Correspondence to:* K. Wapler (kathrin.wapler@dwd.de)

**Abstract.** The development and use of nowcasting systems should inevitably be accompanied by the development and application of suitable verification methods. A thorough verification strategy is needed to adequately assess the quality of the system and consequently to lead to improvements. Different verification methods for thunderstorms and its attributes are discussed along with the importance of observational data sets. They are applied to two radar-based nowcasting algorithms for a convective season using various observation data sets. The results show, that the combination of the two algorithms outperforms a single algorithm.

## 1 Motivation

Severe weather associated with deep convection poses a significant threat to life, property and economy. Fatalities, injuries and damages might be caused by lightning, gusts, hail, heavy precipitation or tornadoes. Therefore the provision of accurate and timely nowcast information, i.e. warnings provided by the national meteorological services, is essential for the general public as well as special users like emergency services and aviation.

Several algorithms exist which detect and nowcast deep convection. Most of them are based on either radar reflectivity measurements, like KONRAD (*KONvektionsentwicklung in RADarprodukten*, convection evolution in radar products, Lang 2001), CellMOS (Cell Model Output Statistics, Hoffmann, 2008) or on satellite measurements like RDT (rapid developing thunderstorm, Morel et al., 2000) and Cb-TRAM (Cumulonimbus TRAcking and Monitoring, Zinner et al., 2008). Some algorihtms rely on the combination of various observational input data (e.g. James et al., 2011; Pierce et al., 2000; Steinacker et al., 2000).

The evolution of forecast systems and the demand for more user oriented verifications has led to a reassessment of traditional verification strategies as well as to the development of new verification methods over the last decades (e.g. Gilleland et al., 2010 and references therein). However, no standard verification methods exist so far to evaluate the performance of nowcasting systems. The verification of convection and severe weather events is especially challenging (e.g. Doswell III, 1996). Due to their small horizontal extent, severe weather phenomena resulting from deep convection are rarely entirely and uniquely captured by current observational systems. However, verification is needed to assess the quality of the algorithms, to determine their strengths and weaknesses and consequently to lead to improvements. Given that several nowcasting systems exist and could be used in the warning process these should not only be verified independently but also comparatively using the same method.

The use of consistent verification methods is crucial to compare the different systems. Of particular interest is the question how these algorithms can optimally be used to issue warnings of thunderstorms as well as accompanying phenomena like gusts or hail. Nowcast verification experiments have been performed e.g. during the World Weather Research Program Forecast Demonstration Projects during the Sydney and Beijing Olympic Games (Ebert et al., 2004; May et al., 2004; Wilson et al., 2010) which compared nowcast provided by different national meteorological services.

In the study presented here we discuss some challenges and approaches of thunderstorm verification. To illustrate these discussions we show some results of verification of the radar based nowcast systems operated at DWD (*Deutscher Wetterdienst*, German Meteorological Service) CellMOS and KONRAD.

The remainder of this paper is organised as follows. In Sect. 2 an overview of the used algorithms and data sets is given. Section 3 discusses the verification methods; starting with a general introduction (Sect. 3.1) followed by a discussion of the specific verification challenges for thunderstorms (Sect. 3.2), convective gusts (Sect. 3.3) and hail (Sect. 3.4). Some verification results are shown in Sect. 4, and Sect. 5 provides a summary and some concluding remarks.

**Table 1.** Overview of DWD warning criteria related to thunderstorm.

| Combination of attributes in addition to lightning | Level |
|---|---|
| Strong gusts (Bft. 7) | moderate |
| Storm force gusts (Bft. 8–10) | strong |
| Heavy rainfall (10–25 $mm\,h^{-1}$) | strong |
| Storm force gusts, heavy rainfall | strong |
| Storm force gusts, heavy rainfall, hail | strong |
| Hurricane force gusts (Bft. 11–12) | severe |
| Very heavy rainfall (25–50 $mm\,h^{-1}$) | severe |
| Storm force gusts, very heavy rainfall | severe |
| Storm force gusts, very heavy rainfall, hail | severe |
| Hurricane force gusts, very heavy rainfall, hail | severe |

## 2 Data

The verification was performed for summer (April to September) 2010. The nowcasting algorithms verified in this study are the radar-based algorithms CellMOS and KONRAD which are based on 5 min 2-D data only. 3-D radar is currently provided only in 15 min interval within the German radar network. For the short lifecycle and rapid development of convection this observation frequency is not sufficient. However, 3-D data will be available at 5 min intervals in the future. KONRAD uses thresholds of 46 dBZ in a 9 $km^2$ area. CellMOS also uses a threshold of 9 $km^2$ but a lower dBZ threshold of 37. Additionally CellMOS uses lightning (at least one stroke had to appear within 10 km of the cell) and GME global model (Majewski et al., 2002) data (several parameters, e.g. wind speed and direction, relative humidity at various heights) applying a model output statistics approach (Klein and Glahn, 1974). Both systems provide as output the location of a cell along with some additional cell information, e.g. hail and gusts. KONRAD provides a hail flag (0, 1 or 2) based on the size of the area of more than 55 dBZ within the detected cell. Within KONRAD it is assumed that the expected convective gusts are equal to the movement speed of the cell. CellMOS provides estimates for the hail size and the gust speed as well as probabilities of hail sizes and gust above certain thresholds (defined in accordance with the warning categories) for up to 2 h. Both systems run operationally every 5 min.

The gusts verified in this study are measured at the about 260 German stations measuring gusts. The dataset provides hourly maximum gusts. Hail is observed at about 60 German stations with visual observations. Additional the hail dataset of the European Severe Weather Database (ESWD, Dotzek et al., 2009; Dotzek and Groenemeijer, 2009; Groenemeijer et al., 2009) was used. It collects events with hailstones having a diameter of 2 cm or more and smaller hailstones that form a layer of 2 cm thickness or more on flat parts of the earth's surface. The database includes information of the location (latitude/longitude) of the event, the time (with an uncertainty), a quality control flag and for some cases some additional information such as hail size. In the ESWD a three-level quality-control is applied. The QC-levels have the following meaning: QC0: "as received", QC0+: "plausibility checked", QC1: "report confirmed" by reliable sources and QC2: "event fully verified" i.e. all information about this event is verified, consistent and comes from reliable sources. For this study, we use data with quality flags QC1 and QC2.

The lightning measurements used are provided by the European LIghtning detection NETwork LINET (Betz et al., 2009). The lightning network consist of 30 antennas in Germany (and many more in Europe) and is considered to have a very high detection efficiency with a quasi continuous spatial and temporal resolution.

## 3 Verification methods for thunderstorms and its attributes

### 3.1 General remarks

One basic option for the verification of a nowcast system is to compare the location (and category) of the nowcast and the analysis. E.g. comparing the location of the cell detected at 12:30 UTC with the location of a cell of the +30 min nowcast based on the 12:00 UTC run. However, these results do not sufficiently indicate whether the system provides adequate support for the warning process since we do not warn cells (defined by dBZ areas) but precipitation, gusts etc. E.g. a nowcasting algorithm could be excellent in tracking a feature however this feature might not be equal to or useful for the event of interest and its warning. Thus, the nowcast system should be verified against and will depend on the warning criteria/categories (see Table 1 for an overview of the DWD thunderstorm warning categories). However, observations of severe convective weather phenomena are rare (see discussion in the following subsections).

Various verification methods are possible (for an overview see Gilleland et al., 2010 and references therein, specifically Davis et al., 2006). An areal approach could measure the

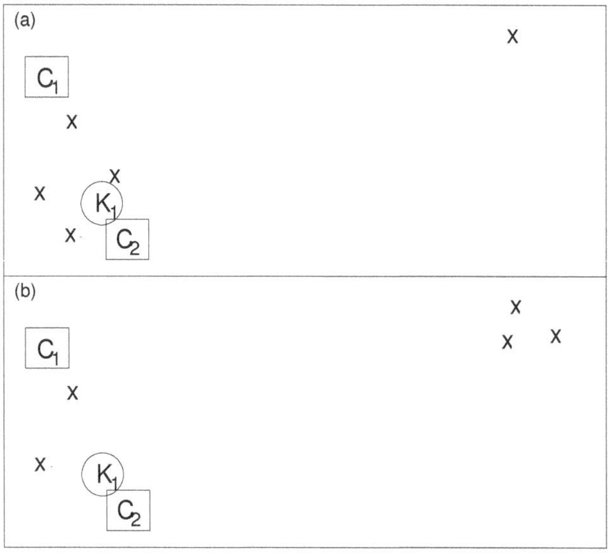


**Figure 1.** Example of two nowcast situations. Crosses indicate a lightning stroke, squares marked with a C indicate CellMOS cells and circles marked with a K indicate KONRAD cells.

areas affected by observed and nowcasted convective events and asses their overlap. A pointwise approach could consider each individual lightning or cell cluster. The results of these two approaches are shown in Sect. 4.2. Standard categorical verification measures are used (as described in Jolliffe and Stephenson, 2011).

## 3.2 Lightning

The verification of thunderstorms without considering accompanying phenomena seems to be much easier due to the high temporal and spatial coverage of lightning data. However, certain aspects have to be addressed when designing a verification methodology.

The most suitable method depends on the definition of a "good forecast". As an example a simple sketch is shown in Fig. 1. It shows two fictitious nowcast situations with the locations of detected lightning strokes and the locations of cells nowcasted by two algorithms (KONRAD (K) and CellMOS (C)). Looking at these figures, the following question should be answered to determine a suitable verification method. Should the result of the verification of the situation in Fig. 1a be the same as that in Fig. 1b? In both examples one of two storms (one storm on the left side of the figure and one on the right side) is detected. Thus, a verification could give a similar score for both cases. However, one could argue that in Fig. 1a the stronger storm (producing more lightning strokes) is detected and a relative weak storm (producing less lightning strokes) is missed; whereas in Fig. 1b the storms seem to have a comparable strength, thus, the miss and the hit could be weighted the same. Thus, the score should be higher in Fig. 1a compared to Fig. 1b. Furthermore, thresholds have to be defined e.g. the maximum distance in space and time to allow for a hit, to be able to calculate scores. Another aspect is the usability by an operational forecaster (not addressed in this study). Probably, the forecaster might prefer to work with the guidance provided by K because the algorithm nowcasts a single cell on the left side of the figures, whereas C nowcasts two cells, even though the detected lightning strokes on the left could be considered to belong to a single event. These discussions indicate that the decision about which of the interpretations of the situation shown in Fig. 1 applies and thus the definition of a "good forecast" might vary for different users of the warning guidance.

As suggested by the discussion above, the calculated scores as well as the possible maximum score depend on the chosen verification method. The chosen thresholds also influence the score. Another factor which influences the maximum possible score of a nowcast is the mean cell lifetime. In current nowcast systems cells usually do not dissolve, but are extrapolated in the future for the duration of the nowcast time frame. As an example of cell lifetimes the KONRAD statistics show that in the summer of 2010 40 % (6802) of all the cells (16 817) were detected just once and 14 % (1972) of the cells lived longer than 30 min. The high number of cells which were detected only once could be due to different reasons. E.g. on a convective day several cells develop, however, several cells might reach the threshold defined by KONRAD only shortly before dissolving and only a few cells intensify further. Additionally, the size of cells which have characteristics close to the thresholds defined in KONRAD might drop below the size threshold. After a renewed increase of the intensity these cells will be detected as new cells.

## 3.3 Gusts

The verification of convective gusts (see Table 1) is hampered by the poor representation of these phenomena by current observation systems. Furthermore, the SYNOP data usually provides information about the highest gust that occurred within an hour; not knowing the exact time of the gust complicates the attribution of the gust to a specific cell at days with numerous cells. Additionally, strong gusts may occur in the surroundings of cells (and not directly close to the highest reflectivity as detected by the algorithms) which further complicates the correlation of observations and nowcast. Not measuring a gust at a station only indicates that no gust occured at exactly this location. It does not mean that a cell which passed this station did not produce a gust. A severe gust might just have developed a few (kilo) meters farther. Furthermore, observations of severe convective gusts are rare (due to low number of events and especially due to relatively low number of stations). E.g. during summer 2010 only five times convective gusts with more than 11 Bft were measured at the about 260 German stations measuring gusts.

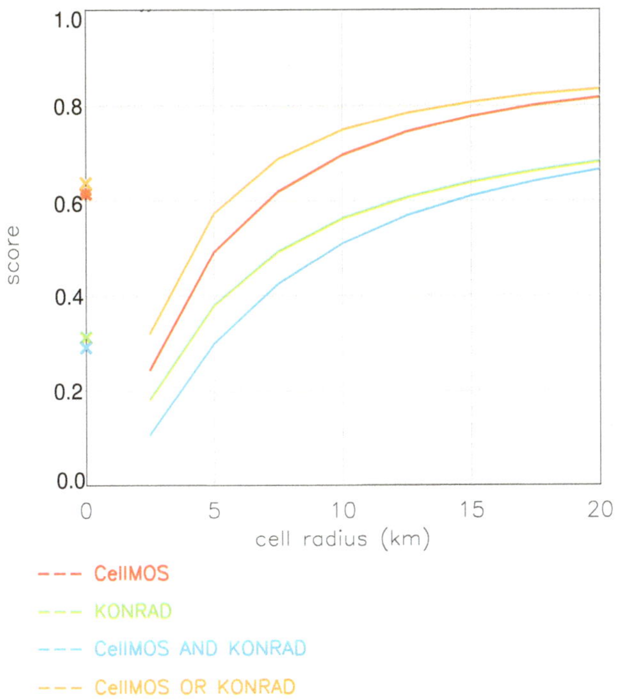


**Figure 2.** Pointwise approach. Probability of detection (percentage of hits of all observed events, 772 347) of lightning strokes that occurred (summer 2010) within the given radius of cells detected by CellMOS, KONRAD, CellMOS and KONRAD, CellMOS or KONRAD analysis. The crosses indicate the results using the cell radius given by the algorithm.

### 3.4 Hail

The problem of low numbers of observations also exists for hail. E.g. during the summer of 2010 at the about 60 German visual observation stations hail was observed just about 20 times. The ESWD provides very useful data. For the summer of 2010 the ESWD contains 82 hail entries with the highest quality flags QC1 or QC2.

However, to derive thorough statistics caution has to be applied when working with this data . It may happen that several data base entries exist for a single event (e.g. a cell in NE Germany had 21 entries (out of 82) in the ESWD, because it was analysed to have hit several villages). Other cells might just have a single entry in the data base which either means that the cell did not leave a long hail path or that the strong extent of the hail event was not observed in its full extent. Most importantly it has to be considered that the hail observations only provide information on "positive events", no entry in the data base or no observation at a station does not mean that no hail occurred, i.e. the hits (event observed and forecasted) and misses (event observed but not forecasted) can be determined but not the false alarms (event forecasted but not observed) nor the correct negatives (event neither forecasted nor observed). Thus a strategy has yet to be developed how to work best with this data in a quantitative way. In this study we show some qualitative characteristics.

**Table 2.** Percentage of lightning strokes that occurred (summer 2010) within 20 km of cells detected and nowcasted by CellMOS, KONRAD, CellMOS and KONRAD ($C \wedge K$), CellMOS or KONRAD ($C \vee K$).

| Nowcast | CellMOS | KONRAD | $C \wedge K$ | $C \vee K$ |
|---|---|---|---|---|
| +0 min | 82 % | 68 % | 67 % | 84 % |
| +30 min | 72 % | 66 % | 58 % | 79 % |
| +60 min | 51 % | 50 % | 34 % | 67 % |

## 4 Results and discussions

### 4.1 Cell tracking

The simple verification of comparing the analysis vs the nowcast of the same algorithm was done for KONRAD. It showed that for all cells during the summer of 2010 which lived at least 30 min, the cell position of the +30 min nowcast was on average 11 km with standard error 11 km away from the location of the analysis. The displacement error of the +60 min nowcast of all cells with a lifetime of at least 120 min was 14 km with standard error 12 km. Such studies could be used for a detailed analysis of the consistency and tracking quality of a system.

### 4.2 Lightning

As an example of thunderstorm nowcast verification, the results of the verification based on the comparison of all observed lightning strokes to detected cells (pointwise approach) is shown in Fig. 2.

These results show that a considerable part of lightning strokes occurred in cells with less than 9 $km^2$ of dBZ values of at least 46 and thus no KONRAD cells were detected. Furthermore, some lightning strokes occurred in cells with less than 9 $km^2$ of dBZ values of at least 37 or were to far away from the cell centre and thus no corresponding CellMOS cells were detected.

This comparison also shows the decrease of the detection rate with a more strict distance thresholds. Furthermore, it is shown that the combination of different algorithms improves the quality of the nowcast. This is due to the different reflectivity thresholds applied by the algorithms as well as due to additional data in CellMOS. Table 2 shows the decrease of the detection rate with longer lead times.

Another example of thunderstorm nowcasting verification is shown in Fig. 3. For this verification, a circular area (with a varying radius given in the figure) was defined to be affected by the event, either a lightning stroke or an analysed cell. Those areas were compared and their overlap calculated (areal approach, Davis et al., 2006). Similar to the

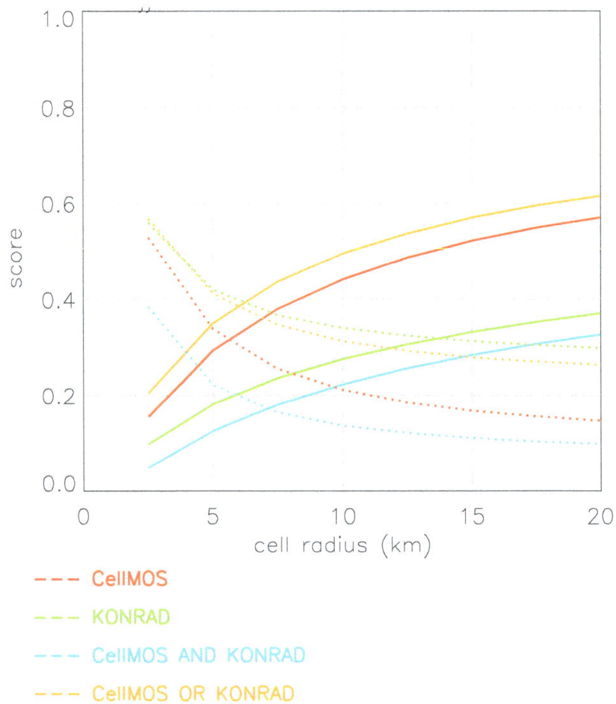


**Figure 3.** Areal approach. Probability of detection (solid, proportion of hits of all observed events) and false alarm ratio (dotted, proportion of false alarms of all forecasted events) for the comparison of areas affected by lightning strokes and areas affected by cells detected by CellMOS, KONRAD, CellMOS and KONRAD, CellMOS or KONRAD analysis. The cell radius determines the size of the area affected by a lightning stroke or a cell detected by one of the algorithms.

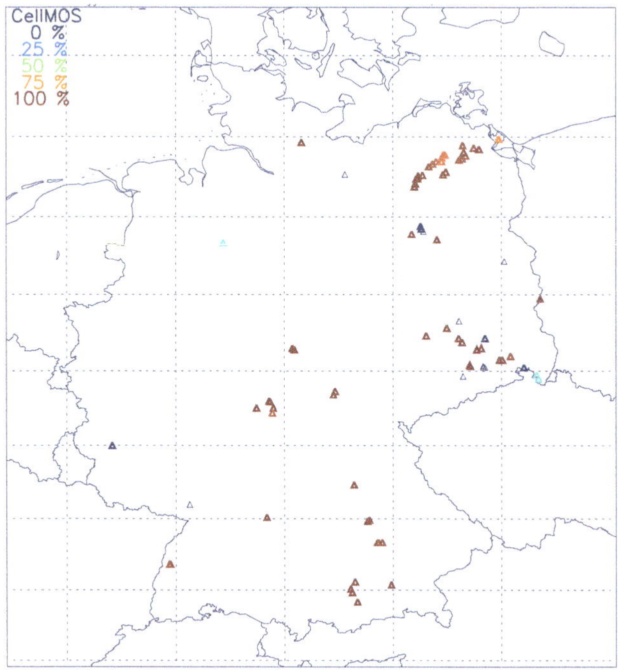


**Figure 4.** CellMOS analysis probability for hail larger than 15 mm (thick blue and red triangles) for all QC1 and QC2 hail entries (summer 2010) in the ESWD. Thin triangles show hail reports with no CellMOS cell within 20 km and ±5 min.

pointwise verification approach (Fig. 2), this method indicates, that the combination of two algorithms outperforms a single algorithm, and that the score improves with larger distance thresholds. However, the comparison of the scores in Figs. 2 and 3 show, that the scores are dependent on the verification method as already discussed in Sect. 3.2. The lower score in the areal approach (Fig. 3) can be explained as follows. The lightning strokes associated with a storm cover a much larger area than a single cell, i.e. area of high reflectivity. This leads to areas near the margin of the storm that are counted as misses which lowers the score.

## 4.3 Gusts

The comparative verification (not shown) for gusts stronger than $14\,m\,s^{-1}$ and gusts stronger than $18\,m\,s^{-1}$ revealed that nowcasts of CellMOS (which uses a model output statistics approach incorporating NWP model output) are superior to KONRAD (which bases its gust nowcast only on the estimated cell movement speed). In general 2-D radar data based algorithms have limited capability in analysing gust speed. Because convective gusts are not (only) determined by the reflectivity in a certain height but parameters such as downdraft convective available potential energy, maximum horizontal momentum and precipitation. The first two could be derived from soundings or models, the latter from vertically integrated liquid (VIL) as derived from 3-D radar measurements.

## 4.4 Hail

Comparison of observations with nowcast show that for most of the summer 2010 ESWD hail reports in Germany a CellMOS (Fig. 4) and a KONRAD (Fig. 5) cell has been detected (within 20 km and ±5 min of the event) which had a hail probability (CellMOS) of more than 75 % or hail warning flag (KONRAD), respectively. As discussed in Sect. 3.4 false alarms and correct negatives cannot be calculated using the existing hail dataset.

For all KONRAD cells detected in summer 2010 64 % had a hail warning flag of 0, 30 % had a hail flag of 1 and only 6 % had a hail flag of 2. Thus, the high number of KONRAD cells with a hail flag of 2 in Fig. 5 is not by chance. An example which shows the KONRAD hail flags for a line of convective cells which crossed Germany on 14 July 2010 is given in Fig. 6. The figure shows that the algorithm analyses a differentiated hail distribution which further supports that the high number of KONRAD cells with hail flag of 1 or 2 near observed hail events is not by chance.

However, in general 2-D radar data based algorithms have limited capability in analysing hail. The potential of

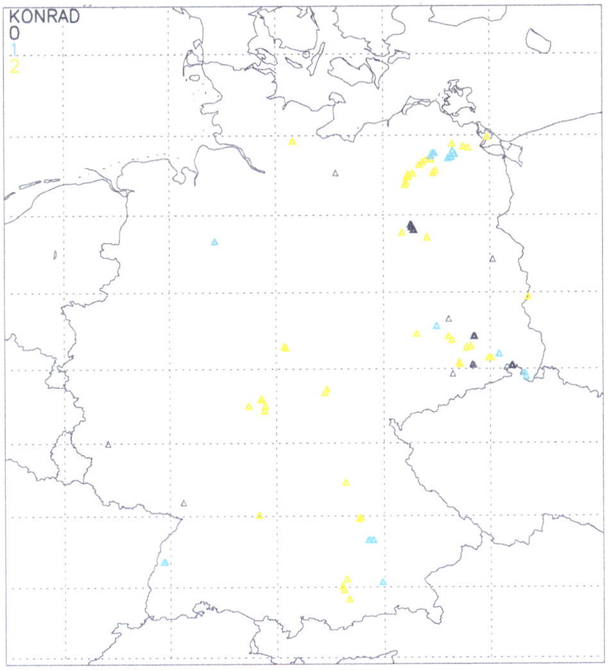


**Figure 5.** KONRAD analysis hailflag (thick dark blue, light blue and yellow triangles) for all QC1 and QC2 hail entries (summer 2010) in the ESWD. Thin triangles show hail reports with no KONRAD cell within 20 km and ±5 min.

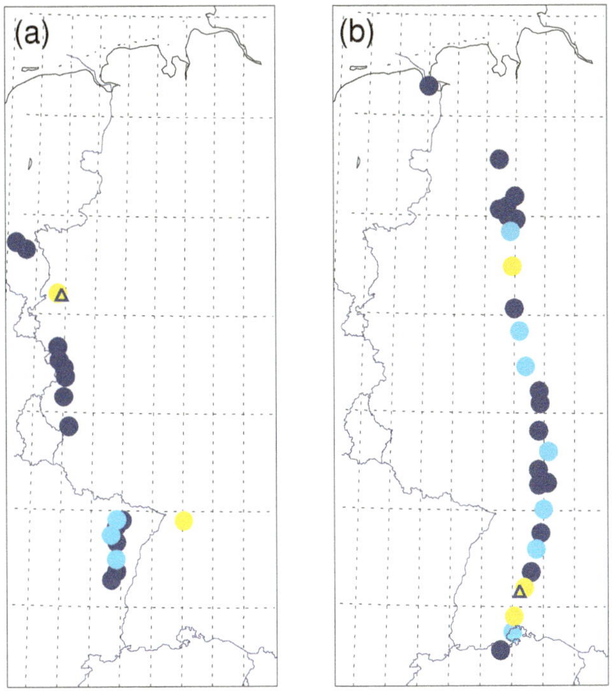


**Figure 6.** KONRAD hailflag (dark blue, light blue and yellow circles) and QC0 eswd hail reports (triangles) within ±5 min of 14 July 2010, 16:15 UTC **(a)** and 18:45 UTC **(b)**.

a convective cell to produce hail is not only determined by a very high reflectivity at a certain height but by high reflectivity values throughout the cell and an intense updraft. Thus, additional information from 3-D radar measurements such as the vertically integrated liquid VIL (Greene and Clark, 1972) provide useful information for the detection of hail.

## 5 Conclusions

In this study the challenges of comparative nowcast verification are discussed and some verification results for cell tracking, thunderstorms, gusts and hail are presented.

It is shown that the combination of different algorithms improves the quality of the nowcast. The low number of observations of rare events, e.g. severe gusts and hail, hampers thorough verification. For one summer season only very few events might be captured by observations. However, nowcasting algorithms need to be verified especially if they are new and no long statistics are available.

The results presented here provide an overview of ongoing work. It is planned to extend this study by using a larger data set, verifying additional phenomena, such as heavy precipitation, and integrating further nowcasting systems in the comparative verification.

**Acknowledgements.** The authors would like to thank Susanne Schroeder for providing the KONRAD cell statistics. We also want to thank the reviewers for their useful comments that helped improve the manuscript. Part of this research was carried out in the Hans Ertel Centre for Weather Research. This research network of Universities, Research Institutes and the Deutscher Wetterdienst is funded by the BMVBS (Federal Ministry of Transport, Building and Urban Development).

Edited by: B. Reichert

## References

Betz, H. D., Schmidt, K., Laroche, P. Blanchet, P., Oettinger, W. P., Defer, E., Dziewit, Z., and Konarski, J.: LINET – An international lightning detection network in Europe, Atmos. Res., 91, 564–573, 2009.

Davis, C. A., Brown, B. G., and Bullock, R. G.: Object-based verification of precipitation forecasts. Part I: Methodology and application to mesoscale rain areas, Mon. Weather Rev., 134, 1772–1784, 2006.

Doswell III, C. A.: Verification of forecasts of convection: Uses, abuses, and requirements, 5th Australian Severe Thunderstorm Conference, Avoca Beach, Australia, http://www.flame.org/~cdoswell/publications/Verification_Oz.pdf, 1996.

Dotzek, N. and Groenemeijer, P.: European Severe Weather Database, ESWD – Data format description ESSL Tech. Rep., 2099-01, 20 pp., available at: http://essl.org/media/publications/essl-tech-rep-2009-01.pdf, 2009.

Dotzek, N., Groenemeijer, P., Feuerstein, B., and Holzer, A. M.: Overview of ESSL's severe convective storms research using the European Severe Weather Database ESWD, Atmos. Res., 93, 575–586, 2009.

Ebert, E. E., Wilson, L. J., Brown, B. G., Nurmi, P., Brooks, H. E., Bally, J., and Jaeneke, M.: Verification of Nowcasts from the WWRP Sydney 2000 Forecast Demonstration Project, Weather Forecast., 19, 73–96, 2004.

Gilleland, E., Ahijevych, D. A., Brown, B. G., and Ebert, E. E.: Verifying forecasts spatially, B. Am. Meteorol. Soc., 91, 1365–1373, 2010.

Greene, D. R. and Clark, R. A.: Vertically integrated liquid water – a new analysis tool, Mon. Weather Rev., 100, 548–552, 1972.

Groenemeijer, P., Kuehne, T. Liang, Z., and Dotzek, N.: New capabilities of the European Severe Weather Database, 5th European Conference on Severe Storms, Landshut, Germany, 311–312, available at: http://www.essl.org/ECSS/2009/preprints/chapter10.pdf, 2009.

Hoffmann, J. M.: Entwicklung und Anwendung von statistischen Vorhersage-Interpretationsverfahren fuer Gewitternowcasting und Unwetterwarnungen unter Einbeziehung von Fernerkundungsdaten, Dissertation, FU-Berlin, 205 pp., 2008.

James, P., Trepte, S., Heizenreder, D., and Reichert, B. K.: NowCastMIX – A fuzzy logic based tool for providing automatic integrated short-term warnings from continuously monitored nowcasting systems, 10th Europ. Conf. Appl. Met., Berlin, 2011.

Jolliffe, I. T. and Stephenson, D. B.: Forecast Verification: A Practioner's Guide in Atmospheric Sciences, Wiley, 292 pp., 2011.

Klein, W. H. and Glahn, H. R.: Forecasting local weather by means of model output statistics. B. Am. Meteorol. Soc., 55, 1217–1227, 1974.

Lang, P.: Cell tracking and warning indicators derived from operational radar products, 30th Int. Conf. Radar Met., Munich, 245–247, 2001.

Majewski, D., Liermann, D., Prohl, P., Ritter, B., Buchhold, M., Hanisch, T., Paul, G., Wergen, W., and Baumgardner, J.: The operational global icosahedral-hexagonal grid point model GME: Description and high resolution tests, Mon. Weather Rev., 130, 319–338, 2002.

May, P. T., Keenan, T. D., Potts, R., Wilson, J. W., Webb, R., Treloar, A., Spark, E., Lawrence, S., Ebert, E. E., Bally, J., and Joe, P.: The Sydney 2000 Olympic Games Forecast Demonstration Project: Forecasting, Observing Network Infrastructure, and Data Processing Issues, Weather Forecast., 19, 115–130, 2004.

Morel, C., Sénési, S., Autones, F., and Labatut, L.: The Rapid Developing Thunderstorms (RDT) product of the nowcasting SAF: Prototyping activities and quality assessment using GOES images. Proceedings, The 2000 Meteorological Satellite Data Users' Conference, Eumetsat and CNR, Bologna, Italy, 698–705, 2000.

Pierce, C. E., Collier, C. G., Hardaker, P. J., and Haggett, C. M.: GANDOLF: A system for generating automated nowcasts of convective precipitation, Meteor. Appl., 7, 341–360, 2000.

Steinacker, R., Dorninger, M., Woelfelmaier, F., and Krennert, T.: Automatic Tracking of Convective Cells and Cell Complexes from Lightning and Radar Data, Meteorol. Atmos. Phys., 72, 101–110, 2000.

Wilson, J. W., Feng, Y., Chen, M., and Roberts, R. D.: Nowcasting Challenges during the Beijing Olympics: Successes, Failures, and Implications for Future Nowcasting Systems, Weather Forecast., 25, 1691–1714, 2010.

Zinner, T., Mannstein, H., and Tafferner, A. Cb-TRAM: Tracking and monitoring severe convection from onset over rapid development to mature phase using multi-channel Meteosat-8 SEVIRI data, Meteorol. Atmos. Phys., 102, 191–210, 2008.

# 6

# First outcomes from the CNR-ISAC monthly forecasting system

**D. Mastrangelo, P. Malguzzi, C. Rendina, O. Drofa, and A. Buzzi**

Institute of Atmospheric Sciences and Climate, National Research Council, Bologna, Italy

*Correspondence to:* D. Mastrangelo (d.mastrangelo@isac.cnr.it)

**Abstract.** A monthly probabilistic forecasting system is experimentally operated at the ISAC institute of the National Council of Research of Italy. The forecasting system is based on GLOBO, an atmospheric general circulation model developed at the same institute. The model is presently run on a monthly basis to produce an ensemble of 32 forecasts initialized with GFS-NCEP perturbed analyses. Reforecasts, initialized with ECMWF ERA-Interim reanalyses of the 1989–2009 period, are also produced to determine modelled climatology of the month to forecast. The modelled monthly climatology is then used to calibrate the ensemble forecast of daily precipitation, geopotential height and temperature on standard pressure levels. In this work, we present the forecasting system and a preliminary evaluation of the model systematic and forecast errors in terms of non-probabilistic scores of the 500-hPa geopotential height. Results show that the proposed forecasting system outperforms the climatology in the first two weeks of integrations. The adopted calibration based on weighted bias correction is found to reduce the systematic and the forecast errors.

## 1 Introduction

Monthly and subseasonal dynamical forecasts are nowadays issued by various meteorological centres. For instance, the European Centre for Medium-Range Weather Forecasts (ECMWF) produces 32-day forecasts (Vitart et al., 2008), the Australian Bureau of Meteorology emits 6-week forecasts (Hudson et al., 2011), and the National Centers for Environment Prediction (NCEP) daily operates the Climate Forecasting System version 2 (Saha et al., 2012) issuing monthly means deduced from seasonal forecasts.

To obtain a successful extended-range forecast, the adopted forecasting system not only must rely on the atmospheric initial conditions but, also, it must be capable to catch the evolution of those slowly varying "boundary" processes whose possible initial anomalous state determine the future atmospheric anomalies (Palmer and Hagedorn, 2006). To accomplish this goal, because of the intrinsic limits of atmospheric predictability, an extended-range forecasting system must be probabilistic: by generating an ensemble of deterministic integrations, the forecasting system samples initial uncertainties to produce an evolution of the probability distribution of the atmospheric state. If the predicted distribution correctly describes an anomalous atmospheric state with respect to the reference climatological distribution, the forecast can be considered skilful.

Initial conditions of land and ocean have been found to positively influence predictability on the monthly time scale, mainly over extratropical and tropical areas, respectively (Chen et al., 2010). Atmospheric initial conditions influence monthly predictability especially when they already feature the signal of a persisting major mode of atmospheric variability (Reichler and Roads, 2003), such as the Madden-Julian Oscillation, one of the most relevant sources of predictability on the monthly time scale over the Northern extratropics (e.g. Vitart and Molteni, 2010).

At the ISAC Institute of the Italian National Research Council (CNR), a new grid-point atmospheric general circulation model, named GLOBO, has been recently developed (Malguzzi et al., 2011). In the framework of a cooperation with the National Italian Civil Protection Agency, GLOBO is currently operated to produce, once a day, medium-range (up to 6 days) deterministic forecasts and, once a month, monthly probabilistic forecasts of atmospheric anomalies. In this paper, we present the monthly forecasting system (Sects. 2 and 3) with the aim of evaluating (Sect. 4) the results obtained from an experience of 13 months of forecasting.

## 2 Numerical setup and forecasting strategy

The GLOBO model integrates the atmospheric equations on a regular latitude/longitude grid that, for the monthly forecast, is set up with a horizontal grid spacing of $0.75 \times 1.0°$ and 50 vertical hybrid levels. A full description of the dynamics and physical parameterizations of the model is given in Malguzzi et al. (2011).

The sea surface temperature (SST) evolution is modelled with a climatological mixed layer ocean with depth $H$. The SST is relaxed to a prescribed field, computed as a sum of the climatological temperature $T_{\mathrm{CLIM}}$ and of the initial observed anomaly $T_{\mathrm{AN}}$ (the latter is assumed to decay slowly in time in the amount of 15 % per month), as follows:

$$\frac{\partial \mathrm{SST}}{\partial t} = \frac{F_{\mathrm{NET}}}{H\rho C} - \gamma(\mathrm{SST} - T_{\mathrm{CLIM}} - T_{\mathrm{AN}}) \qquad (1)$$

where $F_{\mathrm{NET}}$ denotes the sum of turbulent and radiative surface fluxes, $\rho C$ the thermal capacity of sea water, and where the constant $\gamma$ is set to $(2.3\,\mathrm{days})^{-1}$. The sea ice fraction is computed starting from the observed initial state and applying an observed climatological tendency. Both $T_{\mathrm{CLIM}}$ and climatological sea ice cover have been computed from a 20-yr dataset extracted from the ECMWF ERA-Interim reanalysis (Simmons et al., 2007; Dee et al., 2011).

The monthly forecasting system proposed here is based on an ensemble technique aimed at forecasting the probability distribution of atmospheric parameters in the extended range. Forecast anomaly of a single variable is computed by removing from the forecast ensemble mean the modelled climatology of the month obtained through reforecast simulations initialized with ERA-Interim reanalyses (see Sect. 3). The initial conditions for the ensemble forecast are defined using the NCEP Global Forecasting System (GFS) analyses, which are available in real time. Specifically, for a given initialization date of the monthly forecast, meteorological fields from fourteen 00:00 UTC and eighteen 12:00 UTC perturbed analysis of the GFS system are used as initial conditions to produce an ensemble of 32 members.

Over land, the temperature and water content of the deepest, climatological layer are specified from the corresponding climatological data computed from ERA-Interim reanalyses, both for forecasts and reforecasts. This is done to reduce the impact on the surface fields due to the use of different datasets for forecasting and calibration (NCEP and ERA-Interim analyses, respectively).

The model is affected by systematic errors or bias that can be a priori determined and removed from forecast anomalies (calibration). The next section describes how we treat model bias.

## 3 Model calibration and systematic error

Let $B^{\Psi}(\lambda, \vartheta, v, c)$ indicate the model bias for the generic atmospheric parameter $\Psi$ (geopotential height and temperature on standard pressure levels and daily precipitation). The model bias is a function of longitude $\lambda$, latitude $\vartheta$, validity day $v = 0, 1...31$, and calendar day $c = 1, 2...365$. Of course, the bias is zero for $v = 0$.

Our goal is to compute $B^{\Psi}$ from a large number of model simulations of past days (reforecasts). Each simulation is initialized on the ERA-Interim reanalysis corresponding to the 1st and the 15th of each month of the 21-yr period ranging from 1 January 1989 to 15 December 2009. We denote by $C_i$ ($i = 1, 2...24$) the calendar days corresponding to the 1st and the 15th of each month (for the sake of clarity, the problem of leap years is not considered). To increase the statistical ground, a two-member ensemble is performed for each $C_i$ by starting the model integration from both the 00:00 and 12:00 UTC analysis. This gives a total of 1008 (31-day long) reforecasts that can be averaged to provide model climatology as a function of validity day.

Let $\Psi_{i,y,h}(\lambda, \vartheta, v)$ represent the value of $\Psi(\lambda, \vartheta)$ predicted after $v$ days of integration by the reforecast started at the calendar day $C_i$ of year $y = 1, 2...21$ and initial hour $h = 1, 2$. The model error of this particular reforecast is given by the following difference:

$$E_{i,y,h}(\lambda, \vartheta, v) = \Psi_{i,y,h}(\lambda, \vartheta, v) - \Psi^{T}_{C_i+v,y,h}(\lambda, \vartheta), \qquad (2)$$

where $\Psi^{T}$ denotes the verifying ERA-Interim reanalysis of this reforecast. By contracting the previous expression over the index $y$ and $h$ we obtain a first (hereafter referred to as "raw") evaluation of the model bias, which is valid at calendar day $C_i$:

$$B^{\Psi}_{i}(\lambda, \vartheta, v) = \frac{1}{42}\sum_{y=1}^{21}\sum_{h=1}^{2} E_{i,y,h}(\lambda, \vartheta, v), \quad i = 1, ..., 24 \qquad (3)$$

Although we may expect that the model bias changes significantly from season to season, it is reasonable to assume that (possible) large differences between two consecutive $B^{\Psi}_{i}$ are due to unpredictable atmospheric fluctuations that do not average out perfectly (given the limited amount of data available here). Therefore, in order to smooth the annual cycle of the model bias, we introduce a Gaussian weighted running average over the calendar days which also allows us to define the model bias for any arbitrary calendar day $c$ as follows:

$$B^{\Psi}(\lambda, \vartheta, v, c) = \frac{1}{W}\sum_{i=1}^{24} B^{\Psi}_{i}(\lambda, \vartheta, v)\, e^{-(|c-C_i|/\Delta)^2}, \qquad (4)$$

where the normalizing factor $W$ is given by the sum of the Gaussian weights, and where the half-width $\Delta$ is a parameter that can be adjusted to obtain the best results from model calibration. In practice, the raw biases given by Eq. (2) with

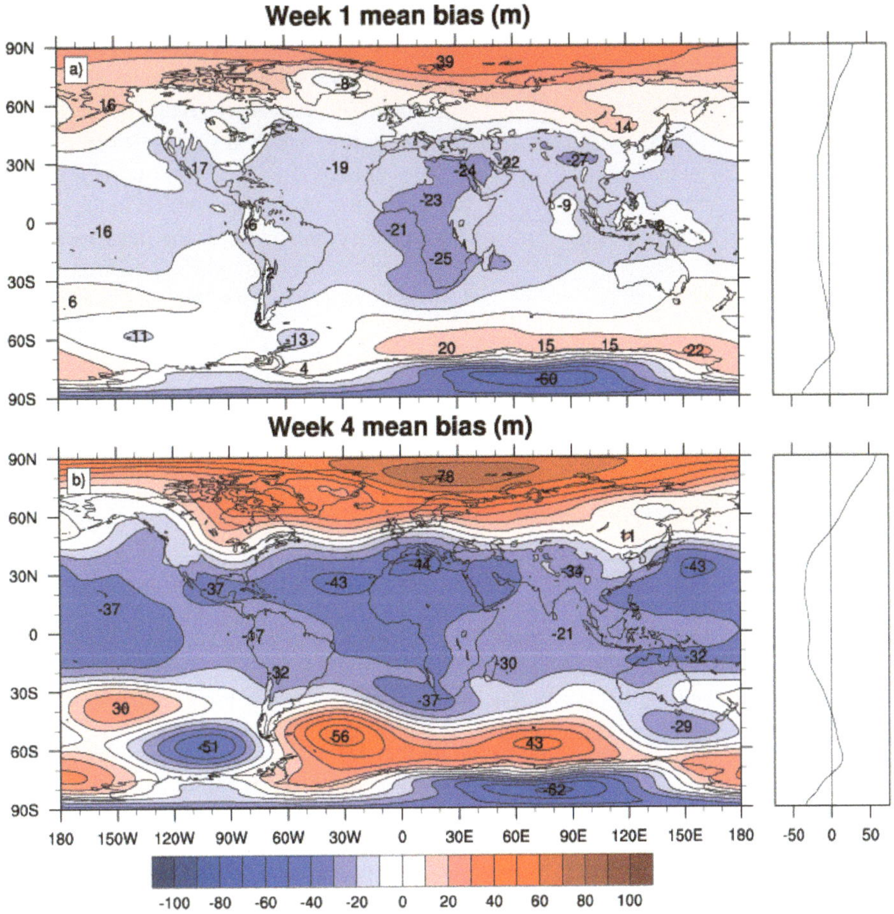


**Figure 1.** 500-hPa geopotential height mean bias (m) computed from reforecasts and corresponding ERA-Interim reanalyses for the 1st **(a)** and 4th **(b)** week of simulation. The zonally averaged bias is shown in side boxes.

$C_i$ close to $c$ give the most significant contribution to the bias at day $c$ (the computation of the distance between two calendar days must take into account the periodic nature of this particular variable).

Expressions (3) and (4) give us the basic elements to calibrate the monthly forecasts for any initial date. In the following, $\Delta = 45$ days is used for 500-hPa geopotential and 850-hPa temperature.

Figure 1 shows the mean model bias (average of Eq. (2) over $i$, $y$, $h$) of the 500-hPa geopotential height for the 1st week (average over $v$ from 1 to 7) and 4th week ($v$ from 22 to 28) of validity. During the first week, the model already develops the main features characterizing the long-term bias: a positive bias centred over the North Pole, a negative one centred over the South Pole, and a weaker negative bias spread over the intertropical belt and adjacent extratropical low-latitude areas. During the model integration, bias values intensify while some pattern differences arise over the southern portion of the Southern Hemisphere (SH) with, in particular, a sign reversal occurring over the area between Southern America and the Weddel Sea. Moreover, a negative bias over the low-latitude Northern Hemisphere (NH) temperate zones extends slightly northward. This implies that our model, in the extended range, reproduces a less intense mid-tropospheric westerly flow than that characterizing the real climate. Furthermore, the model tends to reduce the amplitude of the climatological large-scale wave across the Rocky Mountains.

The amplitude of the spatial pattern of model bias can be quantified by computing its root mean square (RMS) over a spatial domain. Typically, two domains are considered: the NH (latitudes from $\vartheta_1 = 20°$ to $\vartheta_2 = 90°$) and the SH (latitudes from $\vartheta_1 = -90°$ to $\vartheta_2 = -20°$) extratropics. The RMS is defined by the following expression:

$$R^{B^{\Psi}}(v, c) = \left( \frac{1}{2\pi(\vartheta_1 - \vartheta_2)} \int_0^{2\pi} \int_{\vartheta_1}^{\vartheta_2} B^{\Psi}(\lambda, \vartheta, v, c)^2 \cos\vartheta \, d\vartheta \, d\lambda \right)^{1/2} \quad (5)$$

The evolution of the model bias as a function of the validity day is shown in Fig. 2 for the NH and SH extratropics. The dashed curves are the average of the RMS of the raw biases (averaged over $i$ from 1 to 24 of $R^{B_i^{\Psi}}(v)$). Over both domains, the raw RMS steadily grows up to about the day 20, where it reaches the asymptotic value of about 40 m. The solid curves

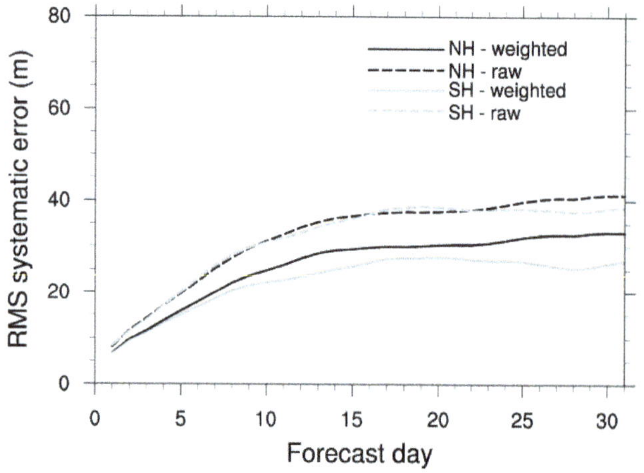


**Figure 2.** RMS (m) spatially averaged over the Northern (black curves) and Southern (grey curves) Hemisphere extratropics. Dashed lines refer to raw biases (see text).

represent the same quantity but for the weighted biases (averaged over $i$ of $R^{B_i^\Psi}(v,C_i)$). Compared to raw RMS, weighted RMS is smaller for all validity days, especially on the SH extratropics. This shows that the weighting procedure described above effectively reduces the estimation of systematic error. This result is in part expected for merely algebraical reasons; in the next section we will show that calibration performed by removing weighted biases effectively improves the model ability to predict anomalies.

## 4 Preliminary evaluation of the monthly forecasting system

The preliminary verification of the monthly forecasts is based on the comparison of 500-hPa geopotential height daily fields with ERA-Interim verifying reanalysis. Both dataset are scaled on a $1 \times 1^\circ$ regular latitude/longitude grid and time averaged 00:00 and 12:00 UTC instants to obtain a single daily field.

The performance of the probabilistic forecasting system is evaluated over 13 monthly outputs (August 2010–August 2011), which are the only months that can be presently verified. Such dataset is neither statistically significant nor allows us to evaluate the forecasting system in a probabilistic framework, which constitutes the proper method to be applied to probabilistic forecasting. However, an overview of the forecast performance can still be drawn by applying non-probabilistic scores to weekly averaged forecast ensemble means. Weekly averaging removes part of the shorter-scales, more directly affected by the chaotic atmospheric nature, leaving the larger-scale signal, which a monthly forecasting system is designed for.

The top panels of Fig. 3 shows weekly time series of uncentred anomaly correlation (AC; Wilks, 2006), while the bottom panels of the same figure report the RMS forecast error for the two extratropical areas. All series are obtained by averaging over the 13 cases available. Blue (red) markers refer to calibrated (uncalibrated) forecasts. On both domains, AC gets lower than 0.6 during the second week of integration. The RMS forecast error reaches the value of the climatological RMS (black markers) by the third week. The calibration method presented in the previous section slightly improves the AC up to the third week, with the highest improvement occurring during the first week over the NH extratropics (this is particularly evident for the first days of forecast, which are not resolved in the figure). The RMS forecast error is improved throughout the whole forecast range in both extratropical zones. If calibration were to be performed by subtracting raw biases instead, worse AC would be obtained (particularly for the first forecast days – not shown). This is a clear indication that reforecasting 21 yr in the past is not enough to separate the model systematic and random errors for a given month, so that we must resort to a weighted average of model errors over nearby months to improve AC by calibration.

## 5 Summary and conclusions

In this paper, the monthly forecasting system run experimentally at the ISAC institute of the National Research Council of Italy has been presented. Ensemble forecasts of 500-hPa geopotential height, 850-hPa temperature, and precipitation anomalies have been produced on a monthly basis through the GLOBO model. Forecast anomalies have been computed based on reforecast and forecast simulations, the former providing the model climatology used to calibrate the forecast ensemble mean.

Through the comparison with ERA-Interim reanalyses, the large number of reforecast simulations has allowed to evaluate the yearly averaged systematic error of the 500-hPa geopotential height field as simulated by GLOBO. The comparison shows that the main structure of the bias pattern is already developed during the first week of integration, and that the model smooths to some extent the main climatological features of the geopotential height.

The 13 monthly forecasts available so far have been analysed to gain some indications on the forecast error, that has been evaluated based on anomaly correlation (AC) and root mean square (RMS) of forecast error. These (typically deterministic) indices have been computed using weekly averages of modelled and observed atmospheric fields to directly compare statistical quantities (model ensemble means) with the single realization available, through reanalysis, of the real atmosphere. Results show that, even if the AC gets lower than 0.6 by the second week, the RMS forecast error outperforms the climatology in the first two weeks of integration. Moreover, the adopted calibration based on bias correction has been found to reduce the systematic and the forecast RMS errors.

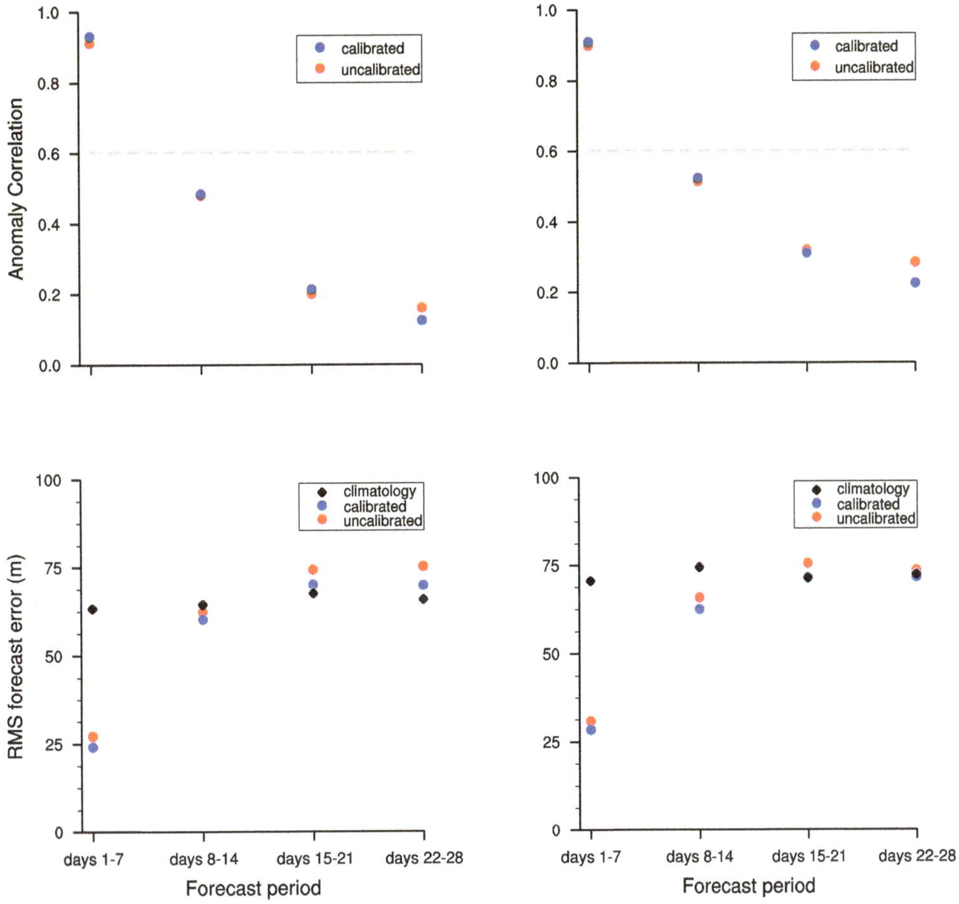


**Figure 3.** Anomaly correlation (top panels) and root mean squared (RMS) error (m, bottom panels) computed from 13 monthly forecast anomalies of 500-hPa geopotential height over the Northern (left panels) and Southern (right panels) Hemisphere extratropics. Blue markers are obtained from calibrated anomalies, red markers from uncalibrated (raw) anomalies. The black markers in the bottom panels represent the RMS computed from 1989–2009 ERA-Interim climatology.

Therefore, preliminary results suggest that the application of the ISAC monthly forecasting system provides some predictive skill in the extended range, in terms of weekly means. Further developments of the GLOBO model, and of the forecasting system, are expected to impact the forecast performance in the next future. Possible evolutions rely on implementing a more sophisticated oceanic module (Rendina, 2011) and different initialization data.

**Acknowledgements.** The authors thank two anonymous reviewers for their constructive comments. ECMWF and NCEP are gratefully acknowledged for providing reanalysis and forecast data.

Edited by: R. Hagedorn

## References

Chen, M., Wang, W., and Kumar, A.: Prediction of monthly-mean temperature: the roles of atmospheric and land initial conditions and sea surface temperature, J. Climate, 23, 717–725, 2010.

Dee, D. P., Uppala, S. M., Simmons, A. J., Berrisford, P., Poli, P., Kobayashi, S., Andrae, U., Balmaseda, M. A., Balsamo, G., Bauer, P., Bechtold, P., Beljaars, A. C. M., van de Berg, L., Bidlot, J., Bormann, N., Delsol, C., Dragani, R., Fuentes, M., Geer, A. J., Haimberger, L., Healy, S. B., Hersbach, H., Hólm, E. V., Isaksen, L., Kållberg, P., Köhler, M., Matricardi, M., McNally, A. P., Monge-Sanz, B. M., Morcrette, J.-J., Park, B.-K., Peubey, C., de Rosnay, P., Tavolato, C., Thépaut, J.-N. and Vitart, F.: The ERA-Interim reanalysis: configuration and performance of the data assimilation system, Q. J. Roy. Meteor. Soc., 137, 553–597, 2011.

Hudson, D., Alves, O., Hendon, H. H., and Marshall, A. G.: Bridging the gap between weather and seasonal forecasting: Intraseasonal forecasting for Australia, Q. J. Roy. Meteor. Soc., 137, 673–689, 2011.

Malguzzi, P., Buzzi, A., and Drofa, O.: The meteorological global model GLOBO at the ISAC-CNR of Italy: Assessment of 1.5 years of experimental use for medium range weather forecast, Weather Forecast., 26, 1045–1055, 2011.

Palmer, T. and Hagedorn, R. (Eds.): Predictability of Weather and Climate, Cambridge University Press, 702 pp., 2006.

Reichler, T. J. and Roads, J. O.: The role of boundary and initial conditions for dynamical seasonal predictability, Nonlin. Processes Geophys., 10, 211–232, doi:10.5194/npg-10-211-2003, 2003.

Rendina, C.: Study of the impact of modelling sea surface temperature in a monthly atmospheric ensemble prediction system, Ph.D. thesis, University of Ferrara, Italy, 2011.

Saha, S., Moorthi, S., Wu, X., Wang, J., Nadiga, S., Tripp, P., Pan, H.-L., Behringer, D., Hou, Y.-T., Chuang, H.-Y., Iredell, M., Ek, M., Meng, J., Yang, R., van den Dool, H., Zhang, Q., Wang, W., and Chen, M.: The NCEP Climate Forecast System Version 2, J. Climate, to be submitted, 2012.

Simmons, A., Uppala, S., Dee, D. P., and Kobayashi, S.: ERA-Interim: New ECMWF reanalysis products from 1989 onwards, ECMWF Newsletter, 110, 25–35, 2007.

Vitart, F. and Molteni, F.: Simulation of the Madden-Julian Oscillation and its teleconnections in the ECMWF forecast system, Q. J. Roy. Meteor. Soc., 136, 842–855, doi:10.1002/qj.623, 2010.

Vitart, F., Buizza, R., Balmaseda, M. A., Balsamo, G., Bidlot, J.-R., Bonet, A., Fuentes, M., Hofstadler, A., Molteni, F., and Palmer, T. N.: The new VAREPS-monthly forecasting system: A first step towards seamless prediction, Q. J. Roy. Meteor. Soc., 134, 1789–1799, 2008.

Wilks, D. S.: Statistical Methods in the Atmospheric Sciences, 2nd Edn., Academic Press, 627 pp., 2006.

# The new WMO RA VI Regional Climate Centre Node on Climate Monitoring

**J. Rapp, H. Nitsche, and P. Bissolli**

Deutscher Wetterdienst, Regional Climate Centre Node on Climate Monitoring (WMO RA VI), Frankfurter Str. 135, 63067 Offenbach am Main, Germany

**Abstract.** Regional Climate Centres are institutions with the capacity and mandate by WMO to develop high quality regional-scale products using global products, national input and incorporating regional information. Recently a pilot network of three Regional Climate Centre consortia was established for the WMO region RA VI (Europe and Middle East). Germany (Deutscher Wetterdienst) has taken the responsibility of the Regional Climate Centre Node on Climate Monitoring. The main basic functions of this centre are the publication of annual and monthly climate diagnostic bulletins, monthly monitoring maps, monitoring of significant events, implementation of a climate watch system, capacity building and offering reference climatologies and trend maps.

## 1 Background

Climate relevant processes have strong inter-scale linkages going beyond borders of individual countries. Therefore, and also to meet the needs of its members, WMO has defined a worldwide three-level infrastructure: Global Producing Centres for Long-range Forecasts (GPCs), Regional Climate Centres (RCCs; WMO, 2009) and National Meteorological or Hydrometeorological Services (NMHSs).

In the light of the enormous challenges climate variability and climate change pose to societies, there is an urgent need to enable WMO Members to provide the best possible climate services in support of climate risk management and adaptation. As global-scale information provided by GPCs (coarse in terms of resolution and reflected features) is typically not sufficient for national-scale services, the concept of WMO Regional Climate Centres was developed to bridge the gap between information at the global and national scales. Moreover, up-to-date climate services require, among others, appropriate computer power, modelling capacities and special expertise, to which not all WMO Members currently have adequate access.

WMO Global Producing Centres (GPCs) for Long Range Forecasts (global seasonal forecasts) are an integral part of the Global Data-Processing and Forecasting System (GDPFS). GPCs adhere to certain well-defined standards – aiding consistency and usability of output: a fixed forecast production cycle, a standard set of forecast products and WMO-defined verification standards (for retrospective forecasts). A comprehensive set of standard verification measures, with which to communicate the skill of forecasts, has also been defined.

Officially designated GPCs of Long Range Forecasts based in Europe are: the European Centre for Medium-Range Weather Forecasts (ECMWF), Toulouse (Météo-France, France), Exeter (Met Office, UK), and Moscow (Roshydromet, Russia). WMO has also designated a Lead Centre for Long Range Forecast Multi-Model Ensemble (LC-LRFMME) and for Standard Verification System of Long Range Forecasts (LC-SVSLRF).

Regional Climate Centres (RCC) are Centres of Excellence that assist World Meteorological Organization (WMO) Members in a given region to deliver better climate services and products including regional long-range forecasts, and to strengthen their capacity to meet national climate information needs. The concept of Regional Climate Centres (RCC) offers excellent opportunities for networking by pooling certain capacities of the National Meteorological and Hydrological Services (NMHSs) and other institutes in a region by an efficient systems approach. Accordingly, each member in a region gets sustained operational access to the most updated climate services. The primary users of an RCC are the NMHSs and other RCCs in the region and in neighbouring areas. NMHSs typically use RCC products to extend or improve their service suite vis-à-vis their end user community.

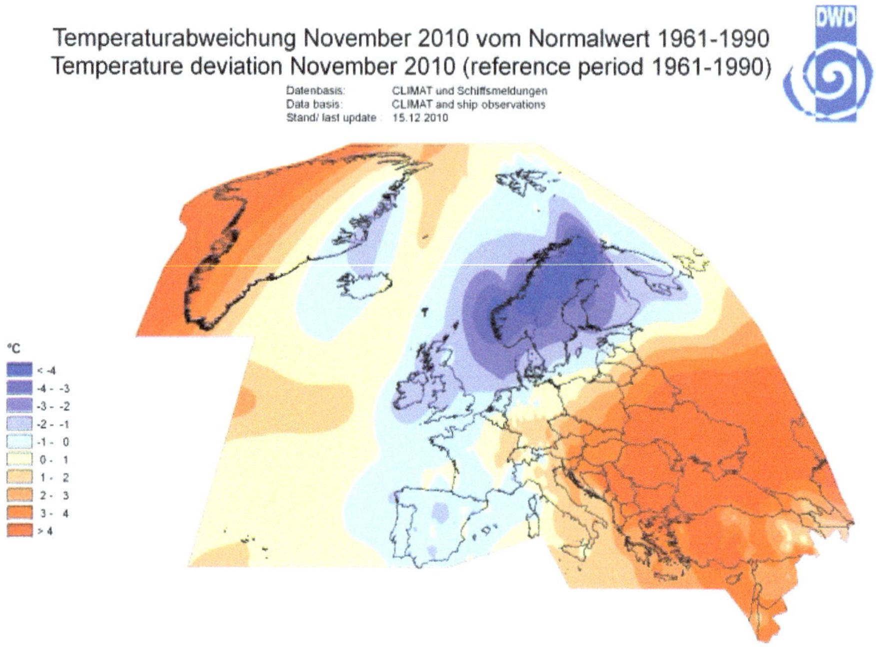


**Figure 1.** Example map of monthly temperature anomaly in November 2010 (Source: Deutscher Wetterdienst).

## 2 Organization

Recently a pilot network of three RCC consortia was established for the WMO region Regional Association (RA) VI (Europe and Middle East) which comprises Europe, Greenland, parts of the North Atlantic and the Middle East (http://www.rccra6.org):

- RCC node on climate data (http://eca.knmi.nl),
- RCC node on climate monitoring (www.dwd.de/rcc-cm),
- RCC node on long-range forecasting (http://neacc.meteoinfo.ru/).

Germany (Deutscher Wetterdienst, DWD) has taken the responsibility of the RCC node on climate monitoring (RCC-CM). The consortium of this node consists of the following NHMSs

- Armstatehydromet/Armenia,
- Deutscher Wetterdienst (Germany) (host),
- Météo-France/France,
- KNMI/The Netherlands,
- RHMS/Serbia, and
- TSMS/Turkey.

RCC-CM supports the South-eastern Europe Climate Outlook Forum (SEECOF). It is the very first regional climate outlook forum in Europe, started in 2008 at Zagreb, Croatia. Regional Climate Outlook Forums (RCOFs), active in several parts of the world, routinely provide real-time regional climate outlook products.

## 3 Basic functions and products

RCCs provide online access to their products and services to National Meteorological and Hydrological Services (NMHSs) and to other regional users (if respective NMHSs agree). Vice versa, RCCs receive data, products, know-how and feedback from the NMHSs as a main source for regional information. By the same time, they provide regional data, products and feedbacks to Global Production Centres and Lead Centres for respective verification and product optimization of the global-scale information.

The RCC-CM will perform basic functions covering the domain of climate monitoring:

- Monthly monitoring maps: Europe (RA VI; see for example Fig. 1; Hogewind and Bissolli, 2011), Eastern Mediterranean, South Caucasus, Southeast Europe,
- Annual and monthly climate diagnostic bulletins (see for example Fig. 2),
- Monitoring of significant events,

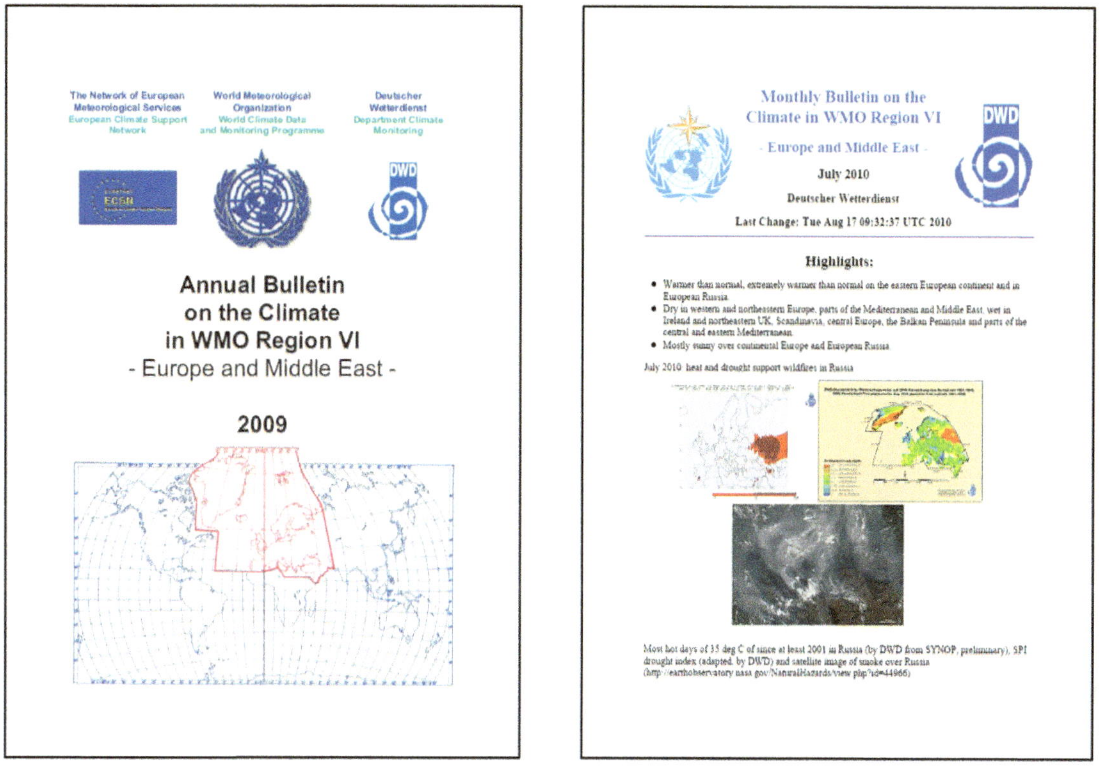


**Figure 2.** Annual and Monthly Bulletin on the Climate in WMO Region VI.

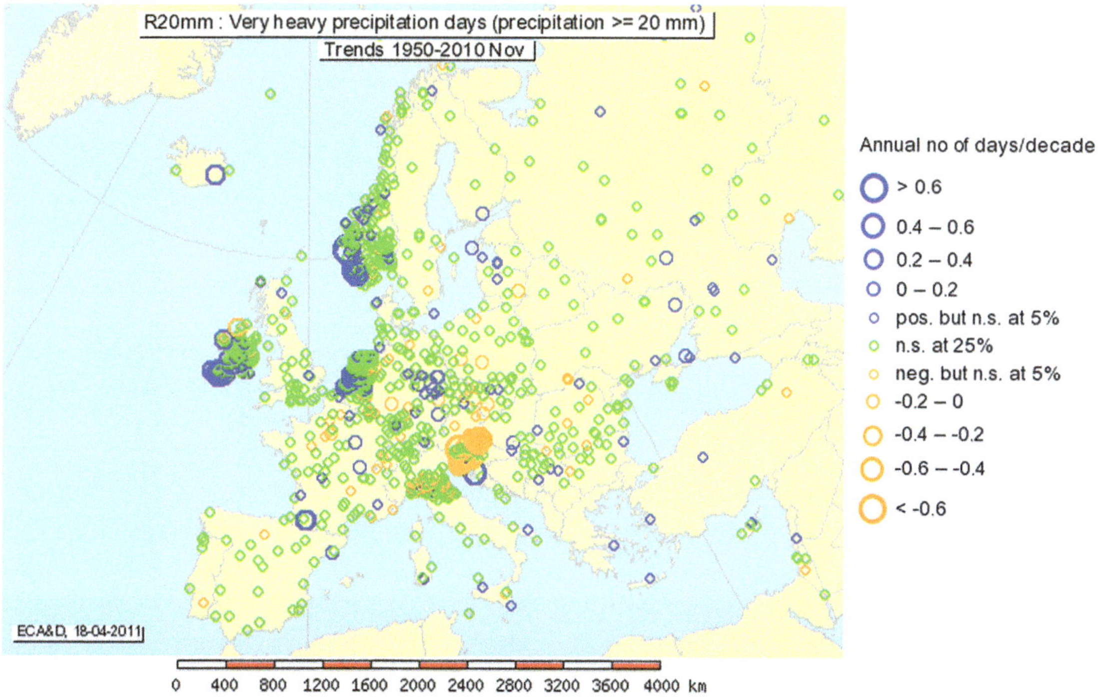


**Figure 3.** Example map of the trend of very heavy precipitation days (≥ 20 mm) in November 1950–2010 (Source: ECA&D, KNMI).

**Table 1.** RCC-CM Product catalogue, Status of 31 December 2010.

**RCC-CM Product catalogue**

| 1 Mandatory Products | |
|---|---|
| **1.1 Monthly and Annual Bulletins and Reports** | |
| Annual Bulletin on the Climate in RA VI (DWD; www.dwd.de/rcc-cm) | Contribution to: Annual 'State of the Climate' Report (BAMS/WMO; www.ncdc.noaa.gov/bams-state-of-the-climate/), European Section (lead: DWD/UK MetOffice) |
| Monthly Climate Diagnostics Bulletin (DWD; www.dwd.de/rcc-cm) | Contribution to: WMO statement on the status of the global climate (www.wmo.int/pages/publications/showcase/index_en.html) |
| **1.2 Monthly Maps** | |
| Monthly maps of air temperature (DWD; www.dwd.de/rcc-cm) | Monthly maps of precipitation (DWD; www.dwd.de/wzn) |
| Monthly maps of snow cover (DWD; www.dwd.de/snowclim) | Monthly maps of sunshine duration (DWD; www.dwd.de/rcc-cm) |
| Monthly maps of cloud coverage (DWD; www.dwd.de/satklim) | Monthly maps of water vapour (precipitable water; DWD; www.dwd.de/satklim) |
| Monthly maps of pressure (DWD; www.dwd.de/rcc-cm) | Monthly maps of temperature, precipitation, pressure, wind, humidity (model based) and anomalies with 3 and 12 months overlapping periods (Météo-France; www.meteo.fr/special/CLIM/clim_model.html) |
| ECA&D based maps, graphs and trends (temperature, precipitation, extremes indices), based on station data and gridded data (KNMI; eca.knmi.nl) | Monthly maps of temperature and precipitation for the Eastern Mediterranean (Turkish State Meteorological Service: emcc.dmi.gov.tr) |
| Monthly anomaly maps of temperature and precipitation for the South Caucasus region (ARMSTATEHYDROMET; www.dwd.de/rcc-cm) | Climate anomalies and extremes indices (Republic Hydrometeorological Service of Serbia; SEE/VCCC; www.seevccc.rs) |
| **1.3 Reference Climatologies** | |
| Temperature (www.dwd.de/rcc-cm) | Precipitation (www.dwd.de/wzn) |
| Snow cover (www.dwd.de/rcc-cm) | Sunshine duration (www.dwd.de/rcc-cm) |
| Cloud coverage (www.dwd.de/satklim) | Water vapour (www.dwd.de/satklim) |
| ECA&D based climatologies (temperature, precipitation, extremes indices), based on station data (KNMI; eca.knmi.nl/utils/mapserver/climatology.php) | |
| **2 Supplementary Products** | |
| Radiation products:<br>Surface incoming short-wave radiation<br>Surface net short-wave radiation<br>Surface outgoing long-wave radiation<br>Surface net long-wave radiation<br>(DWD, www.dwd.de/satklim) | Cloud coverage indices:<br>Number of days with cloud cover < 2/8<br>Number of days with cloud cover > 6/8<br>Duration of days with cloud cover < 2/8<br>Duration of days with cloud cover > 6/8<br>(DWD, www.dwd.de/satklim) |
| Drought index (modified SPI drought index; DWD, www.dwd.de/rcc-cm) | Daily Tmin/Tmax – Temperatur maps (DWD, in development) |
| **3 Event Monitoring** | |
| Significant Weather Event Reports (European scale, www.dwd.de/rcc-cm and http://ecad.knmi.nl/events/selectevents.php) | Significant Weather Event Calendar (European scale - monthly events; www.dwd.de/rcc-cm) |
| Significant Weather Event Reports (countries, www.dwd.de/rcc-cm) | |

- Reference climatologies and trend maps (see for example Fig. 3, Haylock et al., 2008; van Engelen et al., 2008),
- RA VI climate monitoring WebPortal,
- Climate watches (to be performed in future),
- Training (to be performed in future),
- Research and Development (R&D) (to be performed in future).

The pilot RCC-CM generates and distributes various climate monitoring products for the WMO Region VI. Some first products are offered: description of outstanding weather events, maps of various climate quantities and monthly and annual climate reviews ("RA VI Bulletin"). National products for individual countries within the WMO Region VI are provided by the National Meteorological and Hydrological Services of the concerning countries. DWD generates, beside the national products for Germany, also transnational maps and reports for the whole WMO Region VI, based on data and contributions from the countries. A detailed overview of all individual mandatory and supplementary products (e.g. for drought: Pietzsch and Bissolli, 2011) as well as event monitoring information is shown in Table 1.

The climate watch system (CWS; WMO, 2005), which is going to be implemented by RCC-CM in future, will provide advisories and statements for the NHMSs to enable them to inform users, particularly those involved in natural hazards preparedness, mitigation and response, about evolving or foreseen climate anomalies at the regional and national levels, thus allowing them to make informed decisions. The advisories should be based on all available climate information from National Meteorological and Hydrological Services, global producing centres and regional climate centres. The "RA VI Workshop on Climate Monitoring including the Implementation of Climate Watch Systems" (WMO, 2011), held in October 2010 in Offenbach, led to a detailed roadmap for the implementation of a climate watch system.

Updated climate monitoring reports and maps and additional information are available via Internet: www.dwd.de/rcc-cm.

Edited by: W. A. Monna

## References


Haylock, M. R., Hofstra, N., Klein Tank, A. M. G., Klok, E. J., Jones, P. D., and New, M.: A European daily high-resolution gridded data set of surface temperature and precipitation for 1950–2006, J. Geophys. Res., 113, D20119, doi:10.1029/2008JD010201, 2008.

Hogewind, F. and Bissolli, P.: Operational Maps of Monthly Mean Temperature for WMO-Region VI (Europe and Middle East), Idöjaras, 115, 1/2, 31–49, 2011.

Pietzsch, S. and Bissolli, P.: A modified Drought Index for the WMO RA VI Region, 2011, Adv. Sci. Res., submitted, 2011.

van Engelen, A., Tank, A. K., van der Schrier, G., and Klok, L.: European Climate Assessment & Dataset (ECA&D), Report 2008, Towards an operational system for assessing observed changes in climate extremes, KNMI, De Bilt, 68 pp., 2008.

World Meteorological Organization (WMO): Guidelines on Climate Watches, WCDMP-No. 58, 44 pp., 2005.

World Meteorological Organization (WMO) Regional Association VI (Europe): RA VI RCC Implementation Plan, www.wmo.int/pages/prog/dra/eur/documents/RCC%20Network/WMO-RAVI-RCC-Implementation-Plan.pdf (last access: 5 June 2011), 2009.

World Meteorological Organization (WMO): Regional Association VI Workshop on Climate Monitoring including the Implementation of Climate Watch Systems, WCDMP-No. 75, WMO/TD-No. 1565, 154 pp., 2011.

# 8

# Determining the accuracy of gridded climate data and how this varies with observing-network density

**T. P. Legg**

National Climate Information Centre, Met Office, Exeter, UK

**Abstract.** The Met Office National Climate Information Centre regularly produces assessments of mean monthly, seasonal and annual values of weather parameters and their anomalies over the UK. However the gridded values, and corresponding areal-average values, are subject to error. Experiments have been done in an attempt to quantify the mean errors in gridded monthly values and monthly areal averages of temperature and rainfall, and how these errors vary when we artificially thin out the observation network. But there are two additional reasons for this work: firstly we wish to determine how far back we can realistically extend the historical areal series, and secondly, we want to estimate the size of error bars on the historical values. For the UK as a whole, we estimate that error bars of around 0.1 °C would arise for monthly-mean temperatures.

## 1 Introduction

The Met Office National Climate Information Centre regularly produces assessments of mean monthly, seasonal and annual values of weather parameters and their anomalies over the UK[1]. These are based on daily measurements at a network of observing sites, and produced by interpolation to a regular grid as described by Perry and Hollis (2005). Monthly series of gridded temperature and rainfall at 5 km resolution have been produced back to 1910, and of sunshine back to 1929. However the gridded values, and corresponding areal-average values, are subject to error (e.g. due to instrument calibration and reading, site differences, sampling, interpolation). For example, Perry and Hollis (2005) showed that the root-mean-square error for point temperature estimates is around 0.4 °C on average. This paper considers the sampling and interpolation errors, which will vary as a function of network spacing.

Experiments have been done in an attempt to quantify how the mean errors in monthly gridded values and monthly areal averages of temperature and rainfall vary with the size of the observation network. The number of available stations has declined from a peak of around 570 (corresponding to an average horizontal spacing between stations across the UK of approximately 21 km) for temperature and 5400 (7 km) for rainfall in the 1970s, to about 400 (25 km) and 3000 (9 km) respectively in 2010. The objectives of this work were to determine how far back we can realistically extend the historical areal series, and to estimate the size of error bars for the historical values.

## 2 Station data and gridding method

Daily measurements are made of maximum and minimum temperatures, rainfall totals and sunshine amounts. These station observations are then used as the input to our gridding process, which uses regression and interpolation to obtain estimated values on a 5 km × 5 km grid of points to give a better representation of spatial variations than is possible from the station observations alone. There are approximately 10 000 grid-points covering the UK. This process effectively adds information, due to knowledge of orography, exposure etc. even in unobserved areas. Areal averages are then calculated using all grid-point values within each area. The averages should be unbiased overall (a mixture of urban, coastal, upland etc.). Conversion to anomalies is then done using the long-term average for each area.

**Figure 1.** UK station networks for temperature and rainfall; 20 independent verification stations; examples of thinned-out subsets of the network. **(a)** Full station network for temperature observations. **(b)** Full station network for rainfall observations. **(c)** 20 independent verification stations. **(d)** Example of selection of stations left in network when 90% were removed. **(e)** Example of selection of stations left in network when 95% were removed.

## 3 Methodology of this study, and some results

From the full station network (shown in Fig. 1a, b), we take a well-spaced subset of 20 sites as independent verification stations (Fig. 1c). Then we estimate the mean errors at these 20 sites, as interpolated from a gradually decreasing network obtained by artificially removing certain proportions of the original set of stations. Figure 1d and e shows examples of substantially thinned-out subsets of the original network. The network is thinned by removing, in a random manner, fixed percentages of the complete station network (e.g. 4 out of every 10 stations). It is found that the errors increase only slowly as fewer station observations are retained; examples are given in Figs. 2 and 3. This reflects spatial coherence in the data, especially for temperature: when a station is lost, neighbouring stations provide some, though not all, of the missed information.

Biases of the gridded values have been estimated by selecting one independent station at a time and obtaining an estimate of its average error, and relating these errors to location, orography, etc. Individual-station biases tend to be up to ~2 °C, and sometimes (Fig. 4) there is a tendency for similarly-situated stations (e.g. those in valleys) to have similar errors, which constitute elevation-dependent biases.

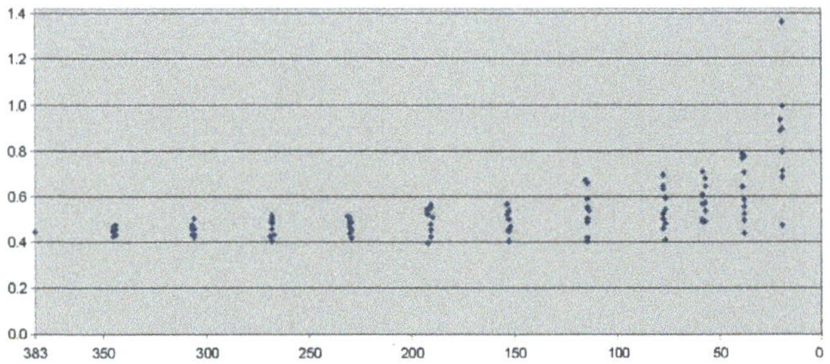


**Figure 2.** Minimum Temperature, April 2007 – Estimated root-mean-square errors (°C) at a subset of 20 independent points, as a function of the number of stations remaining out of the original 403 available (ten runs were made for each proportion of stations removed).

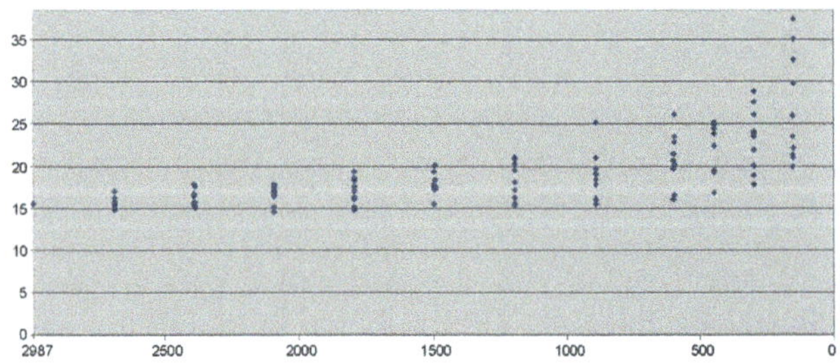


**Figure 3.** Rainfall, July 2007 – Estimated root-mean-square errors (mm) at a subset of 20 independent points, as a function of the number of stations remaining out of the original 3007 available (ten runs were made for each proportion of stations removed).

## 4 Extending our series further back in time

The availability of station data is increasingly limited as we go further back before 1910. But the results presented here show that the mean error of interpolation to the locations of a set of independent stations, increase only slowly as the available station network is artificially thinned out. This is partly because of spatial coherence in the data, especially for temperature. This implies also that errors of areal-average values may increase only slowly as the network is thinned, so that we may be able usefully to extend our series back to earlier dates.

How far back can we extend our series before the errors become unacceptably large? It appears that we need to have approximately 100 temperature stations (average spacing 50 km) (Fig. 2) and 500 rainfall stations (22 km) (Fig. 3) across the UK in order that the estimated mean error at independent sites remains no more than 1.5 times what it is with the current full station network. Currently our historical series for temperatures and rainfall extend back to 1910, but these results, in combination with our information on historical observing networks, suggest that we could reliably extend the series back to at least 1901.

## 5 Errors in areal values

Areal values are calculated for the UK, the four constituent countries, and ten climatological regions. So for temperatures the ten regions each currently contain roughly 40 stations, and for rainfall each region contains about 300 stations.

The areal averages, calculated using all grid-points in each area, effectively smooth out the station errors, so we expect the areal errors to be smaller. Typical point-value errors for monthly-mean temperatures are between 0.3 and 0.6 °C (similar to those in Fig. 2), so errors in areal values will be smaller than this. However individual stations will not be statistically independent of each other, and this is even more true of the grid-points owing to the interpolation. So an error estimate $S/\sqrt{(n)}$, where $S$ is standard deviation and $n$ is the number of independent grid-points, will require a value of $n$ which is substantially fewer than the actual number of grid-points. Noting the assertion by Parker (2010) that errors in monthly-mean Central England Temperature values are around 0.3 °C, based on measurements from three or four stations, errors for our current monthly UK temperature values, based on station density and geographical coherence, are expected to be of the order of 0.1 °C. However, work is continuing to calculate these, because we wish to determine the total error, and the error due to variations in station density is only a part. Errors would be smaller for seasons and years, but larger for individual regions and counties. Rainfall is less straightforward because of the skew nature of the distribution. Typical point-value error estimates for monthly rainfall are around 10 % of the actual rainfall amount, and larger in magnitude for wet months and in wetter locations. Errors in areal values might therefore be considerably less than this, depending on the station density and the geographical coherence, but again these are likely to be larger in wetter areas, such as western Scotland, than drier areas, such as south-east England.

Error bars on our areal values will improve our assessment of their historical context. For example, we will be able to rank values more realistically, e.g. "January 2010 had a mean temperature across the UK of 0.9±0.2 °C, making it the coldest since January 1987 (0.7 ± 0.2 °C)". New graphical ways of illustrating the time-series of areal values could be used, incorporating the error-bars.

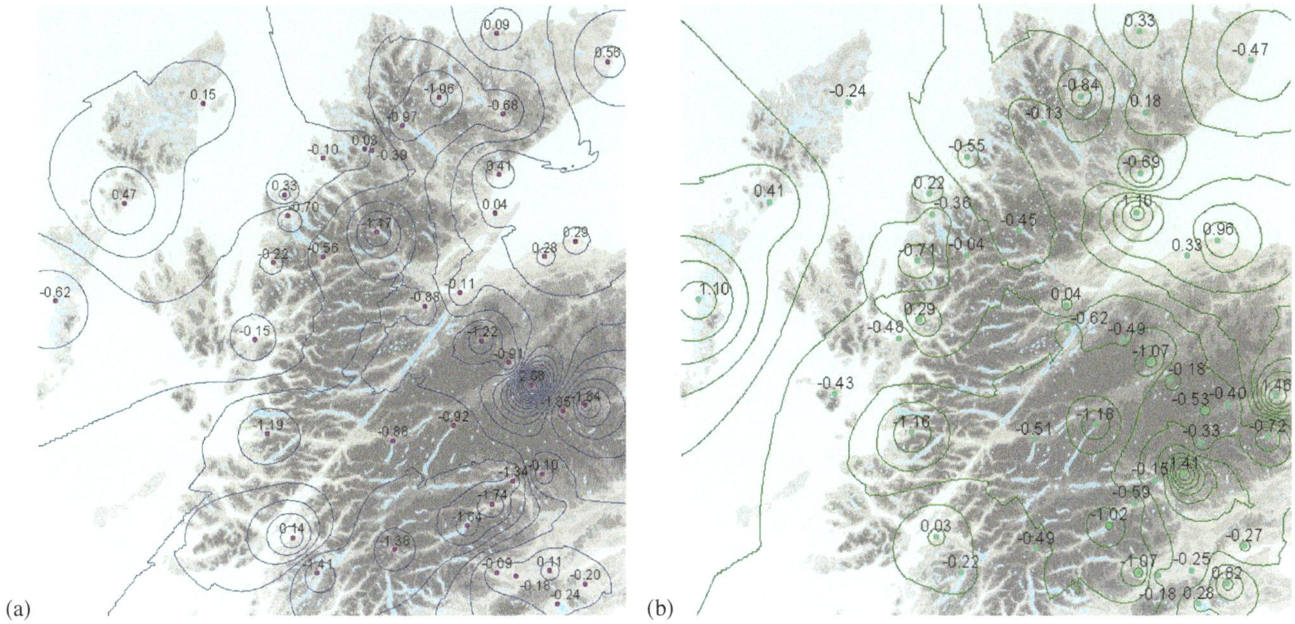


**Figure 4.** (**a**) Maximum Temperature station biases, October 2007. (**b**) Minimum Temperature station biases, January 2009. Values in °C.

Edited by: O. E. Tveito

## References

Parker, D. E.: Uncertainties in early Central England Temperatures, Int. J. Clim., 30, 1105–1113, 2010.

Perry, M. and Hollis, D.: The generation of monthly gridded datasets for a range of climatic variables over the UK, Int. J. Clim., 25, 1041–1054, 2005.

# Quality control procedures at Euskalmet data center

**R. Hernández, M. Maruri, K. Otxoa de Alda, J. Egaña, and S. Gaztelumendi**
Basque Meteorology Agency (EUSKALMET), Vitoria-Gasteiz, Spain
TECNALIA, Meteo Unit. Vitoria-Gasteiz, Spain

*Correspondence to:* R. Hernández (roberto.hernandez@tecnalia.com)

**Abstract.** The Basque Country Mesonet measures more than 130 000 observations daily from its 85 Automatic Weather Stations (AWS). It becomes clear that automated software is an indispensable tool for quality assurance (QA) of this mesoscale surface observing network. This work describes a set of experimental semi-automatic quality control (QC) routines that is applied at Euskalmet data center. It has paid special attention to level validation design and associated flags, as well as to the system outputs, which are used by meteorologist and maintenance staff.

## 1 Introduction

In 1991 the Basque Meteorological Service (Basque Government) began the deployment of an AWS network. Priority was given to the real time observation of water surface level of the rivers. It is noted that flood return periods are quite small in most of the short length hydrologic basins of the Basque Country, especially those oriented to the Cantabrian Sea. In some sense, that objective conditions its current design. Thus much of the stations are gauging or water quality, located in valley bottoms, along the river beds (Fig. 1). In the course of time the network was gaining a more general purpose and was completed with other weather sensors. Nowadays the Mesonet has more than 85 AWS spread over the Basque Country (just over 7000 $km^2$), a quite high density network (Fig. 2) (Gaztelumendi et al., 2003). Its data are used in a wide range of applications related to meteorology – nowcasting, climate, data assimilation, verification –, and with many other fields – transportation, energy, insurance, planning, education, etc. From the beginning it became necessary to perform quality control tasks, both in real time and on recorded data (Navazo et al., 1999; Maruri et al., 2003).

The context in which this work was carried out corresponds to the real time monitoring and nowcasting requirements in Euskalmet (Gaztelumendi et al., 2006). Thus, a set of validation tests have been implemented that try to avoid the inclusion of erroneous data coming from the Mesonet in the visualization system. Moreover, provides information of great use to the QA system.

## 2 Overview of Mesonet QC processes

There are four components that integrate the QC system in the Basque Country Mesonet (Maruri et al., 2010a): (i) laboratory calibration, (ii) maintenance services, (iii) automated routines and (iv) manual inspection. Each component provides valuable information on the operation of the network and their results are shared across the system to ensure the accuracy of the data. Components work to different timescales, from the moment in which data is recorded to the analysis of data quality over time (Table 1).

## 3 Validation levels

The literature about QC methods of meteorological observations is very prolific. It is common to find in it a characteristic sequence of validation procedures: range, step, internal, persistence and spatial. The success of the checks depends largely on the thresholds used. In that sense, it is crucial to adapt them to the specific conditions of the region. In our case, much of the thresholds are based on those proposed by the University of the Basque Country (Maruri et al., 2010a), the WMO guidelines (WMO, 2008) and other meteorological services such as the Oklahoma Mesonetworks (Shafer et al., 2000; Fiebrich et al., 2010; Vejen et al., 2002).

According to Spanish normative (UNE-AENOR, 2004) related to AWS networks, and other operational services, we define six validation levels. Except the visual check, levels are successively applied to the meteorological variables

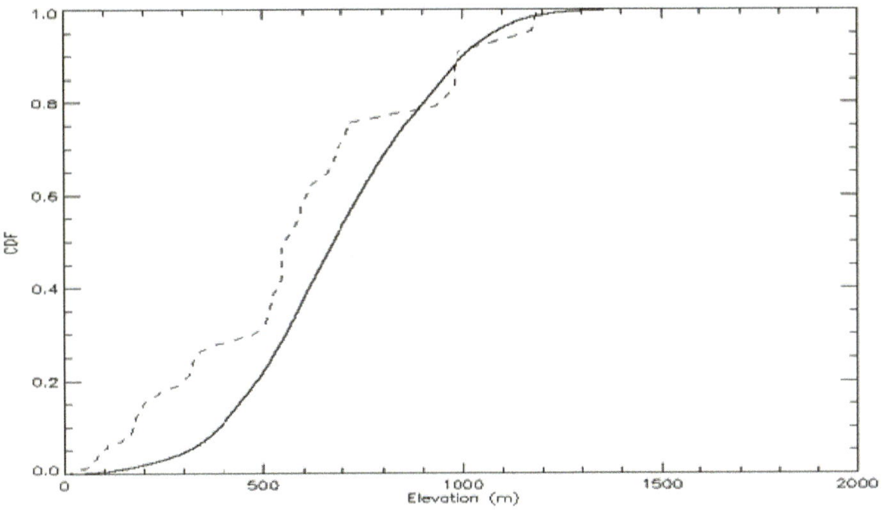


**Figure 1.** Cumulative probability function: Digital Elevation Model and Mesonet sites.

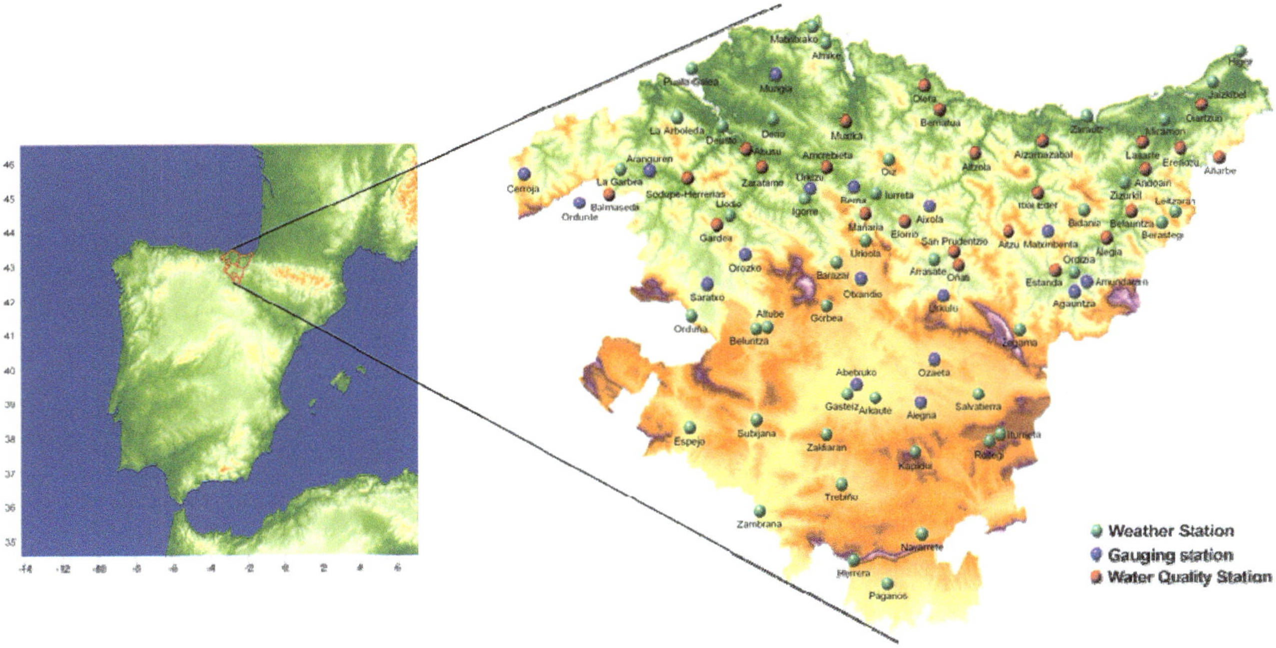


**Figure 2.** Basque Country location and Mesonet map.

(Table 2). The tests are usually generic, but some meteorological variables require certain specifications. The following briefly describe each of them (Table 3).

### 3.1 Validation of the structure of data recorded and the measurement time (level 0)

This section verifies the correct decoding of the data.

### 3.2 Validation of the data according to limits (level 1)

Two types of checks are implemented: (i) sensor-based range test ensures data are between range of sensor hardware specifications or theoretical limits; (ii) climate-based range test ensures data are between certain flexible limits. Currently an observation is compared with the climatological values calculated from representative stations and the expected standard deviation.

**Table 1.** Timescales of QC/QA procedures used by the Basque Country Mesonet (based on Shafer et al., 2000).

| Time interval | Análisis technique |
|---|---|
| Seconds | Data checked at the Campbell datalogger* |
| Minutes | Missing data recovery |
| | Automated QC/QA procedures |
| Hours | Visual inspection (data visualization software and QC/QA reports) |
| Days | Trouble tickets issued to maintenance |
| Weeks | Visual inspection of aggregate data |
| | Missing data report |
| Months | Supervised QC/QA procedures |
| | Preventive or corrective maintenance |
| Year | Sensor calibration |
| | Instrument updating (datalogger, sensors, etc.) |

* This analysis is not yet operational due to the recent update of the dataloggers in the Mesonet.

In the case of solar radiation, the theoretical maximum value is given by the calculation of a clear sky model. The limit is occasionally exceeded under partially covered skies, so the theoretical values are multiplied by a factor of 1.2. Previously, we analyze the existence of noise into the signal.

## 3.3 Validation of the temporal consistency (level 2)

At this level both the consistency of the data and the consistency of the series are analyzed. In the first case, the following checks are performed: (i) step test ensures data do not change more than certain limit in 10 min; (ii) spike-dip test ensures data do not successively increase and decrease (or vice versa) more than certain limit in 20 min.

Regarding the second aspect, the persistence test ensures data change more than certain value in a defined period of time. At this point we do various specifications. For relative humidity, we check if the hygrometers saturate above or below 100 %. For precipitation, we ensure that the rain gauge does not register high rainfall intensities over a period of time questionable.

## 3.4 Validation of the internal consistency of the data (level 3)

The system checks the gust factor, ie the ratio between the mean wind speed and maximum gust, which must exceed a predetermined threshold. We also have established relationships for precipitation, flagging those observations that occur with low humidity or with a high percentage of solar radiation on the clear sky model.

**Table 2.** Meteors and applied levels.

| Meteor | Level 1 | Level 2 | | Level 3 | Level 4 | Level 5 |
|---|---|---|---|---|---|---|
| | | Data | Serie | | Spatial | |
| Wind speed | ✓ | ✓ | ✓ | ✓ | ✓ | ✓ |
| Wind direction | ✓ | x | ✓ | x | x | ✓ |
| Wind gust | ✓ | ✓ | ✓ | ✓ | ✓ | ✓ |
| Air temperature | ✓ | ✓ | ✓ | x | ✓ | ✓ |
| Relative humidity | ✓ | ✓ | ✓ | x | ✓ | ✓ |
| Precipitation | ✓ | x | ✓ | ✓ | x | ✓ |
| Air pressure | ✓ | ✓ | ✓ | x | x | ✓ |
| Solar radiation | ✓* | ✓ | ✓ | x | ✓ | ✓ |

* Based on clear sky model.

## 3.5 Validation of the spatial consistency of the data (level 4)

The test tries to validate the spatial consistency of both absolute data and temporal changes. In the case of absolute data, it performs a cross validation process. The idea lies in removing one datum at a time from the data set and re-estimating this value from remaining data using kriging algorithms. Interpolated and actual values are compared to the standard deviation of the spatial domain:

$$\Delta = \frac{z_K^*(u) - z(u_\alpha)}{\sigma}$$

The observation is flagged when the difference exceeds twice the standard deviation ($\Delta > 2$) or when the error estimation is greater than a certain absolute value. The estimation methods used are ordinary kriging, kriging with external drift and simple kriging with varying local means (Goovaerts, 1997; Hernández, 2001). The last two accounts for secondary information (terrain elevation, etc.) and they are preferably used to estimate the air temperature.

In the case of temporal changes, we perform a reanalysis of the values that have not passed the validation of the temporal consistency. The mechanism is the same as in the previous case, but using simple kriging. In this way we could relax the thresholds used in level 2. The assumption behind is that when a notable temporal variation of a meteorological variable happens in a given station, this should be reflected in the neighbourhood.

Although the high density of the Mesonet is appropriate to carry out this type of spatial tests, the possible spatial anisotropy must be taken into account as far as possible. There are several factors leading to strong gradients in the meteorological variables that influence their effectiveness. One is the existence of distinct climatic barriers. This is the case of Cantabria Mountains, that delimit the comarca of Rioja Alavesa, located in the south of the Basque Country. This represents an additional problem, because it is known that kriging errors grow to the edges of the domain. Other factor is the impact of unique meteorological phenomena: galerna, strong temperature inversions, heat bursts, etc.

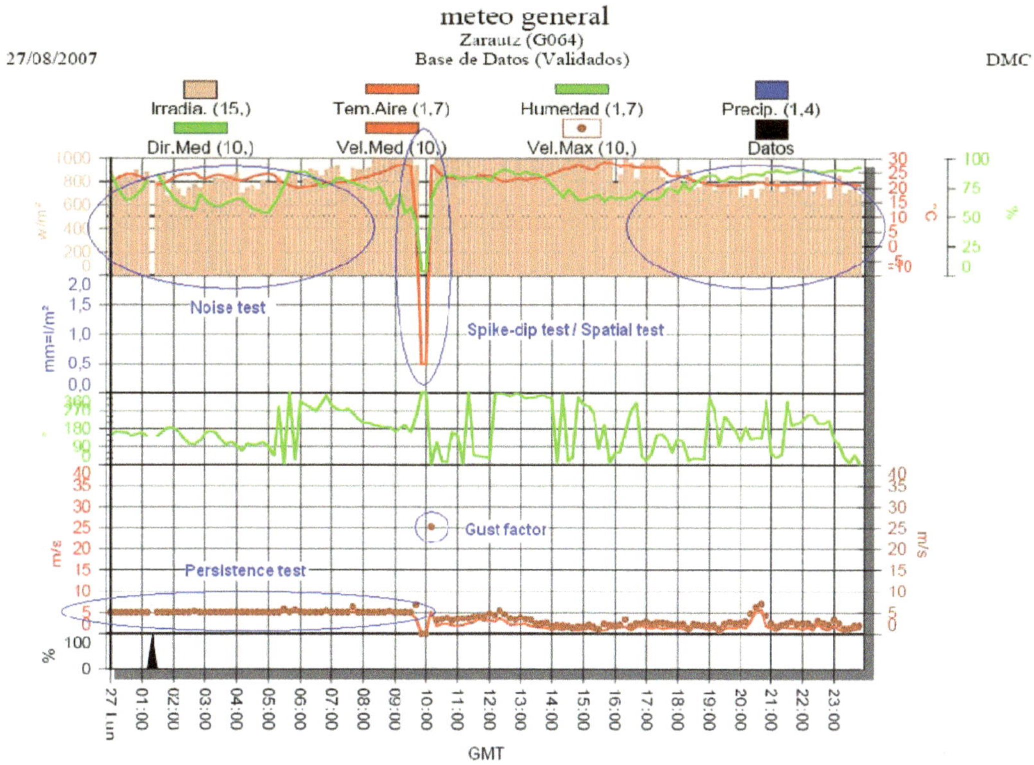


**Figure 3.** This chart shows data from an AWS on a given day with different types of errors in measurement sensors and what kind of test would be capable of detecting them.

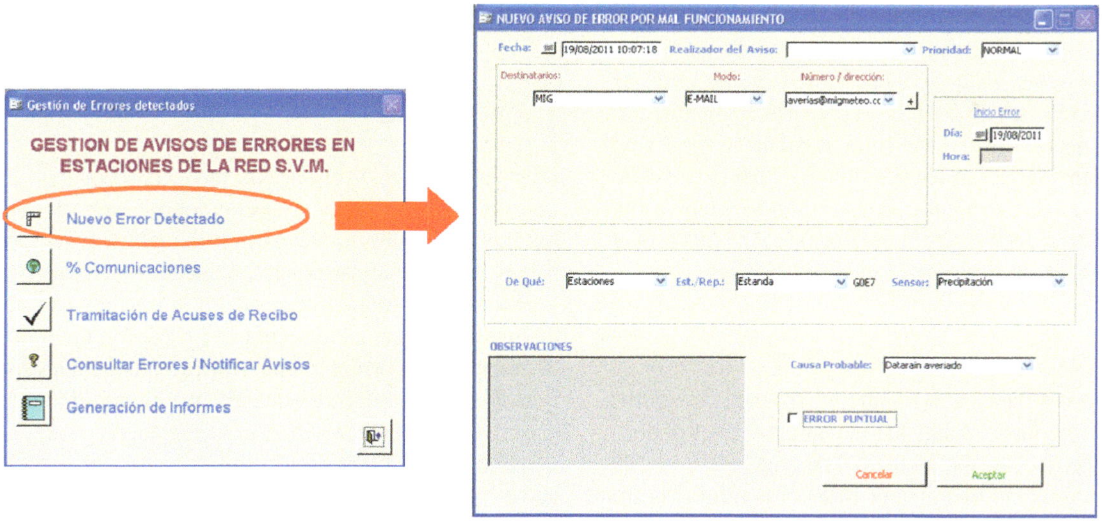


**Figure 4.** The alarm management software.

## 3.6 Visual check (level 5)

We cannot broach the visualization of all information recorded by the Mesonet. Therefore, the time can be reduced to the display of suspect data. At this level it is relatively easy to decide whether the assigned flag is right, but sometimes questions arise. For example, an anemometer stuck in a situation of calms. The test designed to detect the problem is the persistence (level 4), but it is not trivial where to cut.

The casuistry of errors in an AWS is very large. Show all of them here is beyond the scope of this paper. Figure 3 summarizes what happened on a particular day in a given station.

**Table 3.** Validation levels.

| Level | Validation | Name | Test |
|---|---|---|---|
| 0 | Data structure | – | decoding |
| 1 | Limits | sensor/climate range | min < value < max |
| 2 | Temporal consistency | step | (current-previous) < step |
| | | spike-dip | (current-previous) & (current-next) < step |
| | | persistence | $\text{value}\|_{\text{period}}$ > threshold |
| 3 | Internal consistency | variable relation | $\begin{pmatrix}\text{var1}\\ \text{condition}\end{pmatrix} \& \begin{pmatrix}\text{var2}\\ \text{condition}\end{pmatrix} \& \begin{pmatrix}\text{var}N\\ \text{condition}\end{pmatrix}$ |
| 4 | Spatial consistency | kriging | $\text{ObsValue}\|_{\text{AWS}} - \text{EstValue}\|_{\text{neighb.}} < \text{threshold}$ <br> $\left(\text{ObsValue}\|_{\text{AWS}} - \text{EstValue}\|_{\text{neighb.}}\right)/\sigma < 2$ |
| | | | $\left(\text{ObsTemporalChange}\|_{\text{AWS}} - \text{EstTemporalChange}\|_{\text{neighb.}}\right)/\sigma < 2$ |
| 5 | Visual | – | graphics visualization |

We can see at glance the application of different types of test validation.

Subsequently, there is also an adjustment work, which has a great impact on the quality of database. For example, once a step is detected in air temperature, is necessary to define dates that delimit the problem and calculate the magnitude. Other major supervised adjustments try to correct records coming from rain gauges and pyranometers not properly calibrated. Sometimes gauges underestimate the precipitation due to relay failures. Therefore, their amounts are compared and adjusted directly to the volume stored by the totalizer system (collected approximately every month by maintenance). With regard to solar radiation, it is hardly surprising that the constant number of the pyranometer lead to bad data. In this case, we reconstruct the series as best as we possibly could, specifically, fitting the observed data to the theoretical values through a factor calculated in clear sky days.

## 4 Flagging

It is noted that raw data are never altered, instead, all records are coupled with quality flags that indicate the level of confidence assigned by the QC system. The flags are stored in a metadata field, composed of four bytes of control (Maruri et al., 2010b). Each byte deals with (i) origin of the data, (ii) status of the data, (iii) validation levels, (iv) adjustments. On the other hand, the adjusted data series are considered as new variables, so you must define new fields for them.

Data are flagged as erroneous when sensor-based range test are not passed. All others levels (except visual check) qualify the data as suspicious. Also, results from temporal checks are combined with those coming from spatial tests. Thus, if an observation is flagged by both tests data are considered as a failure.

## 5 Automated QC summary report

The automated QC produces a summary report daily that compiles data incidences from the previous day. This report is accessible to the meteorologist on duty responsible for carrying out the surveillance of the network, who determines whether further action is warranted. If so determined by the meteorologist, it sends a new error message for a malfunction to maintenance services. This has an application designed for this purpose (Fig. 4).

## 6 Conclusions

This paper presents very briefly the QC procedures currently used by Euskalmet. As a special contribution on this issue, we highlight the efforts made in the development of algorithms for the analysis of spatial data consistency.

Despite the automation needs, it is important to note that the quality of the data can not fall solely on the application of automatic algorithms. The quality starts with a good location of each station. Subsequently, the information must flow properly between the different components of the QC/QA system. Among other things, this prevents errors are perpetuated over time.

**Acknowledgements.** The authors would like to thank the Emergencies and Meteorology Directorate-Interior Department-Basque Government for public provision of data and operational service financial support. We also would like to thank all our colleagues from EUSKALMET for their daily effort in promoting valuable services for the Basque community.

Edited by: M. Brunet-India

## References

Fiebrich, D. S., Morgan, C. R., McCombs, A. G., Hall, P. K., and McPeherson, R. A.: Quality assurance procedures for mesocale meteorological data, J. Atmos. Ocean. Tech., 27, 1565–1582, 2010.

Gaztelumendi, S., Hernández, R., and Otxoa de Alda, K.: Some aspects on the operative use of the automatic stations network of the Basque Country, 3rd Int. Conf. on Experiences with Automatic Weather Stations, Torremolinos, Spain, February 2003.

Gaztelumendi, S., Otxoa de Alda, K., Gelpi, I. R., and Egaña, J.: An Automatic Surveillance System for Severe Weather Real Time Control in Basque Country Area, 4th International Conf. on Experiences with Automatic Weather Stations, Lisboa, Portugal, May 2006.

Goovaerts, P.: Geostatistics for Natural Resources Evaluation, Applied Geostatistics Series, Oxford University Press, New York, 483 pp., 1997.

Hernández, R.: Comparación de métodos de kriging multivariante para la estimación espacial de la temperatura del aire, Estudios Geográficos, LXII, 243, 285–308, 2001.

Maruri, M., Navazo, M., Alonso, L., García, J. A., Gangoiti, G., Matabuena, M., Iza, J., and Aranda, J. A.: Automatic validation of data from AWS, The experience of Meteorological Basque Service Network, 3rd Int. Conf. on Experiences with Automatic Weather Stations, Torremolinos, Spain, February 2003.

Maruri, M., Lantarón, L., Vallejo, A., Romo, J. A., Serrano, M., and Manso, B.: The new data quality system in the hydrometeorological network of the basque country, WMO Technical Conference on Meteorological and Environmental Instruments and Methods of Observation, Helsinki, Finland, 30 August–1 September 2010a.

Maruri, M., Romo, J. A., Manso, B., and Lantarón, L.: Design of a core of metadata of the Basque Meteorological Service, WMO Technical Conference on Meteorological and Environmental Instruments and Methods of Observation, Helsinki, Finland, 30 August–1 September 2010b.

Navazo, M., Maruri, M., Aranda, J. A., Gezuraga, X., García, J. A., Pérez, G., and Alonso, L.: Sep-up and real time quality control of data from the automatic hydrometeorological network in the Basque Country, 2th Int. Conf. on Experiences with Automatic Weather Stations, Vienna, Austria, 1999.

Shafer, M. A., Fiebrich, C. A., Arndt, D. S., Fredrickson, S. E., and Hughes, T. W.: Quality assurance procedures in the Oklahoma Mesonetwork, J. Atmos. Ocean. Tech., 17, 474–494, 2000.

UNE-AENOR 500540: Automatic weather stations networks: Guidance for the validation of the weather data from the station networks, Real time validation, 2004.

Vejen, F. (Ed), Jacobsson, C., Fredriksson, U., Moe, M., Andresen, L., Hellsten, E., Rissanen, P., Pálsdóttir, T., and Arason, T.: Quality Control of Meteorological Observations, Automatic Methods used in the Nordic Countries, Norwegian Met. Inst., 8/2002 KLIMA, 2002.

WMO: Guide to Meteorological Instruments and Methods of Observation, WMO-No. 8, 2008.

# 10

# On the skill of various ensemble spread estimators for probabilistic short range wind forecasting

**A. Kann**

Central Institute for Meteorology and Geodynamics, Vienna, Austria

*Correspondence to:* A. Kann (alexander.kann@zamg.ac.at)

**Abstract.** A variety of applications ranging from civil protection associated with severe weather to economical interests are heavily dependent on meteorological information. For example, a precise planning of the energy supply with a high share of renewables requires detailed meteorological information on high temporal and spatial resolution. With respect to wind power, detailed analyses and forecasts of wind speed are of crucial interest for the energy management. Although the applicability and the current skill of state-of-the-art probabilistic short range forecasts has increased during the last years, ensemble systems still show systematic deficiencies which limit its practical use. This paper presents methods to improve the ensemble skill of 10-m wind speed forecasts by combining deterministic information from a nowcasting system on very high horizontal resolution with uncertainty estimates from a limited area ensemble system. It is shown for a one month validation period that a statistical post-processing procedure (a modified non-homogeneous Gaussian regression) adds further skill to the probabilistic forecasts, especially beyond the nowcasting range after +6 h.

## 1 Introduction

The analysis and nowcasting system INCA (Integrated Nowcasting Through Comprehensive Analysis), developed at the Austrian national weather service (Central Institute for Meteorology and Geodynamics, ZAMG), provides, among other quantities, short range deterministic analyses and forecasts of the horizontal 10-m wind components on very high resolution in time (60 min) and space ($1 \times 1$ km) with special emphasis on the nowcasting range (usually 0–6 h ahead). In the analysis mode, it combines numerical weather prediction (NWP) forecast fields from the operational limited area model ALADIN-AUSTRIA (Aire Limitée Adaptation Dynamique Développement InterNational) with high-resolution topographic data and integrates about 250 stations within the operational domain. However, although the system shows high skill in the nowcasting range, it is affected by uncertainties mainly due to errors in the initial condition, model formulations and physical parameterizations. The paper compares three methods to quantify these uncertainties and to provide sharp and reliable probabilistic short range 10-m wind speed forecasts. The first method combines the deterministic INCA forecast with the ensemble variance from ALADIN-LAEF (Aire Limitée Adaptation Dynamique Développement InterNational Limited Area Ensemble Forecasting), the operational limited area ensemble prediction system (LAM-EPS) at ZAMG. The second one derives uncertainty estimation from the skill of retrospective INCA forecasts. The third approach uses the uncertainty information of ALADIN-LAEF, but additionally applies a statistical calibration procedure.

## 2 The INCA system

The analysis and nowcasting system INCA, which is in operation at ZAMG, provides improved numerical forecasts especially in the nowcasting range on a very high horizontal resolution ($1 \times 1$ km). A problem for NWP data at all scales is the comparatively low skill within the nowcasting range (0–6 h). To overcome this problem, INCA has been developed to provide temporarily and spatially highly resolved analyses and nowcasts taking into account regional and small scale topographic effects. The basic idea of INCA is to complement and improve NWP direct model output using real-time observations, remote sensing data and high-resolution topographic data (Haiden et al., 2011). The INCA system provides near-real-time analyses and forecasts for the parameters temperature, humidity, wind, precipitation amount, precipitation type, cloudiness, and global radiation.

For wind, z-coordinates with $\Delta z = 125$ m are used in the vertical, where $z$ denotes the height above mean sea level. The three-dimensional wind analyses are based on NWP forecast fields as a first guess and observations to correct them. Operationally, the deterministic limited area model ALADIN-AUSTRIA is used, which runs on 9.6 km horizontal resolution with 60 vertical levels (Wang et al., 2006). ZAMG operates about 250 automated weather stations (Teilautomatisches Wetterstationsnetz, TAWES) in Austria. The wind analysis is constructed by a first guess of the NWP model (ALADIN-AUSTRIA), a 2-D (10-m wind vector) and 3-D (lowest model level wind vector) component. A factor, which transforms a model level wind into a 10-m wind, determines the differences between model-level wind and a 10-m wind observation. After multiplying the observed wind by this factor, differences of the $u$ and $v$ components between the model and the observations are interpolated using a modified inverse-distance weighting (IDW). An iterative relaxation algorithm is applied in order to provide mass consistent fields (Wang et al., 2005; Sherman, 1978). Additionally the generally higher wind speeds over lakes compared the surrounding land due to reduced roughness length is taken into account using boundary-layer similarity theory, as described in Haiden et al. (2010). In Fig. 1 an example of the resulting INCA 10-m wind analysis is shown in comparison to the first guess NWP (ALADIN) 10-m wind field.

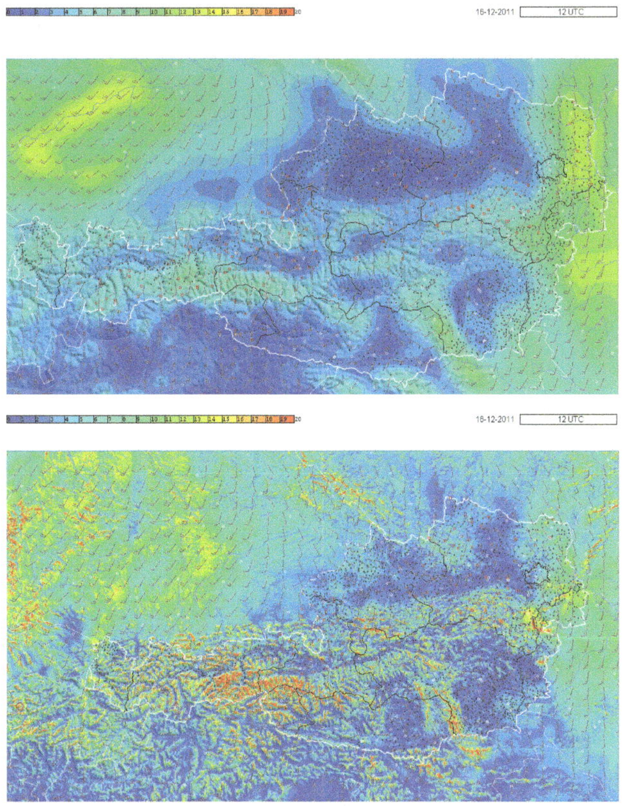

**Figure 1.** Example of a 10-m wind analysis provided by the INCA wind module issued for 16 December 2011, 12:00 UTC (bottom). The top figure shows the corresponding first guess field taken from the ALADIN-AUSTRIA model run initialized at 16 December 2011, 00:00 UTC, +12 h forecast.

## 3 The limited area ensemble system ALADIN-LAEF

The ALADIN-LAEF system, which runs in operational mode since 2007 at ZAMG, provides ensemble forecasts up to +60 h, the horizontal resolution is 18 km with 37 vertical levels. It uses the spectral limited area model ALADIN (Wang et al., 2006) in order to produce 17 ensemble members (16 perturbed members, 1 control). The perturbations for the initial conditions of the 16 ensemble members are composed by a large scale part using 16 ECMWF EPS members which are created with the singular vector method (Leutbecher and Palmer, 2008) and a small scale component derived with the breeding technique (Wang et al., 2011). Both perturbations are finally combined by applying a spectral blending procedure (Brozkova et al., 2001). Model errors are taken into account by a multi-physics approach. A detailed description of the ALADIN-LAEF system can be found in Wang et al. (2011).

## 4 Probabilistic (very) short range forecasting approaches

The basic strategy has been developed and implemented for the purpose of providing sharp and reliable probabilistic short range 2-m temperature forecasts and has been described more detailed in Kann et al. (2011). The main idea is to treat the deterministic INCA 10-m wind speed forecast as ensemble mean and to apply an ensemble variance by different techniques. A statistical method ($EPS_{stat}$), a dynamical one ($EPS_{dyn}$) and a coupled dynamic-statistical approach ($EPS_{dynstat}$) are developed to generate an 18-member ensemble providing probabilistic short range wind speed forecasts up to +36 h.

The statistical method $EPS_{stat}$ uses the deterministic INCA run as ensemble mean and its relative root mean squared error (RMSE) of the past 50 training days as ensemble variance. The ensemble size of 18 members is obtained by extracting the quantile values from a truncated (i.e. cut off at zero) Gaussian cumulative distribution function (CDF) centralized about the mean in such a way that the ensemble spread is determined by a fraction of the relative (i.e. with

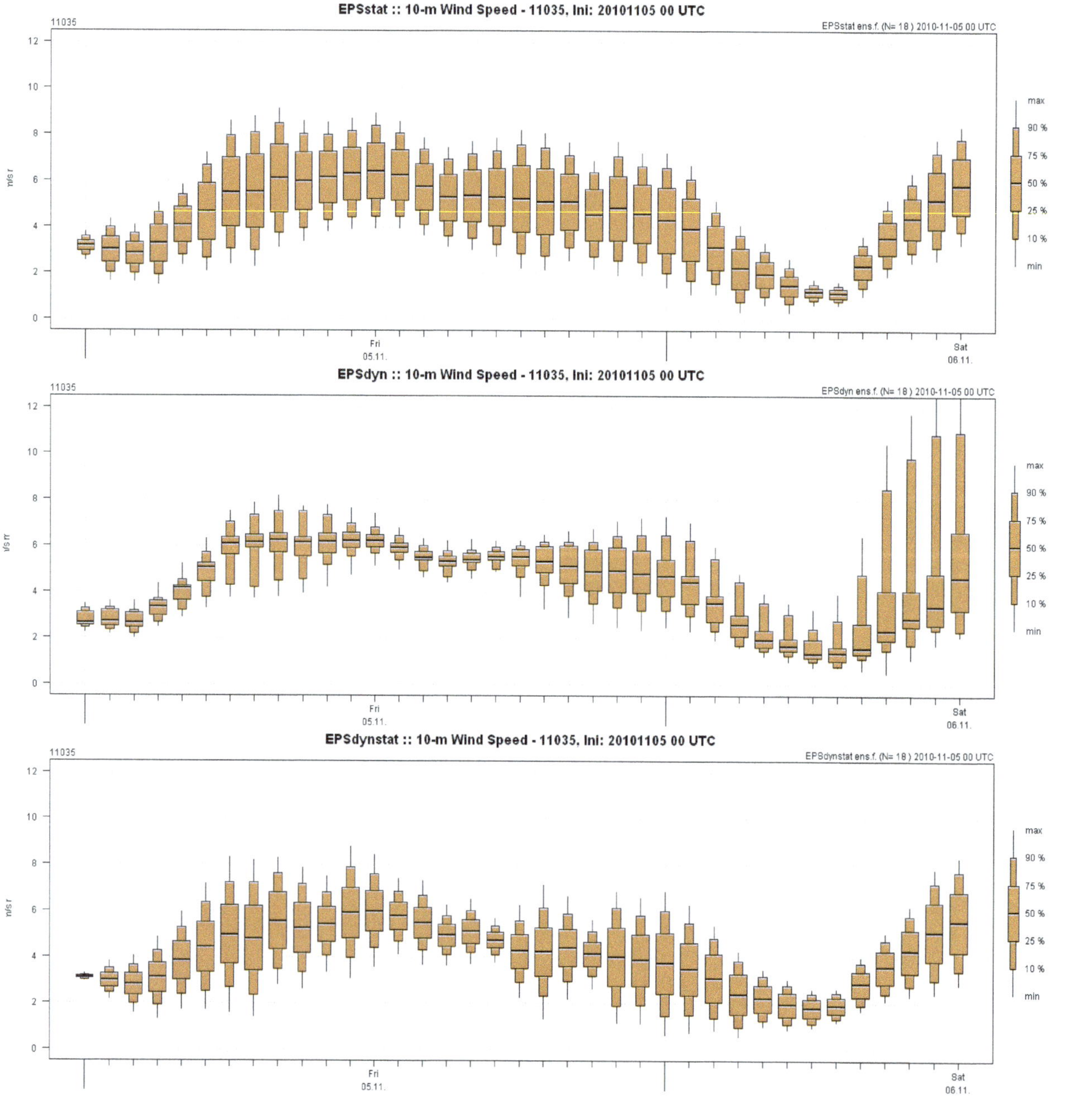


**Figure 2.** EPSgrams for 10-m wind speed, calculated with the three different methods, forecast initialized at 5 November 2010, 00:00 UTC, for station Vienna Hohe-Warte.

respect to the observed average) RMSE of the training data (Kann et al., 2009). Several experiments with varying rescaling factors revealed a value of 3/4 to perform best in terms of the Continuous Ranked Probability Score (CRPS) (not shown).

In the dynamical approach $EPS_{dyn}$, the ensemble variance from ALADIN-LAEF provides the uncertainty information and is superimposed on the deterministic INCA run (Kann et al., 2011). Practically, the differences between the individual ensemble members and the ensemble mean of ALADIN-LAEF are added to the INCA run. Thus, an 18-member ensemble with a high resolution deterministic component and an uncertainty part derived from a limited area ensemble system is created.

The dynamic-statistical method $EPS_{dynstat}$ extends the aforementioned dynamical approach with an additional statistical post-processing application. The $EPS_{dyn}$ is further calibrated by applying a cut-off variant of the non-homogeneous Gaussian regression (NGR) technique (Thorarinsdottir and Gneiting, 2010; Gneiting et al., 2005; Hagedorn et al., 2008; Kann et al., 2009). This technique statistically calibrates the mean and the ensemble variance

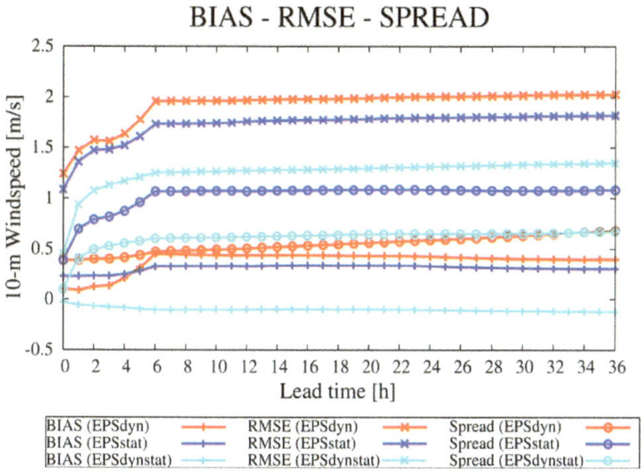


**Figure 3.** Bias, root mean square error (RMSE) of the ensemble mean, and ensemble spread of the 3 experimental INCA ensembles (dynamical: red, statistical: dark blue, dynamic-statistical: light blue).

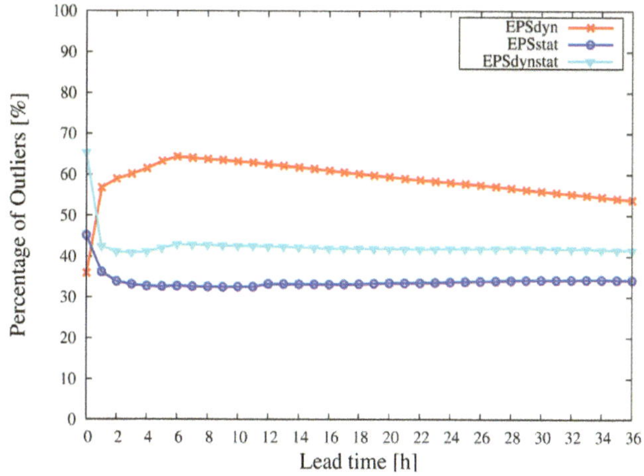


**Figure 4.** Percentage of outliers as a function of lead time, for the dynamical (red), statistical (dark blue) and dynamic-statistical (light blue) approach.

by minimizing the Continuous Ranked Probability Score (CRPS) within a certain training period. The variance of the forecast distribution is a linear function of the ensemble variance. Again, the final 18-member ensemble is created by calculating the quantile values from the Gaussian CDF, centred about the mean with respect to the spread re-scaling factor.

The three sets of ensembles described above are tested for one month from 1 to 30 November 2010. The hourly 10-m wind speed forecasts of INCA and of ALADIN-LAEF are bilinearly interpolated separately for each station and lead time to the locations of the surface weather stations in Austria.

Note that the variance used for the experiments $EPS_{dyn}$ and $EPS_{dynstat}$ is calculated from the latest available ALADIN-LAEF run, valid for the same target time as the INCA run (e.g.: The variance of the INCA ensemble issued at 1 November 2010, 11:00 UTC + 6 h is provided by the ALADIN-LAEF issued at 1 November 2010, 00:00 UTC + 17 h).

Figure 2 shows an example of a probabilistic 10-m wind speed forecast initialized at 5 November 2010, 00:00 UTC at the location of Vienna Hohe-Warte (11035) obtained by the three different methods.

## 5 Validation

To assess the performance of the three probabilistic methods, a comparative evaluation was carried out for the one month period (1–30 November 2010). The 10-m wind speed forecasts are verified against measurements of approx. 250 automatic surface weather stations in Austria. Hourly initializations with lead times up to +36 h are considered. A set of standard ensemble and probabilistic forecast verification methods is consulted to evaluate the performances: ensemble spread, ensemble root-mean square error (RMSE), the continuous ranked probability score (CRPS), percentage of outliers, area under relative operating characteristic (ROC) curve, and the reliability diagram. A detailed description of these verification measures is found in Wilks (2006). In Fig. 3, the error of the ensemble mean in terms of bias and root mean squared error (RMSE) and the ensemble standard deviation is illustrated for the three experiments. Both the statistical and dynamical methods convey a slight positive bias, whereas the combined statistic-dynamical is almost biasfree. The RMSE is substantially reduced to about 1–1.4 m s$^{-1}$ by for the combined approach compared to the pure statistical and dynamical methods, respectively, where the RMSE is about 1.5–2 m s$^{-1}$. Another measure of statistical reliability is the percentage of outliers. It expresses the number of cases where the verifying analysis at any location falls outside the ensemble. A reliable system should have a score close to an expected value of $2/(n+1)$ which corresponds to about 10 % for an 18-member EPS. The pure dynamical ensemble is affected by the highest percentage of outliers (approx. 60 %), but the number of outliers decreases for the dynamic-statistical approach (about 42 %) and the statistical method (35 %) (Fig. 4). The ROC curve and the area below it (Zhu et al., 2002), respectively, offers another methodology for a quantitative evaluation of an ensemble system. With respect to the ROC area, the dynamic-statistical method outperforms the statistical and especially the dynamical approach (Fig. 5). A further assessment of forecast quality with regard to reliability and resolution is given by the reliability diagram which indicates quantitative evidence about the ability of the probability forecasts to reflect the observed relative frequency (Stanski et al., 1989). Figure 6 reveals the high reliability of the dynamic-statistical combination compared to both the dynamical and statistical approach for different lead times (6 h (Fig. 6a) and 24 h (Fig. 6b), respectively). The latter two methods are characterized by the fact that low as well as high observed frequencies are overforecast (apart from the

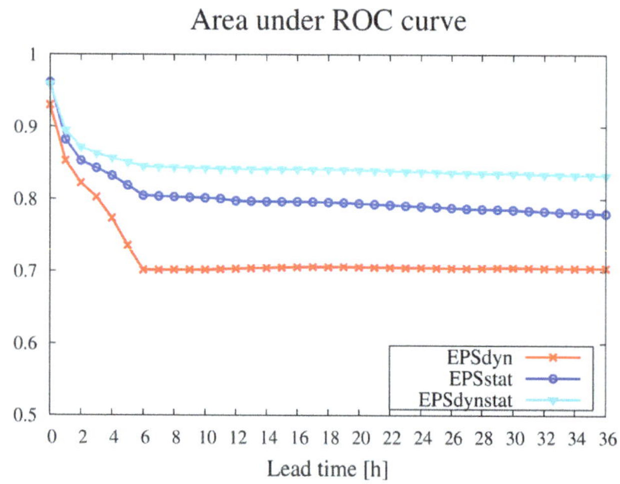


**Figure 5.** Area under ROC curves for 10-m wind speed >2 m $s^{-1}$ as a function of lead time, from the dynamical (red), statistical (dark blue) and dynamic-statistical (light blue) method.

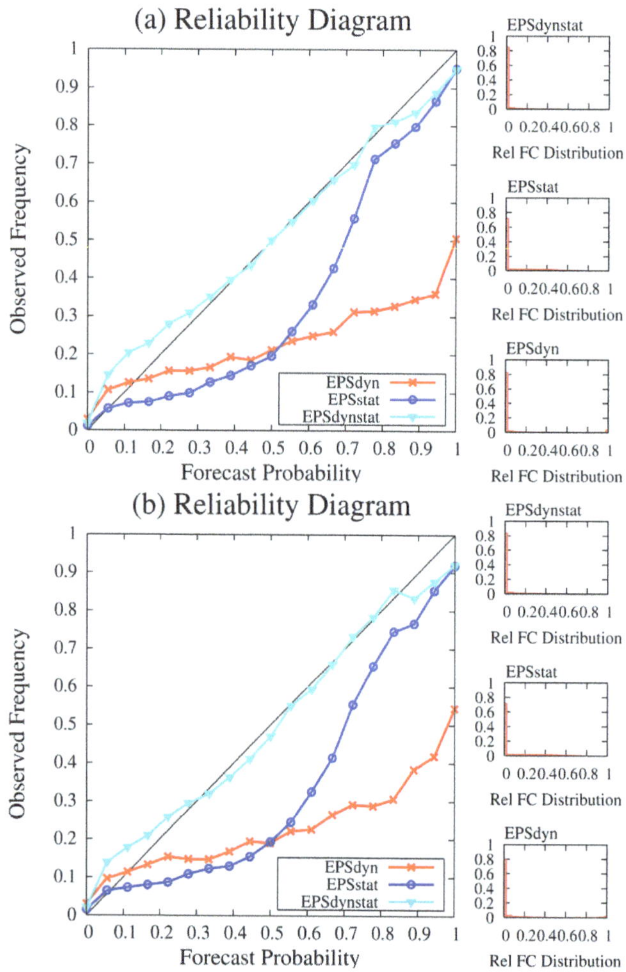


**Figure 6.** Reliability diagram for +6 h **(a)** and +24 h **(b)** from the dynamical (red), statistical (dark blue) and dynamic-statistical (light blue) method. Threshold: 5 m $s^{-1}$.

very low frequencies which are underforecast). The Continuous Ranked Probability Score (CRPS), which can be decomposed into a reliability, resolution and uncertainty term, is the integral form of the discrete ranked probability score over all (possible) thresholds (Hersbach, 2000). It compares a full distribution with the observation, where both are represented as cumulative distribution functions (CDFs). In case of a specific observation value, the corresponding CDF is a single step-function with the step from 0 to 1 at the observed value of the considered parameter (Heaviside function). Figure 7 confirms previous findings that the combination of dynamical and statistical methods outperforms both the pure statistical and dynamical approaches. This is especially true for lead times beyond the nowcasting range, where the CRPS for $EPS_{dynstat}$ is substantially reduced to about 0.7 m $s^{-1}$ compared to the CRPS for $EPS_{dyn}$ (1.2 m $s^{-1}$) and $EPS_{stat}$ (0.9 m $s^{-1}$). However, also within the nowcasting range up to +6 h, the dynamic-statistical method shows highest skill of all three methods. A further elaboration of the statistical approach has been performed by applying the non-homogeneous Gaussian regression to $EPS_{stat}$. Figure 7 confirms that the calibration is able to add further skill to the statistical method (experiment $EPS_{statcal}$, black line), although the error reduction is not pronounced. However, the calibration of the dynamical approach ($EPS_{dynstat}$) allows for a higher skill in terms of the CRPS. In other words, the spread obtained from ALADIN-LAEF provides additional information compared to the RMSE spread estimator.

## 6 Conclusions

Three different methods for constructing an ensemble of 10-m wind speed forecasts are developed and validated for a one month period: a statistical, a dynamical method and a combination of both. The common idea, which has originally been designed for 2-m temperature, is to unite the strengths of a very high resolution nowcasting system with a limited area ensemble system and to derive reliable and sharp probabilistic, site-specific short range 10-m wind speed forecasts. The evaluation conveys that a combined dynamic-statistical approach, where the high resolution deterministic ensemble mean from INCA and the ensemble standard deviation from ALADIN-LAEF are further statistically downscaled, is of superior skill than the pure dynamical and statistical methods. Especially beyond the nowcasting range, the underdispersive behavior is reduced significantly by the statistical adaptation. It has also been demonstrated by a further experiment that the ALADIN-LAEF spread provides additional information compared to the RMSE estimator of spread.

For an operational implementation, some issues have to be still considered: are the results and the conclusion generalizable to other seasons? At least periods in different wind regimes should be studied in order to confirm the results of

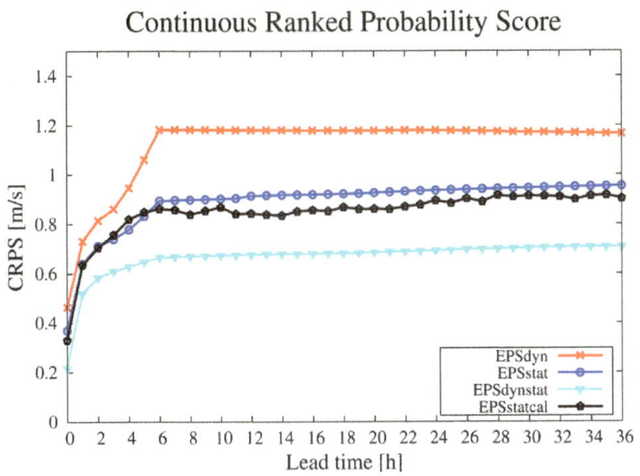


**Figure 7.** CRPS as a function of lead time, from the dynamical (red), statistical (dark blue) and dynamic-statistical (light blue) method. Furthermore, a calibrated-statistical approach indicating the skill of a further calibration of the statistical method is added (black).

the present paper. Is the re-scaling factor used in the experiment $EPS_{dynstat}$ already applicable to other seasons or still improvable?

Both the INCA and the ALADIN-LAEF systems are under permanent development at ZAMG. In INCA, work on the implementation of dynamical and high resolution topographic effects (e.g. by dynamical downscaling techniques) is ongoing and will potentially lead to more realistic wind analyses and short range forecasts. Research activities in ALADIN-LAEF are focusing on the revision of the multi-physics approach and on a stochastic surface perturbation scheme. These activities will have a direct positive impact on such applications which combine these systems algorithmically.

Site-specific, optimized probabilistic wind speed forecasts are of increasing interest in a variety of applications. For example, in the renewable energy sector, precise deterministic and probabilistic wind forecasts for wind farms are of crucial relevance for a detailed planning of the energy production and power management. Thus, improved wind forecasts have a direct impact on a cost-efficient power industry.

**Acknowledgements.** The author is thankful to Christoph Wittmann, Yong Wang and Florian Weidle for fruitful discussions about ALADIN-LAEF and to Thomas Haiden, Georg Pistotnik, Christine Gruber and Benedikt Bica for comments on scientific and operational INCA aspects. The author is also grateful to two anonymous reviewers for their suggestions and constructive comments that led to improvements of the manuscript.

Edited by: R. Hagedorn

## References

Brozkova, R., Klaric, D., Ivatek-Sahdan, S., Geleyn, J. F., Casse, V., Siroka, M., Radnoti, G., Janousek, M., Stadlbacher, K., and Seidl, H.: DFI Blending, an alternative tool for preparation of the initial conditions for LAM, PWRP Report Series No. 31 (CAS/JSCWGNE Report), WMO-TD, No. 1064, 1.7., 2001.

Gneiting, T., Raftery, A. E., Westveld, A. H., and Goldman, T.: Calibrated probabilistic forecasting using ensemble model output statistics and minimum CRPS estimation, Mon. Weather Rev., 133, 1098–1118, 2005.

Haiden, T., Kann, A., Pistotnik, G., Stadlbacher, K., and Wittmann, C.: Integrated Nowcasting through Comprehensive Analysis (INCA) – System description, ZAMG report, 60 pp., available at: http://www.zamg.ac.at/fix/INCA_system.pdf, 2010.

Haiden, T., Kann, A., Wittmann, C., Pistotnik, G., Bica, B., and Gruber, C.: The Integrated Nowcasting through Comprehensive Analysis (INCA) system and its validation over the Eastern Alpine region, Weather Forecast., 26, 166–183, doi:10.1175/2010WAF2222451.1, 2011

Hagedorn, R., Hamill, T. M., and Whitaker, J. S.: Probabilistic forecast calibration using ECMWF and GFS ensemble reforecasts, Part I: temperature, Mon. Weather Rev., 136, 2608–2619, 2008.

Hersbach, H.: Decomposition of the continuous ranked probability score for ensemble prediction systems, Weather Forecast., 15, 559–570, 2000.

Kann, A., Wittmann, C., Wang, Y., and Ma, X.: Calibrating 2 m-temperature of limited area ensemble forecasts using high-resolution analysis, Mon. Weather Rev., 137, 3373–3387, 2009.

Kann, A., Haiden, T., and Wittmann, C.: Combining 2-m temperature nowcasting and short range ensemble forecasting, Nonlin. Processes Geophys., 18, 903–910, doi:10.5194/npg-18-903-2011, 2011.

Leutbecher, M. and Palmer, T. N.: Ensemble forecasting, J. Comput. Phys., 227, 3515–3539, 2008.

Sherman, C. A.: A mass-consistent model for wind field over complex terrain, J. Appl. Meteorol., 17, 312–319, 1978.

Stanski, H. R., Wilson, L. J., and Burrows, W. R.: Survey of common verification methods in meteorology, WMO/WWW Tech. Rep. 8, 114 pp., 1989.

Thorarinsdottir, T. L. and Gneiting, T.: Probabilistic forecasts of wind speed: Ensemble model output statistics by using heteroskedastic censored regression, J. R. Stat. Soc. A. Sta., 173, 371–388, 2010.

Wang, Y., Williamson, C., Garvey, D., Chang, S., and Cogan, J.: Application of a multigrid method to a mass-consistent diagnostic wind model, J. Appl. Meteorol., 44, 1078–1089, 2005.

Wang, Y., Haiden, T., and Kann, A.: The operational Limited Area Modelling system at ZAMG: ALADIN-AUSTRIA, Österr. Beiträge Meteorol. Geophys., 37, 1016–6254, 2006.

Wang, Y., Bellus, M., Wittmann, C., Steinheimer, M., Weidle, F., Kann, A., Ivatek-Sahdan, S., Tian, W., Ma, X., Tascu, S., and Bazile, E.: The Central European limited-area ensemble forecasting system: ALADIN-LAEF, Q. J. Roy. Meteor. Soc., 137, 483–502, doi:10.1002/qj.751, 2011.

Wilks, D. S.: Statistical Methods in the Atmospheric Sciences, 2nd Edn., Academic Press, 627 pp., 2006.

Zhu, Y., Toth, Z., Wobus, R., Richardson, D., and Mylne, K.: The economic value of ensemble based weather forecasts, B. Am. Meteorol. Soc., 83, 73–83, 2002.

# 11

# Climate indicators for Italy: calculation and dissemination

**F. Desiato**[1], **G. Fioravanti**[1], **P. Fraschetti**[1], **W. Perconti**[1], **and A. Toreti**[1,2]

[1]Istituto Superiore per la Protezione e la Ricerca Ambientale (ISPRA), Rome, Italy
[2]Oeschger Centre for Climate Change Research (OCCR) and Institute of Geography, Climatology and Meteorology, University of Bern, Bern, Switzerland

**Abstract.** In Italy, meteorological data necessary and useful for climate studies are collected, processed and archived by a wide range of national and regional institutions. As a result, the density of the stations, the length and frequency of the observations, the quality control procedures and the database structure vary from one dataset to another. In order to maximize the use of those data for climate knowledge and climate change assessments, a computerized system for the collection, quality control, calculation, regular update and rapid dissemination of climate indicators was developed. The products publicly available through a dedicated web site are described, as well as an example of climate trends estimates over Italy, based on the application of statistical models on climate indicators from quality-checked and homogenised time series.

## 1 Introduction

In the context of the United Nations Framework Convention on Climate Change (UNFCCC) and also at the level of European policies, great attention is given to the issue of climate change adaptation. The scientific community recognizes that measures for mitigating greenhouse gases must be accompanied by programs of adaptation in order to reduce the risks and, whenever possible, take advantage of present and future climate change effects. As outlined by the Fourth IPCC Assessment Report (IPCC, 2007), impacts due to global warming, and more generally to climate change, are already evident in many areas of the world. The Mediterranean region is considered a hot spot of climate change (Diffenbaugh et al., 2007; Giorgi, 2006). Due to the specific characteristics of this region, climate change may have wide-ranging impacts that affect socio-economic and production sectors (such as energy, transport, agriculture and tourism) and the environment (including mountainous areas and forests, ecosystems and biodiversity, water resources, coastal areas and marine environment) (e.g. IPCC, 2007; EEA, 2009; EEA-JRC-WHO, 2008). The impacts on human health are expected to be relevant as well (e.g. WHO, 2008). The implementation and evaluation of potential adaptation strategies require accurate knowledge of climate and its variations at the local scale; in addition, the survey of the impacts related to specific features and vulnerabilities of the territory, are needed. This state-of-the-art knowledge should also cover awareness of the uncertainties of the impact estimates, and be continuously updated and clearly communicated to the stakeholders. Instrumental data represent a primary source of information about climate. They are usually archived at daily or sub-daily scales by national and local meteorological networks. Several data sets, which cover different time periods, are useful for the characterization of the recent past climate and its changes over Italy. The identification and estimation of trends in climate time series relies on climate data that meet rigorous criteria of accuracy and temporal continuity. Climate trends may be responsible for different types of impact, already recognized or projected for the near future. Instrumental climate time series are also necessary for the assessment of climate model skill and consequently for tuning the adaptation strategies. In order to address these needs and to build a bridge between climate research and societal sectors involved in climate change impacts, the Italian National Institute for Environmental Protection and Research (ISPRA) has developed a computerized system called SCIA (www.scia.sinanet.apat.it). In the following sections, a description of this project and its related products is provided.

**Table 1.** Main meteorological Italian networks and data sources.

| | | |
|---|---|---|
| Synoptic (GTS dissemination) | | |
| Air Force Weather Service | from '50s–'60s | ~90 stations |
| ENAV-Italian Company for air navigation services | from '80s | ~40 stations |
| National Agrometeorological (Ministry of Agriculture) | | |
| Automatic | from 1991 | ~40 stations |
| Observatories | from 1860 | ~30 stations |
| Regional | | |
| Meteorological services | from '90s | few tens/region |
| Hydrological/Civil Protection Network | from '50s | few tens/region |
| Agrometeorological services | from '90s | few tens/region |

## 2 Data sources and climate indicators

In Italy, the meteorological data necessary and useful for climate evaluations are collected, processed and archived by a wide range of national and regional institutions (Table 1). They range from synoptic observations carried out by the national Air Force weather service and the civil aviation agency, to hourly measurements at automatic stations by regional meteorological, hydro-meteorological and agrometeorological networks. Therefore, the density of the stations, the length and the frequency of the observations, the quality control procedures and the database structure vary from one data source to another. The SCIA system has been developed in order to optimize the use of these data for climate knowledge and climate change assessments. It is dedicated to the collection, quality control, calculation, regular update and dissemination of climate indicators, which reflect the main statistical properties (mean values, intensity and date of occurrence of extreme events, standard deviation, etc.), at different time scales (i.e. 10-daily, monthly and yearly), of a wide range of meteorological variables: temperature, precipitation, humidity, wind, water balance, evapotranspiration, degree-days, cloud cover, sea level pressure and solar radiation. All climate indicators are freely available through the SCIA web site. Climatological values over 30-yr reference periods (e.g. 1961–1990) are calculated and available as well. Today, the SCIA system is fed and updated once a year in collaboration with (and with the data of) the national Air Force weather service, the automatic stations of the national agrometeorological network, the meteorological stations of the national Sea Service and twelve regional environmental protection and agrometeorological agencies. Visitors of SCIA can extract, download and plot: (a) the time series of each climate indicator for each single station; (b) the climate indicators for all the stations which are available in a selected time period; (c) the tables and/or the maps of climate normals over standard climatological 30-yr periods. Through SCIA, users can also access secondary products such as high resolution monthly temperature maps obtained by the application of spatial interpolation techniques (regression kriging). In order to guarantee the integration and comparison of indicators obtained from different data sources, all the indicators undergo a common set of quality controls. The general criteria adopted for the calculation of the indicators follow the recommendations of the World Meteorological Organization (WMO, 2010). Although the validity of input data is the responsibility of each data provider, two classes of formal controls are applied to input data before the calculation of the climate indicators. In the first (weak climatological controls), each value must fall within a range delimited by minimum and maximum thresholds, representing the limits for physically acceptable data. In the second (internal consistency controls), the values of different variables at the same time are checked for consistency; for example, air temperature and dew point temperature are checked in order to verify that the latter is not greater than the former.

## 3 Climate trends estimate

Climate statistics and indicators available through SCIA are used for climate trend recognition and estimation, which are needed for climate impact and vulnerability assessments over Italy, and represent a part of the climate state chapter of the national communications to the UNFCCC. In order to get reliable trends estimates, it is necessary to control and eliminate wrong values from the series and filter out (or at least reduce the influence of) eventual non-climatic signals (i.e. inhomogeneities), such as those due to station relocation or change of meteorological instrumentation. For this reason, time series undergo a number of suitable homogeneity statistical procedures and they are homogenised whenever one or more break points are identified (e.g. Aguilar et al., 2003; Kuglitsch et al., 2009). Then, the variation of climatic variables (in terms of differences or, in the case of

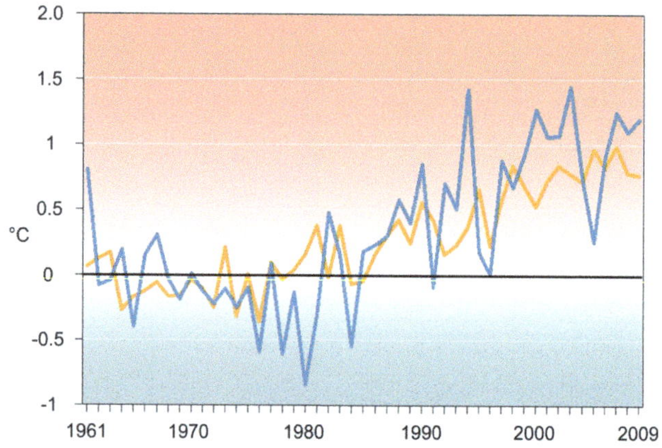


**Figure 1.** Mean temperature annual anomalies. Italian (blue line) and global land (orange line, data from NCDC/NOAA) series.

precipitation, percentage differences) are estimated through the application of statistical models for trend recognition and linear or piecewise-linear statistical models for trends estimate (e.g. Tomé and Miranda, 2004; Seidel and Lanzante, 2004). The estimation of mean temperature trends over Italy was obtained by averaging annual mean temperatures obtained from homogenised time series of 49 synoptic stations with uniform spatial distribution. The application of a sloped-steps trend model (Toreti and Desiato, 2008a) indicated a slightly negative variation of the mean temperature from 1961 to 1981 (−0.6 °C), followed by a positive trend from 1981 to 2008 with an increase of about 1.6 °C. The estimated net variation from 1961 to 2008 is about +1.0 °C. It is confirmed that the mean temperature increase recorded during recent decades in Italy, as in the Mediterranean basin, is higher than the global mean (see Fig. 1). In particular, in 2008 and 2009 the mean Italian temperature anomalies, with respect to 1961–1990, were 1.09 and 1.19 °C, respectively; while the global mean anomalies were about 0.78 and 0.76 °C, respectively. In Italy, 2009 was the 18th consecutive year with a positive anomaly, and its value is ranked 5th from 1961 until now. Besides average climate indicators, quality checked data series are also used for the estimate of indicators and trends of climate extremes. Regarding temperature extremes over Italy, three indices among those defined by the CCl/CLIVAR/JCOMM Expert Team on Climate Change Detection and Indices (Peterson, 2005) have been evaluated: the number of days per year with daily minimum temperature less than or equal to 0 °C (frost days); the number of days per year with daily minimum temperature greater than 20 °C (tropical nights); and the number of days per year with daily maximum temperature greater than 25 °C (summer days). Data processing and the results are described by Toreti and Desiato (2008b). Trend estimates are updated yearly, and recently heat wave indicators (number, duration and intensity) have also been evaluated.

## 4 Conclusions

Besides climate state and trend evaluation purposes, the indicators disseminated through SCIA may be used in several socio-economic contexts, such as energy consumption, water management, agriculture, tourism and health. At the moment, there are several gaps that must be addressed in the near future, in order to better meet the demands for information about the Italian climate. The main limitations concern: (a) the density and the irregular geographical distribution of the observations, which need to include more data sources; (b) the time resolution of climate indicators, which must be increased especially for the evaluation of climate extremes; (c) the frequency and timeliness of updating the climate indicators. Improvement of these aspects would enable better responses to be made to the increasing demand for reliable information on climate and its trends over Italy.

Edited by: J. Prior

## References

Aguilar, E., Auer, I., Brunet, M., Peterson, T. C., and Wieringa, J.: Guidlines on climate metadata and homogenization, WMO TD, 1186, 2003.

Diffenbaugh, N. S., Pal, J. S., Giorgi, F., and Gao, X.: Heat stress intensification in the Mediterranean climate change hotspot, Geophys. Res. Lett., 34, L11706, doi:10.1029/2007GL030000, 2007.

EEA – European Environment Agency: Regional climate change and adaptation, EEA report, 8, 2009.

EEA – European Environment Agency, JRC, and WHO: Impacts of Europe's changing climate – 2008 indicator-based assessment, EEA report, 4, 2008.

Giorgi, F.: Climate change hot-spots, Geophys. Res. Lett, 33, L08707, doi:10.1029/2006GL025734, 2006.

IPCC: Contribution of Working Group II to the Fourth Assessment Report of the Intergovernamental Panel on Climate Change, edited by: Parry, M. L., Canziani, O. F., Palutikof, J. P., van der Linden, P. J., and Hanson, C. E., Cambridge University Press, UK and USA, 2007.

Kuglitsch, F. G., Toreti, A., Xoplaki, E., Della-Marta, P. M., Luterbacher, J., and Wanner, H.: Homogenization of daily maximum temperature series in the Mediterranean, J. Geophys. Res., 114, D15108, doi:10.1029/2008JD011606, 2009.

Peterson, T. C.: Climate Change Indices, WMO Bull., 54, 83–86, 2005.

Seidel, D. J. and Lanzante, J. R.: An assessment of three alternatives to linear trends for characterizing global atmospheric temperature changes, J. Geophys. Res., 109, D14108, doi:10.1029/2003JD004414, 2004.

Tomé, A. R. and Miranda, P. M. A.: Piecewise linear fitting and trend changing points of climate parameters, Geophys. Res. Lett., 31, L02207, doi:10.1029/2003GL019100, 2004.

Toreti, A. and Desiato, F.: Temperature trend over Italy from 1961 to 2004, Theor. Appl. Climatol., 91, 51–58, 2008a.

Toreti, A. and Desiato, F.: Changes in temperature extremes over Italy in the last 44 years, Int. J. Climatol., 28, 733–745, 2008b.

WHO – World Health Organization: Protecting health in Europe from climate change, edited by: Menne, B., Apfel, F., Kovats, S., and Racioppi, F., 2008.

WMO – World Meteorological Organization: Guide to climatological practices, No. 100, 3 Edn., 2010.

# 12

# Processing and analysing an ensemble of climate projections for the joint research project KLIWAS

**F. Imbery, S. Plagemann, and J. Namyslo**

German Weather Service, Offenbach, Germany

*Correspondence to:* F. Imbery (florian.imbery@dwd.de)

**Abstract.** The research programme KLIWAS, funded by the German Federal Ministry of Transport, Building and Urban affairs is focussed on climate change and its impacts on waterways and navigation for Germany in the 21th century. In order to derive sound statements about the range of possible future climate changes, KLIWAS use hydro-meteorological information derived from a wide variety of global and regional climate models.

In the framework of KLIWAS emphasis is taken on the quantification of uncertainties in climate model output. Therefore, a 19-member ensemble of climate model runs was used. On the basis of the SRES-scenario A1B the probabilities of changes in air temperature, precipitation amount, global radiation and several climate indices were computed for near (2021 to 2050) and distant (2071–2100) scenario horizons.

Furthermore, statistical downscaling techniques, including approved bias correction methods, were used to provide a spatial high-resolution sub-ensemble of eight climate model simulations for climate change impact investigations.

## 1 Introduction

The assessment of hydrological impacts of climate change demands accurate and high resolution climate model simulation results. For hydrological applications, the key input variables are typically precipitation, air temperature and radiation, at high spatial resolution or at station scale, and with daily or sub-daily temporal resolution (Kotlarski et al., 2012; Nilson et al., 2010).

Today it is common that only one climate model run can't give the full range of information about uncertainties of possible future climate conditions. Therefore, the use of climate model ensembles is recommended to investigate possible impacts of climate change (Stott and Forest, 2007; Christensen et al., 2010; Deque and Somot, 2010; Giorgi et al., 2009).

Typically, the spatial resolution of global climate models (GCM) is not suitable for hydrological demands on catchment scale (Fowler et al., 2007). Therefore regional climate models (RCM) are used to downscale the simulation results of global climate models. Currently, their typical spatial resolution lies between 10 and 50 km. By using regional climate models, the influence of mountains and also the impact of different land-uses on the local climate, for example, are represented in considerably better detail. Likewise the temporal variability and also the intensity of precipitation amounts are much better captured.

Climate models produce a reduced representation of natural processes such that a systematic error (bias) arises. When using climate model simulation results for impact models, these internal errors must be quantified and corrected. The correction can be realised using the comparison between the control run of the climate model and observed data (e.g. 1961–1990). Depending on the variables to be corrected different methods of bias-correction can be used (Piani et al., 2010a, b; Fowler et al., 2007; Themeßl et al., 2012).

Bias correction procedures assume that model biases do not change over time. Although it is not possible to test whether this assumption is fulfilled for future climate conditions, bias stationarity is a main concern when using bias correction procedures (Maraun, 2012; Teutschbein and Seibert, 2012). On the other side it is important to be aware that the consistency between different variables may get lost after the application of statistical correction terms. Hence, for studies based on climate model output it is important to be aware of possible disadvantages using bias-corrected model data. While corrected RCM datasets are required as input for impact investigations in most cases, studies on climate change signals can also be carried out with uncorrected RCM data.

In this study, we introduce an ensemble of high resolution climate projection runs covering Middle Europe for the time period 1961–2100. Meteorological variables of eight climate projections from different GCM-RCM combinations and three Re-Analysis data driven model runs are downscaled and corrected on a consistent high resolution grid (5 km × 5 km) using different methods based on statistical and physical equations. In this context, existing methods developed by the DWD (German Weather Service) were modified and new methods developed. The results were evaluated and compared to the original model output. This ensemble can be used as input for various climate change impact investigations.

Secondly, a statistical method was developed for analysing climate model ensembles. With this technique it is possible to detect climate change signals as well as uncertainties in the used ensembles. For the investigations on climate change signals in this work we used original model output of 19 different GCM-RCM Combinations.

## 2 Considered area and methods

### 2.1 Domain

The region of interest in this study includes Germany and the hydrological catchment areas of the rivers Rhine, Elbe, Oder and partly Danube (approximately from 45 to 56° N and 3 to 20° E). A Lambert Conformal Conic map projection on a spatial resolution of 5 km with the European Terrestrial Reference System (ETRS-LCC) (Annoni et al., 2003) is the spatial basis for this study.

### 2.2 Data

An ensemble of regional climate models (RCM) is used, which were driven by several GCM on the basis of climate simulations of the years 1961–2000 (control run) and of the years 2001–2100 with SRES A1B climate change scenario (projection run). This multi-model ensemble is composed of different RCM developed in the European framework project ENSEMBLES (Van der Linden and Mitchell, 2009), two so-called consortial runs of CLM (Keuler et al., 2009a, b, c; Lautenschlager et al., 2009) and two additional REMO runs (Jacob et al., 2005a, b, 2009a, b). In addition 3 RCM runs, forced by lateral boundary conditions of the Re-Analysis ERA40 data set (Uppala et al., 2005) are used in this study to assess uncertainties of the models. An overview of the used RCM runs can be found in Table 1. Currently, we regionalised and bias-corrected eight climate projections of this ensemble (so-called KLIWAS ensemble, marked with [x] in Table 1).

Gridded observational data for air temperature and precipitation amount were obtained from the German Weather Service DWD. They were developed in context of KLIWAS and the HYRAS pilot project, funded by the Federal Institute of Hydrology (Bundesanstalt für Gewässerkunde – BfG). These high-resolution (1 km × 1 km) gridded daily data sets are referred to as DWD/BfG-HYRAS v2.0 precipitation data set (HYRAS-PRE) for precipitation amount (Rauthe et al., 2013) and DWD/BfG-HYRAS v0.21 temperature data set (HYRAS-TAS) for the air temperature (Rauthe et al., 2012). Both variables have the same geodetic reference system as the targeted map projection in this study, therefore it is only necessary to aggregate the grid information in 5 km resolution.

Furthermore, for gridded global radiation information the CM-SAF Surface Radiation MVIRI Data Set 1.0 (Posselt et al., 2011) is used and interpolated with distance weighted average.

### 2.3 Regionalisation

#### 2.3.1 Precipitation

The regionalisation of precipitation amount is partly adopted from the observational interpolation method REGNIE (Rauthe et al., 2013). In a first step background fields are constructed by a multiple linear regression and for optimising the explained variance by the Wards- and k-means-clustering. Once the clustered areas in the domain are certain, a multiple Regression with the predictor variables, geographical longitude, latitude, height above sea level and exposition of hillside, could be calculated by the use of the Cholesky decomposition (e.g. Kanzow, 2005). The resulting regression coefficients are applied for the whole cluster in spatial original (e.g. 25 km) and high resolution scale (5 km). The residuals are assessed for each month as a mean of long-term (30 yr) period with a moving average of 10 yr. Afterwards all residuals and predictors are interpolated conservative on the high resolution grid. Consequentially the high resolution values are computed related to the 3 × 3 adjacent RCM grid cells. The required relations are the monthly background fields and are used for the daily precipitation amount by inverse distance weighting.

#### 2.3.2 Air temperature

For air temperature a multiple linear regression is used the spatial high resolution regionalisation. Thereby the potential predictor height above sea level is predefined. Due to a screening method (e.g. Urban and Mayerl, 2008) and cross-validation (e.g. Michaelson, 1987) the other potential predictors (geographical longitude and latitude, sea level pressure, humidity in 700 hPa, air temperature in 850 hPa, thickness 500–925 hPa (geopotential height) and the products of air

**Table 1.** Overview of climate simulations of the years 1961–2000 for the control run (C20), projection runs for the years 2001–2100 based on the scenario A1B and Re-Analysis driven runs (ERA40) used in this study. The symbol X in the last column indicates the projections that are regionalised and bias-corrected.

| Control run/ SRES scenario/ reanalysis driven run | GCM | RCM | KLIWAS ensemble |
|---|---|---|---|
| C20/A1B | HadCM3Q0 (HC) | CLM2.4.6 (ETHZ) | X |
| | | HadRM3Q0 (HC) | X |
| | HadCM3Q16 (HC) | HadRM3Q16 (HC) | |
| | | RCA3 (C4I) | |
| | HadCM3Q3 (HC) | HadRM3Q3 (HC) | |
| | | RCA3 (SMHI) | |
| | BCM2 (NERSC) | RCA3 (SMHI) | X |
| | | HIRHAM5 (DMI) | |
| | ECHAM5-r3 (MPI-M) | RCA3 (SMHI) | |
| | | RegCM3 (ICTP) | X |
| | | HIRHAM5 (DMI) | X |
| | | RACMO2 (KNMI) | X |
| | | REMO5.7 (MPI-M) | X |
| | ECHAM5-r2 (MPI-M) | REMO5.8 (MPI-M) | |
| | | CLM2.4.11 (GKSS) | |
| | ECHAM5-r1 (MPI-M) | CLM2.4.11 (GKSS) | X |
| | | REMO5.7 (MPI-M) | |
| | ARPEGE (CNRM) | HIRHAM5 (DMI) | |
| | | RM5.1 (CNRM) | |
| ERA40 | | CLM2.4.6 (ETHZ) | X |
| | | REMO5.7 (MPI-M) | X |
| | | RM4.5 (CNRM) | X |

temperature in 850 hPa multiply with thickness 500–925 hPa and sea level pressure multiply with thickness 500–925 hPa) are chosen for regression to minimise the mean square error. The resulting regression coefficients apply for the whole cluster in spatial original (e.g. 25 km) and high resolution scale (5 km), which are used for calculation of residuals. Last, all chosen predictors and residuals are interpolated conservative to find the air temperature in high resolution grid.

### 2.3.3 Global radiation

The chosen method for regionalisation of global radiation is not a statistical approach but is based on an empirical procedure. First, for each day of the year and for different Linke turbidity factors, which represent the full spectrum radiation attenuations (Rigollier et al., 2000), the global radiation for clear sky at each grid is calculated. Considering geographical longitude, latitude and height above sea level, solar altitude angle and altitude correction are included (Remund et al., 2003; Whiteman, 2000). The results for the RCM original grid and the targeted high resolution grid are stored in so-called background fields. Then the RCM global radiation dataset is randomized in global radiation for clear sky taking into account cloudiness fraction. These values are compared to the background fields and evaluated. If these computed results are between two global radiation values for clear sky and defined Linke turbidity factors, a simple numeric approximation calculates the correct turbidity. After this, the Linke turbidity factors are interpolated using inverse distance weighting. At last, the procedure is executed in the opposite direction to define the global radiation in the case of the total cloud cover in spatial resolution of 5 km × 5 km.

## 2.4 Bias correction

### 2.4.1 Precipitation

For daily precipitation amounts in climate simulations ($P$) two different bias correction methods were carried out, using model data and observational data (HYRAS) for the time period 1961–1990 for calculating correction terms for future scenarios 2000–2100.

On the one side, a linear scaling approach (Hashino et al., 2007) is used. For every grid $i$ and every month $j$ a correction factor $f$ is computed by calculating the mean ratio between observed reference data ($R$) and simulated data ($C$) of the control period.

$$\begin{aligned} f &= R_j^i \cdot C_j^{i-1} \\ P^* &= f \cdot P \end{aligned} \tag{1}$$

New daily values $P^*$ are calculated with the monthly correction factor $f$.

Also, an empirical quantile mapping approach was carried out. In this case, empirical distributions for observed ($F_{oj}$) and simulated ($F_{sj}$) precipitation amounts were calculated for each grid and month (Hashino et al., 2007; Piani et al., 2010b). With $F_{oj}$ and $F_{sj}$ a transfer function can be obtained, with which a distribution-depending correction

$$P_j^{*i} = [F_{oj}^{-1} F_{sj}](P_j^i) \tag{2}$$

can be carried out.

#### 2.4.2 Air temperature

For simulated daily mean air temperature ($T_a$), a linear scaling approach was carried out. In contrast to precipitation amount, the bias of temperature was calculated as the difference of monthly mean values of simulated control data ($C$) and observed reference data ($R$) for each grid.

$$\begin{aligned} \text{bias} &= R_j^i - C_j^i \\ T_a^* &= T_a - \text{bias} \end{aligned} \tag{3}$$

#### 2.4.3 Global radiation

Values of global radiation were corrected similar to precipitation by computing a correction factor with the ratio of simulated control data and observed reference data (see Eq. 1).

### 2.5 Uncertainties in climate projections

For the consideration of relative changes (trends, climate change signals) of mean values of meteorological variables, a bias-correction is not generally necessary. The use of original simulation data gives the possibility to access more ensemble members which lead to a better assumption of uncertainties for the investigated variables.

To represent the bandwidth of possible future climate conditions, an ensemble of 19 climate projections, based on the SRES emission scenario A1B (IPCC, 2007) was analysed (see Table 1). This ensemble consists of RCM runs in a spatial resolution of 25 km × 25 km that were computed over the time period 1961–2100.

Percentiles of this ensemble were evaluated for different meteorological variables and indices for the time periods 2021–2050 and 2071–2100 relative to the control period 1961–1990, respectively. A conservative lower threshold (15 %) is defined that is almost certainly exceeded, and an upper limit (85 %) is set that expresses unexpectedly drastic changes. The span between these two thresholds covers 70 % of the investigated ensemble and can be understood as a range in which occurrence can be expected in this scenario.

## 3 Results

### 3.1 Post-processing of climate model data

In this section we present some examples of the verification results for the post-processed RCM data. As described in Sect. 2.4.1, for precipitation amounts two different bias-correction methods were used. In Table 1, the processed model runs are indicated with a [x]. In Fig. 1, the results for the linear scaling approach and the quantile mapping method are compared on four different grid points for the ECHAM5-r3 REMO5.7 projection. The selected grid points represent different landscapes in the KLIWAS domain: coastal regions (North Sea, Schleswig-Holstein west coast, 1), hilly landscape, dominated by croplands (Thuringian Forest, 2), mid-latitude mountain regions (Black Forest, 3) and alpine regions (Berchtesgaden National Park, 4). For the time period 1961–1990, all precipitation events are represented in the qq-plots. Dots near the 1 : 1 line represent a perfect accordance to the reference data, dots under this line can be interpret as an underestimation of precipitation amounts.

A slightly improvement can be seen for the simulation data corrected by the linear scaling method. Improved results can be achieved for all regions with the quantile mapping correction. For both methods, limits of improvement can be seen in the highest parts of the distribution (about upmost three percents). This is in agreement with other investigations (e.g. Hashino et al., 2007; Themeßl et al., 2012).

Different metrics are used for the verification of the processed data. For example, the Perkins score (Perkins et al., 2007) is used for testing the quality of the spatial representation of the bias corrected projections. In a first step, for the whole investigated area the probability density functions (PDF) of simulated data and reference data were calculated for each month. Afterwards, the cumulative minimum value of both distributions was compared. This can be expressed as

$$\text{Skill}_{\text{PDF}} = \sum_{1}^{n} \min(Z_m, Z_o) \tag{4}$$

where $n$ is the number of bins used to calculate the PDF for a given region, $Z_m$ is the frequency of values in a given bin from the model data, and $Z_o$ is the frequency of values in a given bin from the observed data. If the simulation data reproduce the reference data perfectly, the skill equals one. In Fig. 2 the simulated air temperature of two ERA40-driven model runs (CLM 2.4.6 and REMO 5.7, both with and without a linear scaling correction) is tested with the Perkins score for the time period 1961–1990.

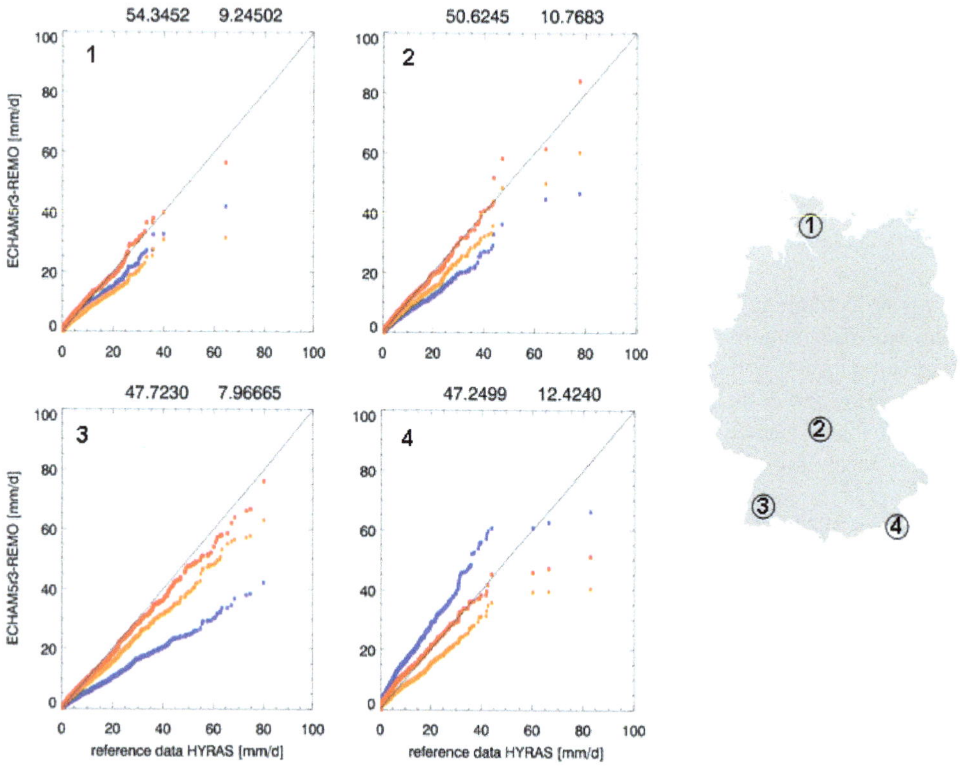


**Figure 1.** qq-Plots for different bias-correction methods for precipitation for the A1B-ECHAM5r3-REMO5.7 run. Uncorrected (blue), linear-scaling method (orange) and quantile mapping approach (red) in comparison with HYRAS-PRE for 1961–1990 on four different regions in the KLIWAS area. (1) Coastal regions (north sea, Schleswig-Holstein west coast), (2) hilly landscape, dominated by croplands (Thuringian Forest), (3) mid-latitude mountain regions (Black Forest) and (4) alpine regions (Berchtesgaden National Park).

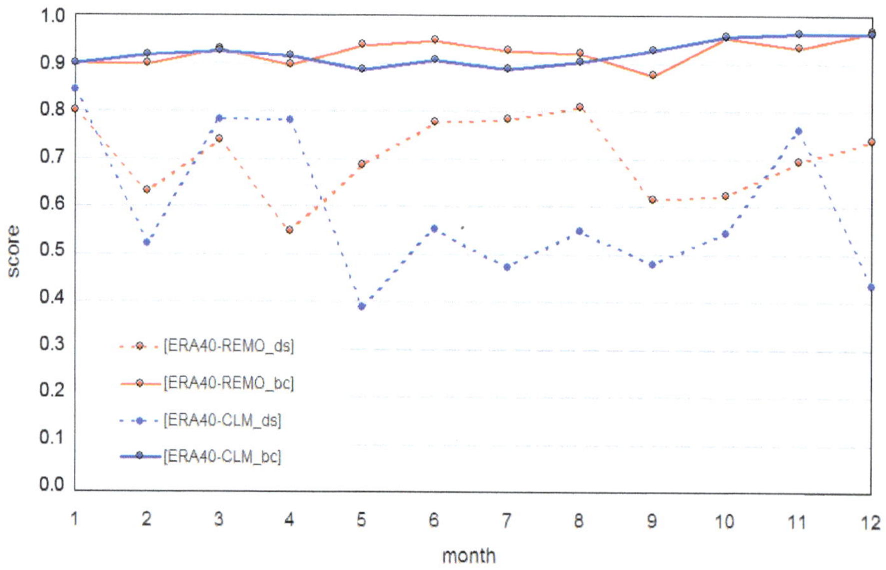


**Figure 2.** Monthly Perkins scores for regionalised and bias-corrected (linear-scaling) mean air temperature for two ERA40-driven model runs (CLM 2.4.6 and REMO 5.7) for the time period 1961–1990. Dashed lines indicate uncorrected regionalised model runs (ds), solid lines represent regionalised and bias corrected (bc) runs.

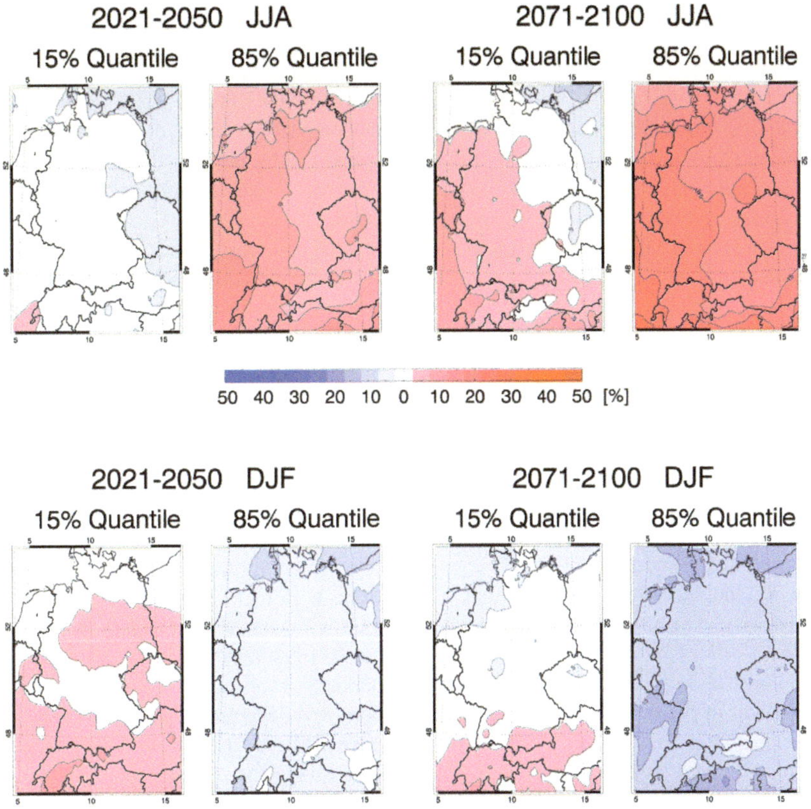


**Figure 3.** Projected relative change [%] of mean summer precipitation (JJA, upper figure), and mean winter precipitation (DJF, below); periods 2021–2050 (left), 2071–2100 (right), with respect to the control run 1961–1990 based on a 19 member ensemble of regional climate projections (see Table 1). Blue colours indicate increasing precipitation, red colours decreasing precipitation.

While the two uncorrected runs show a strong deviation to reference data with values between 0.4 and 0.8, the corrected runs are in good agreement to the observations. The REMO Data have best conformance during summer with a score over 0.95, whereas CLM have highest values during the months October to December.

## 3.2 Climate change signals

As pointed out in Sect. 2.5, we use an ensemble of 19 climate models to analyse projected changes for several variables and indices. Furthermore, with this technique it is possible to quantify the uncertainties within the given ensemble. As an example, Fig. 3 gives an overview of projected relative changes of mean precipitation for a near (2021–2050) and a far (2071–2100) future in relation to the reference period (1961–1990). With respect to precipitation, a differentiation between the seasons is necessary. By the end of the century, winters throughout Germany are likely to become wetter, whereas summers could become drier. Following the investigated climate projections, mean precipitation in the hot season should decrease only weakly until 2050, but in the last quarter of the century it could decrease by nearly 20 percent. However, changes during the first half of the century display regional variation; the amounts of precipitation in summer may even remain nearly unchanged in some regions. In winter, increases in precipitation up to 10 percent in the near future (2021–2050) and 15 percent in the more distant future (2071–2100) are expected.

The DWD has established a "German Climate Atlas" (http://www.deutscher-klimaatlas.de), which allows to analyse more variables and a wide range of climate indices with the presented technique.

## 4 Conclusions

We established a 19 member ensemble of climate projections. With this ensemble it is possible to investigate signals of changes due to possible future climate conditions and the underlying uncertainties of climate projections.

Furthermore we regionalised and corrected air temperature, precipitation and global radiation of a subgroup of eight climate projections of this ensemble. These post-processed climate model runs are an important basis for further investigations on the impacts of climate change, e.g. hydrological modelling. Next steps of our work will be the regionalisation and correction of further variables such as air humidity and wind speed and to expand the ensemble with new projections based on the RCP scenarios (Giorgi et al., 2009).

**Acknowledgements.** This research has been carried out in the project 1.02 "Provision of Application-oriented and Evaluated Climate Projection Data" within the departmental research programme "KLIWAS–Impacts of climate change on waterways and navigation" that is financed by the German Federal Ministry of Transport, Building and Urban Development.

The ENSEMBLES data used in this work was funded by the EU FP6 Integrated Project ENSEMBLES (contract 505539) whose support is gratefully acknowledged.

Edited by: B. Ahrens

## References

Annoni, A., Luzet, C., Gubler, E., and Ihde, J.: Map projections for Europe, available at: http://www.ec-gis.org/sdi/publist/pdfs/annoni-etal2003eur.pdf, 2003.

Christensen, J., Kjellstrom, E., Giorgi, F., Lenderink, G., and Rummukainen, M.: Weight assignment in regional climate models, Clim. Res., 44, 179–194, 2010.

Deque, M. and Somot, S.: Weighted frequency distributions express modelling uncertainties in the ENSEMBLES regional climate experiments, Clim. Res., 44, 195–209, 2010.

Fowler, H. J., Blenkinsopa, S., and Tebaldi, C.: Linking climate change modelling to impacts studies: recent advances in downscaling techniques for hydrological modelling, Int. J. Climatol., 27, 1547–1578, 2007.

Giorgi, F., Jones, C., and Asrar, G. R.: Addressing climate information needs at the regional level: the CORDEX framework, WMO Bulletin, 58, 175–183, 2009.

Hashino, T., Bradley, A. A., and Schwartz, S. S.: Evaluation of bias-correction methods for ensemble streamflow volume forecasts, Hydrol. Earth Syst. Sci., 11, 939–950, doi:10.5194/hess-11-939-2007, 2007.

IPCC: Climate Change 2007: The Physical Science Basis, Contribution of Working Group I to the Fourth Assessment Report of the Intergovernmental Panel on Climate Change, Cambridge University Press, Cambridge, UK, New York, USA, 2007.

Jacob, D.: REMO Climate of the 20th Century Run, UBA Project, 0.088 Degree Resolution, Run No. 006210, 1H Data, World Data Center for Climate, CERA-DB "REMO_UBA_C20_1_R006210_1H", http://cera-www.dkrz.de/WDCC/ui/Compact.jsp?acronym=REMO_UBA_C20_1_R006210_1H, 2005a.

Jacob, D.: REMO A1B Scenario Run, UBA Project, 0.088 Degree Resolution, Run No. 006211, 1H Data, World Data Center for Climate, CERA-DB "REMO_UBA_A1B_1_R006211_1H", http://cera-www.dkrz.de/WDCC/ui/Compact.jsp?acronym=REMO_UBA_A1B_1_R006211_1H, 2005b.

Jacob, D., Nilson, E., Tomassini, L., and Bülow, K.: REMO Climate of the 20th Century Run, BFG Project, 0.088 Degree Resolution, Run No. 2, 1H Data. World Data Center for Climate, CERA-DB "REMO_BFG_A1B_1H", http://cera-www.dkrz.de/WDCC/ui/Compact.jsp?acronym=REMO_BFG_C20_1H, 2009a.

Jacob, D., Nilson, E., Tomassini, L., and Bülow, K.: REMO A1B Scenario Run, BFG Project, 0.088 Degree Resolution, 1H values. World Data Center for Climate, CERA-DB "REMO_BFG_A1B_1H", http://cera-www.dkrz.de/WDCC/ui/Compact.jsp?acronym=REMO_BFG_A1B_1H, 2009b.

Kanzow, C.: Numerik linearer Gleichungssysteme – Direkte und iterative Verfahren, Springer Verlag, Berlin Heidelberg, 77–81, 2005.

Keuler, K., Lautenschlager, M., Wunram, C., Keup-Thiel, E., Schubert, M., Will, A., Rockel, B., and Boehm, U.: Climate Simulation with CLM, Climate of the 20th Century run no.2, Data Stream 2: European region MPI-M/MaD, World Data Center for Climate, doi:10.1594/WDCC/CLM_C20_2_D2, 2009a.

Keuler, K., Lautenschlager, M., Wunram, C., Keup-Thiel, E., Schubert, M., Will, A., Rockel, B., and Boehm, U.: Climate Simulation with CLM, Scenario A1B run no.1, Data Stream 2: European region MPI-M/MaD, World Data Center for Climate, doi:10.1594/WDCC/CLM_A1B_1_D2, 2009b.

Keuler, K., Lautenschlager, M., Wunram, C., Keup-Thiel, E., Schubert, M., Will, A., Rockel, B., and Boehm, U.: Climate Simulation with CLM, Scenario A1B run no.2, Data Stream 2: European region MPI-M/MaD, World Data Center for Climate, doi:10.1594/WDCC/CLM_A1B_2_D2, 2009c.

Kotlarski S., Hagemann S., Krahe P., Podzun R., and Jacob D.: The Elbe River Flooding 2002 as seen by an extended regional climate model, J. Hydrol., 472–473, 169–183, 2012.

Lautenschlager, M., Keuler, K., Wunram, C., Keup-Thiel, E., Schubert, M., Will, A., Rockel, B., and Boehm, U.: Climate Simulation with CLM, Climate of the 20th Century run no.1, Data Stream 2: European region MPI-M/MaD, World Data Center for Climate, doi:10.1594/WDCC/CLM_C20_1_D2, 2009.

Maraun, D.: Nonstationarities of regional climate model biases in European seasonal mean temperature and precipitation sums, Geophys. Res. Lett., 39, L06706, doi:10.1029/2012GL051210, 2012.

Michaelson, J.: Cross-Validation in Statistical Climate Forecast Models, J. Clim. Appl. Meteorol., 26, 1589–1600, 1987.

Nilson, E., Carambia, M., and Krahe, P.: Low flow changes in the Rhine River basin, in: Assessment of climate change impacts on discharge in the Rhine River Basin: Results of the RheinBlick2050 Project, CHR Report No. I-23, edited by: Görgen, K., Beersma, J., Brahmer, G., Buiteveld, H., Carambia, M., de Keizer, O., Krahe, P., Nilson, E., Lammersen, R., Perrin, C., and Volken, D., 115–119, 2010.

Perkins, S. E., Pitman, A. J., Holbrook, N. J., and McAneney, J.: Evaluation of the AR4 Climate Models' Simulated Daily Maximum Temperature, Minimum Temperature, and Precipitation over Australia Using Probability Density Functions, J. Climate, 20, 4356–4376, 2007.

Piani, C., Haerter, J., and Coppola, E.: Statistical bias correction for daily precipitation in regional climate models over Europe, Theor. Appl. Climatol., 99, 187–192, 2010a.

Piani, C., Weedon, G., Best, M., Gomes, S., Viterbo, P., Hagemann, S., and Haerter, J.: Statistical bias correction of global simulated daily precipitation and temperature for the application of hydrological models, J. Hydrol., 395, 199–215, 2010b.

Posselt, R., Müller, R., Stöckli, R., and Trentmann, J.: CM-SAF Surface Radiation MVIRI Data Set 1.0 – Monthly Means/Daily Means/Hourly Means, Sattelite Application Facility on Climate

Monitoring, 2011.
Rauthe, M., Steiner, H., Riediger, U., Mazurkiewicz, A., and Gratzki, A.: HYRAS – Hydrometeorological reference data for German river basins, 12th EMS Annual Meeting & 9th European Conference on Applied Climatology ECAC, MC2/ASI14, EMS2012-87, 2012.
Rauthe, M., Steiner, H., Riediger, U., Mazurkiewicz, A., and Gratzki, A.: A Central European precipitation climatology – Part I: Generation and validation of a high-resolution gridded daily data set (HYRAS), Meteorol. Z., accepted, 2013.
Remund, J., Wald, L., Lefevre, M., Ranchin, T., and Page, J.: Worldwide Linke turbidity information, ISES Solar World Congress, International Solar Energy, 2003.
Rigollier, C., Bauer, O., and Wald, L.: On the Clear Sky Model of the ESRA – European Solar Radiation Atlas – With Respect to the HELIOSAT Method, Sol. Energy, 68, 33–48, 2000.
Stott, P. and Forest, C.: Ensemble climate predictions using climate models and observational constraints, Phil. Trans. R. Soc. A, 365, 2029–2052, 2007.
Teutschbein, C. and Seibert, J.: Is bias correction of Regional Climate Model (RCM) simulations possible for non-stationary conditions?, Hydrol. Earth Syst. Sci. Discuss., 9, 12765–12795, doi:10.5194/hessd-9-12765-2012, 2012.
Themeßl, M. J., Gobiet, A., and Heinrich, G.: Empirical-statistical downscaling and error correction of regional climate models and its impact on the climate change signal, Climatic Change, 112, 449–468, 2012.
Uppala, S. M., KÅllberg, P. W., Simmons, A. J., Andrae, U., Da Costa Bechtold, V., Fiorino, M., Gibson, J. K., Haseler, J., Hernandez, A., Kelly, G. A., Li, X., Onogi, K., Saarinen, S., Sokka, N., Allan, R. P., Andersson, E., Arpe, K., Balmaseda, M. A., Beljaars, A. C. M., Van De Berg, L., Bidlot, J., Bormann, N., Caires, S., Chevallier, F., Dethof, A., Dragosavac, M., Fisher, M., Fuentes, M., Hagemann, S., Hólm, E., Hoskins, B. J., Isaksen, L., Janssen, P. A. E. M., Jenne, R., Mcnally, A. P., Mahfouf, J.-F., Morcrette, J.-J., Rayner, N. A., Saunders, R. W., Simon, P., Sterl, A., Trenberth, K. E., Untch, A., Vasiljevic, D., Viterbo, P., and Woollen, J.: The ERA-40 re-analysis, Q. J. Roy. Meteor. Soc., 131, 2961–2012, 2005.
Urban, D. and Mayerl, J.: Regressionsanalyse: Theorie, Technik und Anwendung, VS Verlag für Sozialwissenschaften – GWH Dachverlage GmbH, 112–114, 2008.
Van der Linden, P. and Mitchell, J. F. B.: ENSEMBLES: Climate Change and its Impacts: Summary of research and results from the ENSEMBLES project, Met Office Hadley Centre, FitzRoy Road, Exeter EX1 3PB, UK, 2009.
Whiteman, C. D: Fundamentals and Applications, Mountain Meteorology, 1, 312–319, 2000.

13

# A cloud-to-ground lightning climatology for north-eastern Italy

**L. Feudale, A. Manzato, and S. Micheletti**

OSMER – Osservatorio Meteorologico Regionale dell'ARPA FVG, Visco (UD), Italy

*Correspondence to:* L. Feudale (laura.feudale@meteo.fvg.it)

**Abstract.** This study analyzes the spatial distribution and temporal characteristics of cloud-to-ground lightnings (C2G) in the North East of Italy and the neighboring areas of Austria, Slovenia and Croatia. The dataset consists of about 6.5 millions C2G flash records, both positive and negative, observed between January 1995 and December 2011 by the "Centro Elettrotecnico Sperimentale Italiano-Sistema Italiano Rilevamento Fulmini" (CESI/SIRF), part of the European Cooperation for Lightning Detection (EUCLID) Network.

The results show that C2G lightnings concentrate in the foothill regions on the southern flank of the Eastern Alps with a maximum of discharge frequency of 10 lightnings per $km^2$ per year. The number of C2G strokes varies with the period of the year: the most active period for lightning starts in April and lasts through November with the highest number of C2G strokes happening during the summer months of July and August, with maximum spatial density slightly moving from the mountain to the coastal area. The least frequency of C2G strokes is observed during wintertime. The mean diurnal C2G lightning activity for the whole domain shows a peak around 16:00–17:00 UTC and reaches a minimum around 07:00–09:00 UTC; the mean spatial distribution follows different patterns depending on the period of the day.

## 1 Introduction

Lightning are one of the most fascinating weather phenomena yet still not completely understood. They usually occur in thunderstorms during a breakdown of electrical fields in the form of cloud-to-cloud and cloud-to-ground (C2G) discharges. C2G lightning is among the major causes of weather-related fatalities. In the United States, for example, some studies assert that C2G lightning events contribute to a higher number of fatalities than tornadoes, hurricanes and strong winds (Ashley and Gilson, 2009); among all thunderstorm-related phenomena, only floods have recorded more average annual fatalities (Curran et al., 2000). Lightning flashes may cause also major property damage, power system breakdowns and forest fires (Rorig and Ferguson, 1999). Therefore lightning spatial distribution becomes of great interest, especially for energy-management and safety purposes. Furthermore, as our society is becoming more and more dependent on information networks, the knowledge of lightning prone areas could help in protecting informatic systems from disruption due to lightning discharges.

OSMER is particularly interested in studying the lightning climatology, because the Eastern Alps have a maximum in lightning spatial density, as it is seen in Schulz et al. (2005) and it will be proved later in this work. For this purpose, this study gives a detailed analysis of the spatial frequency of lightning considering the north-eastern Italy domain and interprets the observed structures in relation with the characteristics of this area. In addition, the frequency of lightning flashes provides a measure of the frequency of intense convection, identifying the areas more prone to severe weather events. The knowledge of lightning distribution may be used as a predictor into very complex weather nowcasting systems, as the system developed in the INCA-CE project (Haiden et al., 2011), and in particular the lightning climatology in terms of thunderstorm probability distribution may give a contribution to nowcasting and alerting in case of severe weather events, such as strong thunderstorms.

In the following sections, data and results regarding both spatial and temporal analysis of C2G lightning distribution in the target area are discussed. Then, it follows the conclusions.

## 2 Data

The C2G lightning data used in this analysis are provided by the "Centro Elettrotecnico Sperimentale Italiano-Sistema Italiano Rilevamento Fulmini" (CESI/SIRF, Bernardi and Ferrari, 2004), part of the European Cooperation for Lightning Detection (EUCLID) Network. The dataset consists of C2G lightning records, both positive and negative, observed between January 1995 and December 2011. These measurements are obtained through a network of Vaisala sensors which detects the return stroke, usually considered the largest discharge transfer during a lightning event. Up to 1999, the CESI/SIRF measurement network included just about 12 sensors located in the whole Italian peninsula. In 1999 four more sensors where added in order to cover the islands of Sicily and Sardinia and part of the sea between them and the mainland. Since 2000, when SIRF entered in the EUCLID Network, the lightning detection system has been strengthened by sensors in the neighboring countries of Austria, Slovenia and Switzerland. Therefore, the records between 1995 and 1999 may be affected by lower localization precision. Figure 1a and b show the C2G lightning density (in lightning number per $km^2$ per hour, multiplied by a factor of $10^4$) for the two periods characterized by a different number of lightning detection sensors: 1995–1999 in Fig. 1a and 2000–2011 in Fig. 1b. Of course, the two C2G lightning density patterns look quite different in areas far from the Italian border, that is in the areas northward (Austria and Switzerland) and eastward (Slovenia and Croatia), revealing an evident influence of the sensor distribution to C2G lightning spatial distribution. Focusing inside the Italian domain (latitude: 44–47° N, longitude: 10–14° E), where, as it will be shown later, the main C2G lightning activity is concentrated, the spatial correlation of these 5 contiguous years against the remnant 12 yr has been computed. In the histogram of Fig. 2, the black bar represents the spatial correlation 1995–1999 vs. 2000–2011 C2G lightning distributions (showed in Fig. 1), which value is 0.78. The grey bars represent the spatial correlation calculated over a running period of 5 contiguous years (starting from 2000 and going up to 2011) against the remnant 12 yr, which have a mean value of 0.88 and a standard deviation of 0.04. So, the natural variability of these latter 5 yr samples (grey bars of Fig. 2) does not seem to justify the low correlation found for the 1995–1999 sample period (black bar); therefore, this low correlation seems to be due also to the lower number of lightning detection sensors.

However, considering the entire available dataset of 17 yr, the effect due to the increase of the sensors number by 2000 should not affect the global results of this work. This statement is supported also by Bernardi and Ferrari (2004) who showed that the CESI/SIRF network efficiency has not changed much in the northern Italy area during this transition.

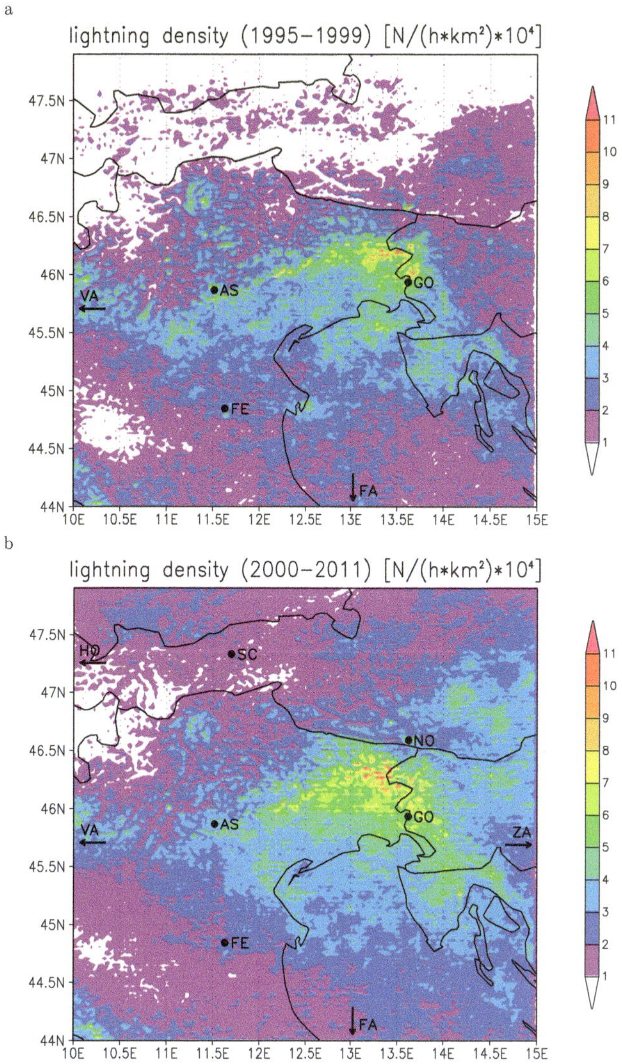


**Figure 1.** Lightning density and location of lightning detection sensors in the studied domain, for the period **(a)** 1995–1999 and **(b)** 2000–2011. Sensors in Italian domain: AS = Asiago, FA = Fano, FE = Ferrara, GO = Gorizia, VA = Varese. Sensors outside the Italian domain: HO = Hohenems (A), NO = Nötsch (A), SC = Schwaz (A), ZA = Zagreb (HR).

The domain of interest is the area included between latitudes 44–48° N and longitudes 10–15° E. The first step of the procedure is a preprocessing aimed at filtering out multiple strokes, considering as a single lightning event all the flash records reported in the same second and in the same area of $0.01 \times 0.01$ degrees. The resulting total number of lightnings is about 6.5 millions. Secondly, the data are regridded into a spatial resolution of $0.03 \times 0.02$ degrees (gridboxes covering

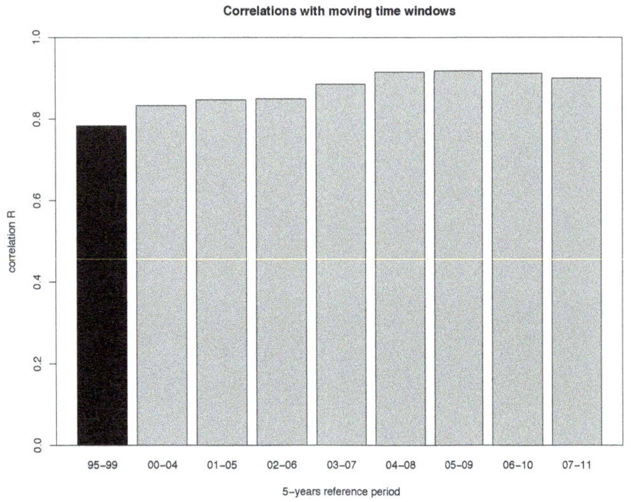


**Figure 2.** Histogram of spatial correlations (domain: LAT 44–47° N, LON 10–14° E) between lightning densities calculated for the 5 yr period 1995–1999 against the 12 yr period 2000–2011 (black bar), and for a running time-window period of 5 yr (listed in the abscissa) against the remnant 12 yr period (grey bars).

an area of about 5.15 $km^2$) and aggregated in the hourly matrices, that will be analyzed in the next section.

## 3 Results

This section discusses the general characteristics of the C2G lightning and how these vary in space and time. The spatial distribution of lightning was computed using all the records available in the CESI/SIRF dataset from 1995 to 2011. The map shown in Fig. 3 displays the mean number of C2G lightnings per $km^2$ per hour, multiplied by a factor of $10^4$. It shows that the C2G lightnings are not homogeneously distributed in the domain: the maximum of C2G density draws an arc-shape on the southern flank of the Carnic and Julian Prealps.

### 3.1 Maximum values

Zooming on the Friuli Venezia Giulia region (FVG), a detailed analysis of the maximum density in the distribution field (Fig. 4a) reveals a peak of $17 \times 10^{-4}$ lightnings per hour per $km^2$ in Austria, on the border with Italy, at the geographic coordinates 46.58° N of latitude and 13.64° E of longitude approximately. Evaluating in details the data, we found that this maximum corresponds to the 165 m high transmission antenna (Fig. 4b) on the summit of Mt. Dobratsch (2115 m a.s.l.). In fact, outliers with a high number of strokes most often correspond to locations with high towers and/or big antennas (Novák and Kyznarová, 2011). The next following maximum values in decreasing order are approximately $11.6{\times}10^{-4}$, $11.3{\times}10^{-4}$, $11.2{\times}10^{-4}$, $10.5{\times}10^{-4}$, $10.5 \times 10^{-4}$, which correspond to the following mountains: Gran Monte (1636 m), Matajur (1641 m), Canin (2587 m), Stol (1673 m, SLO) and Plauris (1958 m), respectively.

The relative maximum in the Eastern Alps confirms what found by Schulz et al. (2005), who analyzed the C2G lightning flashes in Austria for the 10 yr 1992–2001. A part of north-eastern Italy, which includes the Friuli Venezia Giulia region, is on the border of their analyzed domain. They show that the activity in this small area is more than 5 flashes per year per $km^2$, corresponding to the maximum value in their domain. This value agrees qualitatively – but not quantitatively – with our study on 17 yr: the highest density area records about $10 \times 10^{-4}$ lightnings per hour per $km^2$, corresponding to about 9 lightnings per year per $km^2$.

In the following subsections the main temporal characteristics and their spatial evolution are summarized.

### 3.2 Temporal statistics

C2G lightnings show great variability in the domain from year to year, from day to day, as well as they show a diurnal cycle. To better understand the temporal characteristics of lightning activity, which is a representation of convective activity, the mean value in the studied domain of Fig. 3 is computed and analyzed with different time scales.

Figure 5a displays the mean annual value of C2G frequency in the whole target domain and it shows a large variability year by year, with a maximum in 2002, having a mean value above $3.6 \times 10^{-4}$ lightnings per hour per $km^2$ and predominant larger values towards the end of the time-series (2007–2009), indicating a possible positive trend. Nonetheless, no statistically significant trend is observed because a linear fit (dotted line) indicates a low $R^2(0.07)$ with a $P$ value quite high (0.29).

Other studies of lightning in the European area (Holt et al., 2001; Rivas Soriano et al., 2005; Novák and Kyznarová, 2011; Antonescu and Burcea, 2010; Finke and Hauf, 1996; Tuomi and Mäkelä, 2008), in United States (e.g. Orville and Huffines, 2001) and Australia (Kilinc and Beringer, 2007) already evidenced that C2G lightning flashes occur predominantly during the warm season, as expected. Figure 5b depicts the mean value of C2G lightnings on a Julian Day scale; the red line is the ±10 day moving average, which filters higher frequency values. The active period starts approximately in April, reaches its maximum during the months of July and August (around $8 \times 10^{-4}$ lightnings per hour per $km^2$) and fades thereafter until October. Some activity, even if small, is observed also during the month of November, maybe due to unstable conditions often embedded in large precipitating structures (Manzato, 2007, p. 341). Almost 90 % of C2G lightning activity occurs between May and September. The minimum values are recorded during the winter months between December and February.

Figure 5c shows the mean diurnal distribution of C2G lightnings in the north-eastern part of Italy in hourly steps. The values follow an almost sinusoidal-like regime, with the

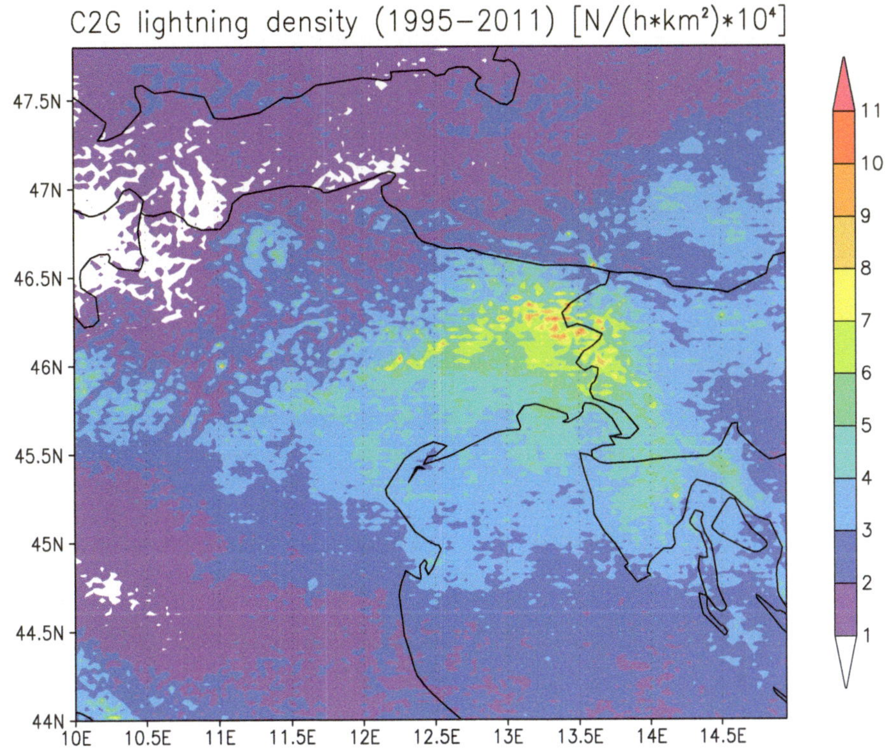


**Figure 3.** Spatial distribution on a grid of $0.03 \times 0.02$ degrees of resolution of the average number of C2G lightning per hour per km$^2$ computed for the 17 yr period 1995–2011. The scale is multiplied by a factor of $10^4$.

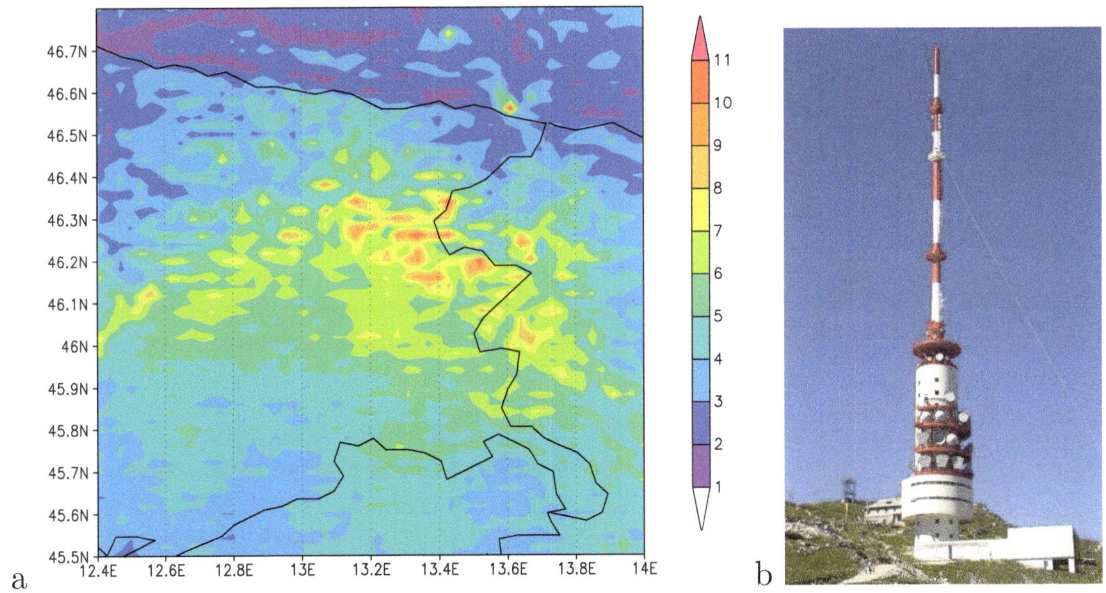


**Figure 4.** **(a)** Zoom on Friuli Venezia Giulia region of C2G lightning distribution of Fig. 3 (in number of lightnings per hour per km$^2$, multiplied by a factor of $10^4$) which shows the detection of an increased number of C2G strokes at the coordinates 46.6° N of latitude and 13.6° E of longitude in the nearby Dobratsch transmitter tower in Austria shown in **(b)**, which is a 165 m high tower on the Dobratsch mountain of 2115 m a.s.l. near Villach.

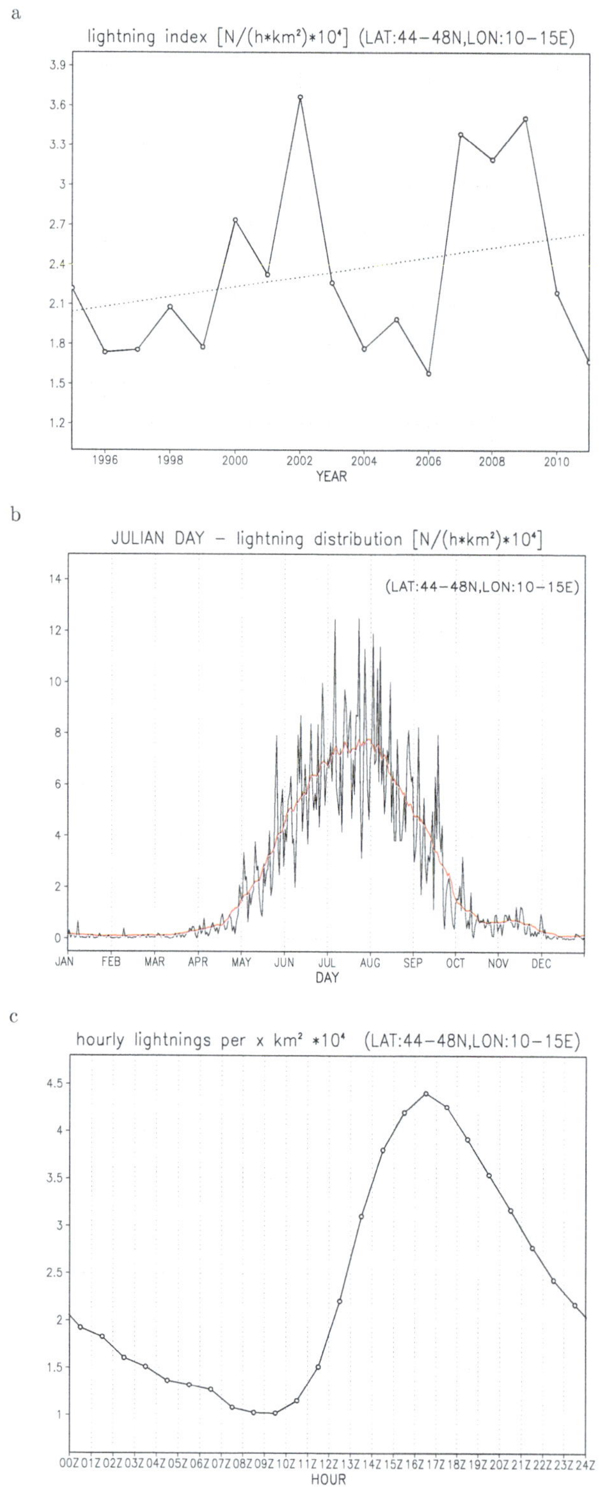


**Figure 5.** **(a)** Domain mean of C2G lightning annual distribution (solid) and the 17 yr linear trend (dotted); **(b)** monthly distribution of C2G lightning with the ±10 day moving average in red; **(c)** daily cycle of the C2G lightning distribution in the target domain. Units are number of flashes per hour per $km^2$ times a factor of $10^4$.

strongest lightning activity recorded during the afternoon, with a peak between 16:00 and 17:00 UTC above $4\times10^{-4}$ flashes per hour per $km^2$, and the weakest lightning activity observed during the morning hours, with minimum of $1\times10^{-4}$ flashes per hour per $km^2$ between 08:00 and 10:00 UTC. The slope from the minimum to the maximum is quite steep since the rise occurs in only 6 h. On the other side, the decrease of lightning activity from 17:00 UTC is slower, occurring in about 14 h, while convection typically moves slightly from the mountain to coastal areas, as it will be described in the next sub-section. Since the frequency of lightnings provides a measure of the probability of the occurrence of convection, the mean diurnal cycle of lightning is in agreement with the 6 h climatology of thunderstorms in the smaller domain, which includes just the Friuli Venezia Giulia region, by Manzato (2007) (his Fig. 2), while the time of the maximum peak found by Gladich et al. (2011) is shifted later (at 19:00 UTC, in their Fig. 3). Thunderstorms being more frequent during the afternoon and less frequent during the morning have also been observed in other European areas like Czech Republic (Novák and Kyznarová, 2011), Romania (Antonescu and Burcea, 2010) and Finland (Tuomi and Mäkelä, 2008).

## 3.3 Spatial distribution

The spatial distribution of C2G lightnings displayed in Fig. 3 is the mean value considering the total number of flashes for all the 17 yr analyzed. From this figure, the lightning distribution seems to be topographically influenced, depicting a decrease in the flash rates with the topographic height. This argument will be further discussed by the authors in a following article. As discussed previously, the "active period" starts in April with some C2G lightnings observed in the Prealpine region (Fig. 6a). By the month of May the flashes become more frequent in the same region and also in the surrounding areas, gradually increasing while approaching the summer season (Fig. 6b–c). The maximum activity occurs during July and August (Fig. 6d–e), extending from the foothills area to the plane and affecting also the coastal areas in the late summer (starting from August). Since September, the main active area moves from land to the sea and on the coastal regions of NE Italy and Croatia's islands, fading thereafter (Fig. 6f–h). In general, it is clear that in spring-summer the largest occurrence of lightnings is over land, whilst in autumn the situation is reversed and the peak values are seen over coastal areas and on the sea, when thunderstorms are influenced by the warm sea and the strong land-sea gradients. These local maxima are mainly due to the Adriatic Sea, which supplies warm and humid air, thus increasing the convection probability, especially during the autumn when this effect is the strongest (as observed in the relative maximum in November in Fig. 5b). This is in agreement with the general study of lightning characteristics in Europe by Holt et al. (2001).

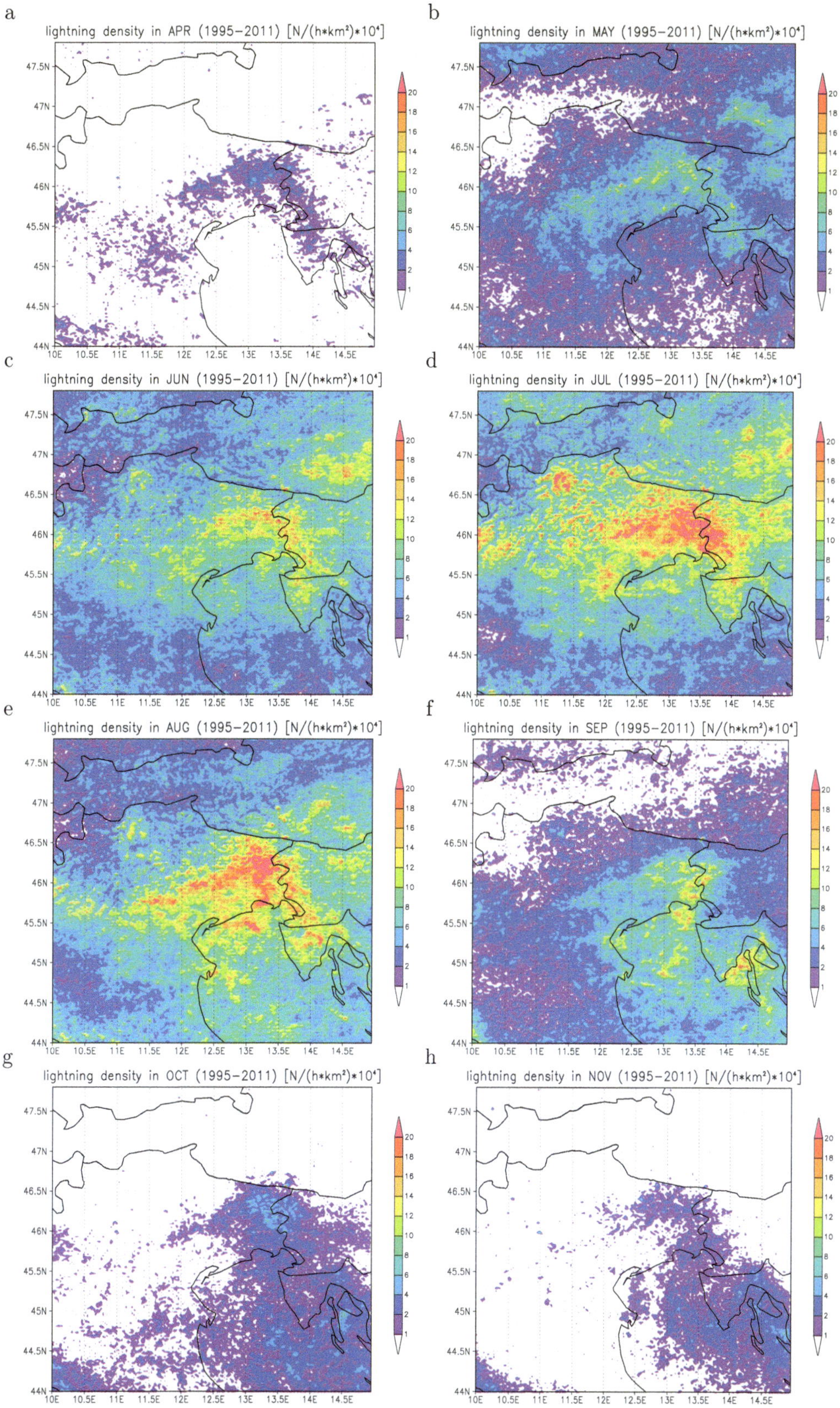


**Figure 6.** Month-by-month C2G lightning spatial distribution starting from April up to November. The scale is number of lightnings per hour per km$^2$ times a factor of $10^4$.

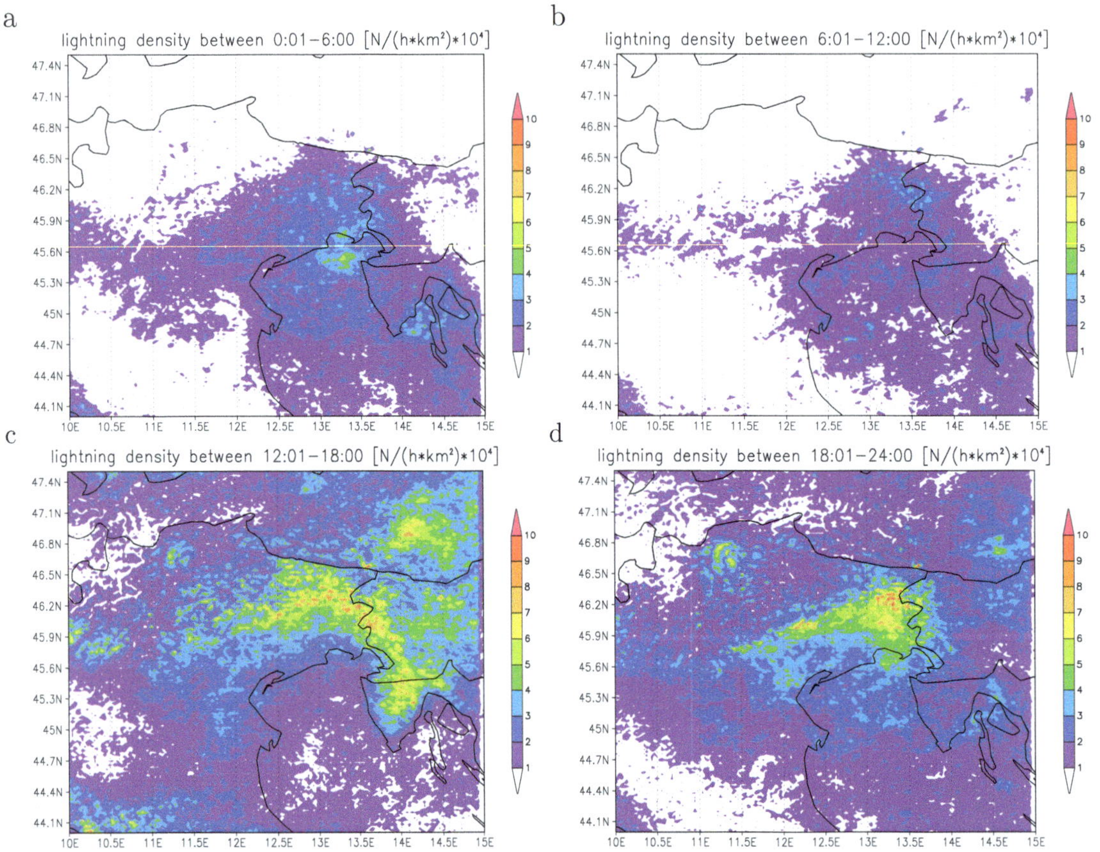


**Figure 7.** Daily cycle spatial distribution of C2G lightning: **(a)** 00:00–06:00 UTC, **(b)** 06:00–12:00 UTC, **(c)** 12:00–18:00 UTC and **(d)** 18:00–24:00 UTC. The peak is in the afternoon over land moving than to the coastal area and the sea during the night. Units are number of flashes per hour per $km^2$ times a factor of $10^4$.

The diurnal cycle of C2G flashes also displays different patterns depending on the time of the day. Because of complex terrain in north-eastern Italy, with mountain-valley areas in the Alps and Prealps, and with the plain being surrounded by the high-terrain of the Alps (with peaks exceeding the 3000 m) in the North and delimited by the Adriatic Sea on the South, the domain is affected by mountain-valley breezes and land–sea breezes, which may trigger convection. Figure 7 shows the migration of thunderstorms, as recorded by the observed flashes, with the diurnal cycle. This figure highlights the different distribution patterns in the four 6 h periods of the day: the peak occurs between 12:00–18:00 UTC (Fig. 7c), with the highest density of flashes in the Julian Prealps down to Istria; from the evening to nighttime (18:00–24:00 UTC) the peak moves to the foothill area of the Carnic–Julian Prealps, and some activity is observed also on the sea area usually more affected by nocturnal thunderstorms (Fig. 7d). Between the night and the first hours of the day (00:00–06:00 UTC) most of the flashes are recorded on the sea and along the coast (Fig. 7a), due to the nighttime inland radiation cooling and corresponding cold drainage air masses. As expected from Fig. 5c, the minimum rate of C2G flashes occurs during the morning, in the 6 h period between 06:00 and 12:00 UTC (Fig. 7b).

## 4 Conclusions

This work focuses on the C2G lightning climatology for the north-eastern part of Italy. Approximately 6.5 million C2G lightnings (both negative and positive), recorded in this area between 1995 and 2011, were analyzed in the spatial and temporal scale, in order to identify the areas prone to high rates of C2G flashes. The main conclusions are that, with respect to the domain investigated in this work and that used by Schulz et al. (2005), it appears that the Friuli Venezia Giulia region is the area in Central Europe with the maximum of lightning density, reaching values as high as 9.6 lightnings per $km^2$ per year. Moreover, the spatial distribution of C2G lightnings is influenced by the topography, such that the gridboxes with maximum lightning density correspond to the tops of some Julian Prealps mountains (or particularly high towers).

C2G lightning activity shows a seasonal cycle with the active period starting in April and continuing up to October, with maximum probability occurring mainly in the Carnic and Julian Prealps. A smaller peak occurs also along the coastal areas in November, due to the unstable conditions induced by the North Adriatic Sea, warmed up during the previous summer season. The seasonal cycle modulates a daily cycle which clearly shows strong lightning activity in the afternoon in the mountain region, fading from the night while moving to the sea. The database analyzed shows that there is an overall tendency to an increase in C2G lightning frequency, but the trend is not statistically significant. An evaluation of the relationship of C2G lightning distribution with the topography will be the topic of a further article by the authors.

**Acknowledgements.** This research was carried out under the support of the project 2CE120P3 INCA-CE, funded by the program ETC Central Europe FESR funds.

Edited by: M. Dolinar

## References

Antonescu, B. and Burcea, S.: A cloud-to-ground lightning climatology for Romania, Mon. Weather Rev., 138, 579–591, 2010.

Ashley, W. S. and Gilson, C. W.: A Reassessment of U.S. Lightning Mortalities, B. Am. Meteorol. Soc., 10, 1501–1518, 2009.

Bernardi, M. and Ferrari, D.: Evaluation of the LLS efficiency effects on the ground flash density, using the Italian lightning detection system SIRF, J. Electrostatics, 60, 131–140, 2004.

Curran, E. B., Holle, R. L., and Lopez, R. E.: Lightning Casualties and Damages in the United States from 1950 to 1994, J. Climate, 13, 3448–3464, 2000.

Finke, U. and Hauf, T.: The characteristics of lightning occurrence in southern Germany, Inst. für Physik der Atmosphäre, 1996.

Gladich, I., Gallai, I., Giaiotti, D., and Stel, F.: On the diurnal cycle of deep moist convection in the southern side of the Alps analysed through cloud-to-ground lightning activity, Atmos. Res., 100, 371–376, 2011.

Haiden, T., Kann, A., Wittmann, C., Pistotnik, G., Bica, B., and Gruber, C.: The Integrated Nowcasting through Comprehensive Analysis (INCA) system and its validation over the Eastern Alpine region, Weather Forecast., 26, 166–183, 2011.

Holt, M., Hardaker, P., and McLelland, G.: A lightning climatology for Europe and the UK, 1990–99, Weather, 56, 290–296, 2001.

Kilinc, M. and Beringer, J.: The spatial and temporal distribution of lightning strikes and their relationship with vegetation type, elevation, and fire scars in the Northern Territory, J. Climate, 20, 1161–1173, 2007.

Manzato, A.: The 6 h climatology of thunderstorms and rainfalls in the Friuli Venezia Giulia Plain, Atmos. Res., 83, 336–348, 2007.

Novák, P. and Kyznarová, H.: Climatology of lightning in the Czech Republic, Atmos. Res., 100, 318–333, 2011.

Orville, R. and Huffines, G.: Cloud-to-ground lightning in the United States: NLDN results in the first decade, 1989–98, Mon. Weather Rev., 129, 1179–1193, 2001.

Rivas Soriano, L., De Pablo, F., and Tomas, C.: Ten-year study of cloud-to-ground lightning activity in the Iberian Peninsula, J. Atmos. Sol.-Terr. Phy., 67, 1632–1639, 2005.

Rorig, M. and Ferguson, S.: Characteristics of lightning and wildland fire ignition in the Pacific Northwest, J. Appl. Meteorol., 38, 1565–1575, 1999.

Schulz, W., Cummins, K., Diendorfer, G., and Dorninger, M.: Cloud-to-ground lightning in Austria: A 10-year study using data from a lightning location system, J. Geophys. Res., 110, D09101, doi:10.1029/2004JD005332, 2005.

Tuomi, T. and Mäkelä, A.: Thunderstorm climate of Finland 1998–2007, Geophysica, 44, 67–80, 2008.

# 14

# An empirical method for estimating probability density functions of gridded daily minimum and maximum temperature

**C. Lussana**

ARPA Lombardia, Milano, Italy

*Correspondence to:* C. Lussana (c.lussana@arpalombardia.it)

**Abstract.** The presented work focuses on the investigation of gridded daily minimum ($TN$) and maximum ($TX$) temperature probability density functions (PDFs) with the intent of both characterising a region and detecting extreme values. The empirical PDFs estimation procedure has been realised using the most recent years of gridded temperature analysis fields available at ARPA Lombardia, in Northern Italy. The spatial interpolation is based on an implementation of Optimal Interpolation using observations from a dense surface network of automated weather stations. An effort has been made to identify both the time period and the spatial areas with a stable data density otherwise the elaboration could be influenced by the unsettled station distribution. The PDF used in this study is based on the Gaussian distribution, nevertheless it is designed to have an asymmetrical (skewed) shape in order to enable distinction between warming and cooling events. Once properly defined the occurrence of extreme events, it is possible to straightforwardly deliver to the users the information on a local-scale in a concise way, such as: $TX$ extremely cold/hot or $TN$ extremely cold/hot.

## 1 Introduction

This paper describes a method for the estimation of daily maximum $TX$ and minimum $TN$ temperature probability density functions (PDFs) from gridded datasets.

The hourly temperature observations have been measured by ARPA Lombardia's meteorological network, a dense surface network of automated weather stations in Northern Italy. The mesonet covers part of the southern alpine ridge and of the Po Plain, where elevation and local advections play important roles; a more detailed description of the network can be found in Lussana et al. (2009a).

As daily temperature extremes are often recorded simultaneously at many stations (Pfahl and Wernli, 2012; Wulfmeyer and Henning-Müller, 2006), it is convenient to use the high-resolution meteorological information provided by the whole surface network through a statistical interpolation procedure.

Lombardy's weather service has implemented an Optimal Interpolation (OI) procedure (Uboldi et al., 2008) to interpolate the sparse hourly temperature observations on a regular grid. The quality of the analysis fields is adequate to resolve the mesoscale features down to the Meso-$\gamma$ scale (Thunis and Bornstein, 1996). A $TX$ and $TN$ analysis field is computed at each grid point as the daily maximum and minimum (respectively) of the hourly averaged temperature.

This work aims to obtain a reliable spatial description of $TX$ and $TN$ extreme events on a local scale. Extreme temperature events, such as heat waves or cold spells, are of great interest for their impact on human activities, and in a warming climate scenario (Schär et al., 2004) they are expected to become more frequent and intense, but the statistical modelling of climate extremes across space remains a challenging issue (Katz, 2010).

The statistical model chosen for $TX$ and $TN$ PDFs is based on the normal distribution, since $TX$ is often observed to be Gaussian (Katz and Brown, 1992), particularly in summer (Mearns et al., 1984). A stationary climate is implicitly assumed, despite climate change (Toreti and Desiato, 2008; Simolo et al., 2010, 2011), because of the short time period considered in the elaborations. In this preliminary study, to enable distinction between extreme warm and cold events, the statistical model is designed to be asymmetrical.

The extreme event definition is usually based on percentile thresholds calculated from observed data (Klein Tank et al., 2009), without any explicit assumption regarding the underlying PDF. In this work, due to the short temporal period covered by the initial datasets, any percentile-based threshold could lead to ambiguous results. To overcame these limitations and, hopefully, to infer more general results, the choice has been to include the a-priori knowledge about the expected temperature behaviour through the PDF.

This paper is structured as follows: Sect. 2 addresses the issues of data quality and data coverage (temporal and spatial); Sect. 3 presents the $TX$ and $TN$ PDF models and the related parameter estimation procedure; Sect. 4 discusses the extreme event definition and briefly presents two case studies.

## 2 Quality of the observations and gridded fields

All new observations of ARPA Lombardia's meteorological network undergo a manual quality control procedure within the network's quality assurance system (Lussana et al., 2012; Ranci and Lussana, 2009), however at present only recent data has been manually verified by meteorologists. To overcame this limitation the OI procedure performs an automated Spatial Consistency Test (SCT), described in Lussana et al. (2010, 2009b). The SCT stops observations likely to be affected by gross measurement errors from entering the OI, and also reduces the effects related to the absence of a data homogenisation procedure because of the filter on large systematic or representativity errors.

The rest of this section concerns the impact of the irregular station distribution on analysis quality. The analysis error's distribution (RMS) at station's location has a median close to 1 °C for all stations but the spread is dependent on station density and is significantly higher for isolated stations. Therefore, though it's reasonable to assume a stable analysis quality for dense stations areas, in areas with poorer station coverage analysis quality is known to be irregular. In the latter case, any long-term statistics is likely to include the nontrivial behaviour of the unstable representativity error (Lussana et al., 2010), and its variability is comparable with the natural variability. It is necessary to reduce the initial dataset in order to use only temporal periods and spatial areas with stationary and adequate station density, thus avoiding ambiguity in the comparisons of the results in different areas or time periods.

A convenient way to quantify station density is through the Integral Data Influence (IDI) parameter described in Uboldi et al. (2008). The IDI field can be intuitively visualised as the analysis obtained by interpolating observations with a constant value of 1 on a constant background field of value 0 (without reference to any particular unit). The IDI analysis combines the effects of irregular station distribution and of OI operational choices, resulting in a field with a straightforward interpretation: dense station areas have IDI values close to 1 while areas with poor station coverage have values close to 0.

In the case of ARPA Lombardia's automated mesonet the bulk of the observations ranges from 2000 to 2011 (with some sparse longer time series). The time series of spatially averaged IDI over Lombardy for this period shows two periods of stationary IDI value: 2000–2002 and 2005–2011. The 2003–2004 period is characterised by a gradual IDI increase corresponding to the mesonet development. Only the latest 7 yr (2005–2011) has been considered in this study.

An IDI-based spatial mask has been computed to include only grid points where the station coverage is sufficiently dense and stationary in time, that is, where the IDI value is mostly close to 1. In Figs. 3 and 5, where the IDI mask is applied, white areas within Lombardy's administrative boundaries represent masked out grid points. The border effects are quite evident, as it's evident the effect of sparse (or variable in time) data coverage in the Alps.

## 3 Estimating probability density functions

The $TX$ or $TN$ time series at a fixed location for a calendar year shows the noisy weather-related variability overlying the typical bell-shaped main annual cycle. To simplify the presentation the following discussion regards $TX$, but it can be easily applied to $TN$.

The $TX$ PDF chosen at any fixed grid point is composed by two normal distributions sharing the same mean value, corresponding to the PDF's location parameter, but with two distinct standard deviations allowing for asymmetry between positive and negative anomalies (i.e. difference between the $TX$ value and the location parameter). The PDF is then completely defined by the three parameters: the location parameter $\langle TX\rangle$, taking into account the main annual cycle, and the two scale parameters $\sigma_+^{TX}, \sigma_-^{TX}$ for positive and negative anomalies, respectively, accounting for the daily variability. The PDF $p_{TX}(x)$ can be written as:

$$p_{TX}(x) = \begin{cases} \left[1/\left(\sqrt{2\pi}\sigma_+^{TX}\right)\right]\exp\left[-\frac{1}{2}\left(\frac{x-\langle TX\rangle}{\sigma_+^{TX}}\right)^2\right], & x \geq \langle TX\rangle \\ \left[1/\left(\sqrt{2\pi}\sigma_-^{TX}\right)\right]\exp\left[-\frac{1}{2}\left(\frac{x-\langle TX\rangle}{\sigma_-^{TX}}\right)^2\right], & x < \langle TX\rangle \end{cases}$$

As a preliminary step in the parameter estimation procedure, each month has been split into three 10-day periods as shown in Fig. 1. The 10-day periods have been used in the selection of the initial statistical datasets for the estimation procedure. It is assumed that each 10-day period is likely to be characterised by similar meteorological conditions. The previous and following 10-day periods show a $TX$ temporal behaviour on longer time scales than the short-term weather variability, but still consistent with the assumptions made in

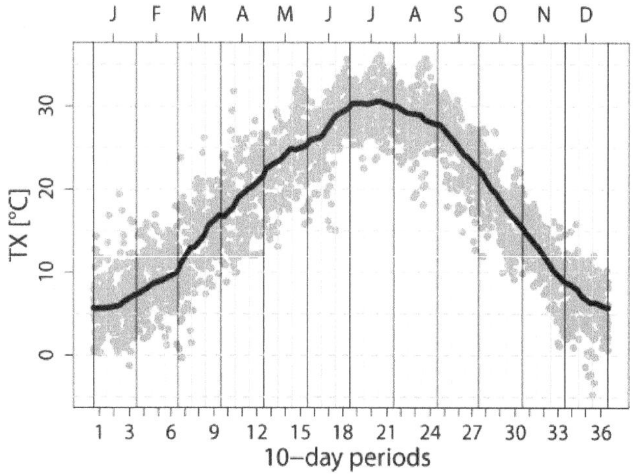


**Figure 1.** Milan grid point. Location parameter estimates for the calendar year: $TX$ values for the period 2005–2011 in grey; $\langle TX \rangle$ in black.

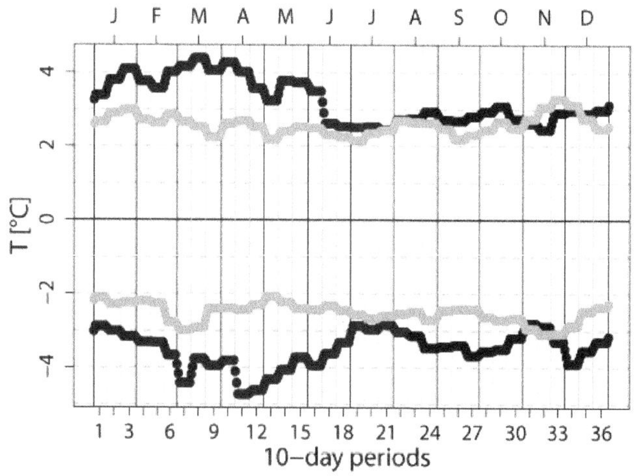


**Figure 2.** Milan grid point. Scale parameters estimates for the calendar year: $\sigma_+^{TX}$, $\sigma_-^{TX}$ in black; $\sigma_+^{TN}$, $\sigma_-^{TN}$ in grey; $\sigma_-^{TX}$ and $\sigma_-^{TN}$ are reported with opposite sign (negative values).

the estimation procedure, such as the linear trend approximation used for $\langle TX \rangle$.

As a first step, the parameter estimation procedure has been applied independently at every grid point. The location parameter for a fixed calendar day is estimated from an initial dataset of about two hundred $TX$ values (the analysis from the 10-day period containing the calendar day and the two adjacent 10-day periods). Then $\langle TX \rangle$ is obtained by choosing the linear trend which best fits the dataset at the calendar day. The method used for the linear trend estimation is the median of pairwise slope regression (Lanzante, 1996), based on computing the slope between pairs of $TX$ values and considering the median of these values.

The two $TX$ scale parameters for a fixed calendar day are estimated from the same dataset using temperature anomalies instead of $TX$ values. As correlations within the anomaly time series could bias the scale parameters estimates (Wilks, 1995), a resampling procedure has been implemented to create uncorrelated synthetic samples of the three 10-day periods under consideration. The standard deviation is estimated separately for the upper and for the lower part –respect to the median– of each of the synthetic samples and their values are assigned to $\sigma_+^{TX}$ and $\sigma_-^{TX}$, respectively. This procedure is an implementation of the asymmetric biweight standard deviation estimation procedure described in Lanzante (1996). The uncertainties in the $\sigma_+^{TX}$ and $\sigma_-^{TX}$ values evaluated from the $\sigma_+^{TX}$ and $\sigma_-^{TX}$ distributions obtained form the synthetic samples are about 0.5 °C for both values (the same holds for $TN$).

The second step in the parameter estimation procedure imposes temporal and spatial consistency by applying a space-time smoothing to the estimates of both location and scale parameters. As the PDF parameters are expected to change slowly in space and in time over the high-resolution gridded domain, for each grid point and fixed calendar day the estimated parameters have been replaced with the average of the estimated values in neighbouring grid points and adjacent days.

Overall the parameter estimation procedure is robust and resistant, based upon simple assumptions for the probability model, and rather insensitive to misbehavior of the data.

As an example, the estimation procedure is shown for Milan's grid point. In Fig. 1, grey dots represent the initial $TX$ dataset (each day has 7 values from years 2005 to 2011) and black dots show the rather smoother $\langle TX \rangle$ time series at the same grid point. Figure 2 shows the $\sigma_\pm^{TX}$ and $\sigma_\pm^{TN}$ time series estimated at this grid point. To avoid confusion, scale parameters for negative temperature anomalies are reported with opposite sign (negative values), and $TX$ and $TN$ scale parameters are shown with black and grey dots, respectively. The estimated values are smaller in summer, around 2.5 °C, and higher in winter, around 4 °C. The $TN$ scale parameters are generally lower than the $TX$ parameters.

The maps of seasonally averaged scale parameter, such as those presented in Figs. 3 and 4, often show a significant (i.e. greater than 0.5 °C) difference between $\sigma_+$ and $\sigma_-$, thus supporting the choice of two distinct scale parameters for positive and negative anomalies. In particular, for $TN$ a significant difference can be found in winter and, for mountainous areas, also in autumn; significant differences for $TX$ have been found in summer, autumn and winter. The spring season is characterised by the highest and more variable values of $\sigma_+$ and $\sigma_-$. In general, the difference between the two scale parameters is more pronounced for $TX$ than for $TN$, due to the greater variability associated with daytime weather conditions. Furthermore, $\sigma_-$ is greater than $\sigma_+$ for all seasons except for winter when the situation is the opposite. In winter, $\sigma_+^{TX}$ is significantly greater than $\sigma_-^{TX}$ in the Apennines, in the south-western part of the Plain and in the alpine valleys, probably due to the frequent foehn episodes; $\sigma_+^{TN}$ is

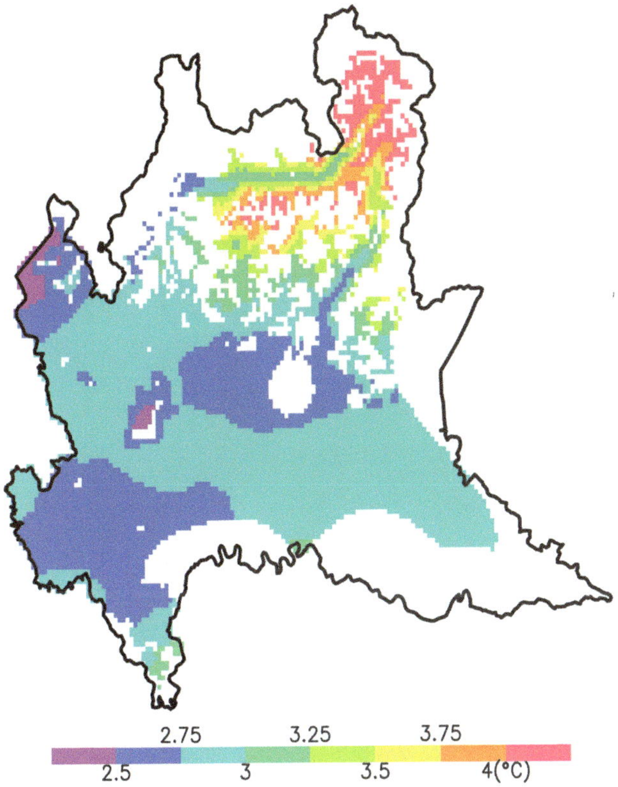


**Figure 3.** Winter season (December-January-February) averaged $\sigma_{-}^{TN}$. The IDI mask has been applied (see Sect. 2).

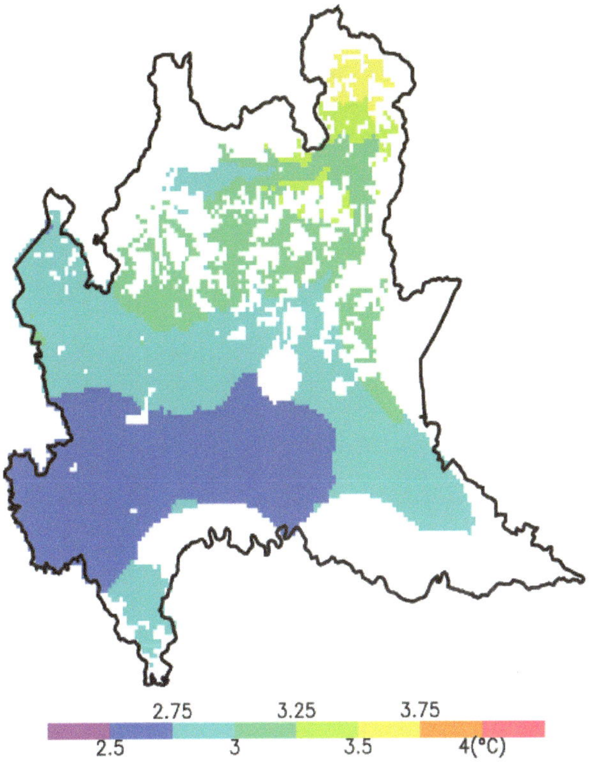


**Figure 4.** Summer season (June-July-August) averaged $\sigma_{+}^{TX}$. The IDI mask has been applied (see Sect. 2).

significantly greater than $\sigma_{-}^{TN}$ in the Plain due to frequent and persistent fog. In the other seasons, $\sigma_{-}^{TX}$ is increased by the greater variability due to cloudiness, while $\sigma_{+}^{TX}$ is associated with the lower temperature variability in clear sky conditions. Figures 3 and 4 show the winter averaged $\sigma_{-}^{TN}$ and the summer averaged $\sigma_{+}^{TX}$ associated with the occurrence of winter cold spells and summer heat waves, respectively. In general, the study of the seasonal averaged maps reveals the strong influence of the elevation, as expected and showed in Fig. 4; nevertheless it is also possible to identify other distinctive features on a very local scale. In the case of the winter averaged $\sigma_{-}^{TN}$, in Fig. 3, the reduced variability of Milan's urban area is clearly evident in the Western Plain, probably related to the constraining effect of the urban heat island on $TN$'s variability.

## 4 Extreme events

Extreme events are by definition rare events, and their probability of occurrence is described by the $TX$ and $TN$ PDFs outermost tails. However, due to the limited temporal period available in this study (only 7 yr), it is more likely that extreme events here detected would be in fact moderate extremes (Klein Tank et al., 2009), that typically occurs several times every year.

To define an extreme event it is convenient to transform the random variable $TX$ (or $TN$) to standard form (i.e. a Gaussian having a sample mean of zero and a sample standard deviation of one (Wilks, 1995)). The resulting standardised random variable $Z$ has values $z$ computed as:

$$z = \begin{cases} |x - \langle TX \rangle| / \sigma_{+}^{TX}, & x \geq \langle TX \rangle \\ |x - \langle TX \rangle| / \sigma_{-}^{TX}, & x < \langle TX \rangle \end{cases}$$

where $x$ represents a $TX$ realisation. The random variable $Z$ follows a standard Gaussian distribution regardless of the calendar day or the grid point considered, therefore this local-scale definition allows for direct comparison both in space and in time.

The $TX$ and $TN$ events are classified according to their estimated probability of occurrence as: unusual cold $P[Z \leq z] \leq 20\,\%$ or warm $P[Z \geq z] \leq 20\,\%$; extreme cold $P[Z \leq z] \leq 5\%$ or warm $P[Z \geq z] \leq 5\,\%$, whichever more restrictive.

The high sensitivity of extreme event definition to the PDF tails suggests avoiding the use of too small probabilities as thresholds because the Gaussian assumption may be problematic.

As an example, in Fig. 5 the $TN$ analysis map is shown for a winter foehn case resulting in extreme warming in separated areas: two areas close to the conjunction between the Plain and the Alpine region and one area in the South corresponding to the mountainous Apennines region. In this case

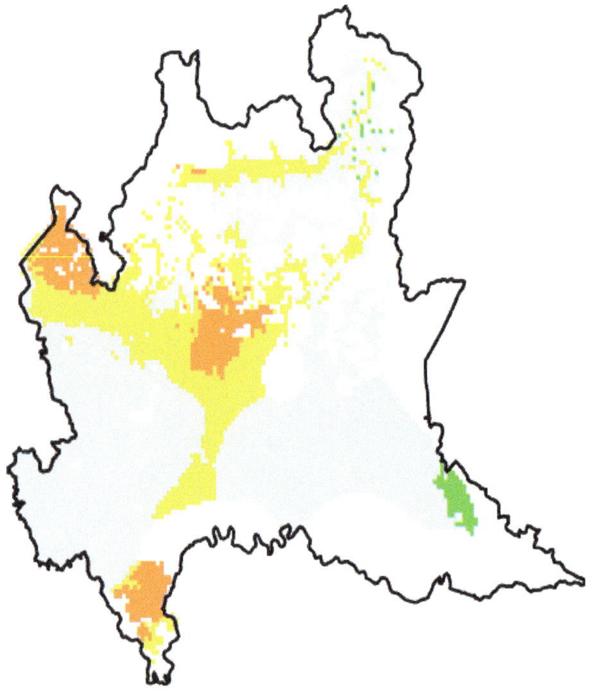

**Figure 5.** 8 January 2012. *TN* extreme events classification. Extreme warming: orange. Unusual warming: yellow. Normal conditions: grey. Unusual cooling: green. The IDI mask has been applied (see Sect. 2).

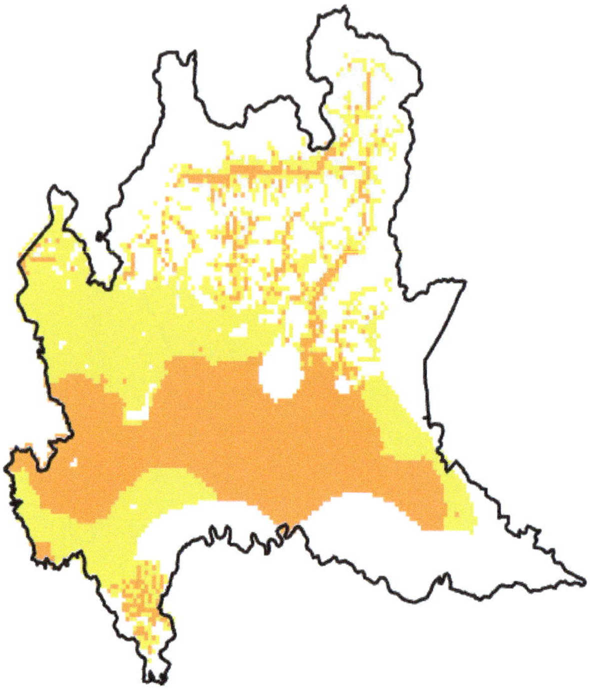

**Figure 6.** 22 June 2012. *TX* extreme events classification. Extreme warming: orange. Unusual warming: yellow. Normal conditions: grey. The IDI mask has been applied (see Sect. 2).

the extreme event occurrence is related to local-scale advection, this scheme is able to correctly detect such events and to provide a reasonable description for their spatial extension. Figure 6 presents a summer heat wave case. Respect to the previous foehn case, in this situation the forcing is at the mesoscale, and results in widespread warming in the whole region. Nevertheless, the map can shows gradients in the probability of occurrence, with extreme warming in the central part of the Po Plain while in the upper part of the Plain *TX* reaches unusual values. In the mountains only the valley floors and only on a local scale show extreme warming.

## 5 Conclusions

The OI hourly high-resolution temperature analysis produced by ARPA Lombardia's weather service is known to provide valuable information on a local scale (Uboldi et al., 2008). This preliminary study investigates the possibility of exploiting these high-resolution analysis to characterise the daily *TX* and *TN* PDFs using an asymmetrical Gaussian-based model.

In studying the daily extremes' statistics it's of crucial importance to use initial temperature analysis fields of adequate and stationary quality. The IDI field can be helpful to locate areas and periods likely to give ambiguous results, which can then be removed from the analysis.

Resistant and robust estimation techniques can be applied to obtain reasonable values for the *TX* and *TN* PDF location and scale parameters, also in the case of sparse data. The significant difference in scale parameter value between positive and negative anomalies clearly supports the use of asymmetrical PDFs.

The *TX* and *TN* PDFs can be used to detect and evaluate extreme events in the daily *TX* and *TN* analysis for each grid point, thus completing the high-resolution spatial information available operationally at Lombardy's weather service.

To improve on this preliminary work the dataset should be enlarged (collecting data from other mesonets in Lombardy and in neighbouring regions), and different choices in statistical model (skew-normal density function, Simolo et al., 2010) and in the parameter estimation procedure (moving-window for the linear trend estimation) should be explored.

**Acknowledgements.** The extensive and important comments of Marta Salvati, Francesco Uboldi and the reviewers are gratefully acknowledged.

Edited by: M. Dolinar

## References

Katz, R. W.: Statistics of extremes in climate change, Climatic Change, 100, 71–76, 2010.

Katz, R. W. and Brown, B. G.: Extreme events in a changing climate: Variability is more important than averages, Climatic Change, 21, 289–302, 1992.

Klein Tank, A. M. G., Zwiers, F. W., and Zhang, X.: Guidelines on Analysis of Extremes in a Changing Climate in Support of Informed Decisions for Adaptation, Climate data and monitoring, WMO, 2009.

Lanzante, J. R.: Resistant, robust and nonparametric techniques for the analysis of climate data: Theory and examples, including applications to historical radiosonde station data, Int. J. Climatol., 16, 1197–1226, 1996.

Lussana, C., Salvati, M. R., Pellegrini, U., and Uboldi, F.: Efficient high-resolution 3-D interpolation of meteorological variables for operational use, Adv. Sci. Res., 3, 105–112, doi:10.5194/asr-3-105-2009, 2009a.

Lussana, C., Salvati, M., and Uboldi, F.: Testing the performance of a spatial consistency test for data quality control, in: 9th EMS Annual Meeting, 9th European Conference on Applications of Meteorology (ECAM) Abstracts, held 28 September–2 October 2009 in Toulouse, France, http://meetings.copernicus.org/ems2009/, id. EMS2009-338., Vol. 1, p. 338, 2009b.

Lussana, C., Uboldi, F., and Salvati, M.: A spatial consistency test for surface observations from mesoscale meteorological networks, Q. J. Roy. Meteor. Soc., 136, 1075–1088, 2010.

Lussana, C., Ranci, M., and Uboldi, F.: A Quality-Control-Oriented Database for a Mesoscale Meteorological Observation Network, in: EGU General Assembly Conference Abstracts, Vol. 14, p. 9942, 2012.

Mearns, L., Katz, R., and Schneider, S.: Extreme High-temperature Events: Changes in Their Probabilities with Changes in Mean Temperature, J. Clim. Appl. Meteorol., 23, 1601–1613, 1984.

Pfahl, S. and Wernli, H.: Spatial coherency of extreme weather events in Germany and Switzerland, Int. J. Climatol., 32, 1863–1874, 2012.

Ranci, M. and Lussana, C.: An automated data quality control procedure applied to a mesoscale meteorological network, in: 9th EMS Annual Meeting, 9th European Conference on Applications of Meteorology (ECAM) Abstracts, held 28 September–2 October 2009 in Toulouse, France, http://meetings.copernicus.org/ems2009/, id. EMS2009-335., Vol. 1, p. 335, 2009.

Schär, C., Vidale, P. L., Lüthi, D., Frei, C., Häberli, C., Lininger, M. A., and Appenzeller, C.: The role of increasing temperature variability in European summer heatwaves, Nature, 427, 332–336, 2004.

Simolo, C., Brunetti, M., Maugeri, M., Nanni, T., and Speranza, A.: Understanding climate change-induced variations in daily temperature distributions over Italy, J. Geophys. Res.-Atmos., 115, D22110, doi:10.1029/2010JD014088, 2010.

Simolo, C., Brunetti, M., Maugeri, M., and Nanni, T.: Evolution of extreme temperatures in a warming climate, Geophys. Res. Lett., 38, L16701, doi:10.1029/2011GL048437, 2011.

Thunis, P. and Bornstein, R.: Hierarchy of Mesoscale Flow Assumptions and Equations, J. Atmos. Sci., 53, 380–397, 1996.

Toreti, A. and Desiato, F.: Changes in temperature extremes over Italy in the last 44 years, Int. J. Climatol., 28, 733–745, 2008.

Uboldi, F., Lussana, C., and Salvati, M.: Three-dimensional spatial interpolation of surface meteorological observations from high-resolution local networks, Meteorol. Appl., 15, 331–345, 2008.

Wilks, D. S.: Statistical Methods in Atmospheric Sciences: An Introduction, in: International Geophysics Series, Acad. Press, 1995.

Wulfmeyer, V. and Henning-Müller, I.: The climate station of the University of Hohenheim: analyses of air temperature and precipitation time series since 1878, Int. J. Climatol., 26, 113–138, 2006.

# 15

# Multivariate testing of spatio-temporal consistence of daily precipitation records

**H. Mächel[1] and A. Kapala[2]**

[1]Deutscher Wetterdienst, Klima und Umwelt, Frankfurter Str. 135, 63067 Offenbach, Germany
[2]Meteorological Institute, University Bonn, Auf dem Hügel 20, 53121 Bonn, Germany

*Correspondence to:* H. Mächel (hermann.maechel@dwd.de)

**Abstract.** The project KLIDADIGI of the German Meteorological Service (DWD) systematically rescues historical daily climate data of Germany by keying and imaging. Up to now, daily nearly gap-free precipitation time series at 118 locations for the period 1901–2000 are collected and extended by digitalization of handwritten protocols. To screen the spatio-temporal consistence of these raw data, we apply principal component analysis (PCA) in S (spatial) mode for daily precipitation records as well as for indices such as the number of rainy days above a certain threshold, intensity and absolute daily maximum in monthly, seasonal or annual resolution. Results of this screening test indicate that the PCA is a useful tool for detection of questionable stations and data preprocessing for further quality control and homogenization.

## 1 Introduction

For an operational quality control of climate records numerous standard methods are used at weather services and in climate research. The common procedural steps are: selection of outlier values using different thresholds (e.g. ±3 or 4 standard deviations), pairwise comparison of the target station with nearest neighbours using differences, ratios, linear regressions and frequency distributions (e.g. Feng et al., 2004; Vincente-Serrano et al., 2010, and their references). However, quality control of historical, especially of precipitation data is still a challenge if only few series with a sparse (irregular) spatial coverage are available.

Up to now, multivariate methods such as principal component analysis (PCA) are not used in the data preprocessing as a first step in quality control and homogenization.

In climatological studies the PCA is commonly applied to data reduction and to detection of leading spatial patterns/regions of similar temporal variability in different climate parameters. Detailed information to the method as well as to the discussion of the interpretation of PCA results can be found in Jolliffe (2002), Wilks (2006), and Compagnucci and Richmann (2008), for example. For analysis of monthly and daily precipitation variability and changes the PCA is used in several studies (e.g. White et al., 1991; Bonell and Sumner, 1992; and Brunetti et al., 2004, 2006a, b).

For example, Widmann and Schär (1997) classified daily precipitation records from Switzerland for the period 1901–1990 by means of unrotated and rotated PCA. The purpose of their study was, in parallel to the trend analysis, the homogenization of the precipitation series. According to the varimax rotation, three subregions in Switzerland were detected. Recently, Brienen et al. (2012) analysed the variability and change in several precipitation indices for winter (DJF) and summer (JJA) in Germany in the period 1901–2000 applying PCA in S (spatial) mode including varimax rotation to identify regions with similar precipitation behaviour.

This contribution will show the applicability of the PCA for identifying questionable stations in raw daily precipitation time series and derived indices by testing all stations of the network simultaneously.

This paper is subdivided in four parts: in Sect. 2 the database is presented. Section 3 shows the results of the PCA classification and some examples of erroneous data are given in Sect. 4. The results are discussed in Sect. 5.

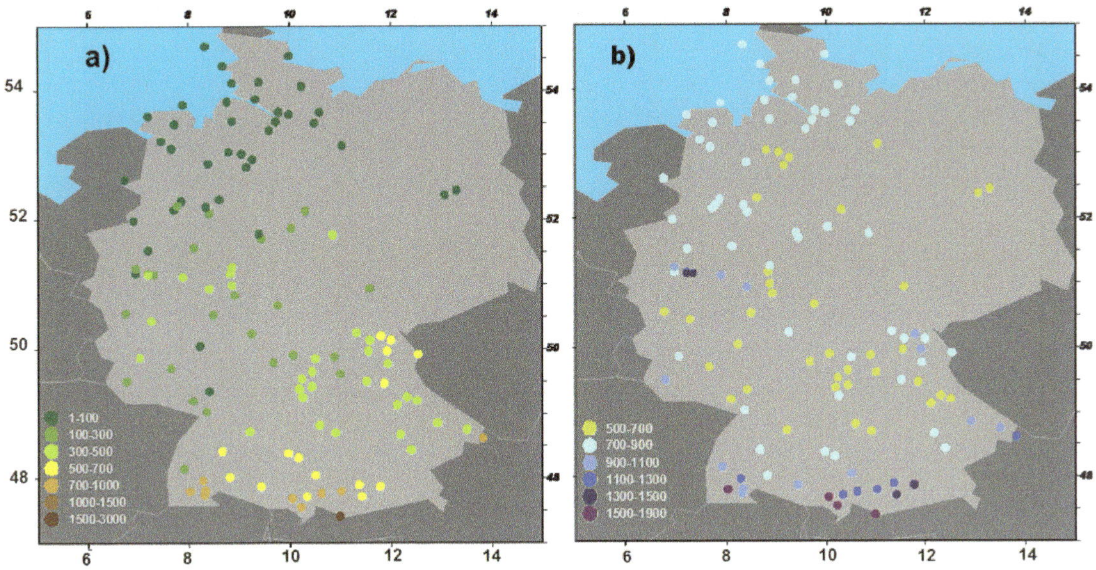


**Figure 1.** Digitized precipitation series: **(a)** location of the digitized records (the colours of the dots indicate the station elevation in metres), **(b)** mean yearly precipitation sum for individual stations for the period 1901–2000 (in mm $yr^{-1}$).

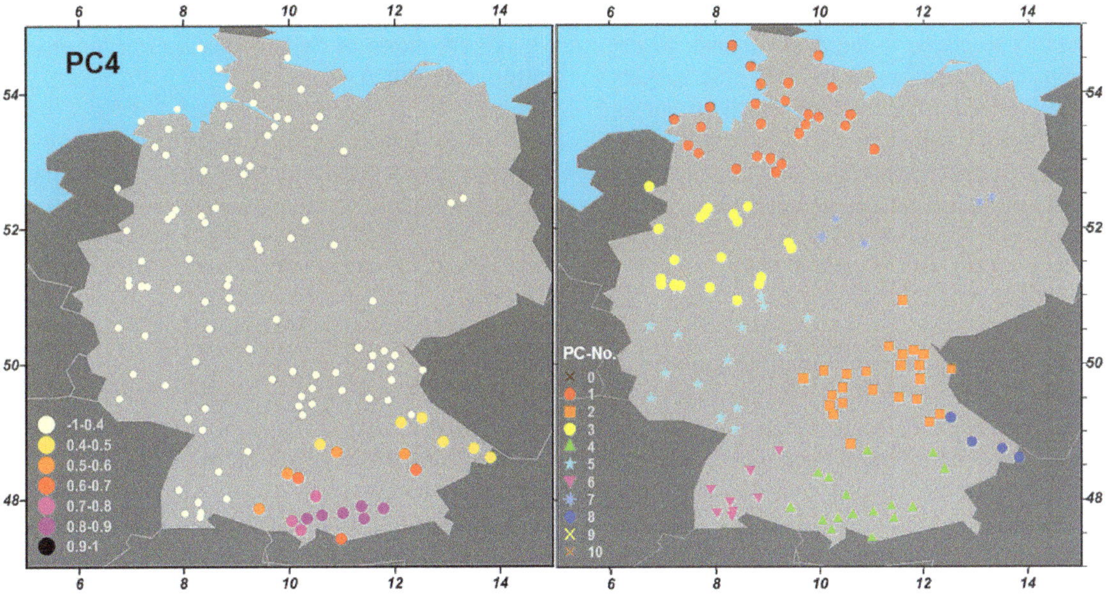


**Figure 2.** PCA of daily precipitation of all 36 525 days in the period 1901–2000. Examples for spatial patterns of the extracted rotated principal components which correspond to the loadings (here: correlation coefficients) of the original records. Left: PC Nr. 4, right: all 8 rotated PCs (PC9 and PC10 contain only the rest variances with loadings < 0.4).

## 2 Database

Most of the digitally available daily precipitation series in Germany start in 1931 or later. Therefore the German Meteorological Service (DWD) decided to digitize hand-written protocols to extend the database. 118 daily precipitation time series for the period 1901–2000 with few gaps (lower than 3 yr or 1095 days, mainly in 1945–1946) are selected to test the PCA for detecting outliers and erroneous data. The location, the station elevation above sea level and the mean yearly precipitation sum for these 118 stations are depicted in Fig. 1. In Eastern Germany the digitally available daily precipitation series starts in 1951 or 1969 and only few series match the above mentioned requirements.

The DWD precipitation stations were equipped since the 1890s with Hellmann rain gauges with a 200 $cm^2$ receiving orifice and brass ring (reading accuracy of 0.1 mm). All rain gauges were protected against evaporation. In the cold season

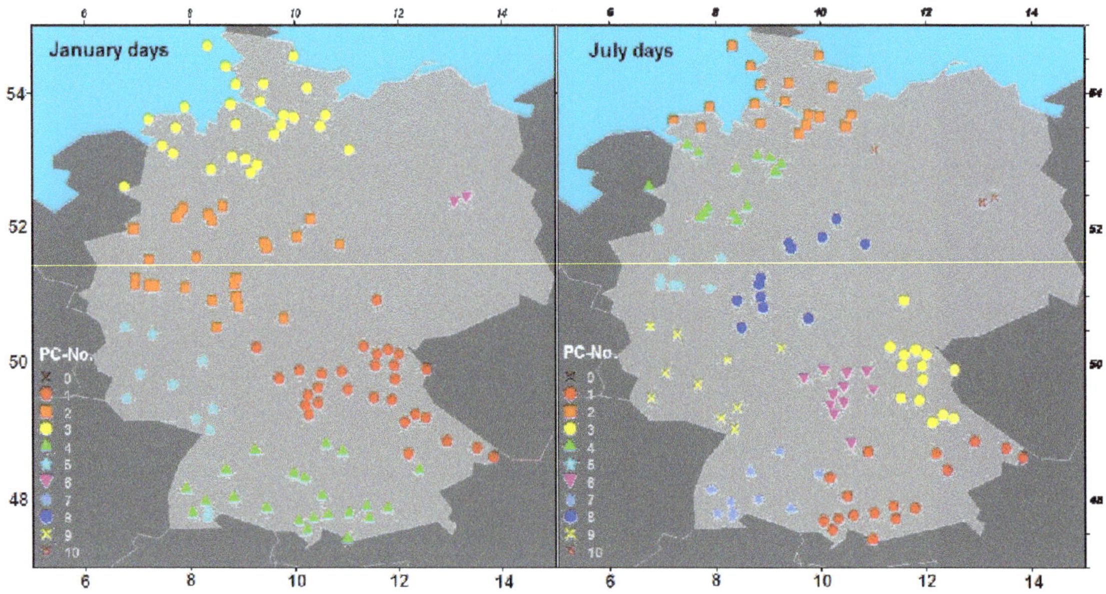


**Figure 3.** Same as Fig. 2, but for all January days (left) and all July days (right) in the period 1901–2000. In January the two stations within the PC4 pattern seems to be questionable and their original protocols should be checked.

a snow cross had to be placed in the gauge funnel. In the 1990s they were replaced at main stations by tipping-buckets combined with drop-counting (for low rain rates) rain gauges with the same orifice of 200 cm$^2$. The accuracy of these automatic precipitation measurements were 0.01 mm min$^{-1}$. The observational time was at the beginning 07:00 LT (local time), from 1 January 1979 the observational time changed to 07:30 CET (Central European Time) (06:30 UTC, Coordinated Universal Time). According to Hellmann (1906), "the changes in observational time and instrumentation are less pronounced than re-locations"; furthermore, "the accuracy of the measurements depends more strongly on the location of the gauges and on the special diligence of the observer than on the used gauge type." On average, the observer changes occur every 10–15 yr often combined with station relocations. A station is continued after relocation with the same identification number if the horizontal distance is less than 5 km, the altitude differs less than 50 m and both locations belong to the same river catchment.

The systematic and documented quality assurance of the data began in 1979 with the introduction of IT (information technology)-supported verification methods (QUALKO, used also in other weather services). With this procedure the daily data was checked automatically for inner, temporal and spatial consistency and flagged. If necessary the erroneous data was manually corrected and flagged accordingly. Since the beginning of the meteorological service in Germany in 1848, the observational protocols are checked manually with neighbouring stations in real time, i.e. in the following month. The newly digitized data is not quality controlled and the following sections should demonstrate a possibility to do this (Mächel et al., 2009).

## 3 Classification of daily precipitation records by means of PCA

To check the spatio-temporal consistence of the precipitation records from the above mentioned 118 stations and to identify questionable stations in this data set the PCA including varimax rotation of the PCs (principle components) is applied to the correlation matrix of different precipitation characteristics (e.g. daily records of different months, single days and total monthly or yearly amount) for the period 1901–2000.

Since the PCA is widely used this method will not be described here, for details see Richman and Gong (1999), Jolliffe (2002), Wilks (2006), Compagnucci and Richmann (2008) and many others.

As shown by Brunetti et al. (2006b), Widman and Schär (1997), and Brienen et al. (2012), for example, the varimax rotation of the PCs leads to a selection of well-separated spatial patterns (regions) of similar precipitation behaviour based on PC-loadings. In this step, single questionable stations can be easily detected. Questionable stations are: outstanding stations within coherent spatial patterns and stations which could not be assigned to one of the extracted PCs.

In the following, the detected outlier stations are candidates for a manual checking of their precipitation records with original reports. The most frequent inconsistencies in the raw precipitation records result from typing errors, inaccurate specification of the days with no precipitation and not-observed (missing) precipitation and measurements over several days (accumulated precipitation amounts).

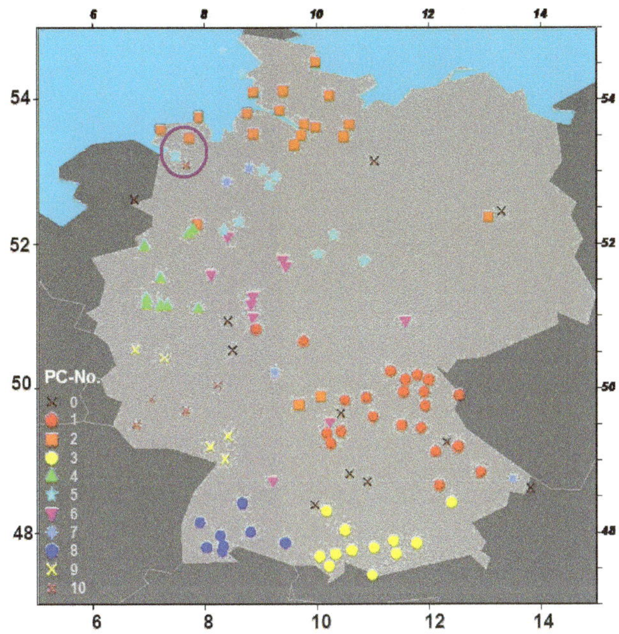


**Figure 4.** Same as Fig. 2 but, for mean precipitation intensity in July of the period 1901–2000.

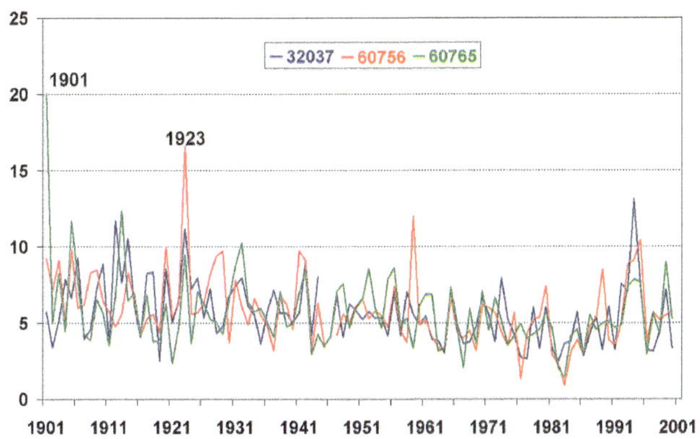


**Figure 5.** Example of 3 neighbouring time series of mean precipitation intensity (mm $day^{-1}$) in July which are classified to different PCs (marked by the circle in Fig. 4).

According to Kaiser's criterion (eigenvalue = 1) and screen plots (Wilks, 2006) we select a maximum of 10 PCs for the varimax rotation of the components. To simplify the interpretation of these patterns the loadings are rescaled according to Wilks (2006) by multiplying them with the square root of the eigenvalue. Accordingly, the loadings are expressed in correlation coefficients of the original data at individual stations with the time series of the corresponding principal components (scores). The single station is assigned to one PC according to the maximum correlation. Figure 2 illustrates this for the extracted 8 PCs for daily precipitation (without annual cycle and standardized) of all 36 525 days in the period 1901–2000. The stations with correlation coefficients above 0.5 in the left panel in Fig. 2 correspond to the green triangles (PC4) of the right panel. In this case we obtain well-separated patterns (regions) which correspond with the topography that influences the precipitation. In some cases, however, neighbouring stations are assigned to different PCs. This seems to suggest that the nearest neighbour is not necessarily adequate for paired comparison of the precipitation records.

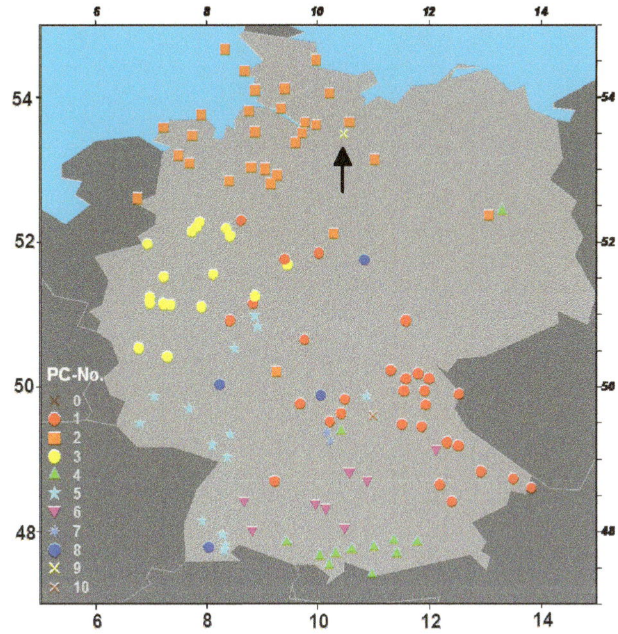


**Figure 6.** Same as Fig. 2, but for precipitation at individual days; here 9 February for the years 1901–2000.

In the following Figs. 2–6 the loadings (correlation coefficients) < 0.4 are cut off (cf. Richman and Gong, 1999). Coloured symbols indicate the PC's rank according to their explained variance. A station is classified to one of the 10 PCs if the PC time series (PC scores) explains at minimum 16 % of the variance of the original time series (loadings/correlation coefficients $r = 0.4$). If a station does not pass this criterion, it is marked by PC-No. = 0.

If the PCA is applied to daily precipitation of individual months the lower variability of the daily precipitation in winter is manifested through only 6 PCs in January (Fig. 3). In July 10 PCs are needed to classify all stations.

## 4 Examples of detected errors

If an individual station does not fit in the well-separated regions, it can be assumed that the data contain either errors or real extraordinary events. In such cases we manually compare the data with original reports.

For the detection of a possible existence of over several days accumulated precipitation measurements, the error sensitive index precipitation intensity, defined as total

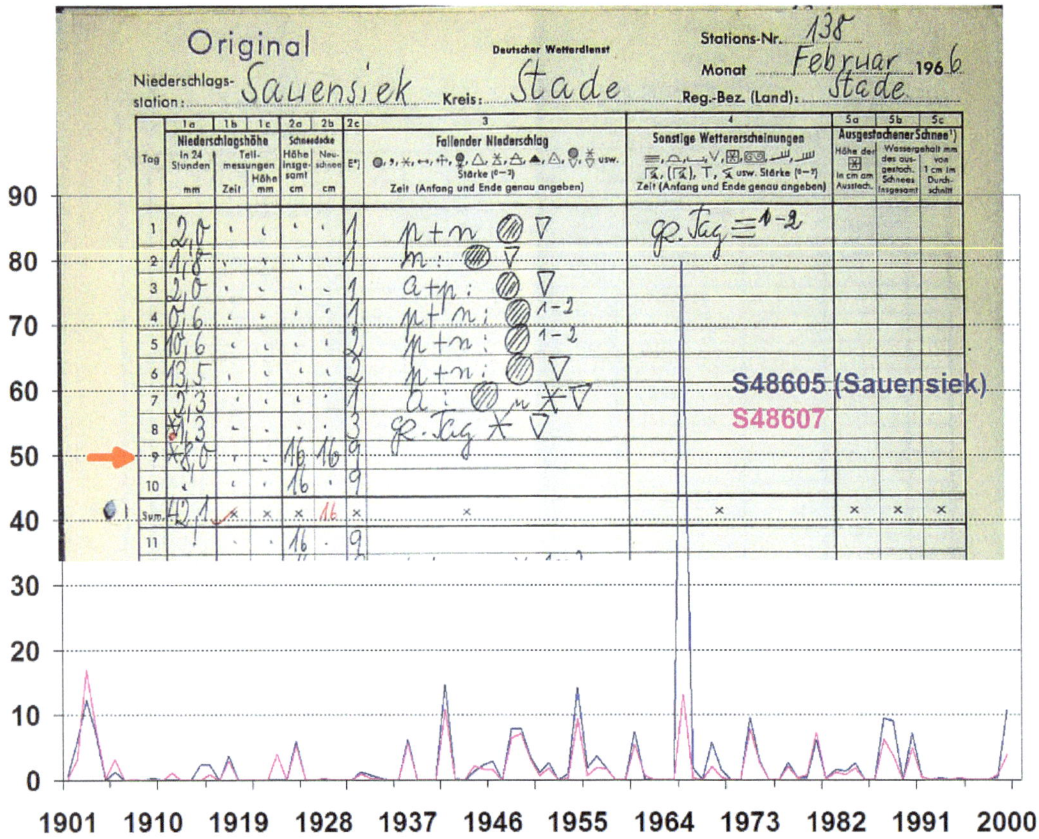


**Figure 7.** Precipitation time series (mm day$^{-1}$) of two neighbouring stations for 9 February for the years 1901–2000 as an example of an outlier station. Station 48605 and its neighbours are classified into different PCs (see arrow in Fig. 6). The image shows the original report of station 48605 for February 1966.

precipitation divided by the number wet days (precipitation amount > 0.0) for single months, seasons or a year, is used.

As can be seen in Fig. 4, the delineation of regions of precipitation intensity in July 1901–2000 is not as clear as in the case of daily precipitation amounts in July (cf. Fig. 3, right). Some of the stations differ from their neighbours and several of the stations have not passed the criterion of 0.4 correlation coefficient. This suggests that the number of wet days and/or the precipitation amounts at some stations may be incorrect. For example, 3 stations are striking in the northwestern corner of Germany marked by the purple circle. Their time series of precipitation intensity show (Fig. 5) some peaks that are verified with the original reports; on 22 July 1901 station 60765 reported a daily value of 78.5 mm, the other two stations only 16 mm (station 32037) and 24 mm (station 60756). Whether the value of 78.5 mm is real or not, cannot be clearly decided without additional information. In fact, some neighbouring stations with shorter time series (not digitally available at this time) also show high values. Therefore the detected peak value of 78.5 mm seems to be real. The high value in 1923 at station 60756 is the result from 2, over 3 days of accumulated values and therefore the number of wet days is lower than at the neighbouring stations. This erroneous amount should be disaggregated in proportion to the neighbouring stations.

The PC-analysis is also applied to a single day in a year for the period 1901–2000. An example for 9 February is shown in Fig. 6. The PCA of individual days enables easy identification of outlier stations and checking of their records if original reports are available. One station in northern Germany, marked by the arrow, is striking. Figure 7 shows the solution. In the time series of daily precipitation on 9 February 1966 an outlier value of 80.0 mm is found which is extraordinary compared to other neighbours. According to the original report only 8.0 mm were measured on this day.

## 5 Conclusions

The quality control of historical daily data is very time consuming. Due to the high daily precipitation variability only large errors can be found. To reduce time and efforts, the screening of the spatio-temporal consistence of the considered data set by means of the PCA can be used as a first step of quality control.

One advantage of the PCA is that this method provides regions with similar behaviour of the data or select stations with similar variability and, therefore, errors could easily be detected. As show in the few examples, the probability of finding questionable stations in the considered data is higher by checking records of every individual day and different indices calculated from the daily records than the precipitation amount itself (cf. Figs. 4 and 5).

An error sensitive index is the precipitation intensity, the ratio between the monthly precipitation sum and the number of precipitation days above a certain threshold (zero or 1.0 mm). By means of this index it is easy to find months with accumulated values, if an observer did not measure continuously every day (perhaps only on work days or not during holidays). Such values can be disaggregated in proportion to their neighbouring stations.

In general, the PCA is an useful tool for a quick check of spatio-temporal coherence of climate records including identification of conspicuous stations. It can be used as a first step in quality control and homogenization of the data. After the first correction of the erroneous data it can be repeated to assess the improvement in the spatio-temporal consistence. Furthermore, the highest loadings of each PC enables one to identify the leading or most representative station in the selected region that can be used as a reference station for a homogenization procedure. However, the disadvantage of PCA is that the data has to be without gaps, in other words, the missing values in one of the series causes that this date be omitted from the analysis. To handle this problem, in a first step monthly data can be analysed for the whole country to find regions. In the second step only stations within such a region can be used for testing the spatio-temporal consistence of daily data or different indices.

However, neither PCA nor other quality control procedure (e.g. regression with a neighbouring station) can guarantee that simultaneously occurred errors/inconsistencies in several time series can be detected.

**Acknowledgements.** The authors would like to thank Jennifer Lenhardt for her assistance during the conference and the reviewers.

Edited by: M. Brunet-India

## References

Bonell, M. and Sumner, G.: Autumn and winter daily precipitation areas in Wales, 1982–1983 to 1986–1987, Int. J. Climatol., 12, 77–102, doi:10.1002/joc.3370120108, 1992.

Brienen, S., Kapala, A., Mächel, H., and Simmer, C.: Regional centennial precipitation variability over Germany from extended observation records, Int. J. Climatol., doi:10.1002/joc.3581, in press, 2012.

Brunetti, M., Maugeri, M., Monti, F., and Nanni, T.: Changes in daily precipitation frequency and distribution in Italy over the last 120 years, J. Geophys. Res., 109, D05102, doi:10.1029/2003JD004296, 2004.

Brunetti, M., Maugeri, M., Monti, F., and Nanni, T.: Temperature and precipitation variability in Italy in the last two centuries from homogenised instrumental time series, Int. J. Climatol., 26, 345–381, doi:10.1002/joc.1251, 2006a.

Brunetti, M., Maugeri, M., Nanni, T., Auer, I., Böhm, R., and Schöner, W.: Precipitation variability and changes in the Greater Alpine Region over the 1800–2003 period, J. Geophys. Res., 111, D11107, doi:10.1029/2005JD006674, 2006b.

Compagnucci, R. H. and Richman, M. B.: Can principal component analysis provide atmospheric circulation or teleconnection patterns?, Int. J. Climatol., 28, 703–726, doi:10.1002/joc.1574, 2008.

Feng, S., Hu. Q., and Qian, W.: Quality control of daily meteorological data in China, 1951–2000: a new dataset, Int. J. Climatol., 24, 853–870, 2004.

Hellmann, G.: Die Niederschläge in den Norddeutschen Stromgebieten, Band I, Dietrich Reimer Verlag, Berlin, 425 pp., 1906.

Jolliffe, I. T.: Principal Component Analysis, Springer, New York, doi:10.1007/b98835, 2002.

Mächel, H., Kapala, A., Behrendt, J., and Simmer, C.: Rettung historischer Klimadaten in Deutschland: das KLIDADIGI-Projekt des DWD, Klimastatusbericht 2008 (Climate Status Report), Deutscher Wetterdienst, Offenbach, 103–118, http://www.ksb.dwd.de/ (last access: 12 January 2013), 2009.

QUALKO: www.dwd.de/nkdz – Quality assurance, last access: 20 April 2013.

Richman, M. B. and Gong, X.: Relationships between the definition of the hyperplane width to the fidelity of principal component loading patterns, J. Climate, 12, 1557–1576, doi:10.1175/1520-0442(1999)012<1557:RBTDOT>2.0.CO;2, 1999.

Vicente-Serrano, S. M., Begueria, S., Lopez-Moreno, J. I., Garcia-Vera, M. A., and Stepanek, P.: A complete daily precipitation database for northeast Spain: reconstruction, quality control, and homogeneity, Int. J. Climatol., 30, 1146–1163, doi:10.1002/joc.1850, 2010.

White, D., Richman, M., and Yarnal, B.: Climate regionalization and rotation of principal components, Int. J. Climatol., 11, 1–25, doi:10.1002/joc.3370110102, 1991.

Widmann, M. and Schär, C.: A principal component and long-term trend analysis of daily precipitation in Switzerland, Int. J. Climatol., 17, 1333–1356, doi:10.1002/(SICI)1097-0088(199710)17:12<1333::AID-JOC108>3.0.CO;2-Q, 1997.

Wilks, D. S.: Statistical Methods in the Atmospheric Sciences, Elsevier, San Diego, 2006.

# 16

# Evaluating multi-scale precipitation forecasts using high resolution analysis

**C. Wittmann, T. Haiden, and A. Kann**

Central Institute for Meteorology and Geodynamics, Vienna, Austria

**Abstract.** The SAL (Structure, Amplitude, Location) method is used for verification of precipitation forecasts at horizontal grid spacings ranging from 2.5 km to 25 km, using a high-resolution 1 km precipitation analysis as a reference. The verification focuses on a summertime period with predominantly convective precipitation. The verification domain contains lowland as well as alpine areas. Evaluation of the individual SAL components shows that with regard to area mean values ($A$) the benefit of high resolutions models becomes apparent only in high impact weather situations. For the summertime period studied, the subjective impression of better structured precipitation fields ($S$) in higher resolution models can generally be confirmed. The most significant improvement appears to be associated with explicit simulation of deep convection.

## 1 Introduction

An increasing number of operational numerical weather prediction (NWP) models are run with horizontal grid spacings in the 1–5 km range, generating highly structured forecast fields. Proper evaluation of the actual benefit of such forecasts compared to those in the more classical 10–20 km range is not straightforward, especially for precipitation. The use of conventional scores, based on point-to-point comparison of gridded data, or for station locations, penalizes the model that generates higher field variances. Spatial up-scaling to a common reference grid solves the variance problem but masks potential structural skill of the high-resolution model. Similarly, temporal up-scaling such as verification of 24-h totals masks potential timing skill concerning, for example, the diurnal cycle of convection, or frontal passages.

Prompted by these issues, several pattern-oriented verification methods have recently been developed (Casati et al., 2008). Gilleland et al. (2009) provide a comprehensive overview and intercomparison of these methods which they categorize into "filtering" and "displacement" types. The primary purpose of this study is to use the displacement-type method SAL (which stands for structure, amplitude, and location) developed by Wernli et al. (2008), to evaluate precipitation forecasts during the convective season over a wide range of model resolutions against a high-resolution precipitation analysis. Previously, raingauge data interpolated to a 5 km grid (Früh et al., 2007) and a 10 km grid (Wernli et al., 2008), or radar rainfall data were used, the latter however only for a 9-day period (Wernli et al., 2009). Also, we compare NWP models over a broader range of resolutions than has been done in the past, with grid spacings ranging from 2.5 to 25 km. Forecasts of 3-hourly accumulated precipitation are considered over a 2 month period in the summer of 2009, for different domains covering lowland and mountain areas.

Section 2 describes the high resolution precipitation analysis of the INCA (Integrated Nowcasting through Comprehensive Analysis) system, and is followed by an overview of the different NWP models and a summary of observed precipitation characteristics during the chosen period in Sect. 3. The verification method SAL is briefly discussed in Sect. 4, and results and conclusions are presented in Sects. 5 and 6, respectively.

## 2 INCA precipitation analysis

The INCA precipitation analysis is a combination of station data interpolation, including elevation effects, and radar data. It is designed to combine the strengths of both observation types, the accuracy of the point measurements and the spatial structure of the radar field. In the following, the analysis method for 15-min precipitation amounts is briefly described. A detailed description of the entire analysis and nowcasting system is given in Haiden et al. (2009).

### 2.1 Interpolation of station data

The irregularly distributed point precipitation values measured at the station locations are interpolated onto the regular $1 \times 1$ km INCA grid using inverse-distance-squared weighting (IDW). To reduce the occurrence of bulls-eyes, only the nearest 8 stations are taken into account in the interpolation. Furthermore, a modification to the classical IDW method has been introduced. It takes into account the inhomogeneous azimuthal distribution of stations around the grid point in question and reduces the relative weight of stations located "behind" (i.e. at a similar azimuth to) nearer stations. The resulting field is denoted by $P_{\mathrm{STAT}}(i, j)$.

### 2.2 Climatological scaling of radar data

The radar data, which are available at 5 min intervals, are aggregated in time and bilinearly interpolated onto the INCA grid. Since the radar field is strongly range-dependent and contains biases due to topographic shielding it must be scaled before use in a precipitation analysis. In a first step, a "climatological" scaling is performed. Operationally, a 3-month temporal average of scaling factors RFC $(i, j)$ centered at the actual month is used. This climatological scaling only partially corrects the reduced radar return in the case of snowfall as compared to rain.

In order to compensate for some of the artifacts in the radar field caused by topographic shielding of the radar beam, the interpolated scaling factor is replaced by a local scaling factor in regions where the radar beam is strongly shielded (indicated by beamlike structures with high local scaling factors). The local scaling factor is the ratio of the monthly accumulated precipitation gained from the interpolated station observations to the accumulated radar precipitation at the respective gridpoint.

### 2.3 Re-scaling of radar data using the latest observations

In a next step the climatologically scaled radar field is rescaled on the basis of a comparison at analysis time of station observations and radar values at the stations. In this comparison, a spatial shift of a maximum of 4 km in either direction between the station and the corresponding radar pixel is allowed to take into account effects due to the finite settling time of hydrometeors, effects of wind-drift, etc. The scaling for a gridpoint is a weighted average of the ratio between station and radar precipitation at the nearest stations, where the weight decreases with increasing distance, with increasing difference in climatological scaling, and with decreasing precipitation at the station (relative to the precipitation at the gridpoint). The resulting field is denoted by $P^{**}_{\mathrm{RADAR}}(i, j)$ (cf. Haiden et al., 2009).

### 2.4 Final combination

The two precipitation fields $P_{\mathrm{STAT}}(i, j)$ and $P^{**}_{\mathrm{RADAR}}(i, j)$ are finally combined to a field $P_{\mathrm{INCA}}(i, j)$ that gives a better estimate of the precipitation distribution than each individual field. The combination is obtained through a weighting relationship

$$P_{\mathrm{INCA}}(i,j) = P_{\mathrm{STAT}}(i,j) + v\left[P^{**}_{\mathrm{RADAR}}(i,j) - P^{**}_{\mathrm{RADSTAT}}(i,j)\right], \quad (1)$$

where the weight $v$ is given by

$$v(i,j) = \begin{cases} 1 & \mathrm{RFC} < \mathrm{RFC}_0 \\ \exp\left[-\ln(2)\left(\frac{\mathrm{RFC}-\mathrm{RFC}_0}{\mathrm{RFC}_H-\mathrm{RFC}_0}\right)^2\right] & \mathrm{RFC} \geq \mathrm{RFC}_0 \end{cases} . \quad (2)$$

In Eq. (2), RFC is the spatially and seasonally varying climatological scaling factor described in Sect. 2.2. The auxiliary field $P^{**}_{\mathrm{RADSTAT}}(i, j)$ is created by interpolating onto the grid, analogous to the station observations, the scaled radar values at the station locations. The threshold value $\mathrm{RFC}_0$, above which the weight of the radar begins to decrease, is 3. The value of $\mathrm{RFC}_H$, at which the radar weight has decreased to one-half, is 5.

Figure 1 shows an example of the stepwise procedure and final analysis from the June 2009 flood event in Austria. Note the large difference between un-scaled radar and station interpolation (top panels), and the importance of the final combination (lower right panel) as a means to smoothly connect areas seen by radar with those only covered by stations. If the final combination step is not made, "edges" remain in the analysis (lower left panel).

The INCA precipitation analysis reproduces (within limits imposed by the grid spacing) the observed values at the raingauge locations. The quality of the analysis as determined by cross-validation based on single-station denial experiments, is superior to raingauge-only IDW interpolation. As expected, the benefit of incorporating radar data is largest for convective precipitation, where the reduction of mean absolute error (MAE) compared to IDW is 20–50%. For stratiform precipitation, the reduction of MAE is typically 5–10%. A paper describing crossvalidation results of the INCA analysis scheme has been submitted. A parameterization of the elevation dependence of precipitation, which was introduced to the system in 2007, is described in Haiden and Pistotnik (2009).

Any precipitation analysis contains measurement errors, sampling and representativeness errors, and uncertainties due to assumptions made in the analysis scheme. Following Pappenberger et al. (2009) we roughly estimate the raingauge measurement errors (undercatch) to be of the order of 10% for lowland and valley stations, and 20% for some of the more exposed mountain stations. Note that we are dealing with cases of rainfall only. According to Skok and Vrhovec (2006), errors of up to 50% for point values can be introduced by the analysis itself. However, the SAL method is equivalent to using spatially aggregated values, which reduces the

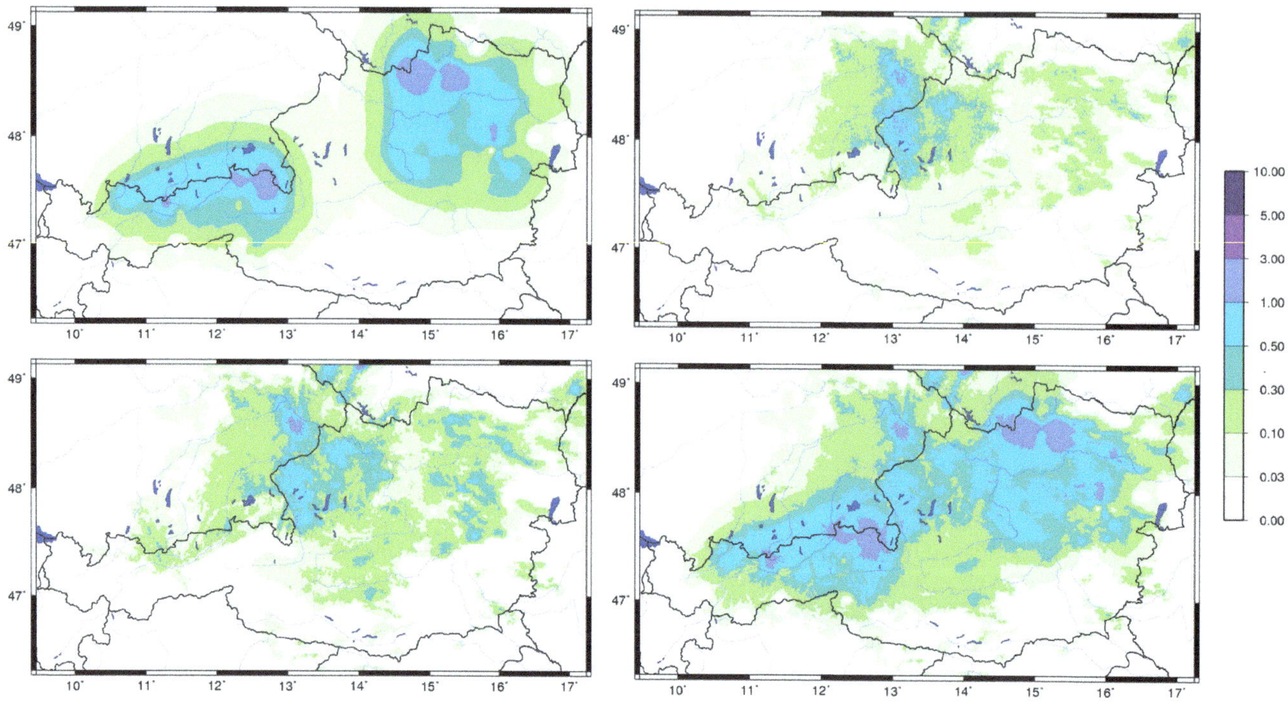


**Figure 1.** Example of a 15-min INCA precipitation analysis (23 Jun 2009, 00:15–00:30 Z) during a flooding event in Central Europe. Upper left panel: pure station interpolation, upper right panel: uncorrected radar field, bottom left panel: corrected radar field, bottom right panel: final INCA precipitation analysis. Units are mm, with color scale on the right.

error. For a comprehensive analysis and discussion of the uncertainty issue we refer to Pappenberger et al. (2009) and references therein.

## 3 Model data

The operational forecast model used at ZAMG is ALADIN (Aire Limitée Adaption Dynamique Developpement International), a spectral Limited Area Model (LAM) which is being developed within the ALADIN consortium, a cooperation of several mainly European national weather services. ALADIN is part of a software library consisting of several models, which allows to choose among various physical schemes and parameterizations. The operational ALADIN version implemented at ZAMG uses the ALARO physics package. It has been developed to address problems in the difficult grid spacing range between 3 and 7 km, where deep convection starts to be at least partly resolved. The ALARO package comprises a prognostic large-scale cloud and precipitation scheme and prognostic convection scheme named 3MT, which stands for Modular Multi-scale Microphysics and Transport. Detailed information about ALARO and 3MT can be found in Gerard et al. (2009).

Apart from the operational ALADIN model (hereafter referred to as ALA-AUT), several other model configurations are used in this verification study. Table 1 gives an overview of the main characteristics of the different models: a 5 km hydrostatic ALADIN version (ALA5) using ALARO physics; a 5 km non-hydrostatic ALADIN version (ALA5-NH); a 6.5 km ALADIN version (ALA-EUR) using ALARO physics; and the 2.5 km AROME (Applications of Research to Operations at Mesoscale) model. These models are used at ZAMG in a pre-operational environment. AROME is a new model which more or less represents the merger of the physical package from the research model Méso-NH (Lafore et al., 1998) and the non-hydrostatic version of ALADIN. The microphysical scheme used in AROME is ICE3 (Pinty and Jabouille, 1998).

In addition to the models listed in Table 1, two more models are considered: The deterministic global model of the European Center for Medium-Range Weather Forecasts (ECMWF, 25 km horizontal resolution) and INCA precipitation forecasts. INCA forecasts are computed on a two dimensional grid with 1 km resolution and start with kinematic extrapolation of analyzed precipitation fields (as described in the previous section) in the nowcasting range (up to +2 h). Outside the nowcasting range the extrapolated fields smoothly merge into a combination of ALA-AUT and ECMWF precipitation forecasts. The combination of these two models uses weights computed on the basis of several years of archived precipitation forecasts. More details about the INCA precipitation forecast are given by Haiden et al. (2009).

**Table 1.** Main characteristics of the different limited area models used at ZAMG.

| | ALA-AUT | AROME | ALA5-NH | ALA5 | ALA-EUR |
|---|---|---|---|---|---|
| Timestep [s] | 415 | 60 | 207 | 207 | 285 |
| Coupling model | ARPEGE | ALA-AUT | ARPEGE | ARPEGE | ECMWF |
| Initialization | ARPEGE | ALA-AUT | ARPEGE | ARPEGE | ECMWF |
| Resolution [km] | 9.6 | 2.5 | 4.9 | 4.9 | 6.5 |
| Grid size | 300×270 | 432×320 | 540×512 | 540×512 | 720×810 |
| Levels | 60 | 60 | 59 | 59 | 45 |
| Forecast Range [h] | 72 | 30 | 48 | 48 | 72 |
| Convection parameterization | Yes (3MT) | No | Yes (3MT) | Yes (3MT) | Yes (3MT) |
| Microphysics | ALARO | ICE3 | ALARO | ALARO | ALARO |
| Kernel H/NH | hydrostatic | non-hydrostatic | non-hydrostatic | hydrostatic | hydrostatic |

The main objective of this study is to evaluate model performance with respect to convective precipitation, so June and July 2009 were chosen as verification period. The number of observed days with precipitation was 76–91% out of the total of 61 days, with higher values found in the more mountainous domains. The corresponding numbers for the forecasts are 87–100%. Only 00:00 UTC runs are considered. This is true for all models except for INCA where the 06:00 UTC runs are taken into account. This is done to allow a comparison of all models available to the forecaster around or shortly after 06:00 UTC in the morning. The common forecast range is defined by the model with the shortest integration time, which in our case is AROME, so all models are compared up to forecast time +30 h, using 3 h accumulated precipitation intervals.

In order to quantify the proportion of days with convective precipitation with respect to the total number of days with precipitation for the chosen period, lightning data was used. Assuming that a day can be classified as "convective" when more than two (cloud-to-ground) lightening strokes are detected within the considered area, the percentage for convective precipitation days ranges from 76–82% with higher values again found in the mountainous domains. The reason for choosing two lightning strokes as a threshold is to reduce erroneous detections. This type of characterization does not distinguish between days with convective cells growing and decaying within the considered area and convective systems moving into and/or over the domain (often related to large scale systems). A subjective, semi-quantitative estimation of this partitioning is based on visual inspection of INCA analyses of 24-h accumulated precipitation. Out of the total number of days with precipitation, the percentage of days with local convection is about 40%, the percentage of days with large scale systems moving over the considered domains is about 20%. The remaining days can not be clearly assigned, showing a mixture of both types. This is valid for the mountainous areas. For the lowlands the proportion of days with local convection is smaller (about 30%), whereas the days with advective convection occur with similar frequency.

## 4 Verification method

In order to allow a fair comparison of the precipitation forecasts coming from models running on different horizontal (and vertical) resolutions it was decided to use the verification package SAL, which stands for Structure, Amplitude and Location. In the following, a short description of the main components of SAL is given. A comprehensive description can be found in Wernli et al. (2008).

SAL is an object-based method which allows to evaluate QPFs on a given geographic domain according to three criteria: structure ($S$), amplitude ($A$) and location ($L$). $A$ is a measure of the deviation of the areal mean QPF relative to the observed value. $S$ gives information on whether the precipitation objects created by the model correspond to the observed objects in terms of size and shape. Finally $L$ yields information about the displacement of the precipitation objects with respect to the observed objects.

To calculate the $S$ and $L$ components it is necessary to identify individual precipitation objects in the forecast and the observation fields. In SAL this is done by using a threshold $R^* = fR_{\max}$, where $R_{\max}$ denotes the maximum precipitation amount found in the domain. Starting from a local precipitation maximum (provided it is larger than $R^*$), all surrounding grid points with values $R_{ij} > R^*$ are considered to be part of the actual precipitation object. Following Wernli et al. (2008) the empirical factor $f$ is set to 1/15. According to their sensitivity analysis, the $S$ and $L$ results are not particularly sensitive to the exact value of this factor. In cases where the threshold $R^*$ happens to be close to the minimum value at the "saddle" between two maxima, the object definition becomes sensitive to $f$, but as these cases comprise only a small fraction of the sample their effect on the overall statistics will be small.

### 4.1 Amplitude *A*

The amplitude component $A$ is defined as the normalized difference of the forecasted and observed domain average precipitation ($D(R_{\mathrm{mod}})$ and $D(R_{\mathrm{obs}})$). Thus, $A$ measures the quality of the total precipitation amount simulated by the model. This normalized difference is computed according to:

$$A = \frac{D(R_{\mathrm{mod}}) - D(R_{\mathrm{obs}})}{0.5(D(R_{\mathrm{mod}}) + D(R_{\mathrm{obs}}))}. \quad (3)$$

Values of $A$ range from −2 to +2. Positive values indicate an overestimation of the domain-averaged precipitation coming from the model, negative values an underestimation, and 0 stands for a perfect forecast.

### 4.2 Location *L*

The location component $L$ is the sum of two components, named $L_1$ and $L_2$. The first part $L_1$ measures the normalized distance between the centers of mass of the modeled and the observed precipitation field ($\boldsymbol{x}(R_{\mathrm{mod}})$ and $\boldsymbol{x}(R_{\mathrm{obs}})$)

$$L_1 = \frac{|\boldsymbol{x}(R_{\mathrm{mod}}) - \boldsymbol{x}(R_{\mathrm{obs}})|}{d}, \quad (4)$$

where $d$ stands for the maximum distance found in the given domain between two boundary points. Values for $L_1$ range from 0 to 1, where for $L_1 = 0$ the centers of mass for model and observed precipitation fields are equal.

The second part $L_2$, ranging from 0 to 1, can be described as the weighted mean distance between the centers of mass and the individual precipitation objects ($r(R_{\mathrm{mod}})$ and $r(R_{\mathrm{obs}})$) and is computed according to

$$L_2 = 2\frac{|r(R_{\mathrm{mod}}) - r(R_{\mathrm{obs}})|}{d}. \quad (5)$$

The total location component $L = L_1 + L_2$ can assume values between 0 to 2, where 0 would only be obtained when the total mass centers *and* the averaged distance between individual objects and the total mass center are identical in the observed and in the model field.

### 4.3 Structure *S*

The structure component $S$ is defined in such a way as to give information about the quality of the forecast with respect to size and shape of the precipitation objects. This is done by calculating a scaled volume $V_n$ for every object $R_n$. Finally a weighted average Volume $V$ is calculated separately for the forecast and observational fields ($V(R_{\mathrm{mod}})$ and $V(R_{\mathrm{obs}})$). The final score for structure is then calculated through

$$S = 2\frac{V(R_{\mathrm{mod}}) - V(R_{\mathrm{obs}})}{V(R_{\mathrm{mod}}) + V(R_{\mathrm{obs}})} \quad . \quad (6)$$

$S$ again takes a continuum of values from −2 to +2. Positive values mean that the predicted forecast objects are too large or widespread with respect to the observed objects. Negative values for $S$ stand for predicted objects being too small or peaked.

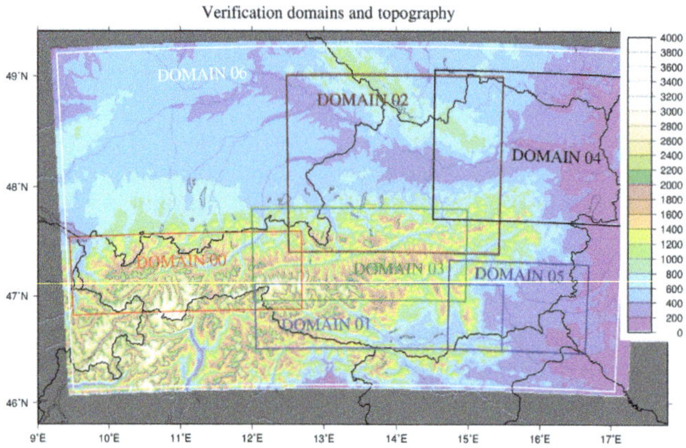


**Figure 2.** Verification domains numbered 00 through 06, and topography. Height in meters with color scale on the right.

### 4.4 Verification domains

The SAL verification is performed for several rectangular domains over Austria covering either Alpine areas or flatland areas. Figure 2 shows the different domains together with the topography. In order to make the various forecasts comparable and usable for the SAL software package, the precipitation fields are interpolated to the same grid, which is the 1 km INCA grid in this case. To discuss the results in Sect. 5 two representative verification domains have been chosen: domain 00, representing an Alpine area in the western part of Austria, and domain 04, an area consisting of mainly rather flat terrain.

## 5 Results

The SAL verification has been performed for the period 20090601–20090731 based on 00:00 UTC model runs (and 06:00 UTC runs for INCA) up to a forecast time of +30 h, considering 3-h accumulations. In the following subsections the results are discussed with regard to each of the individual SAL components. In Sect. 5.4 some verification results using more classical grid-point based scores instead of the SAL components are briefly discussed.

Following the WMO recommendations (WMO, 2009) regarding statistical significance tests for verification results, a bootstrap method is used because the presented scores (and the underlying precipitation fields) do not allow any assumptions concerning theoretical distributions. Further, since we are dealing with fields which are autocorrelated in time, a blocked bootstrap method was chosen. Whenever "significant" or "not significant" is mentioned in the following subsections it is meant in its statistical sense based on a blocked bootstrap and hypothesis test as described in Wilks (1997) using the following test parameters: significance level 0.95; block length $l = 7$; bootstrap sample size $n_b = 5000$. In addition to the bootstrap test a Wilcoxon-Mann-Whitney rank-sum test was performed (Wilks, 2006).

## 5.1 Area mean precipitation

The top panel in Fig. 3 shows the mean amplitude score *A* for each model for the entire period for an Alpine area in the western part of Austria as a function of lead time, so for e.g. lead time +6 h the values in the plot represent the mean amplitude score averaged over 61 forecasts of the 3-hourly precipitation between +3 and +6 h. It is worth noting that all models, including the one with explicit deep convection (AROME), tend to overestimate convective activity in the afternoon. The amplitude score reaches maximum values (the strongest overestimation of area mean precipitation) between 12:00–15:00 UTC. Thus, even if the higher resolution models turn out to have better structural scores, this does not by itself solve the problem of overestimation of alpine afternoon convection. This corresponds to the subjective impression gained from visual inspection of the precipitation forecast fields. All models tend to produce an "envelope" of precipitation over the entire alpine area, whereas in the real atmosphere the precipitation is more selective and clustered into specific regions. The higher resolution models generate more fine-scale structure but still exhibit the envelope characteristics.

Comparing the different models we must keep in mind that the INCA forecast has the advantage of starting by design with values near 0, since in the nowcasting range (up to 4 h) it is an extrapolation of the analysis which is used for verification. For higher lead times the INCA curve is sometimes further from the zero-line than one of its constituent models ECMWF and ALA-AUT. As INCA should represent an optimal combination of ECMWF and ALA-AUT it is concluded that the weights for the combination of these two models are "out-of-date". Several changes have been made to the operational ALA-AUT model and also in the ECMWF model, changing the characteristics of the precipitation forecasts significantly. At the time of writing the weights for the combination have already been updated. It is interesting to see the ECMWF model in Fig. 3 (top panel, yellow curve) as the one showing the fastest decrease of overestimation during late afternoon and evening, whereas the values even become slightly negative (underestimation) in the evening and during night. In contrast to ECMWF, AROME is the model showing the strongest overestimation during the afternoon and evening.

Figure 4 shows the scores for an area in the north eastern part of Austria, representing a predominately flat terrain. The scores exhibit a similar diurnal cycle to the case of the Alpine area, but there are noticeable differences. Most of the models start with an underestimation of the area mean precipitation, which is turning into an overestimation before midday. It is obvious that compared to the results for the Alpine area the phase of the diurnal cycle in terms of the amplitude score is shifted towards shorter lead times. Furthermore it seems that this shift depends on the resolution of the model, that is to say the ECMWF error characteristic is the first to change

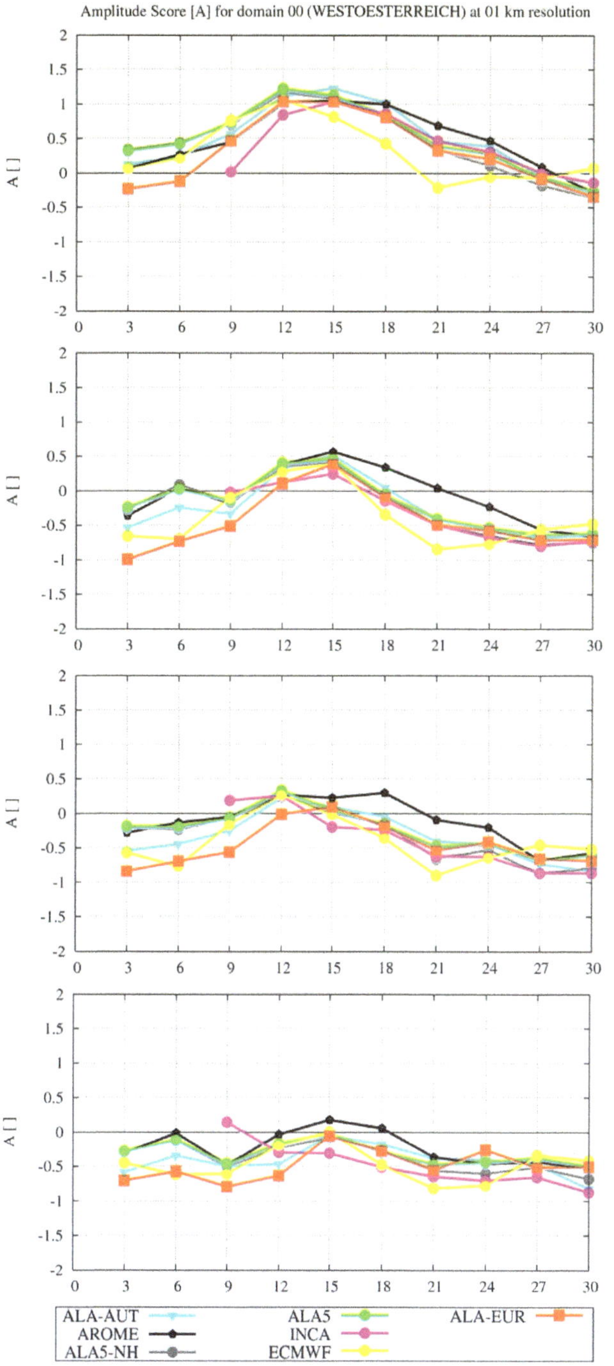


**Figure 3.** Mean values for SAL amplitude component *A* as a function of lead time for various models for different domain average thresholds: (from top to bottom) all events with an observed domain average greater than 0 mm, 0.5 mm, 1.0 mm, 2.5 mm (verification period 20090601–20090731).

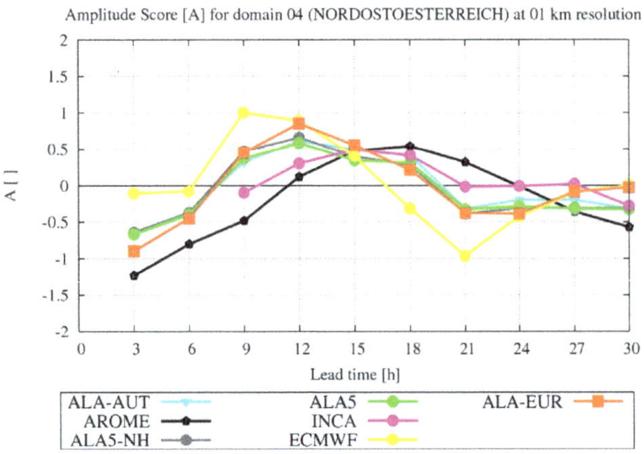


**Figure 4.** Mean values for SAL amplitude component $A$ as a function of lead time for different models (verification period 20090601–20090731).

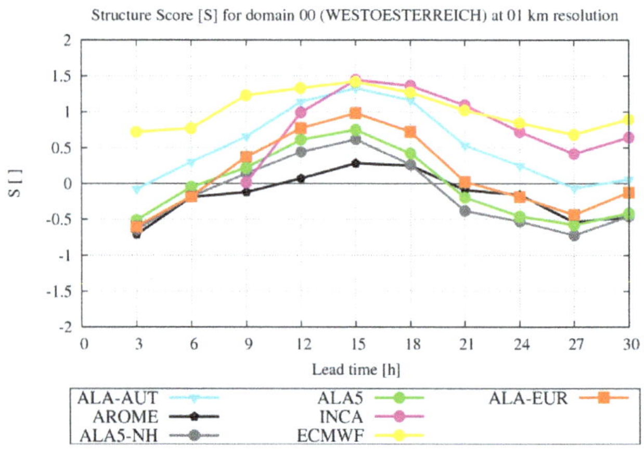


**Figure 5.** Mean values for SAL structure component $S$ as a function of lead time for different models (verification period 20090601–20090731).

from underestimation to overestimation and to reach a maximum before midday. For AROME this is occurring later in the day. As we are dealing mainly with convective precipitation events, the interpretation can be as follows. As the models with higher resolution generally are better capable of simulating the evolution of convective cells, they show higher skills in predicting the spread of convective activity from the mountainous areas to the flatland areas in Austria. But this positive aspect is reduced by the significant negative amplitude scores in the morning, which should be a topic for further investigation.

Two points should be further mentioned. First, from comparing the hydrostatic ALA5 with its non-hydrostatic counterpart ALA5-NH it is evident that in terms of mean scores over a two month period the differences between the two versions are not significant. It is hardly possible to distinguish the curves for ALA5 and ALA5-NH in Figs. 3 and 4. This does not necessarily mean that two versions always produce the same forecasts, but it would need some case studies to point out the more significant differences. The second point is that the INCA precipitation forecast shows consistently higher skill than each of its constituents ALA-AUT and ECMWF from +18 h onwards, and partly also at shorter lead times. Hence, in contrast to the Alpine area the weights for the combination of ALA-AUT and ECMWF over the lowlands still seem to work reasonably well. Forecasting experience has shown that the change of precipitation characteristics of ALA-AUT due to the use of new physics packages was most pronounced in mountainous areas.

Summarizing the results seen from Figs. 3 and 4 one may not see great benefit in running a high resolution model like AROME or ALA5/ALA5-NH. But it is important to point out that the scores shown in the top panel of Fig. 3 represent mean values over a 2 month period without taking into account the intensity of the events. According to Eq. (2) the amplitude score reaches values of 2 for all cases with $D(R_{\text{mod}}) > D(R_{\text{obs}}) = 0$, i.e. when there is precipitation in the model but not in reality. So even when the model domain average is very small, the score is 2 if the observed domain average is 0.

In order to take this important fact into account the scores were recomputed for different domain average thresholds, i.e. different precipitation intensities. Figure 3 also shows the scores separated for events with observed domain averages greater than 0.0 mm, 0.5 mm, 1.0 mm and 2.5 mm respectively (top to bottom). Comparing the curves it can be seen that the higher the intensity of the event gets, the closer the curves for high resolution models (like AROME, ALA5 and ALA5-NH) get to 0 (closer to the observed domain average). This indicates that the benefit of the high resolution models reveals itself for strong precipitation events, i.e. for high impact weather. For events with domain averages greater than 2.5 mm (bottom panel in Fig. 3) it can be seen that the models generally underestimate the amplitude, with AROME showing the best overall A score. During the convectively most active period (12–18 Z) AROME differs from the other models in that it slightly overestimates the amplitude. One needs to keep in mind, however, that the number of considered cases decreases with increasing intensities. With respect to the total number of 61 days the proportions of days with domain averages greater than 0.0 mm, 0.5 mm, 1.0 mm and 2.5 mm are 88%, 37%, 32% and 15%, respectively. This naturally has an effect on the statistical significance of the results, but we expect that the main characteristics of the results do not change when extending the verification period.

## 5.2 Structure and location of precipitation fields

The interpretation of the results addressing the structure of the precipitation forecasts is more straightforward. Figure 5 shows the scores for the structure component for the different models, again as a function of lead time. It can be clearly seen that the higher the resolution of the model, the better the structure of the precipitation forecasts gets in terms of the $S$ component. The structure component is largest for the model with the lowest resolution (ECMWF), implying that predicted precipitation patterns are too big or flat compared to the observed fields. The higher-resolution models ALA-EUR, ALA5, and ALA5-NH show generally significant smaller $S$ values during daytime but apparently underestimate the up-scale development of convective systems in the evening and night hours, as $S$ values become significantly negative there. Averaged over the day, AROME has the smallest $S$ values. Comparing the non-hydrostatic and the hydrostatic 5 km ALADIN version it turns out that the non-hydrostatic version tends to produce smaller or more intense features, as the mean structure scores are noticeable smaller. This is valid for Alpine areas, while the difference is much smaller for areas located in flat terrain (not shown). So in general the benefit of using a high resolution model can be clearly shown with the structure component of SAL.

The results for the location component $L$ do not yield much information when computing mean values for a 2 month verification period, as the values for the various models do not differ significantly. The mean $L$ values (over all lead times) for the different models range from 0.29 to 0.36 for domain 00 and from 0.29 to 0.33 for domain 04. An evolution of $L$ with respect to the lead time is hardly noticeable. Two aspects are worth mentioning. First, the advantage of INCA in the nowcasting range is also visible in the $L$ component. Second, if one had to choose, without considering statistical significance, one model showing the overall lowest $L$ scores it would be ECMWF, in particular for the mountainous domain 00. Taking a look at some case studies while keeping in mind that the $L$ component is constructed as the sum of two components, one gets the impression that the mass centers for the different models are often rather similar. This means that the differences in terms of the $L$ value arise mainly because of the second component of $L$.

### 5.3 Sensitivity of results to grid resolution

Up to now the results were discussed for verification domains with 1 km horizontal resolution. To obtain information about the sensitivity of the SAL results to the resolution of the underlying verification grid, the scores were re-computed for horizontal resolutions of 5 km and 10 km. The up-scaling of the analysis and model fields was performed by thinning the initially created 1 km fields using median values for the final 5 km and 10 km resolution grid, respectively. The median was chosen as it is less sensitive to extreme values than the mean.

Comparing the $A$ scores for the different resolutions for domain 00 (not shown) indicates that the curves for the high-resolution models (AROME, ALA5-NH and ALA5) are slightly shifted towards lower $A$ values for all lead times (0.1 in terms of the mean $A$ value over all lead times) when going to 5 km, and further to 10 km resolution, respectively. For ALA-EUR, ALA-AUT, ECMWF and INCA the differences are smaller, in general even a slight shift towards higher $A$ values may be observed. So it seems that the high resolution model fields lose at least a part of their overestimation characteristics when upscaled to lower resolutions. For domain 04 there is a noticeable shift towards higher values of $A$, this time for all models. The magnitude of this shift is larger (up to 0.3 for ECMWF) for the lowlands than for the mountain areas. The sign of the change is opposite for the high resolution models for the mountainous and the lowland domain. So for the lowlands the upscaling appears to be less beneficial.

The results for the structure component $S$ change in the way that the curves for the different models tend to be located closer together when moving to lower grid resolution. In general the $S$ values decrease for ALA-AUT and ECMWF and increase for the other models. This implies that ALA-AUT and ECMWF precipitation fields gain structural skill with respect to the verifying analysis, while the high resolution model forecasts indeed loose structure in their fields but still retain a better correspondence to the observed field in terms of volume and shape of the precipitation objects. For example, the difference of the mean $S$ value between AROME and ECMWF over all lead times reduces by 0.35 when moving to 10 km grid resolution. But even at 10 km resolution, the $S$ values are significantly lower (and closer to 0) for the high resolution models with respect to ALA-AUT and ECMWF. It may be interesting to extend the verification to even lower resolutions (e.g. the one of ECMWF), but for that it would be necessary to increase the size of the verification domains as the remaining number of grid points within domain 00 and 04 gets too low so that the object identification algorithm within SAL might not work properly anymore.

Finally, when comparing the $L$ component for the different verification grid resolutions, it can be seen that again a shift of the curves is noticeable. The slight but noticeable shift happens towards lower $L$ values. This in turn implies that there is a tendency of the location of the center of mass and the mean distance of individual objects with respect to the center of mass to coincide better when upscaling the analysis and model fields. But as mentioned in the previous section, the differences in terms of $L$ values do not differ greatly between the models.

### 5.4 Grid point verification

In order to evaluate the precipitation forecasts in a more classical way, various grid point scores were computed in addition to the SAL scores: Probability of Detection (POD), False Alarm Ratio (FAR) and Equitable Threat Score (ETS). Definitions and properties of these scores are given in Wilks (2006). For the computation of ETS random chance was chosen as reference forecasts.

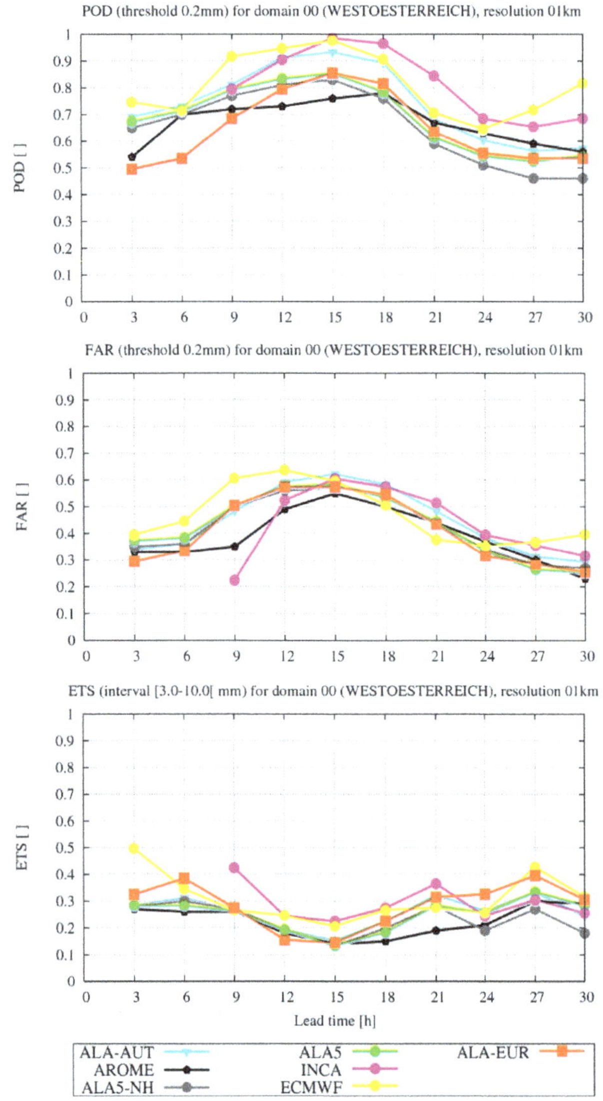


**Figure 6.** FAR, POD and ETS for different models (verification period 20090601–2009731).

Figure 6 shows POD (top), FAR (middle) and ETS (bottom) for the different models as a function of lead time for the mountainous domain 00, again for verification period 20090601-20090731 on a common grid with 1 km horizontal resolution. POD and FAR presented in Fig. 6 just include grid points with precipitation intensities over 0.2 mm, the ETS scores are shown for the intensity interval 3.0–10.0 mm. The diurnal cycle which could be observed in the SAL scores in Figs. 3 and 4 is also visible for the three scores presented in Fig. 6, whereas in contrast to FAR and ETS, POD shows higher values during the period with strongest convective activity. This is consistent with the result for the *A* component in Fig. 3, as a stronger overestimation (in terms of areal means) implies an increase of POD but at the same time also an increase of FAR. Taking a closer look on the curve for POD, one can notice that there is a tendency towards lower POD values for the high resolution models. The difference in terms of POD is around 0.2 during afternoon. This is even better visible in the scores for the lowland domain 04 (not shown), where the difference in terms of POD increases to 0.3. For both domains it is the model running on highest resolution (AROME) showing the lowest POD values for the period 06:00–18:00 UTC, but at the same time the lowest FAR values. This statement is true when taking ALA-EUR and INCA aside for a moment. As the model physics used inside ALA-EUR is the same as it is the case for ALA-AUT and ALA5, the significant difference in the first 12 forecast hours should be related to the initial conditions (ECMWF vs. ARPEGE). This will be a topic for further investigation, as in general the creation of initial conditions for ALADIN/ALARO/AROME models from ECMWF model output is not straightforward, in particular for surface fields like temperature and humidity. Difficulties arise from the differences in the surfaces schemes used in the ECMWF and the ALADIN/ALARO/AROME/ARPEGE models.

In general INCA shows the best scores for ETS and FAR during the first 3 h, as would be expected from a nowcasting system. It must be kept in mind, however, that INCA has the advantage of starting from the analysis which is used for verification. For FAR and ETS the difference between lowlands and mountainous areas is less obvious. In terms of ETS it is ECMWF and INCA showing the highest values during the period with strongest convective activity. When using different intensity intervals to compute the ETS score (e.g. 10–20 mm, > 20 mm) no additional information is obtained, except that in terms of ETS the models show low skill for the forecast period with strongest convective activity.

## 6 Summary

Various models used at ZAMG running on different horizontal grid spacings ranging from 2.5 to 25 km are compared according to their precipitation forecasts for a 2-month period in summer 2009 which is dominated by convective activity. The method used for verification is SAL, a package evaluating the domain average precipitation and also the structure and the location of the precipitation features in the model. On observational side the INCA high resolution precipitation analysis is used, which is a part of the operational analysis and nowcasting system at ZAMG.

It turns out that SAL is a valuable tool to address more aspects than just the quantitative quality of precipitation forecasts. In addition to the *A* component (evaluating the area mean values), the structure component *S* seems to be able to confirm "human" subjective impressions when there is a need to choose the model creating the most realistic precipitation forecast patterns. The location component seems to yield less valuable information due to its rather small variation from model to model (in terms of mean values). *L* should be more valuable when considering single case studies.

The evaluation of the different models shows that the benefit of high resolution models running on grid spacings of 2.5 km or 5 km can be seen when the structure of the model precipitation fields are considered. This is true even when verification grids with lower horizontal resolution are used. In terms of domain average values the benefit is not apparent, unless one concentrates on strong precipitation events. For high impact weather the advantages of models with grid spacing finer than 5 km are visible. Using conventional grid point scores the benefit of high resolution models is not visible.

Among the models used in the present verification study two 5 km model versions are compared, one running with a hydrostatic kernel and one with its non-hydrostatic counterpart. The results show that noticeable differences between the two versions are just visible in the structure component and primarily in mountain areas. In terms of areal mean values the differences are small, but one can expect to reveal more details by performing case studies.

Edited by: D. Giaiotti

## References

Casati, B., Wilson, L. J., Stephenson, D. B., Nurmi, P., Ghelli, A., Pocernich, M., Damrath, U., Ebert, E. E., Brown, B. G., and Mason, S.: Forecast verification: current status and future directions, Meteorol. Appl., 15, 3–18, 2008.

Früh, B., Bendix, J., Nauss, T., Paulat, M., Pfeiffer, A., Schipper, J. W., Thies, B., and Wernli, H.: Verification of precipitation from regional climate simulations and remote-sensing observations with respect to ground-based observations in the upper Danube catchment, Meteorol. Z., 16, 275–293, 2007.

Gerard, L., Piriou J.-M., Brožkova, R., Geleyn, J.-F., and Banciu, D.: Cloud and precipitation parameterization in a meso-gamma scale operational weather rediction model, Mon. Weather Rev., 137, 3960–3977, 2009.

Gilleland, E., Ahijevych, D., Brown, B.G., Casati, B., and Ebert, E. E.: Intercomparison of spatial forecast verification methods, Wea. Forecasting, 24, 1416–1430, 2009.

Haiden, T. and Pistotnik, G.: Intensity-dependent parameterization of elevation effects in precipitation analysis, Adv. Geosci., 20, 33–38, doi:10.5194/adgeo-20-33-2009, 2009.

Haiden, T., Kann A., Pistotnik, G., Stadlbacher, K., and Wittmann, C.: Integrated Nowcasting through Comprehensive Analysis (INCA), System description, available at: www.zamg.ac.at/fix/INCA_system.pdf, 2009.

Lafore, J. P., Stein, J., Asencio, N., Bougeault, P., Ducrocq, V., Duron, J., Fischer, C., Héreil, P., Mascart, P., Masson, V., Pinty, J. P., Redelsperger, J. L., Richard, E., and Vilà-Guerau de Arellano, J.: The Meso-NH Atmospheric Simulation System. Part I: adiabatic formulation and control simulations, Ann. Geophys., 16, 90–109, doi:10.1007/s00585-997-0090-6, 1998.

Pappenberger, F., Ghelli, A., Buizza, R., and Bodis, K.: The skill of probabilistic forecasts under observational uncertainties within the generalized likelihood uncertainty estimation framework for hydrological applications, J. Hydromet., 807–819, 2009.

Pinty, J. P. and Jabouille, P.: A mixed phase cloud parameterization for use in a mesoscale nonhydrostatic model: Simulations of a squall line and of orographic precipitation. Preprints, Conf. on Cloud Physics, Everett, WA, Amer. Meteor. Soc., 217–220, 1998.

Skok, G. and Vrhovec, T.: Considerations for interpolating rain gauge precipitation onto a regular grid, Meteorol. Z., 15, 545–550, 2006.

Wernli, H., Paulat, M., Hagen, M., and Frei, C.: SAL – A novel quality measure for the verification of quantitative precipitation forecasts, Mon. Weather Rev., 136, 4470–4487, 2008.

Wernli, H., Hofmann, C., and Zimmer, M.: Spatial forecast verification methods intercomparison project: application of the SAL technique, Wea. Forecasting, 24, 1472–1484, 2009.

Wilks, D. S.: Statistical Methods in the Atmospheric Sciences, 2nd Edn., Academic Press, 627 pp., 2006.

Wilks, D. S.: Resampling hypothesis tests for autocorrelated fields, J. Climate, 10, 65–82, 1997.

WMO: Recommendations for the verification and intercomparison of QPFs and PQPFs from operational NWP models, World Meteorological Organization, Revision 2, 2009.

# 17

# The French contribution to the Global Climate Observing System

**R. Juvanon du Vachat**

Société Météorologique de France, 1, Quai Branly 75007, Paris, France

**Abstract.** France is participating fully in the Global Climate Observing System (GCOS). It incorporates the following four components: meteorological and atmospheric, oceanic, terrestrial, spatial, which will be briefly presented, especially in relation with the monitoring of the climate. The presentation will give an overview of the general principles governing the GCOS system and particularly the concepts used to maintain efficiently this climate observing system for a long period of time ("from research networks to operational networks"). The presentation will cover all the four components of the GCOS system. The whole report has been published in the Fifth National Communication from France to the UNFCCC (United Nations Framework Convention on Climate Change). The presentation will give an overview of the different networks of these four domains devoted to the monitoring of climate and maintained by France and highlighting the strengths and weaknesses of this climate observing system.

## 1 Introduction and general considerations

This report gives an overview of the ground-based and space-born activities in France in response to the Implementation Plan of the Global Climate Observing System (GCOS) (www.wmo.int/pages/prog/gcos/Publications/gcos-92_GIP.pdf) as specified for the three specific domains (atmosphere, ocean, terrestrial). The GCOS programme in France is mainly driven by Météo-France for the atmospheric domain, but other institutions contribute in the field of climate for specific observations such as of the atmospheric composition, and in the oceanography and the terrestrial domain (glaciers). A number of observations are done for research purposes, and the question of "how to ensure these observations in the long term?" is regularly addressed. Also, in some cases climate observation networks are developed along a European structure (Framework Programme), and the national level is not relevant! Another item for consideration is the question of historical data, which are essential to understand present and future climate (we refer to paleoclimatic research and data rescue programmes). Finally the issue of the contributions to the GCOS programme in developing countries, especially in Africa, has been considered within the international research project and field campaign AMMA concerning the study of West African Monsoon. An inventory of all these observations for climate is compiled every five years for the National Communication under the UN Framework Convention on Climate Change (UNFCCC), in the chapter "Research and Systematic Observation". The last update is the 5th National Communication, which can be found at the UNFCCC website (www.unfccc.int, see National Reports) or at the French website: http://www.developpement-durable.gouv.fr/IMG/pdf/5e_communication_changement_climatique.pdf.

## 2 Atmospheric domain

This is a domain mainly operated by Météo-France. The whole system consists of six GSN stations (surface observations) in mainland France and fourteen ones for overseas territories, all producing monthly CLIMAT messages. For GUAN stations (upper-air observations) the network consists of nine stations in overseas territories (no GUAN in France itself) producing monthly CLIMAT TEMP messages. The ten GCOS monitoring principles (see the GCOS Implementation Plan) have been integrated in the quality management system for observations of Météo-France. In recent years an important programme on data rescue and the reconstitution of long time series have been started, including the African Database (observations of 14 African countries). France is also collaborating with the MEDARE project (Spain) aiming at recovering climate observations in the Mediterranean area.

Concerning the chemistry of the atmosphere the RAMCES network developed by LSCE (Laboratoire des Sciences du Climat et de l'Environnement) is worthwhile for computing the $CO_2$ budget over a given area (region, continent). This network operates at Amsterdam Island (Indian Ocean), Mace Head (Ireland), Puy de Dôme and Biscarosse (mainland France), with some flasks with air samples collected in twelve sites located around the world. Similar measurements are done for other greenhouse gases in the frame of this RAMCES programme. All these measurements contribute to the GAW network. As an example of an application of these measurements, Bousquet et al. (2006) show the contribution of anthropogenic and natural sources to the atmospheric methane variability. In addition, there are four stations in continental France measuring precipitation chemistry, radiation and tropospheric ozone (BAPMON network). Stratospheric and upper tropospheric ozone is measured at three sites overseas (NDACC network) and at different sites located around the globe (Finland, Russia, Greenland and Brazil). The operating agencies are INSU-CNRS for NDACC and LSCE for RAMCES.

Other initiatives in the field of monitoring the atmospheric composition in which France plays a role are the following European projects. The project GEOMON (Global Earth Observation and MONitoring, www.geomon.eu) concerns the measurements of greenhouse gases, atmospheric pollution, aerosols and stratospheric ozone; MACC (Monitoring Atmospheric Composition and Change) follows GEMS (2005–2009), and tries to build high-level monitoring of greenhouse gases, reactive gases and aerosols, including data-assimilation and multi-model forecasting; ICOS (Integrated Carbon Observing Strategy, http://www.icos-infrastructure.eu/) is a pan-European system for observation of greenhouse gases over the European continent and adjacent regions (Siberia, Africa) with various sensors (towers, planes, satellite), with the aim of producing daily maps of concentration of $CO_2$ and other greenhouse gases.

## 3 Oceanic domain

Under the Global Ocean Observation System (GOOS) the French contribution comprises the following measurements: voluntary and occasional observing ships (VOS), ocean gauges (GLOSS network), floating and anchored weather buoys, and sub-surface floaters like ARGO (Coriolis Project). France also maintains the PIRATA observation system (anchored buoys) in the tropical Atlantic Ocean (Bourlès et al., 2008) and contributes to TAO for the Pacific Ocean (IRD). We would like to emphasize the operational approach followed in oceanic observations. The seven French agencies involved in oceanography (CNES, CNRS, IFREMER, IPEV, IRD, Météo-France and SHOM) are joining forces to develop a complete and coherent system of operational oceanography based on three projects: satellite altimetry (JASON), global numerical modelling with data assimilation (MERCATOR) and in situ measurements (Coriolis). The Coriolis project offers the operational structure for acquiring, collecting, validating and distributing world ocean data (temperature, salinity and current profiles) responding to the needs of modellers (MERCATOR) and of the scientific community (CLIVAR programme). For the main results of this project see the Mercator website (http://www.mercator-ocean.fr/).

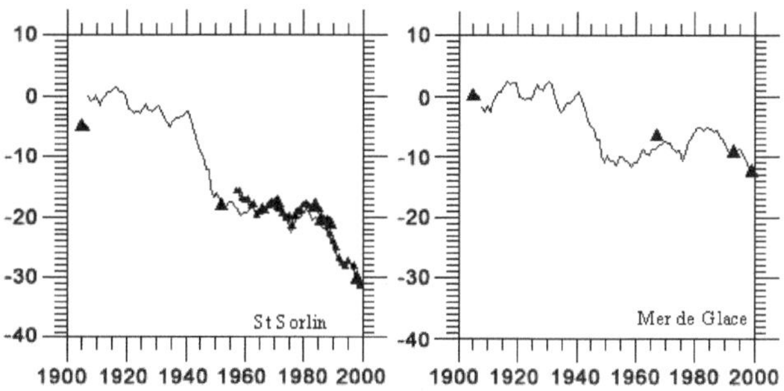


**Figure 1.** Cumulative specific net balance (m water-equivalent) from reconstruction (solid line), from direct measurements (small triangles) and from maps (large triangles) for two glaciers in the Alps (Vincent, 2002).

## 4 Terrestrial observations

Under the Global Terrestrial Network (GTN) the French contribution deals with the observations of mountain glaciers and measurements of carbon fluxes in terrestrial ecosystems (Fluxnet). Mountain glaciers are studied in France and abroad by LGGE, IRD, IPEV and CEMAGREF. The "Glacio-Clim" Observatory, which is labelled as "Observatoire de Recherche en Environnement", consists in monitoring some glaciers at different latitudes (which implies different climate regimes). The sites concerned are in mainland France (five glaciers), in Bolivia and Ecuador, and in Antarctica (two glaciers). Vincent (2002) gives an example of the monitoring of four glaciers in France (evolution of the mass balance during a century, Fig. 1). For the "Glacio-Clim" Observatory we refer to the website http://www-lgge.ujf-grenoble.fr/ServiceObs/contexte-angl.htm.

France is also actively participating in measuring carbon fluxes in terrestrial ecosystems (six sites with different land cover) carried out under the Fluxnet international programme and the various programmes connected with the Carboeurope project. That integrated project aims at determining the role of the European continent in the global carbon cycle. The agencies involved in this programme are: INRA, CNRM, and LSCE. Ciais et al. (2005) give an example of such measurements concerning the heat and drought event of summer 2003 in Europe. More specific experiments have been conducted in the Southwest France in 2005 ("Les Landes", Dolman et al., 2006) with various land cover. Forest ecosystems have been systematically observed by the National Forestry Inventory (IFN) every ten years since 1970, through a survey (sort and volume of trees) put in place after the damage caused by acid rain.

## 5 Space component

Concerning the space observations, we can mention the specific response to the GCOS Implementation Plan done by the Committee on Earth Observing Satellites (CEOS), in which the Centre National d'Etudes Spatiales (CNES) is taking a contributing part.

## 6 GCOS in Africa

After the GCOS regional meeting held in Niamey (March 2003), a number of discussions have arisen on how to restore the deficient networks for climate observations, especially in that part of Africa (West and Central Africa; see Parker et al., 2008). It is important to mention the AMMA experiment (Redelsperger et al., 2006) conducted in that region 2005–2007, leading to restore some radiosoundings and fine scale measurements covering meteorology but also hydrology and aerosols. In the frame of the RIPIECSA experiment the sustainability of the meteorological observational network has been ensured for one more year, but the long term maintenance of that network remains an open question. Finally a new project supported by FFEM (French component of GEF), along with other organizations (as ACMAD) has been proposed, in order to comply with G8 commitments for France ("Climate and observations for Africa"). That project is mainly built on the question of "Climate adaptation for Africa", in the sectors of agriculture, water resources and health, but also the prevention of the consequences of severe weather and climate events. A similar project has been formulated and supported (July 2007) by this FFEM agency in the context of "adaptation", covering the five islands of the west part of the Indian Ocean (Réunion, Mauritius, Madagascar, Comoros, Seychelles). It is a pilot project in the climate change context of "Impacts, Adaptation and Vulnerability".

## 7 Conclusions

As a conclusion we observe that a great number of institutions are involved in climate observations in France, including Météo-France and other agencies connected to the Ministry in charge of Environment and to the Ministry in charge of Research. All these institutions are mentioned below. Coordination is necessary and partially fulfilled under the GEO-GMES French coordination group, at institutional level. We refer again to the French web link to the fifth National Communication, where the detailed inventory of the networks can be found (http://www.developpement-durable.gouv.fr/IMG/pdf/5e_communication_changement_climatique.pdf) with the new reporting format (Chapter 8 and Annex XI for the Tables). Two French publications have also been written for different workshops held in France (Juvanon du Vachat, 2003, 2004) illustrating various aspects of climate and relative to the third National Communication (2001).

## French institutions involved in climate observations

CEMAGREF: Centre d'Etudes du Machinisme Agricole, du Génie Rural et des Eaux et Forêts
CNES: Centre National d'Etudes Spatiales
CNRS: Centre National de la Recherche Scientifique
CNRM: Centre National de Recherches Météorologiques (Météo-France)
IFN: Inventaire Forestier National
IFREMER: Institut Français de Recherche et d'Exploitation de la MER
INRA: Institut National de la Recherche Agronomique
INSU: Institut National des Sciences de l'Univers
IPEV: Institut Paul Emile Victor (Polar Institute)
IRD: Institut de Recherche pour le Développement (former ORSTOM)
LGGE: Laboratoire de Glaciologie et de Géophysique de l'Environnement
LSCE: Laboratoire des Sciences du Climat et de l'Environnement
Météo-France (Direction de la Climatologie)
SHOM: Service Hydrographique et Océanographique de la Marine

## Acronyms not explained in the text

ACMAD: African Centre for Meteorological Applications for Development
AMMA: African Monsoon Multidisciplinary Analysis
ARGO: Array for Real time Geostrophic Oceanography
BAPMON: Background Air Pollution MONitoring
CLIVAR: Climate Variability and Predictability
FFEM: Fonds Français de l'Environnement Mondial
GAW: Global Atmosphere Watch
GCOS: Global Climate Observing System
GEF: Global Environment Fund
GEMS: Global Earth-system Monitoring using Satellite and in-situ data
GEO: Group on Earth Observations
GEOMON: Global Earth Observation and MONitoring of the atmosphere
GLOSS: Global Sea Level Observing System
GODAE: Global Ocean Data Assimilation Experiment
GSN: GCOS Surface Network

GUAN: GCOS Upper Air Network
MACC: Monitoring Atmospheric Composition and Change
MEDARE: MEditerranean climate DAta REscue
NDACC: Network for the Detection of Atmospheric Composition Change
PIRATA: Prediction and Research Moored Array in the Atlantic
RAMCES: Réseau Atmosphérique de Mesure des Composés à Effet de Serre
RIPIECSA: Recherche Interdisciplinaire et Participative sur les Interactions entre les Ecosystèmes, le Climat et les Sociétés d'Afrique de l'Ouest
TAO: Tropical Atmosphere Ocean (Pacific Ocean)
UNFCCC: United Nations Framework Convention on Climate Change

Edited by: W. A. Monna

## References

Bourlès, B., Lumpkin, R., McPhaden, M. J., Hernandez, F., Nobre, P., Campos, E., Yu, L., Planton, S., Busalacchi, A., Moura, A. D., Servain, J., and Trotte, J.: The Pirata program: History, accomplishments and future directions, B. Am. Meteorol. Soc., 89, 1111–1125, 2008.

Bousquet, P., Ciais, P., Miller J. B., Dlugokencky E. J., Hauglustaine, D. A., Prigent, C., Van der Werf, G. R., Peylin, P., Brunke, E.-G., Carouge, C., Langenfelds, R. L., Lathière, J., Papa, F., Ramonet, M., Schmidt, M., Steele, L. P., Tyler, S. C., and White, J.: Contribution of anthropogenic and natural sources to atmospheric methane variability, Nature, 443, 439–443, 2006.

Ciais, P., Reichstein, M., Viovy, N., Granier, A., Ogée, J., Allard, V., Aubinet, M., Buchmann, N., Bernhofer, C., Carrara, A., Chevallier, F., De Noblet, N., Friend, A. D., Friedlingstein, P., Grünwald, T., Heinesch, B., Keronen, P., Knohl, A., Krinner, G., Loustau, D., Manca, G., Matteucci, G., Miglietta, F., Ourcival, J. M., Papale, D., Pilegaard, K., Rambal, S., Seufert, G., Soussana, J. F., Sanz, M. J., Schulze, E. D., Vesala T., and Valentini, R.: Europe-wide reduction in primary productivity caused by the heat and drought in 2003, Nature, 437, 529–533, 2005.

Dolman, A. J., Tolk, L., Ronda, R., Noilhan, J., Sarrat, C., Brut, A., Piguet, B., Durand, P., Butet, A., Jarosz, N., Brunet, Y., Loustau, D., Lamaud, E., Miglietta, F., Gioli, B., Magliulo, V., Esposito, M., Gerbig, C., Körner, S., Glademard, P., Ramonet, M., Ciais, P., Neininger, B., Hutjes, R. W. A., Elbers, J. A., Macatangay, R., Schrems, O., Pérez-Landa, G., Sanz, M. J., Scholz, J., Facon, G., Ceschia, E., and Beziat, P.: The Carbo-Europe Regional Experiment Strategy, B. Am. Meteorol. Soc., 87, 1367–1379, 2006.

Juvanon du Vachat, R.: Le système mondial d'observation du climat, Contribution française, Annales Assoc. Int. de Climatologie, 15, 458–464, www.climato.be/aic, 2003.

Juvanon du Vachat, R.: Le système mondial d'observation du climat, Atelier Expérimentation et Instrumentation, Paris, 6 pp., aei2004.dt.insu.cnrs, 2004.

Parker, D. J., Fink, A., Janicot, S., Ngamini, J.-B., Douglas, M., Afiesimana, E., Agusti-Panareda, A., Beljaars, A., Dide, F., Diedhiou, A., Lebel, T., Polcher, J., Redelsperger, J.-L., Thorcroft, C., and Wilson, G.A.: The AMMA radiosonde programme and its implications for the future of atmospheric monitoring over Africa, B. Am. Meteorol. Soc., 89, 1015–1027, 2008.

Redelsperger, J.-L., Thorncroft,C. D., Diedhiou, A., Lebel, T., Parker, D. J., and Polcher, J.: African Monsoon Multidisciplinary Analysis : An international research project and field campaign, B. Am. Meteorol. Soc., 87, 1739–1746, 2006.

Vincent C.: Influence of climate change over the 20th century on four French glacier mass balances, J. Geophys. Res., 107(D19), ACL4, 1–12, 2002.

18

# The future climate characteristics of the Carpathian Basin based on a regional climate model mini-ensemble

**I. Krüzselyi[1], J. Bartholy[2], A. Horányi[1], I. Pieczka[2], R. Pongrácz[2], P. Szabó[1], G. Szépszó[1], and Cs. Torma[2]**

[1]Hungarian Meteorological Service, Budapest, Hungary
[2]Department of Meteorology, Eötvös Loránd University, Budapest, Hungary

**Abstract.** Four regional climate models (RCMs) were adapted in Hungary for the dynamical downscaling of the global climate projections over the Carpathian Basin: (i) the ALADIN-Climate model developed by Météo France on the basis of the ALADIN short-range modelling system; (ii) the PRECIS model available from the UK Met Office Hadley Centre; (iii) the RegCM model originally developed at the US National Center for Atmospheric Research, is maintained at the International Centre for Theoretical Physics in Trieste; and (iv) the REMO model developed by the Max Planck Institute for Meteorology in Hamburg. The RCMs are different in terms of dynamical model formulation, physical parameterisations; moreover, in the completed simulations they use different spatial resolutions, integration domains and lateral boundary conditions for the scenario experiments. Therefore, the results of the four RCMs can be considered as a small ensemble providing information about various kinds of uncertainties in the future projections over the target area, i.e., Hungary. After the validation of the temperature and precipitation patterns against measurements, mean changes and some extreme characteristics of these patterns (including their statistical significance) have been assessed focusing on the periods of 2021–2050 and 2071–2100 relative to the 1961–1990 model reference period. The ensemble evaluation indicates that the temperature-related changes of the different RCMs are in good agreement over the Carpathian Basin and these tendencies manifest in the general warming conditions. The precipitation changes cannot be identified so clearly: seasonally large differences can be recognised among the projections and between the two periods. An overview is given about the results of the mini-ensemble and special emphasis is put on estimating the uncertainties in the simulations for Hungary.

## 1 Introduction

Climate change is an essential environmental challenge, which is of great interest for the society. Climate models are the most appropriate tools available for the provision of estimates for the future evolution of the climate system, consequently, their use is indispensable for the adaptation to the (positive and negative) impacts of climate change. Evidently, climate projections possess various kinds of uncertainties, therefore, different simulations have different results over the same domain. Uncertainties, among others, can be originated from the emission scenarios, from the model formulations (different numerical schemes and physical parameterisation packages for instance) and the internal variability also existing in the absence of any external radiative forcing. In order to estimate these uncertainties the evaluation has to be realised in an ensemble way, which means considering several model projections together. This multi-model approach was applied within the international ENSEMBLES project (http://ensembles-eu.metoffice.com), where numerous regional climate simulations (RCMs) mainly with 25 km horizontal resolution were conducted and analysed together.

In Hungary four RCMs (ALADIN-Climate, PRECIS, RegCM and REMO) were adapted in the past few years for the dynamical downscaling of global climate projections over the Carpathian Basin giving the opportunity to create a Hungarian mini-ensemble. In this paper the evaluation of this small ensemble is discussed in order to identify the climate uncertainty range over Hungary. It is especially interesting in the case of precipitation change since (as it was also

**Table 1.** Main characteristics of the simulations.

| | ALADIN-Climate | PRECIS | RegCM | REMO |
|---|---|---|---|---|
| Horizontal resolution | 10 km | 25 km | 10 km | 25 km |
| Domain | Carpathian Basin | Central Europe | Carpathian Basin | Central and Eastern Europe |
| LBCs | ARPEGE-Climat/OPA | HadCM3 | ECHAM5/MPI-OM → RegCM_25 | ECHAM5/MPI-OM |
| Scenario runs | 2021–2050: A1B<br>2071–2100: A1B | 2071–2100: B2 | 2021–2050: A1B<br>2071–2100: A1B | 2001–2100: A1B |

identified already by the PRUDENCE project; Christensen and Christensen, 2007) Hungary is situated in an transition zone between the northern regions characterised by annual precipitation increase and the southern regions by decrease.

The following models are applied in Hungary and their results are examined in this paper: (i) ALADIN-Climate model (Spiridonov et al., 2005) developed by Météo France within an international cooperation; (ii) PRECIS model (Wilson et al., 2005) available from the UK Met Office Hadley Centre; (iii) RegCM model (Giorgi et al., 1993a, b) originally developed at the US National Center for Atmospheric Research and improved at the International Centre for Theoretical Physics in Trieste (it has to be mentioned here that a tuned and calibrated version of the model was applied in order to reduce the model biases for the Carpathian Basin; Torma et al., 2008); and (iv) REMO model (Jacob and Podzun, 1997) developed by the Max Planck Institute for Meteorology in Hamburg.

After validation of (monthly) mean temperature and precipitation, their annual and seasonal changes (including their statistical significance) have been evaluated focusing on two 30-year future periods (2021–2050 and 2071–2100) with respect to the past reference (1961–1990).

In the following sections first the details of the regional climate simulations are introduced, which is followed by the validation and projection results, and finally conclusions are drawn.

## 2 Simulations

First, some validation experiments were performed for the past (1961–1990) using on the one hand, ERA-40 re-analysis data (Uppala et al., 2005) provided by the European Centre for Medium-Range Weather Forecasts, and on the other hand, the results of global climate models (GCMs) as lateral boundary conditions (LBCs). The simulations were accomplished on 10 and 25 km horizontal resolutions covering the Carpathian Basin (Fig. 1).

After identification of the strengths and weaknesses of the models, simulations for the future were carried out.

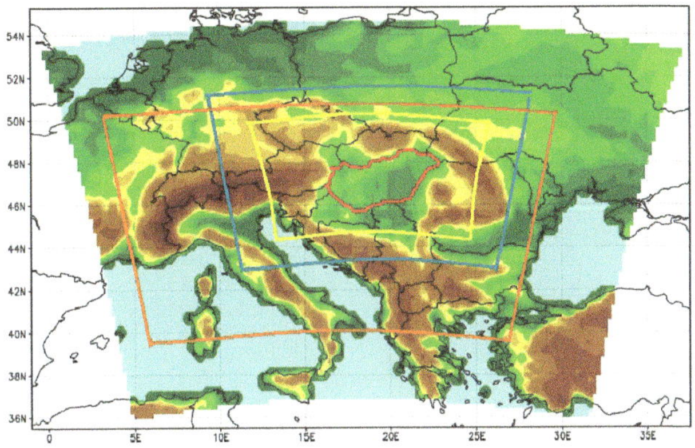


**Figure 1.** Model integration domains: yellow, blue and orange curves represent the domains covered by ALADIN-Climate, RegCM and PRECIS, respectively; the largest domain and its orography belongs to REMO.

Scenario simulation for REMO was a transient run for 2001–2100 while the other models were accomplished in time slices focusing on two 30-year periods (2021–2050 and 2071–2100). The anthropogenic forcings were taken into account according to A1B SRES scenario (Nakicenovic and Swart, 2000) except for PRECIS, where B2 SRES scenario was available. The main characteristics of the simulations are presented in Table 1.

## 3 Validation

Figure 2 shows the normalised Taylor diagram (Taylor, 2001) for monthly mean temperature and precipitation values over Hungary for the period 1961–1990 using model simulations and the 10-min horizontal resolution Climate Research Unit (CRU, Mitchell et al., 2004) observational data set. (Monthly mean calculations were performed on the own grid of each simulation data set in order to avoid the interpolation of the raw model data.) It can be seen that models describe better the temporal characteristics for temperature than for precipitation, especially for REMO driven by GCM. This

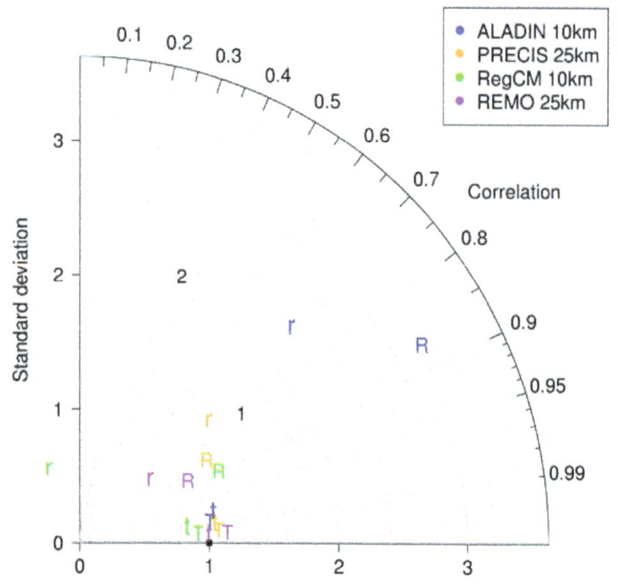


**Figure 2.** Taylor diagram (normalised by the reference standard deviation) for monthly mean temperature (T: ERA-40-driven case, t: GCM-driven case) and precipitation values (R: ERA-40-driven runs, r: GCM-driven runs) over Hungary for 1961–1990. CRU data are used as observational dataset (represented by black dot).

latter simulation provides the smallest bias for temperature. Otherwise ERA-40 and GCM-driven runs significantly differ in case of precipitation: one can generally conclude that ERA-40 runs have better performance in terms of correlations and that is particularly true for ALADIN-Climate and RegCM. Precipitation of GCM-driven RegCM has negative correlation with observations, which indicates that this run cannot reproduce the annual cycle. The fact, that the REMO simulation using the same GCM forcings as RegCM does not possess this deficiency, implies the question whether the same calibration used to eliminate the bias from the ERA-40 run can be applied in the GCM-driven case, when the simulation biases consist of the RCM and GCM errors simultaneously. Models usually overestimate monthly precipitation, particularly ALADIN-Climate in summer.

It can be seen from the validation process that the applied models carry different kind of errors. More details about the validation results can be found in Csima and Horányi (2008); Torma et al. (2008); Szépszó and Horányi (2008); Bartholy et al. (2009). This information is essential for the further development of these RCMs. For future projections the delta-approach is used, which computes the deviations of the model values between two periods (future and past) in order to eliminate systematic errors (which are supposed to be unchanged).

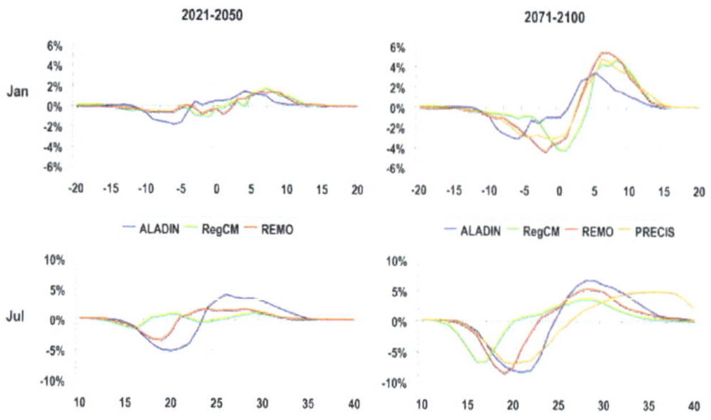


**Figure 3.** Changes of probability density function for daily temperature values over Hungary for 2021–2050 and 2071–2100 relative to 1961–1990.

## 4 Projections

Two future periods (2021–2050 and 2071–2100) are analysed. In 2021–2050 three models were accomplished, which are forced by A1B scenario, while for the further future four models are used, among them one is using B2, the others are using A1B scenario. Uncertainties coming from different (global and regional) model formulations are quantified for the first period and additionally the scenario-uncertainties, which is expected to be more robust for temperature in the larger lead time (i.e., after 2050; Hawkins and Sutton, 2009), are also addressed for the end of the 21st century.

### 4.1 Temperature

The annual and seasonal mean temperature values are projected to increase over Hungary during the 21st century. These statistically significant changes are larger for the far-future; annual temperature rises are 1.1–1.9 °C by 2021–2050 and 3.1–4.0 °C by 2071–2100. The largest changes, 0.7–2.6 °C and 3.5–6 °C and related uncertainties are expected to occur in summer. The 10 km resolution ALADIN-Climate simulation indicates some fine-scale spatial details: the temperature change has a northwest-southeast gradient, i.e., the warming has larger magnitude over the southeastern part of Hungary. The uncertainty is growing for the end of the century (not shown), which can arise from the use of more RCMs and different emission scenarios.

Change of probability density function for temperature (Fig. 3) indicates that the mean temperature increase is accompanied by the decreasing frequency of lower and increasing frequency of higher temperature values. This tendency is even more intensified at the end of the century, when probability density functions are shifted towards warmer temperature values.

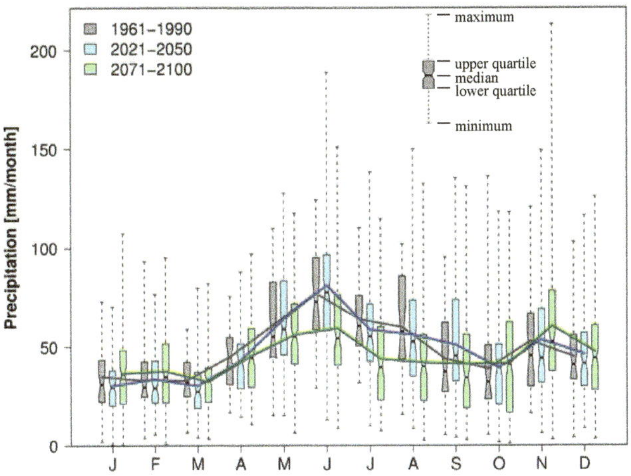


**Figure 4.** Box diagrams of the simulated monthly precipitation (mm/month) area-averaged over Hungary for 2021–2050 (blue) and 2071–2100 (green) and the observations (CRU, grey) for the period of 1961–1990. The future precipitation amounts are calculated by adding the simulated relative changes to the reference mean observed in 1961–1990. Curves are indicating 30-year average values, while lower and upper whiskers of the plot represent the minimum and maximum of the time-series of the given month.

## 4.2 Precipitation

Precipitation change is more ambiguous, i.e., in certain cases models disagree even in the direction of change, and in contrast to temperature values, the changes in precipitation are mostly non-significant. Annual, spring, and summer mean precipitation will probably slightly decrease, among them the summer decrease is clearly projected by all models. Annual changes are (−10)–0% and (−21)–(+3)% for the near- and far-future periods, respectively. In autumn and winter the precipitation amounts are likely to rather increase, however, the sign of the winter precipitation change is very uncertain in the near-future period, with higher chance for decrease. The uncertainty grows larger towards the end of the 21st century (not shown).

Figure 4 shows box diagrams of simulated area-averaged monthly precipitation compared to observations for 1961–1990 over Hungary. The future monthly values are not purely the model output values, they are calculated by adding the simulated monthly changes to the reference mean as observed in 1961–1990. Curves representing the corresponding 30-year average values indicate that the annual cycle may slightly change in the far-future. In the past, there was a primary maximum in June and a secondary one in November, while in 2071–2100 these months seem to become equally wet. Moreover, in some months (January, February, October), especially in the far-future, it can happen that almost no rain occurs. June in 2021–2050 and November in 2071–2100 can become the months with the largest inter-annual variability.

Probabilities of winter precipitation change exceeding certain thresholds can be seen in Fig. 5. In 2021–2050 it is probable that the decrease will be less than 20% (not shown), but there are some chances for both increase and decrease larger than 10%, as well. In the far-future this tendency changes: precipitation increase is more likely than decrease. It is suspected that this important change of direction is not due to the use of the fourth model with another scenario, since one of the three RCMs available also for the near-future period projected precipitation change with different sign in near- and far-future, which is the indication of the non-linear change between the two periods.

## 5 Conclusions

In this paper four RCMs adapted for the Carpathian Basin were studied in order to evaluate the uncertainty of their climate projections over Hungary. First, their validation was presented, then, projection results were demonstrated concerning mean temperature and precipitation changes.

Past simulations showed that model performance depends on the LBCs (ERA-40 or GCM), the considered variables (temperature or precipitation), the resolution (not shown). The validation results provide solid basis for the model developments, however, until the improved RCMs are available the delta-approach is used to eliminate systematic model errors (which are considered to be constant in time).

The Hungarian mini-ensemble projections provide one important addition with respect to previous results: while winter precipitation is decreasing in 2021–2050 it is increasing in 2071–2100 in opposition to the negative trend for the end of the 21st century discussed in earlier studies (Bartholy et al., 2008) based on the RCM results of the PRUDENCE project. Increasing mean and extreme temperature and decreasing annual and summer precipitation are the main characteristics of the results. For the annual precipitation, the mini-ensemble projects a slightly decreasing trend. However, there are some uncertainties over Hungary in the magnitude of summer warming and winter precipitation. Note that this mini-ensemble has only four members, hence, it is hard to determine the entire uncertainty range of climate change valid for the Carpathian Basin, therefore, this limitation should be considered when evaluating the results. Finally, we are planning to place our mini-ensemble into a larger uncertainty field based on the 25 km resolution model runs for Hungary provided by the ENSEMBLES project (van der Linden and Mitchell, 2009).

**Acknowledgements.** The authors are very grateful to Gabriella Csima and Adrienn Hunyady for their contribution to this study. This work was supported by the European Commission's 6th Framework Programme in the framework of CECILIA project (contract number 037005) and in the framework of CLAVIER project (contract number 037013). Research leading to this paper has been supported by the following additional sources: the

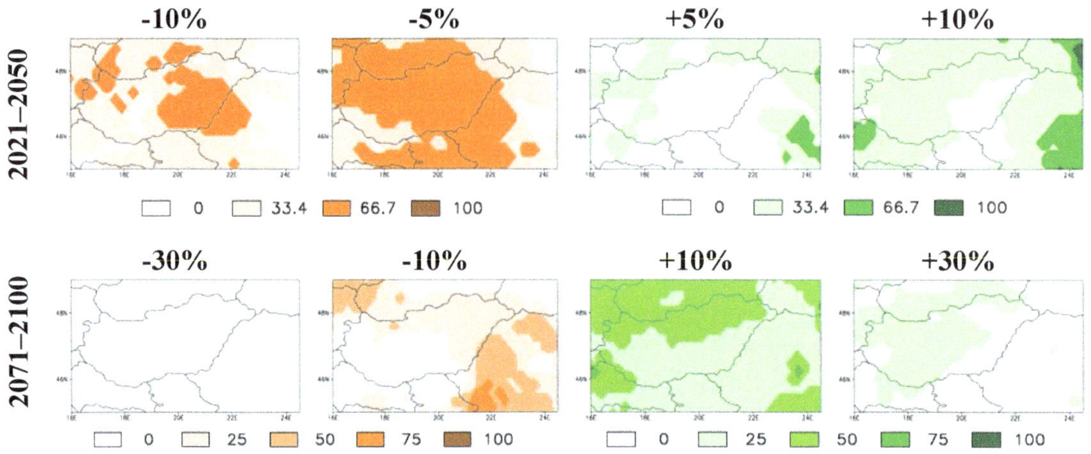


**Figure 5.** Probability (%) of winter precipitation change larger than 5, 10 and 30 percent for 2021–2050 and 2071–2100 relative to 1961–1990. Colors indicate in percentage how many out of three/four models are projecting a signal exceeding a selected threshold.

Hungarian National Science Research Foundation under grants K-78125, and the European Union and the European Social Fund (TÁMOP-4.2.1/B-09/1/KMR-2010-0003).

Edited by: S. Gualdi

## References

Bartholy, J., Pongrácz, R., Gelybó, Gy.,and Szabó, P.: Analysis of expected climate change in the Carpathian Basin using the PRUDENCE results, Időjárás, 112, 3–4, 249–264, 2008.

Bartholy, J., Pongrácz, R., Torma, Cs., Pieczka, I., Kardos, P., and Hunyady, A.: Analysis of regional climate change modelling experiments for the Carpathian basin, International Journal of Global Warming, 1, 238–252, 2009.

Christensen, J. H. and Christensen, O. B.: A summary of the PRUDENCE model projections of changes in European climate by the end of this century, Climatic Change, 81, 7–30, 2007.

Csima, G. and Horányi, A.: Validation of the ALADIN-Climate regional climate model at the Hungarian Meteorological Service, Időjárás, 112, 3–4, 155–177, 2008.

Giorgi, F., Marinucci, M. R., and Bates, G. T.: Development of a second generation regional climate model (RegCM2). Part I: Boundary layer and radiative transfer processes, Mon. Weather Rev., 121, 2794–2813, 1993a.

Giorgi, F., Marinucci, M. R., Bates, G. T., and DeCanio, G.: Development of a second generation regional climate model (RegCM2). Part II: Convective processes and assimilation of lateral boundary conditions, Mon. Weather Rev., 121, 2814–2832, 1993b.

Hawkins, E. and Sutton, R.: The potential to narrow uncertainty in regional climate predictions, BAMS, 90, p. 1095, doi:10.1175/2009BAMS2607.1, 2009.

Jacob, D. and Podzun, R.: Sensitivity studies with the regional climate model REMO, Meteorol. Atmos. Phys., 63, 119–129, 1997.

Mitchell, T. D., Carter, T. R., Jones, Ph. D., Hulme, M., and New, M.: A comprehensive set of climate scenarios for Europe and the globe, Tyndall Centre Working Paper 55, 2004.

Nakicenovic, N. and Swart, R. (Eds.): Emissions Scenarios, A Special Report of IPCC Working Group III, Cambridge University Press, Cambridge, UK, 2000.

Spiridonov, V., Déqué, M., and Somot, S.: ALADIN-CLIMATE: from the origins to present date, ALADIN Newsletter 29, 2005.

Szépszó, G. and Horányi, A.: Transient simulation of the REMO regional climate model and its evaluation over Hungary, Időjárás, 112, 3–4, 203–231, 2008.

Taylor, K. E.: Summarizing multiple aspects of model performance in single diagram, J. Geophys. Res., 106(D7), 7183–7192, 2001.

Torma, Cs., Bartholy, J., Pongrácz, R., Barcza, Z., Coppola, E., and Giorgi, F.: Adaptation of the RegCM3 climate model for the Carpathian Basin, Időjárás, 112, 3–4, 233–247, 2008.

Uppala, S. M., Kallberg, P. W., Simmons, A. J., Andrae, U., da Costa Bechtold, V., Fiorino, M., Gibson, J. K., Haseler, J., Hernandez, A., Kelly, G. A., Li, X., Onogi, K., Saarinen, S., Sokka, N., Allan, R. P., Andersson, E., Arpe, K., Balmaseda, M. A., Beljaars, A. C. M., van de Berg, L., Bidlot, J., Bormann, N., Caires, S., Chevallier, F., Dethof, A., Dragosavac, M., Fisher, M., Fuentes, M., Hagemann, S., Hólm, E., Hoskins, B.J., Isaksen, L., Janssen, P. A. E. M., Jenne, R., McNally, A. P., Mahfouf, J.-F., Morcrette, J.-J., Rayner, N. A., Saunders, R. W., Simon, P., Sterl, A., Trenberth, K. E., Untch, A., Vasiljevic, D., Viterbo, P., and Woollen, J.: The ERA-40 re-analysis, Q. J. Roy. Meteor. Soc., 131, 2961–3012, 2005.

van der Linden, P., Mitchell, J. F. B. (Eds.): ENSEMBLES: Climate Change and its Impacts: Summary of research and results from the ENSEMBLES project, Met Office Hadley Centre, Exeter EX1 3PB, UK, 2009.

Wilson, S., Hassell, D., Hein, D., Jones, R., and Taylor, R.: Installing and using the Hadley Centre regional climate modelling system, PRECIS, Version 1.3. Met Office Hadley Centre, Exeter, UK, 2005.

# A data portal for regional climatic trend analysis in a Peruvian High Andes region

**M. Schwarb[1], D. Acuña[2], Th. Konzelmann[3], M. Rohrer[1], N. Salzmann[4], B. Serpa Lopez[2], and E. Silvestre[2]**

[1]Meteodat GmbH, Zurich, Switzerland
[2]SENAMHI, Lima, Peru
[3]Federal Office of Meteorology and Climatology MeteoSwiss, Zurich, Switzerland
[4]Department of Geography, University of Zurich, Switzerland

**Abstract.** In the frame of a Swiss-Peruvian climate change adaptation initiative (PACC), operational and historical data series of more than 100 stations of the Peruvian Meteorological and Hydrological Service (SENAMHI) are now accessible in a dedicated data portal. The data portal allows for example the comparison of data series or the interpolation of spatial fields as well as download of data in various data formats. It is thus a valuable tool supporting the process of data homogenisation and generation of a regional baseline climatology for a sound development of adequate climate change adaptation measures. The procedure to homogenize air-temperature and precipitation data series near Cusco city is outlined and followed by an exemplary trend analysis. Local air temperature trends are found to be in line with global mean trends.

## 1 Introduction

The impacts of climatic changes are unequally distributed over the globe. The reason for this is on the one hand because climatic trends have a spatial variability (e.g. Groves et al., 2008) and on the other hand because different regions and systems are dissimilarly vulnerable and sensitive to changes (Schneider et al., 2007). With regard to ongoing international efforts (e.g. Adaptation Fund (AF) and the new Climate Fund under the United Nations Framework Convention on Climate Change, UNFCCC) for developing and implementing adequate adaptation measures, reliable climate data baselines on a regional – or even local – level become increasingly important. Successful and sustainable adaptation measures require knowledge on regional climatic trends and must be embedded in their regional and local context, that is, they must be developed in close collaboration with the people affected and considering their cultural, political and environmental context (Yohe et al., 2007).

Within the frame of a Peruvian-Swiss climate change adaptation initiative (PACC; Salzmann et al., 2009) for the Central Peruvian Andes, substantial efforts have been undertaken to provide a climatological baseline for the region Cusco-Apurimac in the Central Peruvian Andes (Fig. 1). It is aimed at completing and homogenizing available data records and consequently provide reliable, long-term climate time series in order to (i) derive and analyse climatic trends in the target region, to (ii) compute scenarios (statistical downscaling of GCM results), and eventually to (iii) serve as the climatological basis for developing and implementing adequate adaptation measures. Additionally, the performance of alternative data sets has been analysed for the region, such as satellite precipitation estimations from TRMM (Scheel et al., 2010).

So far, there are two widely known meteorological data portals publicly accessible that allow for download, graphical representation and analysis of climate data worldwide: (i) the KNMI Climate Explorer (2010) and (ii) the Goddard Institute of Space Studies (GISS) global surface temperature analysis data portal (NASA-GISS, 2010), which is partly based on the Global Historical Climatology Network (GHCN; Peterson et al., 1997). However, these data portals only provide a handful of stations for the Peru's Central Andes region. Here, we present an approach for collecting, pre-analysing and distributing climate data on a regional level through a data portal specifically developed for the Cusco-Apurimac Region (Fig. 1). The aim is to include as many hydro-meteorological data as possible for the target region.

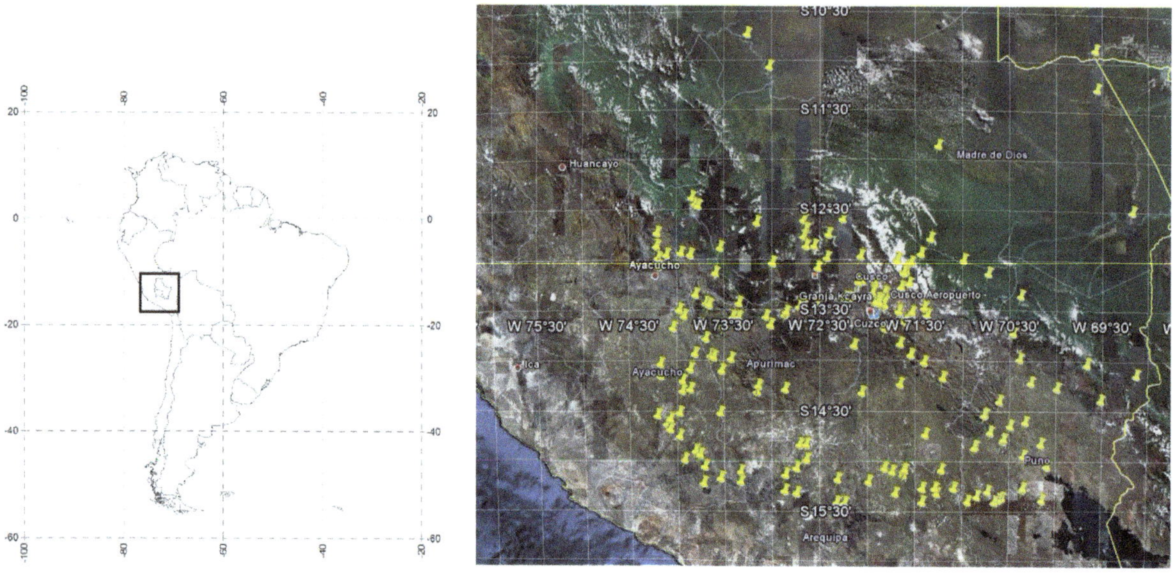


**Figure 1.** Region Cusco-Apurimac in the Peruvian Central Andes and the respective available climate stations. The plots, as shown, are provided in the PACC data portal.

Moreover, a strategy is described on how to generate reliable long-term daily climate time series in a remote mountain region with typically relatively low spatial and temporal data coverage. In an exemplary manner, results for the station Granja Kcayra (near Cusco) are shown and analyzed.

## 2 Methods

In Peru, measuring and maintaining hydrological and meteorological data is under the responsibility of the National Service for Meteorology and Hydrology (SENAMHI). Data records are stored as hard copies, and the process of digitizing this data is currently ongoing. SENAMHI publishes daily data of the past few years on their webpage (www.senamhi.gob.pe). For access to longer historical time series, SENAMHI must be contacted personally.

For the purposes and goals of PACC, a data portal has been established which is described in the following section.

### 2.1 Data portal

In remote mountain regions proper measurement of meteorological data and station maintenance is typically challenging, and human as well as financial resources are often very limited, data series are incomplete, implausible and inhomogeneous. Prior to use these data records for climatic analyses, it is required to first quality check, treat, complete and homogenize these data (Begert et al., 2005; Appenzeller et al., 2008). Currently, there is an increasing demand of reliable climate time series by scientists and decision makers alike for the purposes of past climate trend analyses, scenario construction and impact assessment, and eventually as a basis for adequate development of adaptation measures. Therefore, tools and strategies must be developed to ease data homogenization processes, even and particularly in remote mountain regions, where spatial climate variability is high and often complex due to strong topographic influences. The set-up of a data portal can thereby be a critical initial step and a helpful tool. It allows for an overview and initial exploration of the available data in a specific region, and for transparent and centrally managed data update. That is, a data portal is a management and initial exploring tool, rather than a substitute for scientific data analysis tools.

The initiation of a data portal goes naturally along with easy access of many valuable data, which opens many possibilities for new studies. Communication and declaration about data policy, including access, use and eventual publication of data is thus critical and appropriate attention should be given to this issue at a very early stage of the implementation.

The data portal created within the frame of PACC is based on a kernel programmed in FORTRAN, and provides a user-interface realised in HTML. For the relatively small project region (Cusco-Apurimac; Fig. 1), a considerable amount of climate data is actually available. Figure 2 shows the frequency distribution per altitude range of the stations included in the data portal. From Fig. 2 it is evident, that the number of stations has decreased from more than 120 stations between 1960 and 1970 to about 80 stations. Between 1980 and late 1990, the station count reduction was particularly pronounced. During the last two decades, the number of station again increased moderately. At altitudes between 3000 m and 4000 m a.s.l., there are currently about 40 stations available.

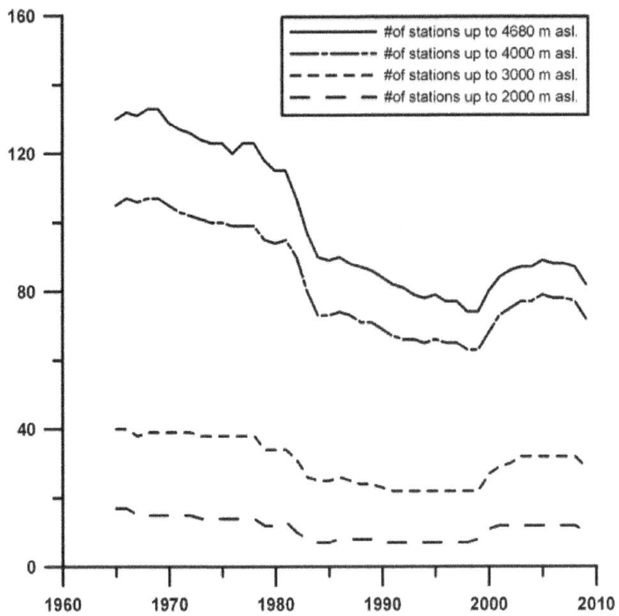


**Figure 2.** Frequency distribution per altitude class of climatological stations in the Cusco-Apurimac region that are accessible in the PACC data portal.

Furthermore, the PACC data portal also includes data from alternative sources like satellites (Tropical Rainfall Measuring Mission; TRMM) or hydrological run off data and provides several useful functionalities to screen, explore, visualize, export, etc. data (Fig. 3). In the following some of the most important features are described.

### 2.1.1 Metadata access

Proper metadata description and access is a critical element of a data portal, in particular when a portal is aimed at supporting data homogenization processes. The PACC data portal allows for easy access to all available meta data for each data record (Fig. 4). For each station, geographical longitude, latitude and altitude is indicated. A click on the station name leads to a outline map of the station. If a station relocation is known, then the station-ID changes, the station name is indexed and a separate metadata-file is provided. For each station the range of each variable is listed. The lower and upper limits for a variable can actively be set in the portal and outliers can optionally be turned into missing values.

### 2.1.2 Temporal and spatial comparison of data records

The PACC portal enables temporal and spatial comparison of time series through several plotting possibilities.

*The function xy-plot* allows for comparison of two stations (Fig. 5). The example in Fig. 5 shows a good correlation, with a root mean square deviation (RMS) of 27 mm for monthly precipitation of the stations La Angostura and Tisco. These two stations are about 20 km apart and differ in elevation by some meters only.

The xy-plot function (Fig. 5) may be used for finding stations with similar climatic characteristics in order to fill data gaps, or for getting a rapid impression about the meteorological variability within a region.

*The function matrix-plot* (Fig. 6) allows (i) for a rapid overview of available data (daily or temporal aggregated, see Sect. 2.1.3) within a defined time period (in red: missing values) and (ii) for a visual impression about the seasonality and inter-annual variability of precipitation sum.

*The sumplot function* (Fig. 9) allows for enhanced double-mass curve homogeneity testing (Craddock-test). In the data portal enhanced double-mass curve functionality allows for a first homogeneity test. The double-mass curve is a popular tool to check homogeneity (e.g. Buishand, 1982). This curve is obtained by plotting the cumulative amounts of the considered station, against the cumulative amounts of one or a set of neighbouring stations. The plotted points tend to fall along a straight line under conditions of homogeneity. Instead of the double-mass curve one can also plot the cumulative deviations from some average value, as it is implemented in the portal. The cumulative deviations have the advantage that changes in the mean amount of rainfall are easier recognized (Craddock, 1979). The graph of the cumulative deviations is sometimes called a residual mass curve. A more current application of the "craddock" method can be found e.g. in Brunetti et al. (2006).

### 2.1.3 Temporal data aggregation

The data portal allows for temporal data aggregation. The data in the portal is stored as daily values, which can be temporally aggregated to monthly or annual means or sums. The temporal aggregation can then be visualized with the tools presented above and may be exported in aggregated form (see also Sect. 2.1.5).

### 2.1.4 Spatial interpolation of daily/monthly precipitation fields

Another useful function is the spatial interpolation of daily and monthly precipitation fields (works only for precipitation). Within the portal, the inverse-distance weighting (IDW) interpolation method is implemented, intended for a first, rapid and explorative overview of the precipitation regime in the target region (Cusco and Apurimac, Peru). Because the availability of precipitation stations is variable in time, the interpolation includes only the available stations, which can optionally be plotted on the map. Plots can easily be generated for daily or monthly values over a user-defined time period. Moreover, the generated plots can be animated and in this manner provide an even better impression of the dynamics of the precipitation regime.

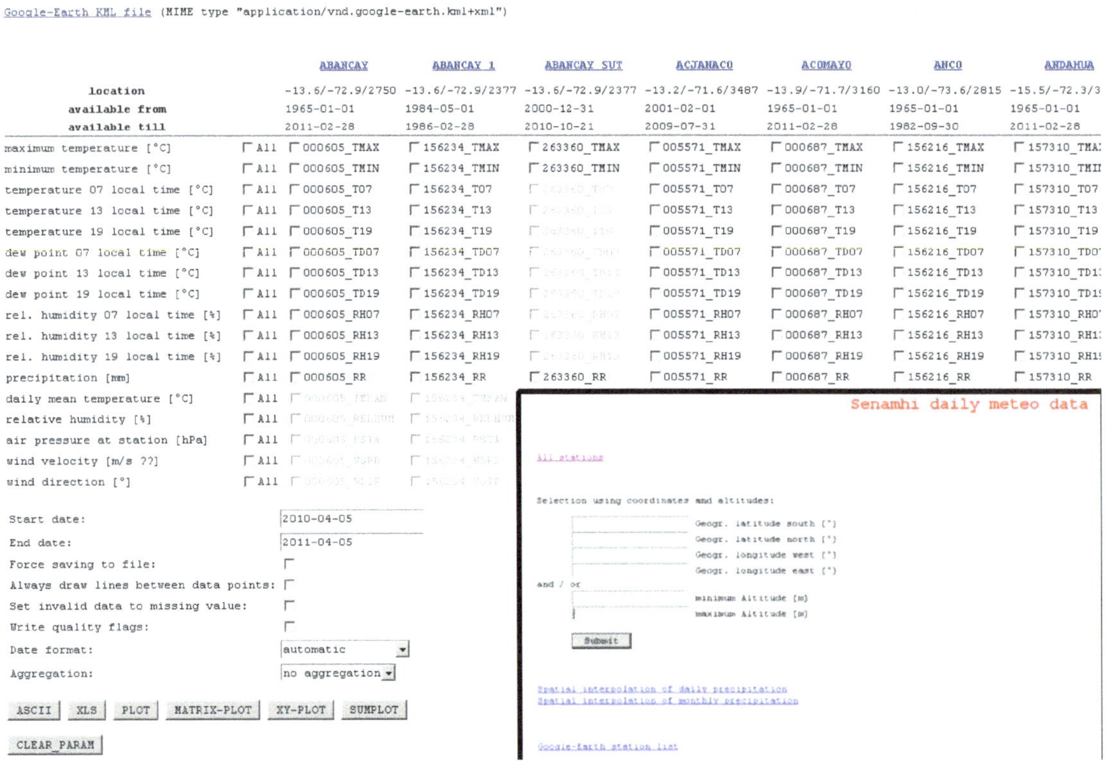


**Figure 3.** Data portal of the PACC project with the possibility of pre-selecting stations by a geographical criterion, as a latitude-longitude and/or an altitude range (lower right corner). In the main screen of the data portal, a station-parameter combination can be selected for download, analysis or graphical representation of meteorological data.

**Figure 4.** Metadata in the PACC data portal, here for the Station Granja Kcayra. The metadata includes a link to Google Earth, with the geographic location of each station. The thresholds for plausible values are also indicated.

For the application of more sophisticated interpolation methods in order to produce final interpolation maps, the data can be exported in order to apply specific and more sophisticated interpolation methods (e.g. PRISM (Parameter-elevation Regressions on Independent Slopes Model); see Daly et al., 1994)

### 2.1.5 Export and download functionalities

All of the aforementioned products generated within the functionalities provided by the data portal may be exported and downloaded. Time series (daily or temporally aggregated) can be downloaded in ASCII or CSV format, plots exported in the PNG format.

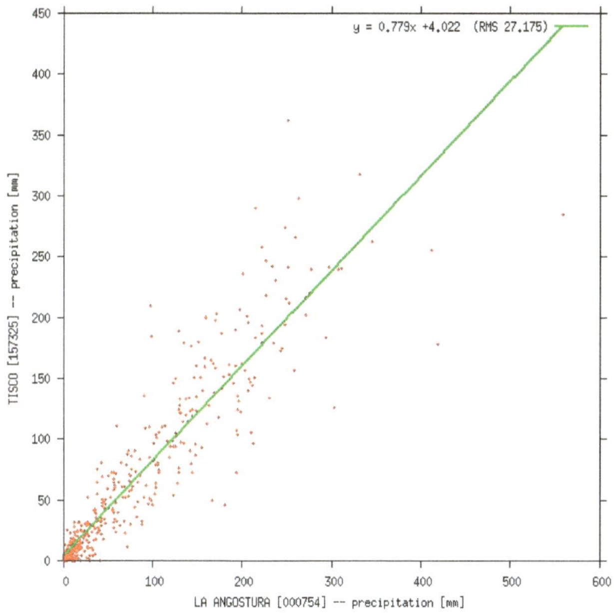


**Figure 5.** XY-plot for the two high-altitude precipitation stations La Angostura (15.18° S, 71.65° W, 4150 m a.s.l.) and Tisco (15.35° S, 71.45° W, 4175 m a.s.l.). Shown are monthly precipitation sums (red crosses), the equation of the line of best fit (green line) and the RMS value.

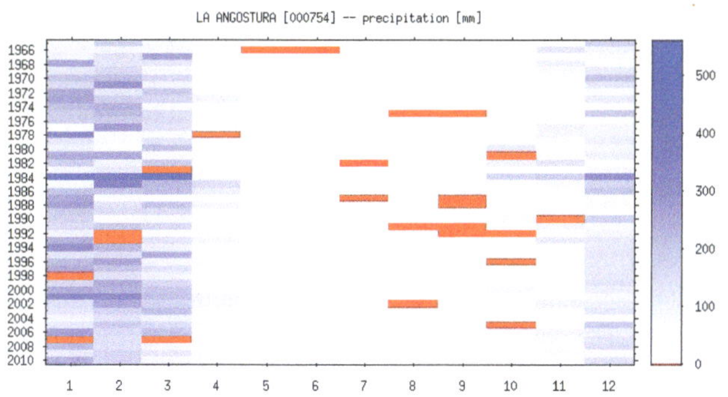


**Figure 6.** Matrix plot of monthly (aggregated from daily values) precipitation sums for the precipitation station La Angostura (15.18° S, 71.65° W, 4150 m a.s.l.) from January 1965 to December 2010. The red bars are for monthly values having less than the predefined minimum number of plausible daily values required for the calculation of a monthly mean or sum.

### 2.1.6 Data plausibility

The process of data homogenization (see next section) includes the check of data series regarding their plausibility and the detection and replacement of implausible or erroneous values. The above presented data portal tools provide several ways to check data records regarding their plausibility. Figure 7 shows an example of a detected implausible value. While precipitation sums of the three stations Janacancha, La Angostura and Tisco have similar monthly sums between 1970 and 1988, it is evident from the comparison with the other two stations that for La Angostura in February 1984 the monthly sum is implausibly high. This error may be introduced by technical problems with the measurement station, misreading of the value by the person who transmitted the data etc. In a first step, this value may be turned into "missing value" in the data portal. In a second step, the reason for the implausible value either is found and corrected or a plausible value is estimated through interpolation.

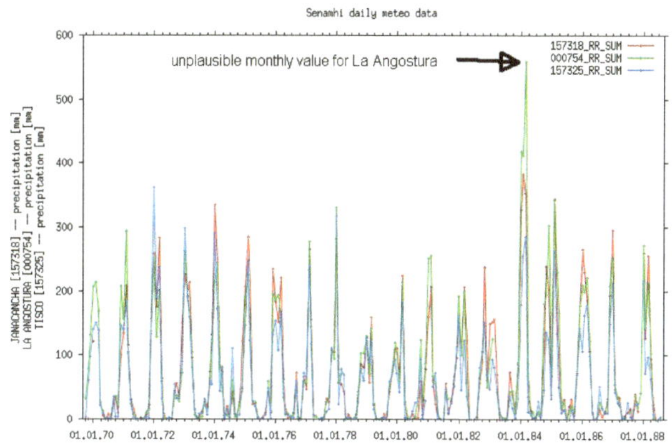


**Figure 7.** Detection of an unplausible value in a precipitation data series of the stations Janacancha (15.18° S, 71.77° W, 4320 m a.s.l., in red), La Angostura (15.18° S, 71.65° W, 4150 m a.s.l., in green) and Tisco (15.35° S, 71.45° W, 4175 m a.s.l., in blue).

### 2.2 The homogenization approach

As stated above, plausible, complete and homogeneous climate data series are an absolute prerequisite for any climate related trend analysis (e.g. Appenzeller et al., 2008). Despite this importance, quality-proofed and homogenized data series are often not available, particularly for mountain regions. A reasonable explanation herefore is probably that data quality control, completion and homogenization is generally a challenging and time consuming task, particularly in remote high mountain regions where data is scarce and topography raises additional complexity (Begert et al., 2005; Auer et al., 2007).

In the region of Cusco-Apurimac, there is a significant number of data series available (see above and Fig. 1) and many of them with data records starting as early as 1965. On the one hand, however, station records and station history are often fragmentary or lost, among other reasons also as a result of armed conflicts in the region, particularly in the 1980s, and during the past decades, many stations also have been relocated. On the other hand, methods and measuring techniques have remained the same in the Cusco-Apurimac region during the entire measuring period.

In order to improve data quality despite the limitations outlined above, a strategy was developed to enhance data quality and reliability significantly and to complete data series where

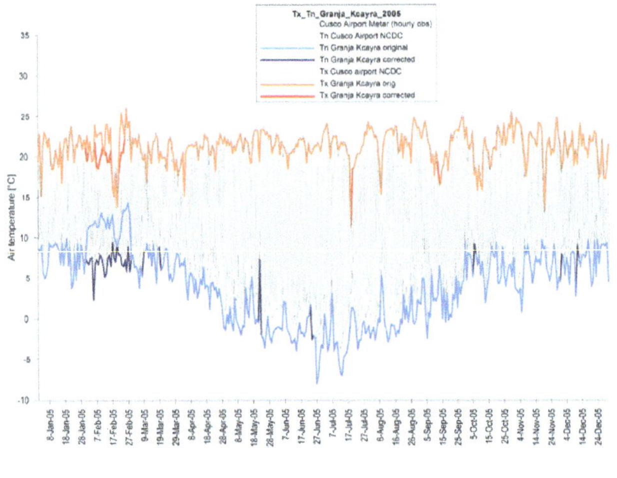


**Figure 8.** Comparison between daily minimum (Tn, in light blue) and maximum Temperature (Tx, in orange) of Granja Kcayra station (13.56° S, 71.88° W, 3219 m a.s.l., near Cusco city), hourly METAR air temperature observations (in light grey) and NCDC Tx and Tn data records of the nearby station of Cusco-Airport (13.56° S, 71.88° W, 3219 m a.s.l., in grey), and the respective corrections made for Granja Kcayra (in dark red and dark blue) for the year 2005.

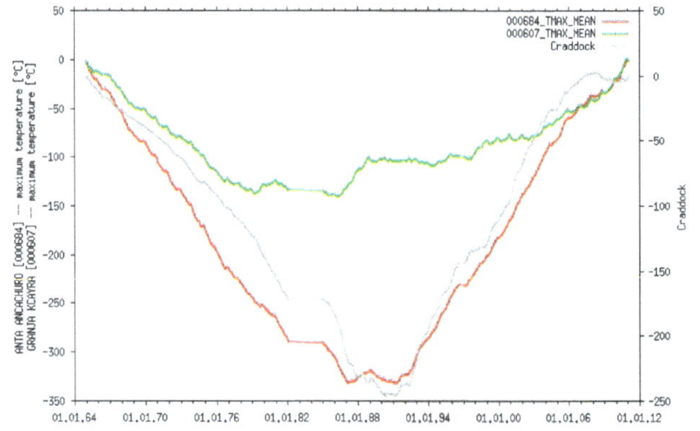


**Figure 9.** The craddock test (Craddock, 1979), as available in the PACC data portal shows a strong inhomogeneity in the Anta Ancachura-Granja Kcayra relation of monthly mean maximum temperature.

possible. The strategy followed here is based on known regional climatic trends and thus requires complete, long-term time series, which, however, are not available for the Cusco-Apurimac region. To overcome this limitation, instead of using a single long-term record as a reference station, clusters/ensembles of available station series where used which then served as a substitute for one single, "reference" station. This is similar to the procedure described in Brunetti et al. (2006), where the a priory existence of a homogeneous reference series is also not necessary. In the following, our approach is outlined in more detail for the example for the station cluster around Granja Kcayra, near Cusco (13.56° S, 71.88° W, 3219 m a.s.l.; data record back to 1965).

In the vicinity of the station Granja Kcayra, two additional data sources provide time series of climate variables; the Meteorological Aerodrome Records (METARs) and data from the National Climate Data Center (NCDC) for Cusco Airport (13.54° S, 71.94° W, 3249 m a.s.l.). METAR-records can be downloaded e.g. from NOAA (NOAA, 2011). These two additional data sources can be used for homogenization and data correction. As shown in Fig. 8, e.g. erroneous minimum and maximum temperatures can be corrected. The METARs of Cusco Airport, unlike to most other airports worldwide, provide additional variables, such as hourly precipitation sums. With this basis, e.g. daily precipitation values wrong by the factor 10 can be detected and corrected.

Four more stations are situated within 50 km horizontal distance and with less than 200 m elevation difference: Acomayo (13.92° S, 71.68° W, 3160 m a.s.l.), Anta Ancachuro (13.47° S, 72.22° W, 3340 m a.s.l.), Paruro (13.77° S, 71.84° W, 3084 m a.s.l.) and Paucartambo (13.32° S, 71.59° W, 3042 m a.s.l.).

The homogeneity of the station cluster can than be checked with the data portal by forming station couples and applying the craddock test (Craddock, 1979). Figure 9 shows the result of the Craddock test for maximum temperature of the station couple Granja Kcayra vs. Anta Ancachuro. A strong inhomogeneity for the values before and after March 1988 is indicated by the strong deviation of the cumulative deviation from average. As the station couple Paucartambo vs. Anta Ancachuro also shows this inhomogeneity, but not Paucartambo vs. Granja Kcayra, or Acomayo vs. Granja Kcayra, it is very likely that maximum temperature at Anta Ancachuro is not homogeneous before and after March 1988. Consequently, the time series of Anta Ancachuro is homogenized by adding the monthly mean differences between March 1988 and 2010 to the respective mean maximum temperature of the other stations. Precipitation is homogenized in a similar way using the respective quotients of the monthly sums between the stations to be homogenized and the reference station cluster. As also stated by Brunetti et al. (2006), it is problematic to homogenize small breaks (craddock curve deviates only slightly from straight line) with no indication from metadata. Therefore, small breaks are not homogenized here, neither.

## 3 Trend analysis for the station Granja Kcayra, Cusco

Trends for the station Granja Kcayra were analyzed based on the records previously treated as outlined above. Linear trends and gaussian filtered trends are shown (Figs. 10 and 11). Mean yearly air temperature show a rather small year-to-year variation of about 2 °C at most. For mean yearly air temperature, a clear positive trend is found for the period 1965 to 2010, which is somewhat more pronounced between

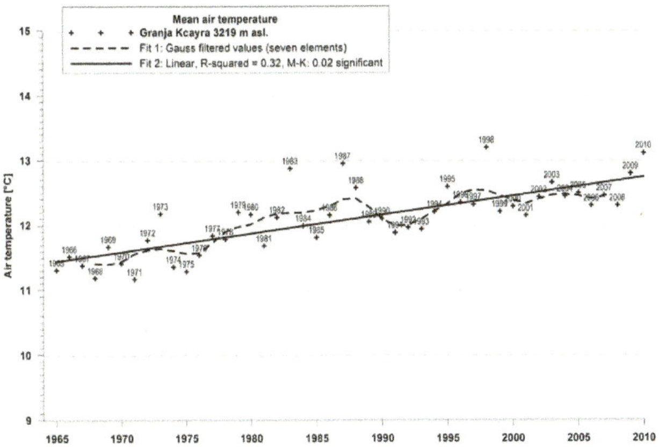


**Figure 10.** Air temperature and linear and gaussian low-pass filtered trends for the homogeneous time series of Granja Kcayra (−13.56° S, −71.88° E, 3219 m a.s.l.). Linear trend is 0.3 °C decade$^{-1}$ between 1965 and 2010. This trend is significant at the 0.02 level with Mann-Kendall test.

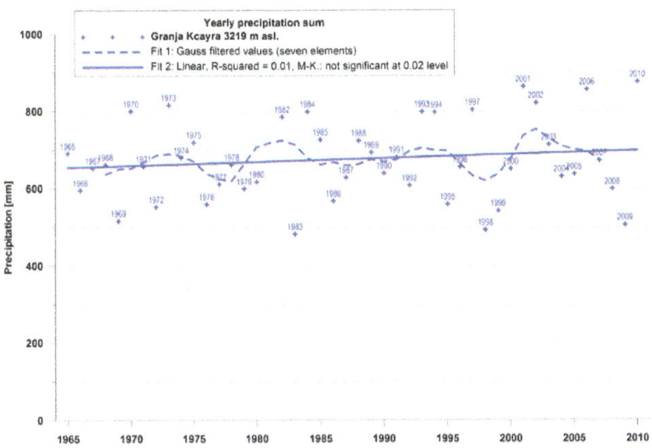


**Figure 11.** Yearly precipitation sum and linear and gaussian low-pass filtered trends for the homogeneous time series of Granja Kcayra (−13.56° S, −71.88° E, 3219 m a.s.l.). Linear trend between 1965 and 2010 is not significant at the 0.02 level with Mann-Kendall test.

1965 and 1988 (Fig. 10). Yearly precipitation sums show virtually no trend for the mentioned period, but a considerable year-to-year variation of up to about 400 mm, the long-term mean being around 700 mm (Fig. 11).

Putting these trends in a more continental/global context, we can state that from visual comparison the linear air temperature trends found so far found for the Cusco area are in line with the large-scale global air temperature trends (IPCC, 2007). Compared to observed trends e.g. in Europe, the trends for Cusco are however lower. The trend of the mean yearly air temperature in Granja Kcayra is about 0.30 °C decade$^{-1}$ for 1965 to 2010, whereas e.g. for Zurich-Fluntern (Switzerland) the respective trend is 0.43 °C decade$^{-1}$ for the same period (MeteoSchweiz, 2011).

The lower trend of Granja Kcayra air temperature may be related the so called "global dimming" or "decreasing clear sky visibility over land" in the last few decades, as described e.g. in Wild (2009) or in Wang et al. (2009). While there was a "European brightening" since the mid-80s, global dimming continued, particularly over Asia, South America, Australia and Africa.

Regarding precipitation sums, it seems that there is a slight increase since mid 1960s for Cusco (Fig. 11). However, this very weak trend is not significant with a Mann-Kendall test at 0.02 level. In the central Andes region, wet season precipitation is rather depending on large scale wind conditions in the upper troposphere than on ENSO as stated e.g. by Lenters and Cook (1999); Garreaud and Aceituno (2001) and Rohrer et al. (2010). So, the rather pronounced year-to-year variations in precipitation may be correlated to the continental upper air wind regime.

## 4 Conclusion and perspectives

The challenges of adapting to adverse impacts of climatic changes require solid data baselines on regional to local scales. In mountain regions, where impacts are expected to be particularly high and the adaptive capacity of the people is often low, reliable long-term climate time series are rare and spatial and temporal data coverage is typically low. Therefore, straightforward tools and methods are needed, to generate a data baseline, that is, reliable long-term climate time series. Here, we presented a data portal tool, developed in the frame of the PACC program. We found the data portal a particularly valuable tool for data collection, screening, homogenization and distribution. The study also shows that search for alternative data (e.g. METAR) can be well worth as they can serve as important data sources in a data quality control, interpolation and homogenization process.

Our experience has furthermore shown, that a common climate data platform such as the described data portal can fulfil an important role for interdisciplinary research in climate change adaptation. It also stimulates an enhanced open interaction between various government agencies involved in planning adaptation measures. In this sense, our data portal approach could be a model case for other adaptation projects at the local to national level, especially in developing countries.

Currently, the portal is open for PACC collaborators only, for ongoing studies in the fields of hydrology, food security, and disaster risk reduction. In future, however, it might become accessible also for a wider scientific community.

First trend analysis for temperature and precipitation show a slight positive linear trend.

**Acknowledgements.** We thank the Swiss Agency for Development and Cooperation (SDC) and the Ministry of Environment of Peru for enabling the studies on which this work is based through the Programa de Adaptacion al Cambio Climatico (PACC) in Peru.

Edited by: M. Brunet-India

## References

Appenzeller, C., Begert, M., Zenklusen, E., and Scherrer, S. C.: Monitoring climate at Jungfraujoch in the high Swiss Alpine region, Sci. Total Environ., 391, 262–268, 2008.

Auer, I., Böhm, R., Jurkovic, A., Lipa, W., Orlik, A., Potzmann, R., Schöner, W., Ungersböck, M., Matulla, C., Briffa, K., Jones, P. D., Efthymiadis, D., Brunetti, M., Nanni, T., Maugeri, M., Mercalli, L., Mestre, O., Moisselin, J. M., Begert, M., Müller-Westermeier, G., Kveton, V., Bochnicek, O., Stastny, P., Lapin, M., Szalai, S., Szentimrey, T., Cegnar, T., Dolinar, M., Gajic-Capka, M., Zaninovic, K., Majstorovic, Z., and Nieplova, E.: HISTALP – Historical Instrumental Climatological Surface Time Series of the Greater Alpine Region, Int. J. Climatol., 27, 17–46, 2007.

Begert, M., Schlegel, T., and Kirchhofer, W.: Homogeneous temperature and precipitation series of Switzerland from 1864 to 2000, Int. J. Climatol., 25, 65–80, 2005.

Brunetti, M., Maugeri, M., Monti, F., and Nanni, T.: Temperature and precipitation variability in Italy in the last two centuries from homogenised instrumental time series, Int. J. Climatol., 26, 345–381, 2006.

Buishand, T.: Some methods for testing the homogeneity of rainfall records, J. Hydrol., 58, 11–27, 1982.

Craddock, J. M.: Methods of comparing annual rainfall records for climatic purposes, Weather, 34, 332–346, 1979.

Daly, C., Neilson, R. P., and Phillips, D. L.: A Statistical-Topographic Model for Mapping Climatological Precipitation over Mountainous Terrain, J. Appl. Meteor., 33, 140–158, 1994.

Garreaud, R. D. and Aceituno, P.: Interannual rainfall variability over the South American Altiplano, J. Clim., 14, 2779–2789, 2001.

Groves, D. G., Yates, D., and Tebaldi, C.: Developing and applying uncertain global climate change projections for regional water management planning, Water Resour. Res., 44, W12413, doi:10.1029/2008WR006964, 2008.

IPCC: The Physical Science Basis. Contribution of Working Group I to the Fourth Assessment Report of the Intergovernmental Panel on Climate Change, edited by: Solomon, S., Qin, D., Manning, M., Chen, Z., Marquis, M., Averyt, K. B., Tignor, M., and Miller, H. L., Cambridge University Press, Cambridge UK, 235–336, 2007.

KNMI Climate Explorer: http://climexp.knmi.nl/, last access: 21 December 2010.

Lenters, J. D. and Cook, K. H.: Summertime precipitation variability over South America: Role of large-scale circulation, Mon. Weather Rev., 127, 409–431, 1999.

MeteoSchweiz: http://www.meteoschweiz.admin.ch/web/de/klima/klima_heute/homogene_reihen.html, last access: 20 April 2011.

NASA-GISS: http://data.giss.nasa.gov/, last access: 21 December 2010.

NOAA: ftp://tgftp.nws.noaa.gov/data/observations/metar, last access: 12 July 2011.

Peterson, Th., Russell, C., and Vose, S.: An Overview of the Global Historical Climatology Network Temperature Database, B. Am. Meteorol. Soc., 78, 2837–2849, 1997.

Rohrer, M., Acuña, D., Calanca, P., Huggel, C., Konzelmann, T., Salzmann, N., Schwarb, M., and Silvestre, E.: Dependence of wet season precipitation in the Central Andes on Pacific-SST anomaly patterns, tropospheric temperature, pressure and humidity fields, Poster presented during the American Geophysical Union Meeting of the Americas, 8–12 August 2010, Foz de Iguassu, Brasil, http://www.meteodat.ch/pdf/AGUFOZ.pdf, 2010.

Schneider, S. H., Semenov, S., Patwardhan, A., Burton, I., Magadza, C. H. D., Oppenheimer, M., Pittock, A. B., Rahman, A., Smith, J. B., Suarez, A., and Yamin, F.: Assessing key vulnerabilities and the risk from climate change, Climate Change 2007: Impacts, Adaptation and Vulnerability, Contribution of Working Group II to the Fourth Assessment Report of the Intergovernmental Panel on Climate Change, edited by: Parry, M. L., Canziani, O. F., Palutikof, J. P., van der Linden, P. J., and Hanson, C. E., Cambridge University Press, Cambridge, UK, 779–810, 2007.

Salzmann, N., Huggel, C., Calanca, P., Díaz, A., Jonas, T., Jurt, C., Konzelmann, T., Lagos, P., Rohrer, M., Silverio, W., and Zappa, M.: Integrated assessment and adaptation to climate change impacts in the Peruvian Andes, Adv. Geosci., 22, 35–39, doi:10.5194/adgeo-22-35-2009, 2009.

Scheel, M. L. M., Rohrer, M., Huggel, C., Santos Villar, D., Silvestre, E., and Huffman, G. J.: Evaluation of TRMM Multi-satellite Precipitation Analysis (TMPA) performance in the Central Andes region and its dependency on spatial and temporal resolution, Hydrol. Earth Syst. Sci. Discuss., 7, 8545–8586, doi:10.5194/hessd-7-8545-2010, 2010.

Wang, K., Dickinson, R. E., and Liang, S: Clear Sky Visibility Has Decreased over Land Globally from 1973 to 2007, Science, 323, 1468–1470, 2009.

Wild, M: Global dimming and brightening: A review, J. Geophys. Res., 114, 1–31, 2009.

Yohe, G. W., Lasco, R. D., Ahmad, Q. K., Arnell, N. W., Cohen, S. J., Hope, C., Janetos, A. C., and Perez, R.T.: Perspectives on climate change and sustainability, Assessment Report of the Intergovernmental Panel on Climate Change, edited by: Parry, M. L., Canziani, O. F., Palutikof, J. P., van der Linden, P. J., and Hanson, C.E., Cambridge University Press, Cambridge, UK, 779–810, 2007.

# 20

# Improving short-term forecasting during ramp events by means of Regime-Switching Artificial Neural Networks

**C. Gallego[1], A. Costa[1], and A. Cuerva[2]**

[1]Wind Energy Forecasting Group, CIEMAT. Avd. Complutense 22, 28040 Madrid, Spain
[2]IDR/UPM, E.T.S.I. Aeronáuticos, Universidad Politécnica de Madrid, Pza. de Cardenal Cisneros 3, 28040 Madrid, Spain

**Abstract.** Ramp events are large rapid variations within wind power time series. Ramp forecasting can benefit from specific strategies so as to particularly take into account these shifts in the wind power output dynamic. In the short-term context (characterized by prediction horizons from minutes to a few days), a Regime-Switching (RS) model based on Artificial Neural Nets (ANN) is proposed. The objective is to identify three regimes in the wind power time series: Ramp-up, Ramp-down and No-ramp regime. An on-line regime assessment methodology is also proposed, based on a local gradient criterion. The RS-ANN model is compared to a single-ANN model (without regime discrimination), concluding that the regime-switching strategy leads to significant improvements for one-hour ahead forecasts, mainly due to the improvements obtained during ramp-up events. Including other explanatory variables (NWP outputs, local measurements) during the regime assessment could eventually improve forecasts for further horizons.

## 1 Introduction

Since wind energy cannot be neither scheduled nor large-scale stored, wind power forecasting is required in order to minimize the impact of the variability of the wind. In particular, short-term forecasting is currently required by energy producers (in a daily electricity market context) and the transmission system operators (TSOs) in order to keep the balance of an electrical system. Within the short-term context, time-series based models (i.e. statistical models) have shown a better performance than Numerical Weather Prediction (NWP) models for horizons up to few hours (Giebel, 2003; Costa, 2005). These models try to learn and replicate the dynamic shown by a certain time series, for instance the power output time series of a wind farm.

Wind power ramp events are characterized by large gradients in the time series during relatively short time periods (few hours). Ramp events occur not very often, but they may have a real impact on an electrical system due to the unexpected variation in the generation side. Additionally, energy traders incur penalties in energy markets due to deviations from the scheduled energy (see Greaves et al., 2009, and references therein). These impacts could be reduced with forecasting models focused on ramp events (Potter et al., 2009). Ramp events are usually motivated by specific meteorological processes (crossing fronts, fast changes in the local wind direction) involving several scales (synoptic, mesoscale, microscale). In other cases, wind power generation may experience drops related to other causes, such as voltage dips. A ramp event can be then considered as an "unexpected/atypical" dynamic due to a change in the underlying causes of the wind power conversion process. Consequently, traditional statistical models considering only one dynamic for the whole wind power time series may be inadequate. Section 2 describes the implementation of statistical models based on Artificial Neural Networks (ANNs) for the case of wind power short-term forecasting. In particular, Sects. 2.1 and 2.2 deal with a non-regime and a regime-switching strategy respectively. Main results of the study are gathered and discussed in Sect. 3.

## 2 Methodology

Let $\{y_t\}$, $t = 1,...,N$ be a discrete time series with $N$ observations of hourly averaged wind power. Time series based models traditionally states that $\{y_t\}$ follows a stochastic process like (Peña, 1987; Madsen, 2007):

$$y_{t+k} = f(y_{t-d+1},...,y_t,\Theta) + \varepsilon_{t+k} \tag{1}$$

where $k$ is the prediction horizon. $f$ represents the deterministic component of $y_{t+k}$ as a function of a set of $d$ previous values $[y_{t-d+1},...,y_t]$ and a certain set of parameters $\Theta$. $\{\varepsilon_t\}$ is the noise of the stochastic process, assumed to follow a white noise process $\{\varepsilon_t\} \sim N(0,\sigma^2)$.

The purpose of time series models is to estimate the unknown function $f(\cdot,\Theta)$ in order to provide accurate forecasts. In this way, an appropriate window size $d$ and an optimized set of parameters $\Theta$ have to be found from historical observations. A criterion based on the minimization of the Root Mean Squared Error (RMSE) is usually employed.

### 2.1 The single-ANN model

Non-linear models are usually required in order to estimate the unknown function $f(\cdot,\Theta)$ introduced in Eq. (1). It is mainly due to the fact that there are many non-linear effects involved in the wind power conversion process, such as the behaviour of the atmosphere and the power curve of wind turbines. Multilayer Feedforward Networks, a kind of ANNs, represent a powerful tool to approximate any function (Hornik et al., 1989). In particular, we employ Multilayer Perceptrons to estimate $f(\cdot,\Theta)$ based on the backpropagation learning algorithm (Lippmann, 1987).

The implementation of an optimal ANN requires an appropriate choice of:

- The ANN architecture, which comprises the number of layers and the number of neurons per layer. The set of parameters $\Theta$ gathers weights and bias of the connections between neurons from one layer to the next one.

- The window size $d$ (Kil et al., 1997).

In this work, the selection of the optimal ANN is based on a cross-validation criterion. First, several ANNs with different architectures (from 1 to 3 layers) and window size (from 1 to 10 lags) are trained over a historic dataset (training-set). The mentioned range of values are based on empirical experience. Then, the ANN with the lowest RMSE over the validation dataset is considered to provide the best approach of $f(\cdot,\Theta)$, being the validation dataset different from that dataset used for training the ANN. The performance of the selected ANN is evaluated over a third dataset (test-set), different from the previous ones.

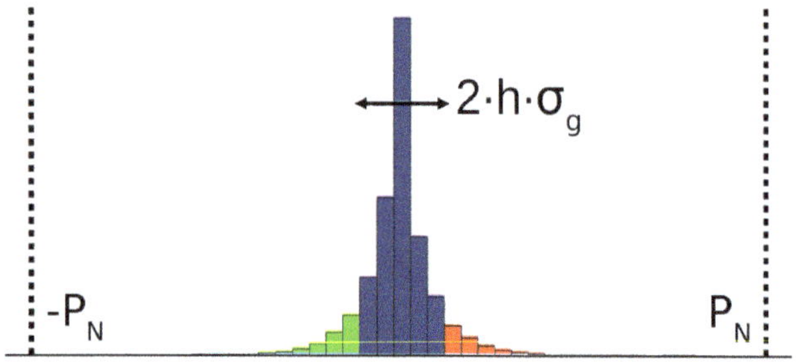


**Figure 1.** Histogram of local gradient time series $g_t$ (case for $h = 1$). $P_N$ stands for the rated power of the wind farm. Local gradients triggering Ramp-up and Ramp-down regimes are in red and green colour respectively.

### 2.2 A Regime-Switching model based on ANNs

Regime-Switching (RS) models consider that a certain time series evolves shifting between a number $r$ of different dynamics. In this case, the stochastic process can be written as follows:

$$y_{t+k} = \begin{cases} f^{(1)}(y_{t-d_{(1)}+1},...,y_t,\Theta^{(1)}) + \varepsilon^{(1)}_{t+k}, & \text{if } s_t = 1; \\ f^{(2)}(y_{t-d_{(2)}+1},...,y_t,\Theta^{(2)}) + \varepsilon^{(2)}_{t+k}, & \text{if } s_t = 2; \\ \dots & \\ f^{(r)}(y_{t-d_{(r)}+1},...,y_t,\Theta^{(r)}) + \varepsilon^{(r)}_{t+k}, & \text{if } s_t = r; \end{cases} \tag{2}$$

where $\{s_t\}$, $s_t \in \{1,2,...,r\}$ provides the current regime at time $t$. This approach permits to consider different generating processes $f^{(r)}$ with different window sizes $d^{(r)}$ and white noise processes $\{\varepsilon^{(r)}_t\} \sim N(0,\sigma_r^2)$ , which is consistent with the idea that the underlying causes of each regime are different. This work considers three different regimes: Ramp-up, Ramp-down and No-ramp regime. Each regime is modelled by means of an ANN.

Additionally, a criterion have to be defined so as to determine $s_t$ at each time step. Assuming that ramp events are motivated by specific conditions at certain meteorological scales (Cutler et al., 2007), meteorological expertise could provide $s_t$ as a function of local observations, NWP outputs, etc. This work proposes an on-line regime assessment based on the observed wind power local gradient. The gradient time series $\{g_t\}$ can be defined as:

$$g_t = y_t - y_{t-1} \tag{3}$$

Then, the regime assessment is based on the following proposed criterion:

$$\begin{array}{lll} g_t > h\cdot\hat{\sigma}_g & \Longrightarrow s_t = 1 & \text{(Ramp-up)} \\ |g_t| \leq h\cdot\hat{\sigma}_g & \Longrightarrow s_t = 2 & \text{(No-Ramp)} \\ g_t < -h\cdot\hat{\sigma}_g & \Longrightarrow s_t = 3 & \text{(Ramp-down)} \end{array} \tag{4}$$

where $\hat{\sigma}_g$ is an estimate of the standard deviation of $\{g_t\}$ obtained from historical data. $h$ is a scalar factor related to the ramp event definition. For example, higher values of $h$ provide few but large ramp events (see Fig. 1). A suitable value for $h$ has to be found during the training process.

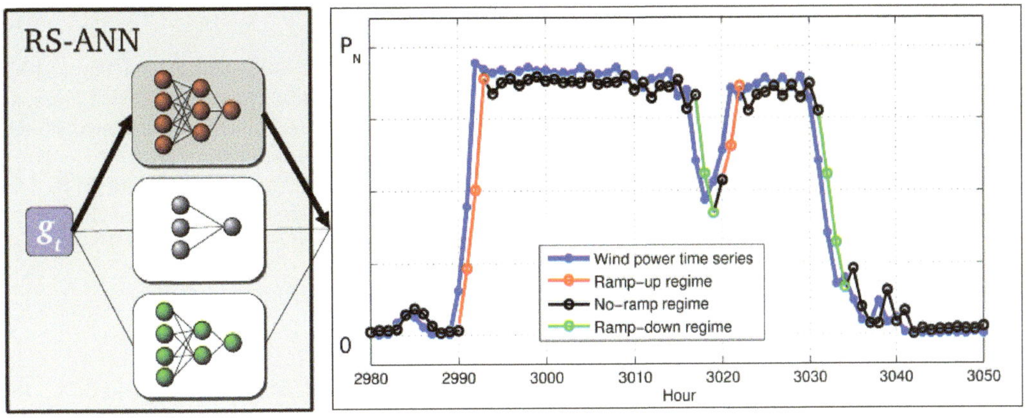


**Figure 2.** RS-ANN model (left). One-hour ahead forecasting (right). The architecture of each ANN was optimized for each prediction horizon considered.

## 3 Results and discussion

The case of a wind farm located in the complex terrain of Alaiz (north of Spain) has been considered. Three years of available power output data with an hourly resolution have been divided into three sets, the length of each set being one year: training-set (2001), validation-set (2002) and test-set (2003). In this sense, considering that the underlying dynamics in the wind power time series is strongly conditioned by atmospheric processes, it is important to remark that the one-year period was selected for each set so as to take into account monthly seasonality in the wind behaviour. Additionally, the three mentioned periods showed a similar relative frequency of ramp events (around the 5–10% of the time).

For a given prediction horizon $k$, a single-ANN model was performed following the methodology described in Sect. 2.1. Then, the RS-ANN (Eq. 2) was implemented by combining three ANNs with the regime assessment criterion given by Eq. (4). The optimal ANN architecture and the window size was tracked for each regime. Several values for $h$ were also considered, varying from 0.2 to 2 in 0.2 steps. Figure 2 illustrates the proposed RS-ANN model.

The performance of the models have been obtained based on the Improvement over Persistence (IoP%). Persistence is the most common reference framework for time series forecasting models (Giebel, 2003), which states that the forecasting at time $t+k$ is the last observation at time $t$. The IoP% of a certain model is defined as follows:

$$\text{IoP\%} = 100 \cdot \frac{\text{RMSE}_{\text{persistence}} - \text{RMSE}_{\text{model}}}{\text{RMSE}_{\text{persistence}}} \tag{5}$$

Figure 3 illustrates the results obtained for different prediction horizons. The improvement over Persistence tend to be modest since Persistence is difficult to improve on for few-hours ahead (Madsen et al., 2005). It can be seen that RS-ANN improved the single-ANN only for the case of one-hour ahead. It may be due to the fact that the on-line regime

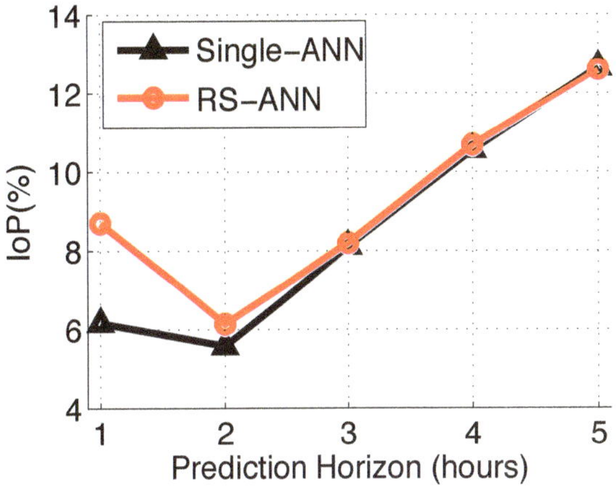


**Figure 3.** Improvement over Persistence for different prediction horizons (case for $h = 1$).

assessment was based on previous observations of the wind power time series. Hence, the ramp regime is triggered when the observed local gradient is already significant enough. We speculate that off-site measurements and NWP outputs could lead to relevant improvements also for further horizons. Another open question is how different time series resolution (for example, 10-min averaged data) may modify the conclusion here attained.

The case of one-hour ahead was analysed in depth. It was found that the optimal thresholds between regimes were given by a scalar factor $h = 1$ (the estimated standard deviation of $\{g_t\}$ during the training-set is $\hat{\sigma}_g = 10.08\%$ of the rated power $P_N$). However, the performance of the RS-ANN models did not show a smooth trend with respect to $h$. This may be due to the fact that the training process involves a numerical learning algorithm. This algorithm may evolve in a different manner depending on several factors, for instance, the initial conditions of the parameters.

**Table 1.** IoP% for the case of one-hour ahead forecasting.

| Model | Globally | Ramp-up | Ramp-down |
|---|---|---|---|
| Single-ANN | 6.16% | 17.60% | 18.55% |
| RS-ANN | 8.66% | 22.25% | 19.55% |

Additionally, a locally-weighted evaluation was carried out to obtain the IoP% during ramp-up and ramp-down events for both models (see Table 1). It was found that the improvement during ramp-up events was clearly higher than during ramp-down events. This fact suggest that ramp-up patterns are more regular (hence, better captured by the model) than ramp-down patterns. It could be due to the fact that ramp-down events may be motivated by a broader number of causes, including yaw-misalignments and wind turbine shut-down due to high wind speeds. If further research reinforces this effect, special attention should be paid to understand the underlying processes during ramp-down events in order to get more reliable predictive models.

Edited by: M. Sprenger

## References

Costa, A.: Mathematical/Statistical and Physical/Meteorological Models for Short-term Prediction of Wind Farms Output, Ph.D. thesis, Escuela Técnica Superior de Ingenieros Industriales (Universidad Politécnica de Madrid), 2005.

Cutler, N., Kay, M., Jacka, K., and Nielsen, T. S.: Detecting, categorizing and forecasting large ramps in wind farm power output using meteorological observations and WPPT, Wind Energy, 10, 453–470, 2007.

Giebel, G.: The state of the art in short-term prediction of wind power – A literature overview, Tech. rep., ANEMOS EU project, 2003.

Greaves, B., Collins, J., Parkes, J., and Tindal, A.: Temporal Forecast Uncertainty for Ramp Events, Wind Engineering, 33, 309–320, 2009.

Hornik, K., Stinchcombe, M., and White, H.: Multilayer feedforward networks are universal approximators, Neural Networks, 2, 359–366, 1989.

Kil, R. M., Park, S. H., and Kim, S.: Optimum window size for time series prediction, Annual International Conference of the IEEE Engineering in Medicine and Biology – Proceedings, 4, 1421–1424, 1997.

Lippmann, R. P.: An introduction to computing with neural nets, IEEE ASSP magazine, 4, 4–22, 1987.

Madsen, H.: Time series analysis, Chapman & Hall, CRC, 2007.

Madsen, H., Pinson, P., Kariniotakis, G., Nielsen, H. A., and Nielsen, T. S.: Standardizing the performance evaluation of short-term wind power prediction models, Wind Engineering, 29, 475–489, 2005.

Peña, D.: Estadística Modelos y métodos, Vol. 2, Alianza Editorial, 2nd Edn., 1987.

Potter, C. W., Grimit, E., and Nijssen, B.: Potential Benefits of a Dedicated Probabilistic Rapid Ramp Event Forecast Tool, in: 2009 IEEE/PES Power Systems Conference and Exposition, Seattle, Vol. 1–3, 409–413, 2009.

# A quantitative evaluation of the high resolution HARMONIE model for critical weather phenomena

**E. V. van der Plas, B. Wichers Schreur, and K. Kok**

Royal Netherlands Meteorological Institute, De Bilt, The Netherlands

*Correspondence to:* E. V. van der Plas
(emiel.van.der.plas@knmi.nl)

**Abstract.** The high resolution non-hydrostatic Harmonie model (Seity et al., 2012) seems capable of delivering high quality precipitation forecasts. The quality with respect to the European radar composite is assessed using the Model Evaluation Tool, as distributed by the NCAR DTC (Developmental Testbed Center, 2012), and compared to that of the reference run of Hirlam (Unden et al., 2002), the current operational NWP model at KNMI. Both neighbourhood and object-based verification methods are compared for a week with several high intensity precipitation events in July 2010. It is found that Hirlam scores very well in most metrics, and that in spite of the higher resolution the added value of the Harmonie model is sometimes hard to quantify. However, higher precipitation intensities are better represented in the Harmonie model with its higher resolution. Object-based methods do not yet yield a sharp distinction between the different models, as it proves difficult to construct a meaningful and distinguishing metric with a solid physical basis for the many settings that can be varied.

## 1 Introduction

The interest in high resolution numerical weather prediction is mainly driven by the presumed ability to skillfully predict extremes in critical weather situations. Heavy precipitation and strong wind may be very local phenomena, for which even a 2.5 km grid spacing could be too coarse to resolve, but present day numerical models try to improve upon the forecasts of these events nevertheless.

The higher resolution of the models also poses a challenge for the verification of the forecasts. When the timing or location of a shower is only a few minutes or kilometers off, a pixel-per-pixel comparison will see this as a double mismatch: a shower is forecasted where it is not observed, giving rise to a false alarm, and the observed shower is not forecasted, counting as a “miss”. Nevertheless, these may be very useful forecasts, giving relevant information to most of the end-users.

The IMPACT project aims to evaluate the high resolution non-hydrostatic model HARMONIE (Seity et al., 2012), being developed by the Hirlam/Aladin consortium, for a series of cases in which the weather was critical to the operations of Schiphol Airport (Amsterdam, The Netherlands).

In this paper a period with several events with convective precipitation will be studied using two verification approaches that try to circumvent the double penalty problem.

## 2 Methods

Various approaches to the problem of verification of high resolution precipitation forecasts have been proposed. In this paper two of those will be studied: neighbourhood methods, such as the Fractions Skill Score (FSS) (Roberts and Lean, 2008), and the object-based MODE (Davis et al., 2006). The software to apply these methods is contained in the Model Evaluation Tool (MET), developed by the developmental testbed center (DTC) at NCAR (Developmental Testbed Center, 2012).

### 2.1 Neighbourhood methods

The most widely used method to take into account that higher resolution forecasts may introduce localisation errors is the *fractions skill score* (FSS). This method attributes merit to a forecast if the criterion (e.g. 1 h accumulated precipitation > 5 mm) is met in a *neighbourhood* of the observed event.

So, following Roberts and Lean (2008), the observed and forecasted rainfall distribution $O_r$ and $M_r$ are converted into binary fields $I_O$ and $I_M$, set to 1 when a threshold $q$ is exceeded and zero otherwise.

Subsequently, for every grid point in the binary fields $I_O$, $I_M$, the fraction of points with a value 1 within a square of (odd) length $n$ are computed, $O_n$ and $M_n$. This is effectively *smoothing* the respective fields.

The mean squared error (MSE) for the observed and forecast fractions is then given by

$$\mathrm{MSE}_{(n)} = \frac{1}{N_x N_y} \sum_{i=1}^{N_x} \sum_{j=1}^{N_y} [O_{n,ij} - M_{n,ij}]^2.$$

This score depends highly on the frequency of the event itself, so by defining a reference MSE,

$$\mathrm{MSE}_{(n),\mathrm{ref}} = \frac{1}{N_x N_y} \left[ \sum_{i=1}^{N_x} \sum_{j=1}^{N_y} O^2_{n,ij} + \sum_{i=1}^{N_x} \sum_{j=1}^{N_y} M^2_{n,ij} \right],$$

a skill score relative to a low-skill forecast can be defined: the fractions skill score (FSS):

$$\mathrm{FSS}_{(n)} = \frac{\mathrm{MSE}_{(n)} - \mathrm{MSE}_{(n),\mathrm{ref}}}{\mathrm{MSE}_{(n),\mathrm{perfect}} - \mathrm{MSE}_{(n),\mathrm{ref}}} = 1 - \frac{\mathrm{MSE}_{(n)}}{\mathrm{MSE}_{(n),\mathrm{ref}}}.$$

This score is minimal for a neighbourhood size of 1, i.e. pixel-per-pixel comparison, and tends asymptotically to a ratio of the frequencies of the observed and forecasted events, as:

$$\mathrm{FSS}_{(n)} \rightarrow 1 - \frac{(f_O - f_M)^2}{f_O^2 + f_M^2} \quad \text{for } n \rightarrow N,$$

where $N = N_x N_y$ is the total number of available grid points, and $f_O$ and $f_M$ are the fractions of the observed and forecasted points exceeding the threshold over the whole domain.

Using the neighbourhood aggregation over an area around the pixel, it is also possible to compute other well-known contingency table statistics (CTS, see Table 1, and e.g. Ebert (2012)). The MET suite provides most of the relevant CTS scores, but the score that will be used in this paper is the Hanssen-Kuiper discriminant (HK). Another well-known quantity in this respect is the Gilbert Skill Score (GSS), also known as the Equitable Threat Score (ETS), but the same overall behaviour emerges as for the HK, and is therefore omitted here for brevity.

The Hanssen-Kuiper discriminant is defined as

$$\mathrm{HK} = \frac{a}{a+c} - \frac{b}{b+d},$$

which gives a measure of how well the areas with precipitation are distinguished from the areas without.

**Table 1.** Contingency table

| | | Observed | |
|---|---|---|---|
| | | yes | no |
| Forecast | yes | $a$ | $b$ |
| | no | $c$ | $d$ |

## 2.2 Object-based verification: MODE

The method for object-based diagnostic evaluation (MODE, Davis et al., 2006), has a quite different approach. Again a binary field ($I_O$, $I_M$) is constructed depending on the exceedance of the threshold $q$. Only now, this field is convoluted by a circular kernel $K$ of $r$ pixels wide to make the field more contiguous, and filter out small and potentially uninteresting features,

$$I_{O,C} = \int I_O(\mathbf{x}) K(\mathbf{x}, r) \mathrm{d}^2 x,$$

thus constituting the *objects*. The combination of threshold $q$ and convolution radius $r$ determine the distribution of the resulting objects.

To associate a forecasted object to an observed object, an interest function $F$ prescribes, on a scale from 0 to 1 (1 being perfect), how closely an attribute of the forecasted object matches the same attribute of the observed object. For any forecast and observed object pair, the total interest $C$ is then defined as

$$C = \frac{\sum_{i=1}^{M} c_i w_i F_{i,j}}{\sum_{i=1}^{M} c_i w_i}.$$

Here $w$ is the weight assigned to a certain attribute $i$ of an object, e.g. the location of its center of mass, and $c$ is a function of attributes that describes the confidence in a partial interest value obtained from $w_i F_{i,j}$. Total interest assumes a value between 0 and 1, and it is used to associate the objects to one another. When the total interest is larger than 0.65 objects are considered "matched". When several objects match to each other above this threshold the objects are called a "cluster".

As an example, the result of one particular time and model is shown in Fig. 1. Here, e.g. the forecasted objects 19, 20 and 21 are compared to observed object 15 (lower right corner, in green) because they are not too far apart, they have a similar intensity, the difference in the angle is not too large etc. This amounts to a total interest larger than 0.65, and hence they are matched.

# 3 Data

The objective of this paper is to compare the forecast produced by a non-hydrostatic high-resolution model to a current operational standard, in this case HARMONIE and the hydrostatic Hirlam RCR model, respectively. These are then

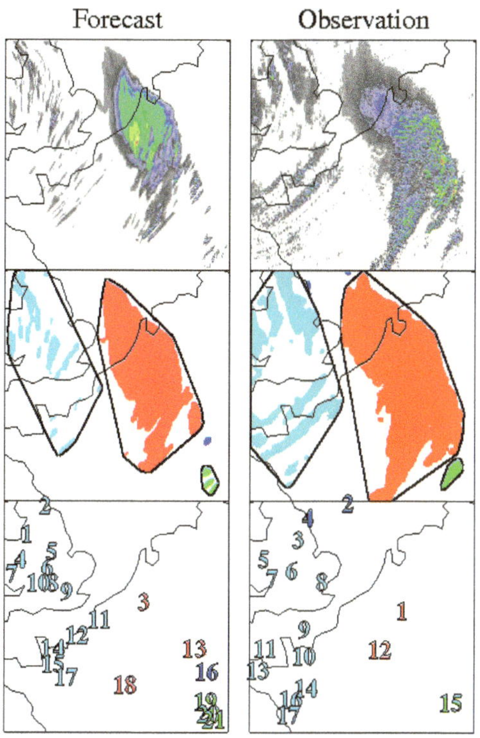


**Figure 1.** Typical output of the MODE algorithm, here for Harmonie with Hirlam boundaries, 14 July, 18:00 UTC: on the left the forecasted precipitation field (top: raw data, middle: thresholded and convoluted objects, bottom: objects numbered and colored by matching criteria with observations), on the right the radar data.

compared to the European radar composite at 4 km resolution.

In this study we have chosen a 10 day period from 6 to 15 July 2010. In this period weather alerts have been issued by KNMI for extreme precipitation on 10, 12 and 14 July. The latter case also developed a phenomenon which was believed to be a *micro-downburst*: the passage of the front induced a very strong yet highly localised wind field, resulting in substantial damage and the loss of life.

### 3.1 Model data

The HARMONIE model is the latest high resolution model that is used within the Hirlam/ALADIN community, using the AROME non-hydrostatic dynamical core (cycle 36h1.3) which was developed by the ALADIN community. It runs on a limited area, getting initial conditions and boundaries from either the Hirlam RCR run or the ECMWF operational analysis and forecast. Runs are executed on a $400^2$ grid with 2.5 km resolution using both boundary strategies.

The HARMONIE runs are initialised by either the Hirlam RCR or the ECMWF analysis. However, as some of the HARMONIE prognostic variables, e.g. vertical velocity for convection and the hydrometeors, are not initialised, this will effectively be considered as a "cold start". The starting times are 00:00, 06:00, 12:00 and 18:00 UTC, with 12 h forecasts. A spin-up time of 3 h is taken into account, so the data of T+003 until T+009 are considered in this particular study.

**Table 2.** Model configuration

| | Harmonie | Hirlam RCR | Radar |
|---|---|---|---|
| Vertical levels | 40 | 60 | n.a. |
| Domain | 400 × 400 | 582 × 446 | 512 × 512 |
| Resolution | 2.5 km | 0.15° | 4 km |
| Projection | Lambert | rot. lat-lon | polar stereogr. |
| Assimilation | none | 3DVAR | |

The Hirlam RCR run is performed on a substantially larger domain (see Table 2), and performs an analysis using 3DVAR data assimilation.

### 3.2 Observations

The choice for the European composite radar product RADNL23 can be motivated by the assertion that in this pilot study we primarily concern ourselves with the distribution of (extreme) precipitation. It is recognised that this data is aggregated from various different radar installations, giving rise to quantitative differences for the different areas. Also, over the North Sea radar clutter can give rise to complications for the verification methods, especially the object-based methods. As the scatter has very little spatial extent but may have considerable intensity, the convolution step in MODE tends to overestimate these areas. A more reliable radar product may one day be available as a result of the OPERA project, and by cross-validating with MSG satellite data during daytime.

### 3.3 Grids

To compare the gridded observations and the model output it is necessary to transform the data to a uniform grid. To prevent the introduction of artificial values all data was resampled unto the highest resolution grid available, i.e. the HARMONIE Lambert-conformal grid, using a nearest-neighbour method.

## 4 Results

First of all, the total precipitation over the 10 day period can be compared, see Fig. 2. The radar shows the largest amount of rain, also with the largest amount of variation. The Hirlam RCR data gives a considerably smoother picture. Here we point out that the data shown here is already resampled to the Harmonie grid, giving rise to some obvious resampling artefacts. The Harmonie data resembles the radar data more closely, where the run nested in the Hirlam

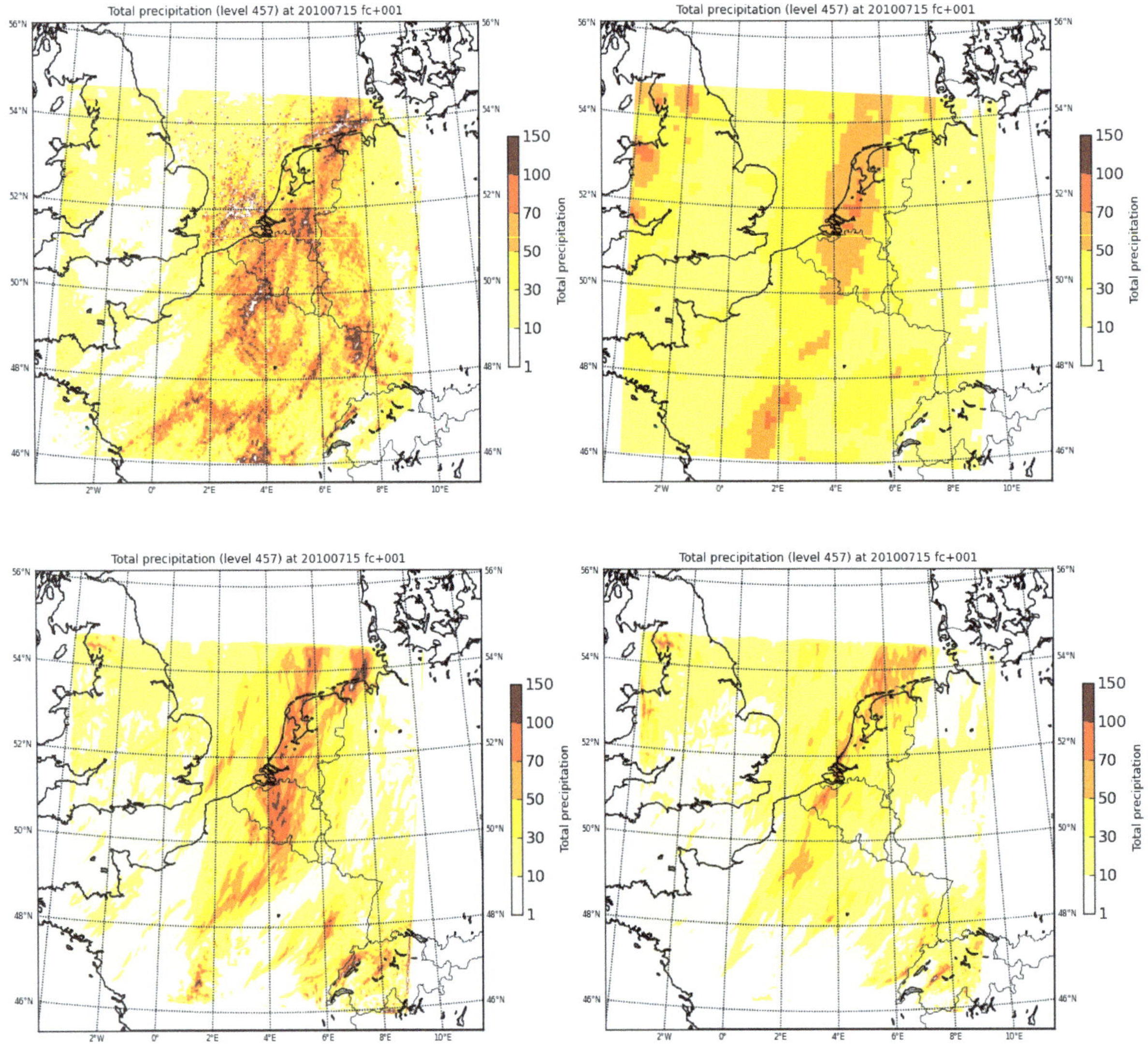


**Figure 2.** Total accumulated precipitation for the period 6–15 July 2010, using the resampled data. Top left: radar data, right: Hirlam RCR. Bottom left: Harmonie with Hirlam boundaries, right: Harmonie with ECMWF boundaries.

model gives higher quantities, especially in the band of intense rain that crossed Belgium and the Netherlands. These figures can be summarised in the histogram in Fig. 3 (top). The distribution of the radar data is shifted towards higher intensities than the model data, whereas the Harmonie runs have more dry pixels, reflecting the tendency of the model to underrepresent light rain. The maximum in the distribution in the Hirlam RCR precipitation around 20 mm can probably be attributed to the coarser resolution of the model set-up: the extremes (> 100 mm) are underrepresented, because generally high intensity precipitation does not extend over large areas. On the other hand the amount of "dry" pixels is reduced because of the same effect: if there is precipitation in the grid box, the whole box will give non-zero precipitation. This leads to a shift of the distribution towards its mean, in this case approximately 35 mm over the whole period.

In the bottom panel of Fig. 3 the histogram of the three hour precipitation sums in the same periodis shown. Again we see how the Hirlam RCR run tends to overestimate and smear out low intensities due to the coarser resolution. This might also lead to the overestimation of the 60–80 mm bins in the total accumulation in the top figure: many low-intensity precipitation events could lead to accumulative amounts in this range (10 days times 8 3-h accumulations times 1 mm/3 h yields 80 mm).

Harmonie however clearly underrepresents low precipitation intensities. This is in line with the experience that the forecasted showers tend to be too localised: stratiform, light rain is not allways well captured in the model.

A time series of the FSS is shown in Fig. 4. The three events with heavy precipitation are clearly distinguishable.

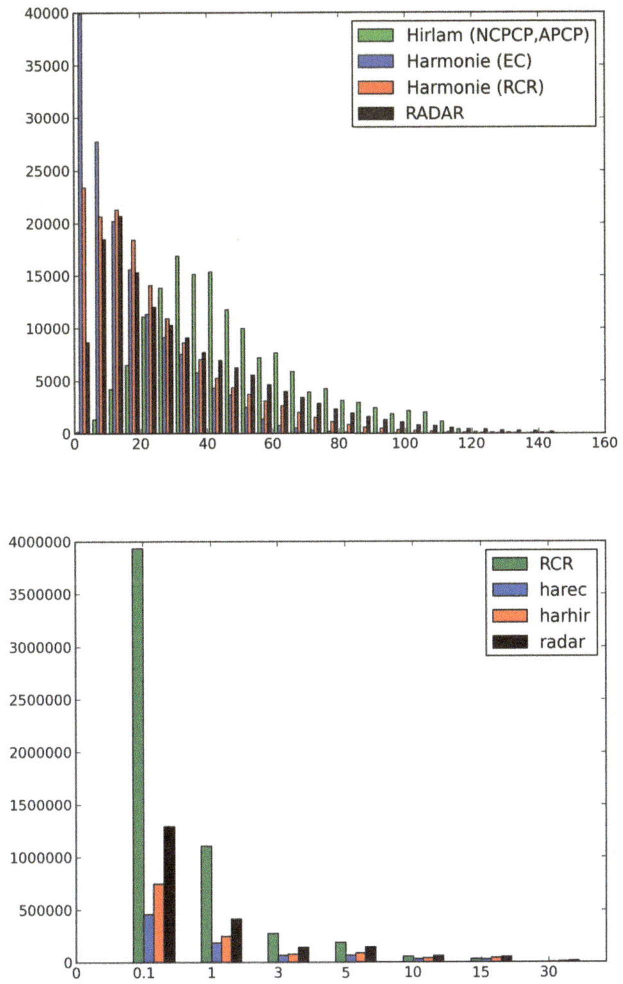


**Figure 3.** Top panel: histogram of the total accumulated precipitation for the period 6–15 July 2010. Here, the resampled data was binned in 5 mm bins (the interval [0,5] is depicted at 0, etc.). The bottom panel shows the accumulative histogram for each three hour period, binned between 0.1, 1, 3, 5, 10, 15, 30 and 100 mm per three hours (the interval (0.1,1] is depicted at 0.1, etc).

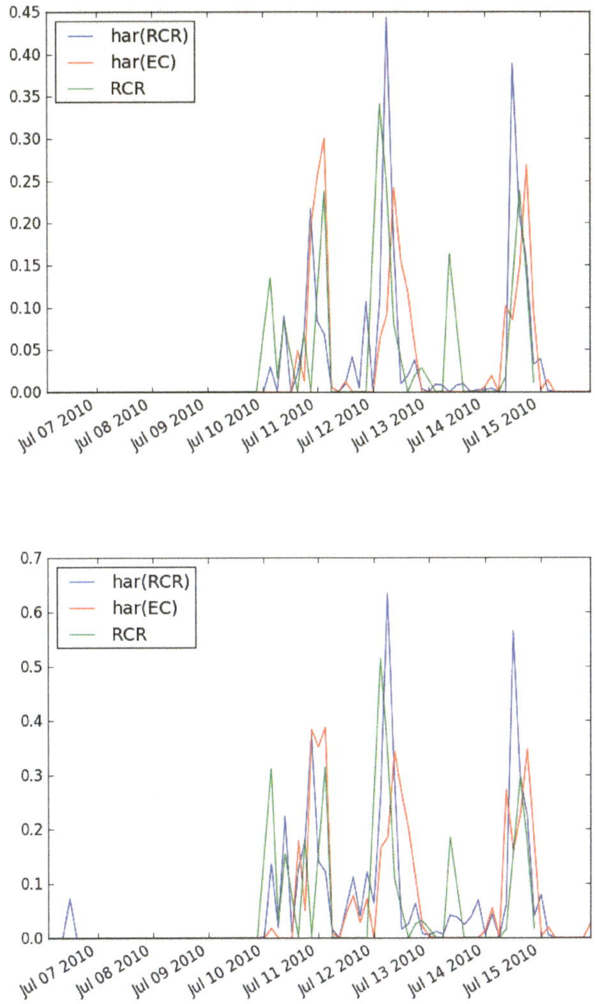


**Figure 4.** Time series of the FSS for precipitation over 5 mm/3 h, with a neighbourhood size of 1 pixel or 2.5 × 2.5 km. (top) and $15^2$ pixels or 37.5×37.5 km (bottom). The green line represents the RCR run data, blue Harmonie using Hirlam boundaries and red Harmonie using ECMWF boundaries.

We compare the different runs taking either a neighbourhood size of 1 by 1 pixel, the classical CTS score, or smoothing the data over an area of between 3 by 3 (7.5 by 7.5 km) to 75 by 75 (187.5 by 187.5 km) pixels. This has been taken as the upper limit, as the amount of areas within the computed domain as well as the amount of counted events drops significantly for larger areas.

In this figure the scores for precipitation of more than 5 mm per three hours is shown. The scores are comparable for the three models. For the 1 pixel neighbourhood size, i.e. pixel-per-pixel comparison, shown on top, Hirlam scores generally slightly higher for the large scale events, and also picks up an event on 9th July (not shown here). This holds for both the FSS and the HK discriminant score. Increasing the neighbourhood size, to 15 by 15 pixels, shown in the bottom panel, results obviously in increasing scores as generally the number of counted events over larger areas will increase, but the same qualitative picture remains. The scores of the Harmonie runs increase a bit more than those of the Hirlam RCR run.

The behaviour of the FSS and HK scores as a function of the neighbourhood size is depicted in Fig. 5. The FSS for the Harmonie model runs increase monotonically and visibly faster than the FSS for the Hirlam model, where the median does not reach beyond a FSS of roughly 0.2.

The HK score is less pronounced in this respect. For smaller regions (1 by 1 to 9 by 9 gridboxes) the scores are of the same order. Harmonie with ECMWF boundaries scores slightly higher (median at HK ≈ 0.1 vs. HK ≈ 0.05). For areas of 15 and 25 gridboxes wide Hirlam scores more consistently, but from 35 gridboxes upward this advantage is reversed again.

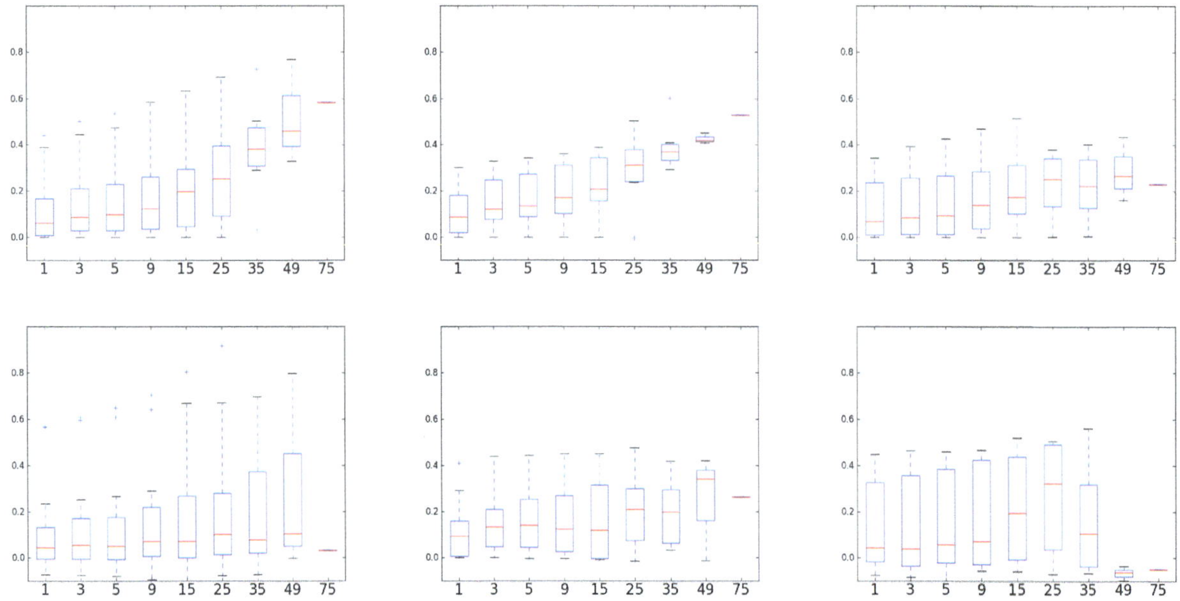


**Figure 5.** Box plots of the FSS (top) and HK (bottom) for rain over 5 mm per three hour as a function of the neighbourhood size for Har(RCR), Har(EC) and RCR (left, middle and right). The red line represents the median of the data, and the box extends from the 25th to the 75th percentile. The whiskers extend to 1.5 times the distance from the mean to the 25th and 75th percentile range. Points outside the whiskers are denoted with a blue +, and considered outliers from a visualisation perspective.

The outliers, corresponding to the events with the highest precipitation intensities, show a considerable edge to Harmonie with RCR boundaries. The neighbourhood method averages over a square of increasing size, and this may yield unfavourable results for the typically highly localised precipitation forecasts of Harmonie at the intermediate areas.

If we look at the MODE analysis, we first observe that the process of grouping precipitation features into objects of a certain minimum size makes it easier to do visual, subjective verification. An example of the visual output of the thresholding and convolution process as performed by MET is shown in Fig. 1. Matched events are coloured accordingly (dark blue means not matched), and one can easily perform a quick subjective assessment of the situation considered. Per case or time step one may compare certain attributes of objects one is interested in, such as total interest or centroid distance. However, to condense this information into a single score over the entire time range is less straightforward.

Furthermore, the configuration of MODE has many degrees of freedom. Some settings, such as the radius of the convolution kernel, can make a substantial difference in how the objects come out.

Statistics of the individual objects may give interesting information for model intercomparison studies (different physics, initial and boundary conditions etc.), but generally have very little meaning in the context of (longer) timeseries. One might follow e.g. the largest matched object or identified cluster, but the objects or clusters are not tracked in time, so this naive attribute does not tell whether the largest cluster in one time step is related to the largest cluster in the next. Also, the grouping of objects into a cluster is not always very consistent in the sense that storms that belong to different clusters in one time step may coalesce into the same cluster and vice versa. In Fig. 1 one sees how grouping may seem a bit arbitrary.

One more consistent method to construct a score was proposed in Davis et al. (2009), using the median of the maximum interest (MMI) of the whole domain. The matching procedure computes the interest between all features, and considers it a match when this number is above a certain, user defined threshold. By considering the median of all these interest values, we have a measure that reflects how well the forecast performed for a given moment, and that can be used to compare different models and different (dryer and wetter) periods. The results for this particular case are shown in the right panel of Fig. 6. It is remarkable that the the high-resolution Harmonie model performs comparable to the Hirlam model for this particular score, with the exception of the large lower tail for Harmonie using ECMWF boundaries.

## 5 Discussion

A few different verification methods have been applied to Hirlam and HARMONIE model data, using the European composite radar product to verify against.

The high level of detail in the forecast would suggest that classical methods, such as (neigbourhood) contingency table statistics, are less well equipped to attribute the forecasted

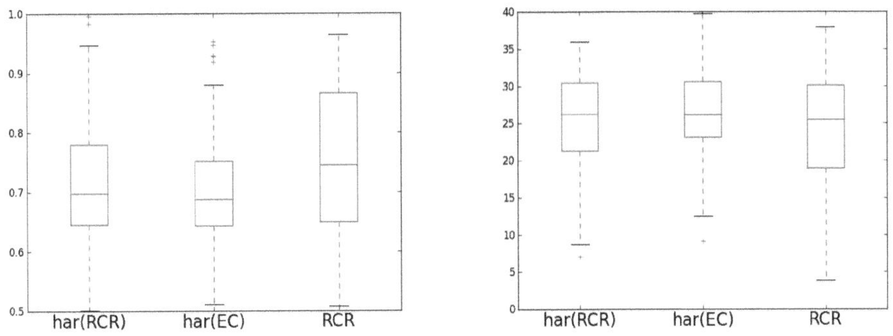


**Figure 6.** Box plot of the median of the maximum interest (MMI, left) and scores and the median of the centroid distance between two matched objects (right) for Harmonie using Hirlam and ECMWF boundaries and the Hirlam RCR run over the 10-day period. Box definitions as in Fig. 5.

precipitation features to its observed counterparts than the more flexible object-based methods, such as MODE. However, the qualitative picture that arises when comparing the scores of Harmonie and Hirlam shows that using the former CTS-based methods, with e.g. the fractions skill score or the HK discriminant, give relatively more credit to the higher resolution forecasts than for this specific choice of MODE results.

The higher resolution of the Harmonie model does give a better representation of the more extreme events (> 5 mm/3 h) than Hirlam. This is one of the expected advantages from a high-resolution model that properly takes into account the dynamics on the smaller scales. These higher precipitation intensities were hardly present in the Hirlam data as a result of the coarser resolution. It is noted (not shown) that the events with the highest precipitation intensities generally correspond to positive outliers in the scores, though more investigation and a larger dataset are needed. These points hardly contribute to the value of the median of the distribution, even if the median should be regarded as the score for a certain measure.

Also, the Harmonie results shown here are obtained with a model set-up that does not use data-assimilation of any kind. This means that especially quantities that are not being initialised by the analysis on which they are based, such as cloud cover and precipitation, suffer from spin-up. Experiments show that for the atmosphere to display a properly dsitributed state it takes the model between 4 to 6 h. Longer lead times show better scores for these parameters when cold starts are considered.

It is stressed that this is just a preliminary study with a very modest amount of data. Furthermore, the output of the MODE algorithm is so rich, that obviously more effort should be invested into combining the attributes in such a way as to produce a score that gives intuitive results and can be compared over a variety of cases.

**Acknowledgements.** Model Evaluation Tools (MET) was developed at the National Center for Atmospheric Research (NCAR) through a grant from the United States Air Force Weather Agency (AFWA). NCAR is sponsored by the United States National Science Foundation.

Edited by: H. Böttger

## References

Seity, Y., Brousseau, P., Malardel, S., Hello, G., Bénard, P., Bouttier, F., Lac, C., and Masson, V.: The AROME-France convective scale operational model, Mon. Weather Rev., 139, 976–991. 2012,

Roberts, N. M. and Lean, H. W.: Scale-Selective Verification of Rainfall Accumulations from High-Resolution Forecasts of Convective Events, Mon. Weather Rev., 136, 78–97, 2008,

Davis, C. A., Brown, B. G., and Bullock, R.: Object-Based Verification of Precipitation Forecasts. Part I: Methodology and Application to Mesoscale Rain Areas, Mon. Weather Rev., 134, 1772–1784, 2006.

Davis, C. A., Brown, B. G., Bullock, R. and Halley-Gotway, J.: The Method for Object-Based Diagnostic Evaluation (MODE) Applied to Numerical Forecasts from the 2005 NSSL/SPC Spring Program, Weather Forecast., 24, 1252–1267, 2009.

Ebert, E.: Forecast Verification: Issues, Methods and FAQ, http://www.cawcr.gov.au/projects/verification, 2012.

Developmental Testbed Center, MET: version 3.0.1 Model Evaluation Tools Users Guide, http://www.dtcenter.org/met/users/index.php, 2010.

Unden, P., Rontu, L., Järvinen, H., Lynch, P., Calvo, J., Cats, G., Cuxart, J., Eerola, K., Fortelius, C., Garcia-Moya, J. A., Jones, C., Lenderink G., Mcdonald, A., Mcgrath, R., Navascues, B., Woetman Nielsen, N., Degaard, V., Rodriguez, E., Rummukainen, M., Sattler, K., Hansen Sass, B., Savijarvi, H., Wichers Schreur, B., Sigg, R., and The, H.: HIRLAM-5 Scientific Documentation, 2002.

# 22

# Statistical processing of forecasts for hydrological ensemble prediction: a comparative study of different bias correction strategies

**I. Zalachori[1], M.-H. Ramos[1], R. Garçon[2], T. Mathevet[2], and J. Gailhard[2]**

[1]IRSTEA, Hydrology Group, UR HBAN, 1 rue Pierre-Gilles de Gennes CS 10030, 92761 Antony, France
[2]EDF-DTG, Electricité de France, Direction Technique de Grenoble, France

*Correspondence to:* I. Zalachori (ioanna.zalachori@irstea.fr)

**Abstract.** The aim of this paper is to investigate the use of statistical correction techniques in hydrological ensemble prediction. Ensemble weather forecasts (precipitation and temperature) are used as forcing variables to a hydrologic forecasting model for the production of ensemble streamflow forecasts. The impact of different bias correction strategies on the quality of the forecasts is examined. The performance of the system is evaluated when statistical processing is applied: to precipitation and temperature forecasts only (*pre-processing* from the hydrological model point of view), to flow forecasts (*post-processing*) and to both. The pre-processing technique combines precipitation ensemble predictions with an analog forecasting approach, while the post-processing is based on past errors of the hydrological model when simulating streamflows. Forecasts from 11 catchments in France are evaluated. Results illustrate the importance of taking into account hydrological uncertainties to improve the quality of operational streamflow forecasts.

## 1 Introduction

Probabilistic information is of special importance for users vulnerable to climatic and hydrological hazards at different scales (agriculture and irrigation, navigation, public safety, energy companies, etc.). In hydrology, a common approach to produce probabilistic information is the use of ensemble-based streamflow forecasting systems (see review by Cloke and Pappenberger, 2009). The key advantage of these systems is that they can provide future scenarios of streamflow evolution in time, with information on the uncertainty of the predictions, which can be potentially more useful at longer forecast lead times, notably in terms of increasing preparedness for severe flood events and reducing losses (Bartholmes et al., 2009; Boucher et al., 2012; Verkade and Werner, 2011). However, model output predictions may lack precision and reliability due to several reasons like imperfect numerical representation of physical process or insufficiency account of all sources of uncertainty involved in the system being modelled (e.g. Thirel et al., 2008; Jaun and Ahrens, 2009; Randrianasolo et al., 2010; Velazquez et al., 2011). To improve the quality of probabilistic forecasts and provide reliable estimates of uncertainty, statistical processing of forecasts is recommended (Schaake et al., 2010). The aim is to remove forecast biases and to improve ensemble dispersion. Several techniques have been proposed in meteorology and hydrology, mainly based on empirical dressing techniques, Bayesian methods or regression analysis (e.g. Krzysztofowicz, 1999; Raftery et al., 2005; Fortin et al., 2006; Hashino et al., 2007; Olsson and Lindström, 2008; Brown and Seo, 2010; Zhao et al., 2011).

In hydrologic forecasting systems, statistical correction techniques can be applied to the forecast input of the hydrological model (meteorological variables like precipitation and temperature), to the forecast output of the hydrological model (streamflows) or to both. As shown in Fig. 1, from the hydrological model point of view, a forecasting system can comprise *pre-processing* approaches (statistical correction applied previously to the hydrologic modelling) and *post-processing* approaches (statistical correction applied to flow predictions). In all cases, calibration against

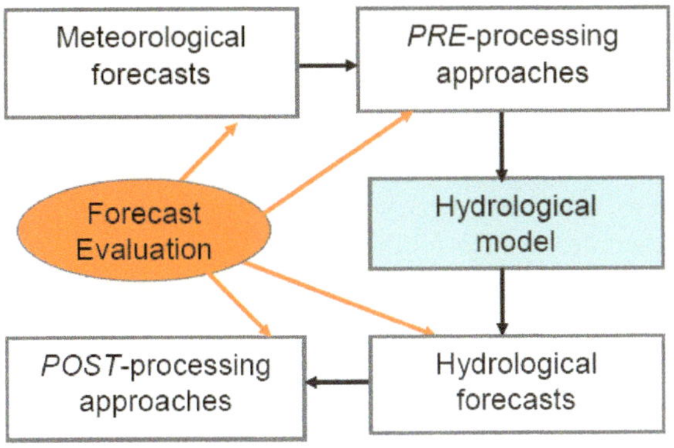


**Figure 1.** Schematic of a hydro-meteorological forecast chain.

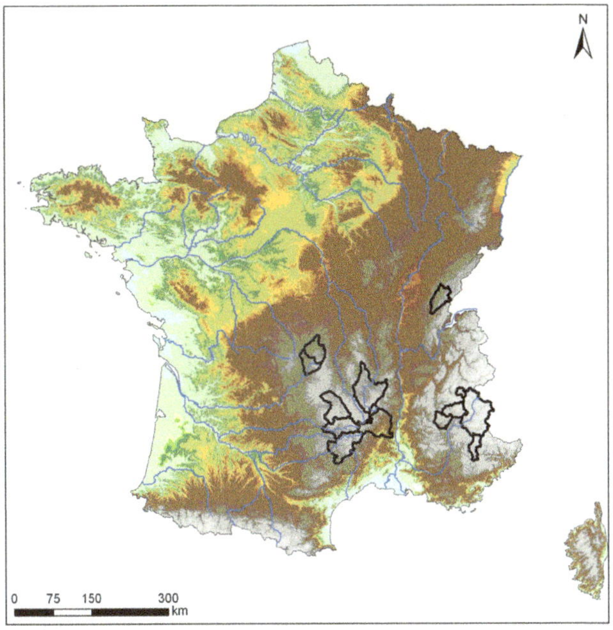


**Figure 2.** Location of the studied catchments in France.

observations and extensive testing over different hydrologic conditions are usually required to develop an operationally robust system. In order to optimize the implementation of post-processing techniques in real-time operational forecasting systems, a better understanding of the propagation of uncertainty from weather forecasts through hydrologic models and on the impact of non-linear hydrological transformations and hydrological updating on the ensemble streamflow forecasts is needed.

The aim of this paper is to investigate the use of statistical correction techniques in hydrological ensemble forecasting. We focus on the evaluation of different correction strategies (*pre-processing*, *post-processing* or both) and on their impact on the quality of operational streamflow forecasts. The context of the study and the modelling framework, including data and model used, are presented in Sect. 2; methodology and verification measures are described in Sect. 3; Sect. 4 presents the results, and in Sect. 5, conclusions are drawn.

## 2 Study context and modelling framework

The study is based on a modelling framework set up at the French electricity company (EDF) for the forecast of streamflows in France. EDF has produced hydrological forecasts for the past 60 yr (Lugiez and Guillot, 1960). Their operational interest includes flood forecasting (for human safety and dam security), short-term forecasting of water inflows to reservoirs, long-term prediction and reservoir management. For the last decade, EDF has invested into ensemble-based hydrological forecasting, comprising the acquisition of real-time meteorological forecast data and the set up of an appropriate hydrological modelling framework. Moreover, special attention has been paid to the performance of experiments on the use and communication of uncertainty in decision-making (Ramos et al., 2010). EDF forecasting chain follows the schematic description in Fig. 1. In this study, meteorological forecasts come from the 50 perturbed members of the ensemble prediction system produced by the European centre of medium-range weather forecasts (ECMWF-EPS). Operationally, EDF also uses the ECMWF-EPS control forecast, as well as deterministic forecasts produced by Météo-France. Forcing data (temperature and precipitation) are spatially aggregated at the catchment scale and evaluated by forecasters before being used as input to the hydrological model. The hydrological model is the MORDOR model. It is a lumped soil-moisture-accounting type rainfall-runoff model developed at EDF (Garçon, 1999). MORDOR has four reservoirs representing the physical processes in a river basin and a snow module that accounts for snow storage and melting in the catchment. The model version used in this study has 11 free parameters that were calibrated against observed data.

This study focuses on 11 catchments located in France, with areas ranging from 220 to 3600 $km^2$ (Fig. 2). Meteorological input fields are available at a horizontal resolution of $0.5 \times 0.5$ degree in latitude/longitude, ca. 50 km over France. The number of grid points falling within each catchment (at different percentages of coverage) varies from 2 to 42, with a median value of 15 grid points inside a catchment. Although some catchments are much smaller than the available meteorological grid scale, they were kept in the study and the sensitivity of the results to the catchment size was evaluated. At each catchment, forecasts are run at the daily time step and for a maximum forecast horizon of 7 days. Verification is performed over a 48-month forecast evaluation period (2005–2008). Forecasts are evaluated against observed daily areal precipitation and daily discharge data available at the outlet of the catchments.

## 3 Methodology and performance measures

### 3.1 Statistical correction strategies

Four main scenarios corresponding to different strategies for the processing of forecasts within the hydrologic modelling framework are tested:

- Scenario 1 – *Raw forecasts*: no pre- or post-processing of forecasts is performed and raw model outputs are evaluated;

- Scenario 2 – *Pre-processing of meteorological ensemble forecasts*: meteorological forecasts of temperature and precipitation are corrected using a technique developed at EDF, which is based on the search for analog situations previously archived in a database. At a given forecast day, for each perturbed member of the ECMWF-EPS system, the 50 most analog situations (according to the forecast fields of geopotential height at 700 and 1000 hPa) are retrieved from the database (for details on the use of analog techniques, see Zorita and von Storch, 1999 or Obled et al., 2002). A total of 2500 scenarios (50 ECMWF-EPS × 50 analogs) is obtained. Each scenario consists of a pair of an analog situation (precipitation and temperature) and its corresponding ECMWF-EPS forecast. A corrected forecast scenario is calculated for each lead time $t$ as indicated in Eq. (1):

$$V'_t = k \cdot V_t^{\mathrm{ECMWF}} (V_t^{\mathrm{ANALOG}} / \bar{V}_t^{\mathrm{ANALOG}})^{\alpha} \quad (1)$$

$V'_t$: corrected forecast scenario for lead time $t$;
$V_t^{\mathrm{ECMWF}}$: ECMWF-EPS raw forecast for lead time $t$;
$V_t^{\mathrm{ANALOG}}$: analog forecast for lead time $t$;

$\bar{V}_t^{\mathrm{ANALOG}}$: mean value of the 50 selected analog forecasts for lead time $t$;
$\alpha, k$: parameters to be calibrated.

This combination of ECMWF-EPS and analog-based forecasts aims at perform an ensemble dressing of ECMWF forecasts (to improve reliability and to remove biases). The values of the parameters $\alpha$ and $k$ can vary according to the studied catchment and lead time. In this study, they were considered constant and equal to 0.3 and 1.2, respectively. The 2500 scenarios obtained were sorted in ascending order and 50 scenarios (equidistant quantiles) were selected to be used as input to the hydrological model.

- Scenario 3 – *Post-processing of hydrological ensemble forecasts only*: the statistical correction technique used here takes into account only the errors from the hydrological model. It is thus independent of the meteorological forecasts (raw or pre-processed) used during the forecast evaluation period. Based on the past performance of the model, empirical errors are evaluated by taking the logarithm of the ratio between the observed and the simulated streamflows. Simulations are obtained using observed precipitation and temperature available in the period 1970–2000 as input to the hydrological model. Subsamples of the errors are defined according to 20 classes of streamflow values (corresponding to a discretization of the cumulative distribution function at steps of 8 % between the 10 % and 90 % quantiles and 2 % for the tails of the distribution) and for each lead time (Mathevet, 2010). During forecasting, to "dress" each hydrological ensemble member, error values are drawn from the subsamples, according to the category to which the forecast discharge belongs, and added to the raw forecast value.

- Scenario 4 – *Pre-processing of meteorological forecasts and post-processing of hydrological ensemble forecasts*: the last scenario tested is the combination of the two correction approaches described above: the pre-processing of meteorological forecasts (scenario 2) and the post-processing of hydrological forecasts (scenario 3).

### 3.2 Forecast evaluation methods

Forecasts were evaluated against observations for a 48-month period from 2005 to 2008. Various measures are available in the literature to evaluate probabilistic forecasts and some have already been applied for the evaluation of hydrological forecasts (Wilks, 2011; Casati et al., 2008; Laio and Tamea, 2007). The scores used here are briefly presented below (see the references for details):

- Normalized RMSE: the root-mean-square error of the ensemble mean is normalized by the mean value of the observations during the forecast evaluation period to allow comparison among catchments of different sizes. Although the RMSE is not a score adapted to ensemble or probabilistic forecasts, we included it here as it is a score commonly used in hydrology.

- Brier Score (BS): one of the most common accuracy measure for forecast verification, the BS is essentially the mean squared error between the predicted probabilities for a set of events and their outcomes (= 1, if the event occurs and = 0 if it does not occur). The score takes values between 0 and 1; the lower the score, the higher the accuracy. In this study, we focus on the evaluation of severe events given by predictions exceeding the 80 % quantile of the empirical distribution of observed values.

- Rank Probability Score (RPS): the RPS is an extension of the BS to the many-event situation, computed, however, with respect to the cumulative probabilities in the forecast and observations vectors. The score takes values between 0 and 1; a "perfect forecast" receives RPS = 0. Here, we used 10 forecast categories to define

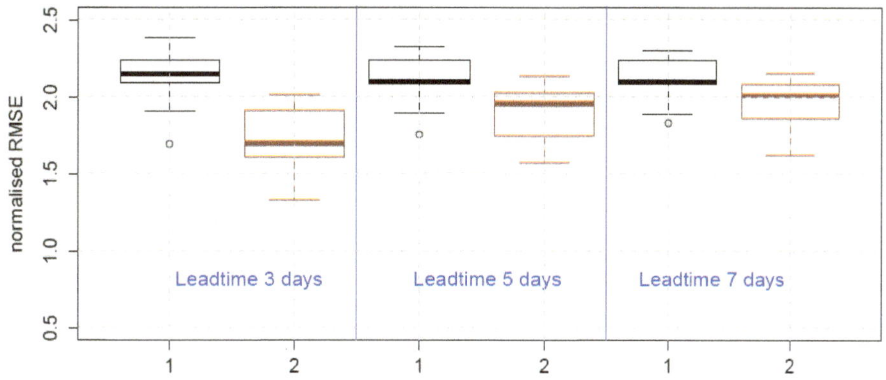


**Figure 3.** Boxplots of mean normalised RMSE of daily areal precipitation forecasts for lead times of 3 days (left), 5 days (centre) and 7 days (right) using: (1) raw forecasts, (2) pre-processed meteorological forecasts. Forecast evaluation over 48 months (2005–2008) and 11 catchments.

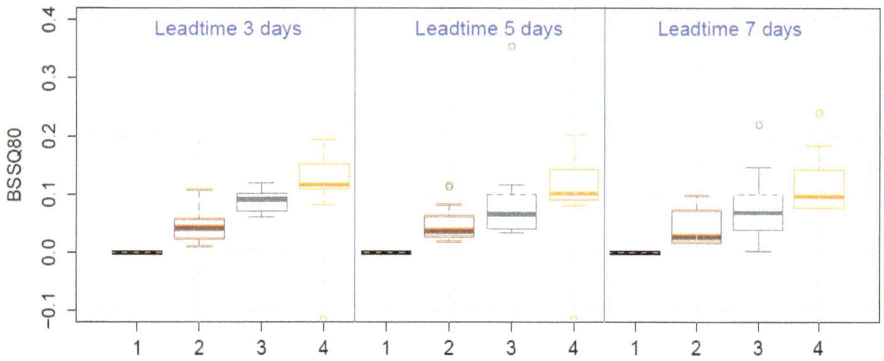


**Figure 4.** Boxplots of BSS (with reference to raw forecasts) for forecast streamflows exceeding the 80 % percentile of observed flows for lead times of 3 days (left), 5 days (centre) and 7 days (right) using: (1) raw forecasts, (2) pre-processed meteorological forecasts, (3) post-processed hydrological forecasts, (4) pre-processed meteorological forecasts and post-processed hydrological forecasts. Forecast evaluation over 48 months (2005–2008) and 11 catchments.

the events, determined by the values of the 10 % to the 90 % quantiles of the empirical distribution of the observed values during the forecast evaluation period.

- Skill Scores (BSS or RPSS): the BS and the RPS are compared to a reference, which in this study corresponds to the raw forecasts (scenario 1). A skill score of 0 indicates a forecast with skill similar to the reference, while a forecast which is less (more) skilful than the reference will result in negative (positive) skill score values.

- Probability Integral Transform (PIT) histogram: the PIT histogram is a continuous analog of the rank histogram (Gneiting et al., 2005; Wilks, 2011), frequently used to verify the consistency of the forecasts, i.e. if the ensemble members of a forecast and the corresponding observations are samples from the same population (Wilks, 2011). If the ensemble consistency condition is satisfied, the relative frequencies given by the ensembles should estimate the actual (observed) probability. In this case, the PIT histogram shows as a uniform histogram, giving an indication of reliable forecasts. Under-dispersed forecasts will give U-shaped PIT histograms, while over-dispersed forecasts show relative frequencies concentrated in the middle ranks (arch-shaped). Asymmetrical histograms are an indication of over- or under-forecasting bias.

## 4 Results

The main results obtained from the comparative study of different bias correction strategies applied to meteorological and hydrological forecasts are illustrated in Figs. 3 to 6. For each studied catchment, mean values of the scores RMSE, BSS and RPSS are obtained by averaging the daily scores over the entire evaluation period (2005–2008). In Figs. 3 to 5, results from all catchments are displayed using the boxplot representation: the 75th and the 25th percentiles are represented at the top and the bottom of the box, respectively, while the top and the bottom of the tail indicate the 95th and

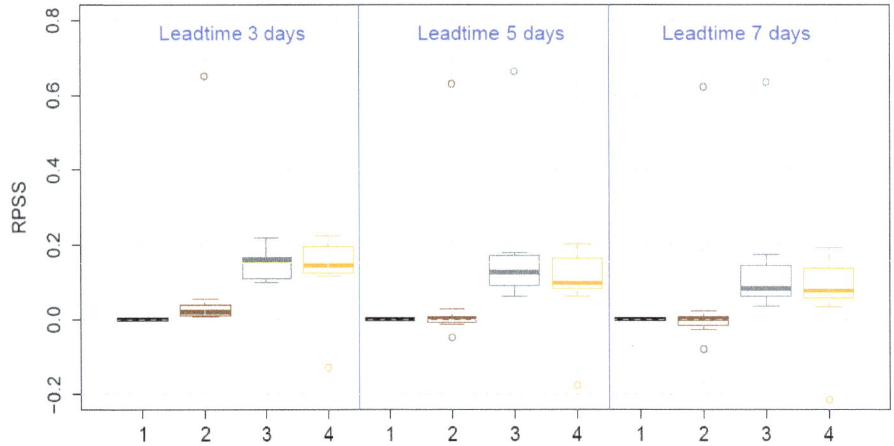


**Figure 5.** Boxplots of RPSS (with reference to raw forecasts) for forecast streamflows and lead times of 3 days (left), 5 days (centre) and 7 days (right) using: (1) raw forecasts, (2) pre-processed meteorological forecasts, (3) post-processed hydrological forecasts, and (4) pre-processed meteorological forecasts and post-processed hydrological forecasts. Forecast evaluation over 48 months (2005–2008) and 11 catchments.

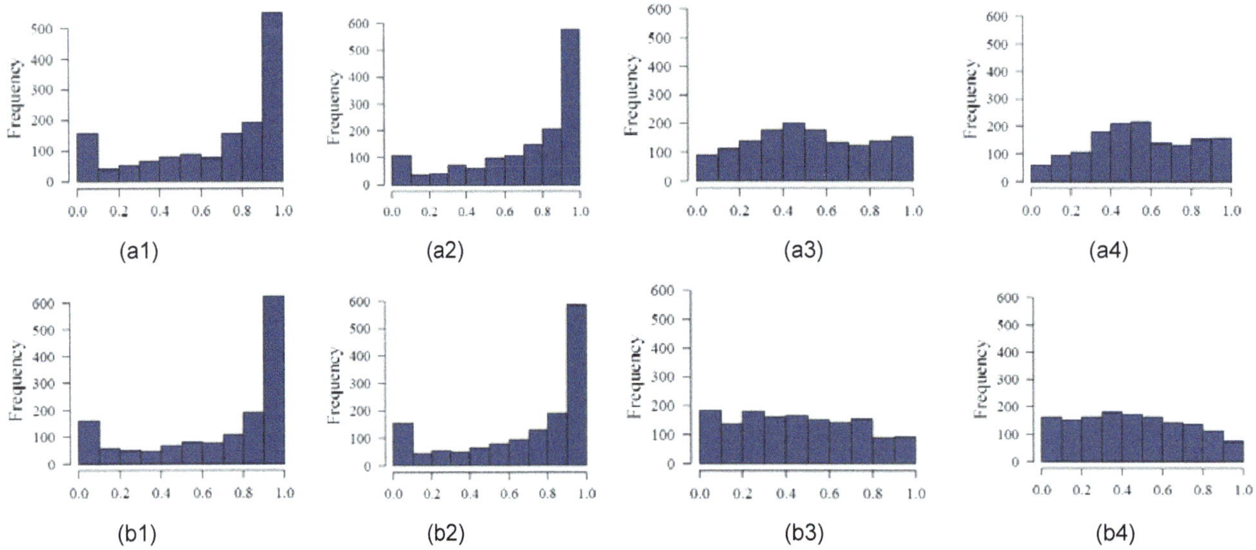


**Figure 6.** PIT histograms of forecast streamflows for lead time 7 days and the catchments **(a)** Tarn river at Milau (2138 $km^2$) and **(b)** Durance river at Serre Ponçon (3580 $km^2$) using: (1) raw forecasts, (2) pre-processed meteorological forecasts, (3) post-processed hydrological forecasts, and (4) pre-processed meteorological forecasts and post-processed hydrological forecasts. Forecast evaluation over 48 months (2005–2008).

the 5th percentile, respectively. PIT histograms are evaluated for each catchment. While Fig. 3 illustrates the results for precipitation (main meteorological input to the hydrological model), the other figures focus on the evaluation of streamflows.

Figure 3 shows the normalized RMSE values obtained from the evaluation of daily areal precipitation forecasts against observed precipitation data. The quality of raw forecasts (scenario 1) is compared to the quality of statistically processed forecasts (according to scenario 2), for leadtimes of 3, 5 and 7 days. The statistical correction applied reduces significantly the forecast errors and improves forecast precision, especially for short lead times. The results from the other scores (not shown here) indicate the same tendency, confirming the efficiency of the applied statistical correction technique to improve the precision and the reliability of the meteorological input to the hydrological model.

Concerning the impact of statistical correction on the quality of streamflow forecasts, Figs. 4 and 5 show, respectively, the Brier Skill Scores for flow forecasts exceeding the 80 % percentile of observed flows and the Rank Probability Skill Scores for the four scenarios of correction strategies studied.

Both scores show that the use of a pre-processing technique (scenario 2) generally improves the quality of ensemble streamflow forecasts in the studied catchments: BSS and RPSS values are higher than 0, showing an improvement in forecast skill with respect to the raw forecasts (scenario 1). This positive impact of pre-processing meteorological forecasts on the quality of hydrological forecasts is greater at shorter lead times.

Furthermore, results for the scenarios 3 and 4 illustrate the added value of implementing also a post-processing correction approach of the hydrological model outputs. For the prediction of events of high flows (Fig. 4), the implementation of pre- and post-processing techniques together (scenario 4) conducts to the highest score values (better forecast quality). When considering the RPSS values (Fig. 5), differences between scenario 3 and scenario 4 are more significant for shorter lead times. For longer lead times, skill scores achieved for scenario 3, where raw meteorological forecasts are used and only hydrological outputs are post-processed, are basically equivalent to those achieved when using scenario 4, where both pre-processing and post-processing are performed. The analysis of BSS and RPSS as a function of catchment size (not shown) did not indicate a clear sensitivity of the results to the catchment area. Only at longer lead times, hydrological forecasts based only on pre-processed meteorological forecasts showed negative values of skill scores for the largest catchments, i.e. less skilful forecasts comparatively to the reference (raw forecasts). Further analysis, with a larger sample of catchments, would be necessary to better detect any general tendency.

The PIT histograms in Fig. 6 illustrate the impact of statistical correction strategies on the reliability of streamflow forecasts for two catchments representative of the studied sample and for forecast lead time of 7 days. For both catchments, scenario 1 with no bias correction strategy (raw forecasts) displays biased under-dispersive streamflow ensemble forecasts. The use of scenario 2 (statistical correction applied only to meteorological forecasts) does not improve the PIT histograms of streamflow forecasts. Since the PIT histograms of statistical corrected precipitations (not shown here) do not display significant under-dispersion problems, the examples shown in Fig. 6 illustrate the impact of the rainfall-runoff transformation on the spread of the ensemble streamflow forecasts. It is possible that the added value of pre-processing meteorological input forcings in these cases has been obscured by the mixed evaluation of high and low (more frequent) streamflow periods. Biases in the modelling of the recession part of hydrographs can be of different nature from biases in the modelling of high flows. It would be interesting to separate the evaluation of hydrological forecasts by considering separately flood and recession periods. This is part of an ongoing study and is beyond the scope of this paper. Furthermore Fig. 6 presents also the impact of applying the post-processing approaches described in scenarios 3 and 4. For these scenarios a significant improvement of forecast reliability is observed for both catchments studied.

## 5 Conclusions

This paper investigates the use of statistical bias correction techniques in hydrological ensemble forecasting. Our main focus is on evaluating of the impact of different strategies of statistical bias correction on the quality of operational streamflow forecasts. From the hydrological model point of view, forecasters can use *pre-processing* approaches (statistical corrections applied prior to the hydrologic modelling, i.e. on the meteorological forcing), *post-processing* approaches (statistical corrections applied only to the output of the hydrological model, i.e. streamflow predictions) or both. We compared performance measures obtained for 11 catchments in France during a 48-month evaluation period (2005–2008) according to four scenarios of statistical bias correction: raw forecasts, only pre-processed meteorological forecasts, only post-processed hydrological forecasts and with statistical processing applied to both meteorological and hydrological forecasts.

Results show that even though correcting the meteorological uncertainties is of high importance to obtain precise and reliable inputs to the hydrological model, the errors linked to hydrological modelling remain a key-component of the total predictive uncertainty of hydrological ensemble forecasts. Statistical corrections made to precipitation forecasts can lose their effect when propagated through the hydrological model. As a result efforts to also implement a post-hydrological model correction may be necessary. In this paper we showed that even a relatively simple empirical post-processing approach can be useful to achieve reliable hydrological forecasts for operational needs. Future work should include the application of other statistical correction techniques and the use of other hydrological models and performance measures on a larger set of catchments.

**Acknowledgements.** The authors would like to thank Massimiliano Zappa and the anonynous referee for their constructive comments on this paper.

Edited by: B. Ahrens

## References


Bartholmes, J. C., Thielen, J., Ramos, M. H., and Gentilini, S.: The european flood alert system EFAS – Part 2: Statistical skill assessment of probabilistic and deterministic operational forecasts, Hydrol. Earth Syst. Sci., 13, 141–153, doi:10.5194/hess-13-141-2009, 2009.

Boucher M.-A., Tremblay, D., Delorme, L., Perreault, L., and Anctil, F.: Hydro-economic assessment of hydrological forecasting systems, J. Hydrol., 416–417, 133–144, 2012.

Brown, J. D. and Seo, D.-J.: A nonparametric post-processor for bias-correction of hydrometeorological and hydrologic ensemble forecasts, J. Hydrometeorol., 11, 642–665, 2010.

Casati, B., Wilson, L. J., Stephenson, D. B., Nurmi, P., Ghelli, A., Pocernich, M., Damrath, U., Ebert, E. E., Brown, B. G., and Mason, S.: Forecast verification: current status and future directions, Meteorol. Appl., 15, 3–18, 2008.

Cloke, H. and Pappenberger, F.: Ensemble Flood Forecasting: A Review, J. Hydrol., 375, 613–626, 2009.

Fortin, V., Favre, A. C., and Eriem, S.: Probabilistic forecasting from ensemble prediction systems: Improving upon the best-member method by using a different weight and dressing kernel for each member, Q. J. Roy. Meteorol. Soc., 132, 1349–1369, 2006.

Garçon, R.: Modèle global Pluie-Débit pour la prévision et la prédétermination des crues (Lumped rainfall-runoff model for flood forecasting and design flood estimation), La Houille Blanche, 7/8, 88–95, 1999 (in French).

Gneiting, T., Raftery, A. E., Westveld, A. H., and Goldman, T.: Calibrated Probabilistic Forecasting Using Ensemble Model Output Statistics and Minimum CRPS Estimation, Mon. Weather Rev., 133, 1098–1118, 2005.

Hashino, T., Bradley, A. A., and Schwartz, S. S.: Evaluation of bias-correction methods for ensemble streamflow volume forecasts, Hydrol. Earth Syst. Sci., 11, 939–950, doi:10.5194/hess-11-939-2007, 2007.

Jaun, S. and Ahrens, B.: Evaluation of a probabilistic hydrometeorological forecast system, Hydrol. Earth Syst. Sci., 13, 1031–1043, doi:10.5194/hess-13-1031-2009, 2009.

Krzysztofowicz, R.: Bayesian theory of probabilistic forecasting via deterministic hydrologic model, Water Resour. Res., 35, 2739–2750, 1999.

Laio, F. and Tamea, S.: Verification tools for probabilistic forecasts of continuous hydrological variables, Hydrol. Earth Syst. Sci., 11, 1267–1277, doi:10.5194/hess-11-1267-2007, 2007.

Lugiez, F. and Guillot, P.: Dix années de prévisions d'apports à Electricité de France, IAHS-AISH P., 51, 558–566, 1960.

Mathevet, T.: Erreur empirique de modèle, Note technique internet EDF-DTG D4165/NT/2010-00395-A, Internal Technical Report, p. 11, 2010.

Obled, C., Bontron, G., and Garçon, R.: Quantitative precipitation forecasts: a statistical adaptation of model outputs though an analogues sorting approach, Atmos. Res., 63, 303–324, 2002.

Olsson, J. and Lindström, G.: Evaluation and calibration of operational hydrological ensemble forecasts in Sweden, J. Hydrol., 350, 14–24, 2008.

Raftery, A. E., Gneiting, T., Balabdaoui, F., and Polakowski, M.: Using Bayesian model averaging to calibrate forecast ensembles, Mon. Weather Rev., 133, 1155–1174, 2005.

Ramos, M. H., Mathevet, Th., Thielen, J., and Pappenberger, F.: Communicating uncertainty in hydro-meteorological forecasts: mission impossible?, Meteorol. Appl., 17, 223–265, 2010.

Randrianasolo, A., Ramos, M.-H., Thirel, G., Andréassian, V., and Martin, E.: Comparing the scores of hydrological ensemble forecasts issued by two different hydrological models, Atmos. Sci. Lett., 11, 100–107, 2010.

Schaake, J., Pailleux, J., Thielen, J., Arritt, R., Hamill, T., Luo, L., Martin, E., McCollor, D., and Pappenberger, F.: Summary of recommendations of the first workshop on Postprocessing and Downscaling Atmospheric Forecasts for Hydrologic Applications held at Météo-France, Toulouse, France, 15–18 June 2009, Atmos. Sci. Lett., 11, 59–63, 2010.

Thirel, G., Rousset-Regimbeau, F., Martin, E., and Habets, F.: On the impact of short-range meteorological forecasts for ensemble stream flow predictions, J. Hydrometeorol., 9, 1301–1317, 2008.

Velazquez, J. A., Anctil, F., Ramos, M. H., and Perrin, C.: Can a multi-model approach improve hydrological ensemble forecasting? A study on 29 French catchments using 16 hydrological model structures, Adv. Geosci., 29, 33–42, 2011, http://www.adv-geosci.net/29/33/2011/.

Verkade, J. S. and Werner, M. G. F.: Estimating the benefits of single value and probability forecasting for flood warning, Hydrol. Earth Syst. Sci., 15, 3751–3765, doi:10.5194/hess-15-3751-2011, 2011.

Wilks, D. S.: Statistical Methods in the Atmospheric Sciences, 3rd Edn., Academic Press, Amsterdam, 627 pp., 2011.

Zhao, L., Duan, Q., Schaake, J., Ye, A., and Xia, J.: A hydrologic post-processor for ensemble streamflow predictions, Adv. Geosci., 29, 51–59, 2011, http://www.adv-geosci.net/29/51/2011/.

Zorita, E. and von Storch, H.: The analog method as a simple statistical downscaling technique: Comparison with more complicated methods, J. Climate, 12, 2474–2489, 1999.

# 23

# Observations and WRF simulations of fog events at the Spanish Northern Plateau

**C. Román-Cascón[1], C. Yagüe[1], M. Sastre[1], G. Maqueda[2], F. Salamanca[3], and S. Viana[4]**

[1]Dpto. Geofísica y Meteorología, Universidad Complutense de Madrid, Spain
[2]Dpto. Astrofísica y Ciencias de la Atmósfera, Universidad Complutense de Madrid, Spain
[3]Lawrence Berkeley National Laboratory (LBNL), Berkeley, CA, USA
[4]AEMET, Barcelona, Spain

*Correspondence to:* C. Román-Cascón (carlosromancascon@fis.ucm.es)

**Abstract.** The prediction of fogs is one of the processes not well reproduced by the Numerical Weather Prediction (NWP) models. In particular, the role of turbulence in the formation or dissipation of fogs is one of the physical processed not well understood, and therefore, not well parameterized by the NWP models. Observational analysis of three different periods with fogs at the Spanish Northern Plateau has been carried out. These periods have also been simulated with the Weather Research and Forecasting (WRF) numerical model and their results have been compared to observations. The study includes a comparison of the skill of different planetary boundary layer (PBL) parameterizations, surface layer schemes and a test of the gravitational settling of clouds/fogs droplets option. A statistical analysis of this comparison has been evaluated in order to study differences between the periods and between the various parameterizations used. The model results for each PBL parameterization were different, depending on the studied period, due to differences in the features of each fog. This fact made it difficult to obtain generalized conclusions, but allowed us to determine which parameterization performed better for each case. In general, judging from the models results of liquid water content (LWC), none of the PBL schemes were able to correctly simulate the fogs, being Mellor-Yamada Nakanishi and Niino (MYNN) 2.5 level PBL scheme the best one in most of the cases. This conclusion is also supported by the root mean square error (RMSE) calculated for different meteorological variables.

## 1 Introduction

The adverse effects of fog on human activities, especially on transport, are widely known. The physical processes affecting the formation or dissipation of fogs are still not well understood and these processes are not always well parameterized by the Numerical Weather Prediction (NWP) models. The chosen planetary boundary layer (PBL) scheme is very important to obtain appropriate simulations of fogs. Several studies comparing different PBL schemes for the Weather Research and Forecasting (WRF) model have been done in the last years (concerning fogs or not), as Shin and Hong (2011) who did an intercomparison of various PBL parameterizations for a single day, and found differences between unstable and stable conditions. While Welch and Welicki (1986) support the theory that turbulence acts favouring the formation of fog, other authors, such as Roach et al. (1976), established that turbulence favours its dissipation. It seems that there exists a threshold on the relation between turbulence and fog development (Zhou and Ferrier, 2008). Several recent studies have been carried out investigating how NWP models simulate fogs under different conditions. Van der Velde et al. (2010) studied the ability of the High-Resolution Limited-Area Model (HIRLAM) and WRF model to simulate a case of fog under frost conditions and they compared different microphysics and PBL options.

In the present preliminary work, the effect of turbulence on fogs is studied through the use of different PBL parameterizations in order to draw some conclusions and improve fog forecasting. Three different periods with fogs (mainly radiation fogs) have been studied and simulated with the WRF-ARW (Advanced Research WRF solver) in order to evaluate the ability of the model to simulate these events. For this purpose, three planetary boundary layer (PBL) parameterizations have been compared, as well as three different surface layer (SL) parameterizations. Additionally, an option of gravitational settling of cloud/fog droplets has been tested. Differences between simulated and observed values (bias) and RMSE (root mean square error) for different meteorological parameters have been calculated for each period and for each model option.

## 2 Data collection

This study uses data from the Research Centre for the Lower Atmosphere (CIBA in Spanish, 41°49′ N 4°56′ W, 840 m a.s.l.) (Cuxart et al., 2000), located near Valladolid (Spain). It is a place situated over an extensive (800 km$^2$) and homogenous plateau, being a prosperous place for the development of radiation fogs in autumn and winter. The main instrumentation used for this study is installed on two towers (10 and 100 m). The 10 m tower was equipped with two fast-response (20 Hz) sonic anemometers at 1.5 and 10 m and with conventional cup anemometers, vanes thermometers and hygrometers at 1.5 and 10 m. The 100 m tower was equipped with standard anemometers and thermometers at 10, 20, 35 and 97 m and with hygrometers at heights of 10 and 97 m among other instrumentation. For further information about the CIBA towers, Yagüe et al. (2009) can be consulted. Additionally, visibility information from Valladolid Airport METAR reports was used to support the existence of fogs (human observations made manually). This airport is located 14.5 km SE from CIBA, but since the terrain is very homogeneous, it was supposed that the conditions in both places were approximately the same. However, sometimes there exists a doubt in the possible local nature of the fog, so for this reason in our study, fogs were considered with relative humidity higher than 95 % at CIBA, which was supported by METAR reports on visibility. Visibilities below 1000 m are considered fogs by METAR reports; visibilities between 1000 and 5000 m are considered mists by METAR reports.

## 3 WRF-ARW 3.3

The Weather Research and Forecasting (WRF) Model is a next-generation mesoscale NWP system designed to serve both operational and atmospheric research needs (Skamarock et al., 2008). In this work, WRF-ARW 3.3 version was used. Three turbulent kinetic energy (TKE) PBL parameterizations were chosen: Mellor-Yamada-Janjic (MYJ) (Janjic, 1990), Quasi-Normal Scale Elimination (QNSE) (Sukorianski et al., 2005) and Mellor-Yamada-Nakanishi-Niino Level 2.5 (MYNN) (Nakanishi and Niino, 2004). These PBL parameterizations have been chosen among others because it has been demonstrated that TKE closure schemes are better at simulating cases dominated by stable conditions, which is what was expected during these periods, at least before the fog onset.

MYJ is the Eta operational scheme. It has a one-dimensional prognostic TKE scheme with local vertical mixing. QNSE is an extended version of MYJ based on MYJ PBL code and uses a TKE-prediction option that uses a theory for stably stratified regions (considering anisotropy for diffusivity as well as the effect of internal-gravity waves). It was designed to improve the results in stably stratified conditions. MYNN predicts a more realistic entrainment at the top of the PBL (moreover it forecasts other second order moments besides TKE). It has a new equation for the master mixing length, it parameterizes the buoyancy effects on the pressure covariances and it includes a new condensation physic with respect to the older MYJ (NCEP, 2011).

MYNN allows the use of three different surface layer (SL) parameterizations, which were also compared: MM5 similarity, Eta similarity and MYNN surface layer. MM5 scheme is based on Monin-Obukhov with Carlson-Boland viscous sub-layer and similarity functions from look-up tables. Eta similarity was used in the Eta model and it is based on Monin-Obukhov theory with Zilitinkevich thermal roughness length and standard similarity functions from look-up tables (NCEP, 2011). MYNN surface layer is the Nakanishi and Niino PBL's surface layer scheme and it uses the Monin-Obukhov theory to calculate the surface layer length scale.

Four two-way nested domains with a horizontal resolution of 27, 9, 3 and 1 km, and fifty vertical levels were used (28 levels below 1 km and 8 levels below 100 m). The boundary conditions were taken from NCEP (1° resolution, each 6 h) and a time step of 90 s and a spin up of 12 h were used. Dudhia was the SW radiation scheme chosen, Rapid Radiative Transfer Model (RRTM) the LW radiation and WRF Single-Moment 3-class scheme (WSM3) the microphysics package. A first comparison of three PBL parameterizations was done using MYJ, QNSE and MYNN schemes. Then a comparison of three surface layer (Eta, MM5 and MYNN) parameterizations was done using MYNN PBL scheme. Finally, gravitational settling of cloud/fog droplets option was tested. Since no observed liquid water content (LWC) is available for the studied site, a comparison of the LWC simulated by WRF for the different parameterizations will be shown together with the observed relative humidity.

## 4 Observational analysis

Three different periods of fogs have been studied. Each period is composed of 3 days (72 h), but the number of fog hours was different for each period. Most of the periods were characterized by favourable synoptic situations for the development of radiation fogs (see Supplement information 1), i.e. high pressures with low pressure gradient and clear skies allowing the nocturnal radiative cooling required to saturate the air. However, in some cases it is difficult to conclude that they are purely radiation fogs. Differences between periods are shown in Table 1, which shows mean values of different

**Table 1.** Mean observed values for different parameters for complete periods (3 days) and during strictly fog moments (in brackets).

| | 10-11-12 November 2009 | 10-11-12 December 2009 | 4-5-6 November 2010 |
|---|---|---|---|
| Total hours of fog (72 h) | 13 | 44 | 29 |
| 1.5 m Relative Humidity (%) | 81.16 (96.38) | 87.96 (96.3) | 88.76 (97.39) |
| 1.5 m Temperature (°C) | 9.56 (6.91) | 3.62 (1.75) | 9.78 (6.96) |
| 1.5 m Mixing Ratio ($g\,kg^{-1}$) | 6.56 (6.58) | 4.74 (4.59) | 7.32 (6.71) |
| 10 m Wind Speed ($m\,s^{-1}$) | 4.17 (3.5) | 2.05 (1.15) | 1.82 (1.6) |
| Sensible Heat Flux ($W\,m^{-2}$) | 10.35 (1.36) | 12.9 (11.92) | 20.78 (29.35) |
| Friction Velocity ($m\,s^{-1}$) | 0.76 (0.57) | 0.25 (0.09) | 0.14 (0.15) |

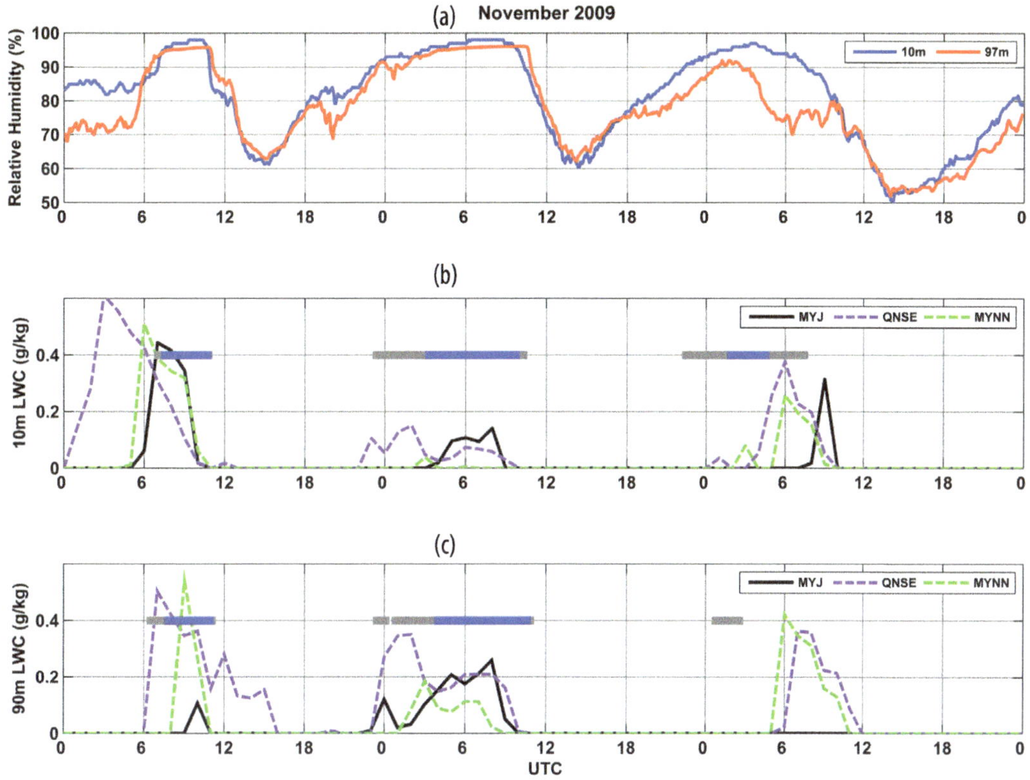


**Figure 1.** Analysis for 10, 11 and 12 November 2009. **(a)** Observed relative humidity at 10 m (blue line) and 97 m (red line). **(b)** Simulated LWC at 10 m using MYJ (black line), QNSE (dashed purple line) and MYNN (dashed green line). Blue and grey rectangles indicate observed relative humidity at 10 m larger than 95 % and larger than 90 %, respectively. **(c)** Same as **(b)** for 90 m. Blue and grey rectangles indicate the same as in **(b)** but for 97 m.

parameters for the periods. Mean values for strictly fog moments are also shown in the same table in brackets. It should be underlined the different degree of turbulence (evaluated from the friction velocity) that is present for the fog moments.

First period corresponds to the period 10-11-12 November 2009. It is composed by a total of 13 h of fog, characterized by short morning fogs (Fig. 1a). Fogs were present from 07:00 to 10:30 UTC approx. on 10 November. The main feature of this day was the relatively moderate wind speed (not shown), remaining around 2.5 $m\,s^{-1}$ before the fog onset and increasing from 3 to 4 $m\,s^{-1}$ when the fog was present. Probably, this increase in wind contributed to the evaporation of the previously condensed dew over the surface during the night and allowed the fog formation. Later, the wind continued increasing and contributed to the fog dissipation. It can be observed how during this day, there is one moment from 06:00 to 07:00 UTC when the relative humidity at 97 m was greater than at 10 m.The mixing ratio at the height of 97 m was higher (5.5 $g\,kg^{-1}$) than those at 10 m (4 $g\,kg^{-1}$) (not shown). This increase in mixing ratio at 97 m was caused by a change in wind direction from SW to S at

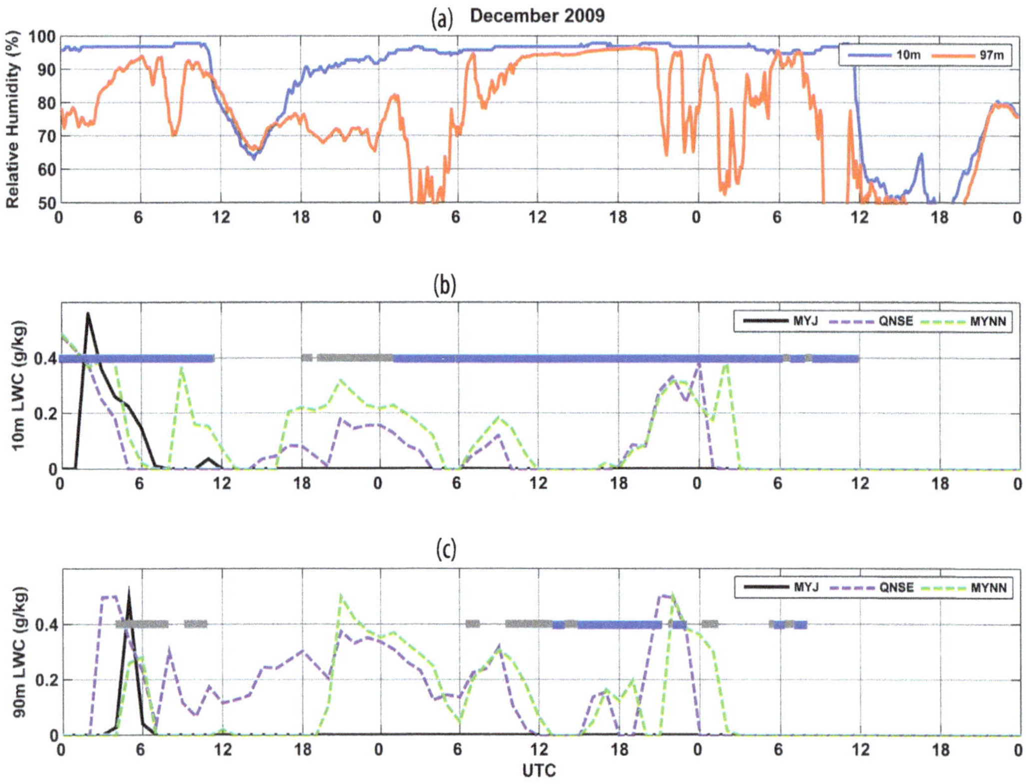


**Figure 2.** Analysis for 10, 11 and 12 December 2009. **(a)** Observed relative humidity at 10 m (blue line) and 97 m (red line). **(b)** Simulated LWC at 10 m using MYJ (black line), QNSE (dashed purple line) and MYNN (dashed green line). Blue and grey rectangles indicate observed relative humidity at 10 m larger than 95 % and larger than 90 % respectively. **(c)** Same as **(b)** for 90 m. Blue and grey rectangles indicate the same as in **(b)** but for 97 m.

this height during these times (which can represent a source of water vapour at 97 m). The second day (day 11), fogs were present from 03:00 to 09:30 UTC; and from 02:00 to 06:00 UTC on day 12, in accordance with relative humidity records. METAR information support this fog schedule except for day 12, when the fog could have a patchy behaviour, being difficult to affirm that a relative humidity higher than 95 % could indicate fog formation on this day. Relative humidity at 97 m shows how the fog progressed upward and reached this level during days 10 and 11, but not on day 12.

The second period corresponds to the period 10-11-12 December 2009. It is composed by a total of 44 h of fog and can be divided in two fog events (Fig. 2a). The first one is an event of dense fog with visibilities around 100 m (information from METAR reports, not shown). The onset was around 00:00 UTC on day 10 and the dissipation at 11:00 UTC approximately. The second event was a case of a very persistent fog without dissipation during the daytime, established approximately at 00:00 UTC on day 11 and dissipated the next day (12 December at 12:00 UTC), i.e. remaining for at least 36 h. Relatively low values of friction velocity and wind velocity (not shown) contributed to this persistent fog.

The third case corresponds to the period 4-5-6 November 2010, with a total of 29 h of fog. It is a clear case of pure radiation fogs developed during the early hours in the night and dissipated in the afternoon (Fig. 3a). Fogs were established at 05:00 UTC and dissipated at 09:00 UTC on day 4, with a possible progressive transformation into low clouds (decrease in relative humidity at 10 m and increase at 97 m). Day 5 was characterized by fogs formed at 01:00 UTC approximately and dissipated at 12:00 UTC, with a higher relative humidity at 97 m during the last 3 h of the event (from 09:00 to 12:00 UTC). On day 6, the fog developed around 23:00 UTC of the previous day and dissipated at 13.30 UTC of day 6, possibly reaching the height of 97 m from 09:00 to 13:30, as relative humidity records indicate.

## 5 WRF Simulation analysis

The studied periods were simulated with WRF-ARW model using the three PBL parameterizations previously commented on Sect. 3.

Figure 1b and c show simulated liquid water content (LWC) at 10 and 90 m, respectively, using the different PBL parameterizations for the first studied period (November

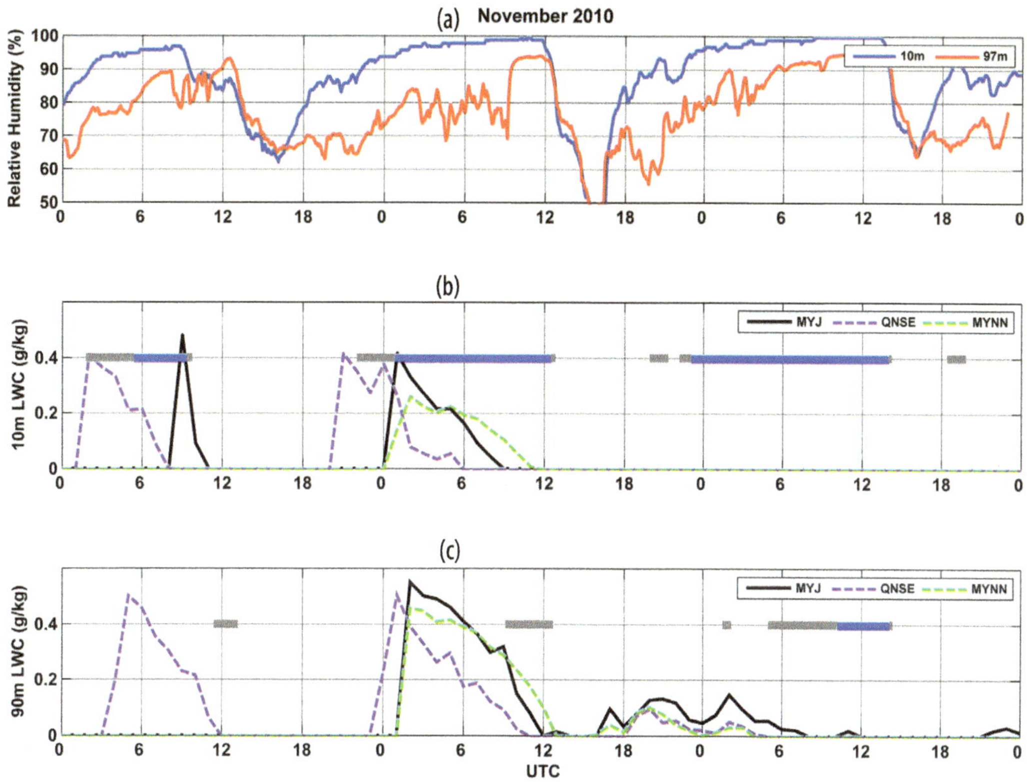


**Figure 3.** Analysis for 4, 5 and 6 November 2009. **(a)** Observed relative humidity at 10 m (blue line) and 97 m (red line). **(b)** Simulated LWC at 10 m using MYJ (black line), QNSE (dashed purple line) and MYNN (dashed green line). Blue and grey rectangles indicate observed relative humidity at 10 m larger than 95 % and larger than 90 %, respectively. **(c)** Same as **(b)** for 90 m. Blue and grey rectangles indicate the same as in **(b)** but for 97 m.

2009). This can be compared with observed relative humidity at 10 and 97 m. For the first day (day 10), the three PBL parameterizations simulated the fog relatively well, but with an earlier dissipation and an earlier formation for QNSE, giving also a wrong transformation of the fog into low clouds (see LWC at 90 m). MYJ predicted a correct fog at 10 m but not at 90 m. MYNN seems to be the best parameterization simulating LWC at 10 and 90 m. For day 11, MYJ was the best PBL scheme simulating the fog at both levels, while for day 12, QNSE and MYNN obtained more realistic results, but slightly delayed on time.

Regarding the second period (December 2009), Fig. 2b and c shows how MYJ was able to simulate the fog during some hours for the fog event 1 but it was not able to simulate the persistent fog event. MYNN was probably the best PBL scheme simulating the fog during the first event and also during the second fog event, but it was not able to remain the fog during the daytime. QNSE gave a similar behaviour of MYNN. The thickness of the fog was not correctly simulated by any of the parameterizations during this period.

Finally, for the third period (November 2010) (Fig. 3b and c), on day 4 MYJ and QNSE predicted morning fogs, but not MYNN. In any case, the predicted fog was not totally correct on time. QNSE seems to fail in the thickness of the fog, as can be seen looking at LWC simulated at 97 m. Fog on day 5 was relatively well simulated by the three parameterizations; MYNN was the best one, while QNSE and MYJ failed on the onset and dissipation of the fog. Regarding relative humidity at 97 m, all of them overestimate the thickness of the fog. The third day (day 6), none of the PBL parameterizations were able to simulate LWC at 10 m, while in reality the fog was present for more than 12 h and reached 97 m during the last hours of the fog.

A similar comparison but for the different surface layer parameterizations is shown in Supplement information 2. From this comparison is deduced that no important differences can be found using different SL schemes, only small differences.

Additionally, a testing of the Gravitational Settling (GS) option is also shown in Supplement information 2. The effect of using this option was mainly to produce more realistic results in these cases, with less LWC at high levels (it produces a shallower fog and closer to the ground). For November 2009 period, the effect was to produce no LWC at 90 m on the first day. During days 11 and 12, the effect of using this option did not change significantly the results. For December 2009, the effect of using the GS option was to obtain more realistic results at 90 m, consequently producing a shallower fog and closer to the ground. For November 2010 period,

**Table 2.** RMSE and bias calculated for each period (3 days) and during fog moments (indicated by "fogs") for different parameters and for different PBL parameterizations, MYJ (Mellor-Yamada Janjic), QNSE (Quasi-Normal Scale Elimination) and MYNN Eta (Mellor Yamada Nakanishi and Niino 2.5 level). Lowest values are indicated in bold.

| 10-11-12 Nov 2009 | MYJ | | QNSE | | MYNN Eta | |
|---|---|---|---|---|---|---|
| | RMSE | Bias | RMSE | Bias | RMSE | Bias |
| 2 m Temperature | **3.03** | **−1.51** | 3.24 | −2.13 | 3.03 | −1.69 |
| 10 m Relative Humidity | **13.00** | **9.60** | 15.33 | 12.85 | 14.21 | 10.74 |
| 10 m Wind Speed | 2.55 | 0.97 | **2.37** | **0.59** | 2.54 | 1.19 |
| Nov 2009 (fogs) | | | | | | |
| 2 m Temperature | 2.22 | 0.77 | 2.08 | **0.38** | **1.89** | 0.74 |
| 10 m Relative Humidity | **2.55** | 2.09 | 3.02 | 2.88 | 2.80 | **1.92** |
| 10 m Wind Speed | 3.42 | 2.34 | **3.08** | **2.20** | 3.82 | 2.94 |
| 10-11-12 Dec 2009 | MYJ | | QNSE | | MYNN Eta | |
| | RMSE | Bias | RMSE | Bias | RMSE | Bias |
| 2 m Temperature | 2.87 | 1.87 | 2.73 | −1.01 | **2.38** | **−0.89** |
| 10 m Relative Humidity | 19.03 | −12.38 | 10.49 | 4.78 | **8.08** | **3.94** |
| 10 m Wind Speed | 1.08 | 0.28 | **0.92** | **0.17** | 0.94 | **0.17** |
| Dec 2009 (fogs) | | | | | | |
| 2 m Temperature | 3.54 | 3.18 | **1.32** | 0.36 | 1.35 | **0.25** |
| 10 m Relative Humidity | 23.61 | −18.93 | 4.50 | **0.11** | **4.34** | 0.78 |
| 10 m Wind Speed | 0.79 | 0.45 | 0.66 | 0.33 | 0.66 | **0.21** |
| 4-5-6 Nov 2010 | MYJ | | QNSE | | MYNN Eta | |
| | RMSE | Bias | RMSE | Bias | RMSE | Bias |
| 2 m Temperature | 3.28 | 0.42 | 3.75 | −1.15 | **3.20** | **0.10** |
| 10 m Relative Humidity | **10.08** | 0.20 | 10.61 | 4.56 | 10.91 | **0.08** |
| 10 m Wind Speed | **1.06** | **0.29** | 1.09 | 0.51 | 1.35 | 0.87 |
| Nov 2010 (fogs) | | | | | | |
| 2 m Temperature | 2.98 | 2.13 | **2.58** | **1.57** | 2.77 | 1.68 |
| 10 m Relative Humidity | 5.85 | −3.62 | **5.08** | **−3.05** | 5.91 | −3.29 |
| 10 m Wind Speed | 1.15 | 0.53 | **0.91** | **0.55** | 1.47 | 1.23 |

the effect of using the GS option was to obtain better results for LWC for day 6, with a better simulated fog at 10 m but mistakenly simulated at 90 m.

## 6 WRF simulations statistics

In the previous section it has been shown how, depending on the different fog event analysed, sometimes a PBL parameterization is better than others. In order to have an overall view, a statistical analysis comparing simulated to observational results has been carried out. A total of 216 h were analyzed and simulated (72 h for each period). The results of these simulations for different parameters were compared with observations. For this purpose, RMSE and bias were calculated for each PBL parameterization. The model produced instant values for each hour, but a comparison with the observed averaged values was done after checking that the temporal variability of the model in the different studied parameters was not very high. Table 2 indicates RMSE and bias for temperature, relative humidity and wind speed for the whole periods. These statistical values have also been calculated only for those moments with determined observed fogs and are labelled in the tables as "fogs". In Supplement information 3, a complete statistical analysis is shown, including variables as friction velocity, mixing ratio and sensible heat flux. Differences between the values for the different periods can be observed.

According to the November 2009 period, lower values for RMSE and bias are observed during fog times, except for wind speed, although it should be taken into account the lower number of fog hours (only 13 h). MYNN obtained slightly better results and QNSE for the wind speed prediction. Nevertheless, there were not too many differences between the 3 parameterizations.

For the December 2009 period, MYJ produced unsatisfactorily results. MYNN was the best scheme with the lowest values of RMSE and bias and QNSE gave the lowest values for wind speed. The model overestimated the temperature during this period because it was not able to predict the persistent fog during the daytime.

For November 2010, QNSE obtained the lowest values for bias and RMSE during fog times. Despite this, the results for the three PBL parameterizations used were quite similar.

During fog moments, all of the parameterizations overestimated the temperature. This could be one of the reasons why the model did not predict the fog. These overestimated temperatures did not allow relative humidity to increase sufficient to produce condensation. In general, these temperature biases were positive during the whole period because the model did not predict the fog, and a higher temperature was simulated during daytime, a consequence of a higher downward short wave radiation at surface than in the reality. These results agree with Shin and Hong (2011), who found how the PBL parameterizations in WRF tended to underestimate the surface cooling rates during nighttimes (in this case before the fog onset).

The predicted wind can be also an important factor affecting the fog forecasting: with positive bias values, the higher wind speed predicted could give an earlier dissipation of fogs.

The first problem is that a comparison done with only 72 h is not sufficient to obtain reliable results. It has also been deduced that a complete statistics, using the three periods together, cannot be done since the studied periods and their statistical values are different from each other, depending on the period. The fact of using different types of radiation fogs and trying to use them as the same type should affect the results of the simulation with different PBL parameterizations. The evolution and the type of fog are mainly driven by the phase changes of the fog water and its microphysics, radiation budget, advection, turbulence and effects of the terrain (Cuxart and Jiménez, 2012). That is, the relative importance of turbulence over fog would not be the same, depending on the other factors and on the type of the fogs. For instance, it would not be the same in a case of persistent, dense and deep fog as in a case of short, not very dense and shallow fog. A classification of different types of radiation fogs has to be done in order to correctly calculate this type of statistics using different PBL parameterizations. Despite these uncertainties, MYNN scheme seems to be the best PBL parameterization for most cases, with the minimum RMSE values. Also, the use of the gravitational settling option produced slightly better results (see Supplement information 3).

There were well known discrepancies in the formation and dissipation of fogs by MYJ PBL scheme (see Nakanishi and Niino, 2004), and the improvements added to MYNN and QNSE (see Sect. 3) seem to work during some of these cases. Nevertheless, these improvements are not sufficient in some cases, as for the case of persistent fog in December 2009 period and the radiation fog observed on 6 November 2010.

## 7 Conclusions

Three periods with fogs at the Spanish Northern Plateau were analysed and simulated with WRF-ARW model. The features of the fogs were different for each period, and consequently, the results of the simulations depended on the period more than on the option used to simulate, making it difficult to obtain generalized conclusions from this study. Despite this, MYNN obtained slightly better results simulating LWC during fog moments. The surface layer scheme used was not very important in terms of LWC simulated by the model, with only local differences between the different schemes. Additional gravity settling of cloud/fog droplets produced less LWC at high levels, a shallower fog, and it usually served to improve the simulation. The relative importance of the different physical processes (turbulence among them) depends on the features and the type of radiation fogs. A radiation fog classification is needed to compare the differences between observations and simulated values with different PBLs parameterizations. Despite the improvement obtained using MYNN and QNSE parameterizations, it has been demonstrated how they have several uncertainties concerning fog simulations during specific events.

**Acknowledgements.** The authors wish to thank Javier Peláez (CIBA) for his technical support and help as well as J. L. Casanova, Director of the CIBA. We are also very grateful to the editor and two anonymous referees for their constructive suggestions, which helped to improve this paper. This research has been funded by the Spanish Ministry of Science and Innovation (project CGL2009-12797-C03-03). The GR35/10 program (supported by BSCH and UCM) has also partially financed this work through the Research Group "Micrometeorology and Climate Variability" (No 910437).

Edited by: G.-J. Steeneveld

## References


Cuxart, J. and Jiménez, M. A.: Deep radiation fog in a wide closed valley: Study by numerical physics: its design and verification, Pure Appl. Geophys., doi:10.1007/s00024-011-0365-4, in press, 2012.

Cuxart, J., Yagüe, C., Morales, G., Terradellas, E., Orbe, J., Calvo, J., Fernandez, A. Soler, M. R., Infante, C., Buenestado, P., Espinalt, A., Joergensen, H. E., Rees, J. M., Vilá, J., Redondo, J. M., Cantalapiedra, I. R., and Conangla, L.: Stable Atmospheric Boundary-Layer Experiment in Spain (SABLES 98): A report, Bound.-Lay. Meteorol., 96, 337–370, 2000.

Janjic, Z. A.: The step-mountain coordinate: physics package, Mon. Weather Rev. 118, 1429–1443, 1990.

Nakanishi, M. and Niino, H.: An improved Mellor-Yamada level-3 model with condensation physics: its design and verification, Bound.-Lay. Meteorol., 112, 1–31, 2004.

NCEP: Mesoscale & Microscale Meteorology Division, Weather Research & Forecasting ARW Version 3 Modeling System User's Guide, 2011.

Roach, W. T., Brown, R., Caughey, S. J., Garland, J. A., and Readings, C. J.: The physics of radiation fog I – a field study, Q. J. Roy. Meteor. Soc., 102, 313–333, 1976.

Shin, H. H. and Hong, S.-Y.: Intercomparison of Planetary Boundary Layer Parameterizations in the WRF Model for a Single Day from CASES-99, Bound.-Lay. Meteorol., 281, 139–261, 2011.

Skamarock, W. C., Klemp, J. B., Dudhia, J., Gill, D. O., Barker, D. M., Duda, M. G., Huang, X. Y., Wang, W., and Powers, J. G.: A description of the advanced research WRF version 3, NCAR Technical note, NCAR/TN-475+STR, 113 pp., 2008.

Sukorianski, S., Galperin, B., and Perov, V.: Application of a new spectral theory of stable stratified turbulence to the atmospheric boundary layer over sea ice, Bound.-Lay. Meteorol., 117, 231–257, 2005.

Van Der Velde, I. R., Steeneveld, G. J., Wichers Schreur, B. G. J., and Holtslag, A. A. M.: Modeling and Forecasting the Onset and Duration of Severe Radiation Fog under Frost Conditions, Mon. Weather Rev., 138, 4237–4253, 2010.

Welch, R. M. and Welicki, B. A.: The stratocumulus nature of fog, J. Appl. Meteorol., 25, 101–111, 1986.

Yagüe, C., Sastre, M., Maqueda, G., Viana, S., Ramos, D., Vindel, J. M., and Morales, G.: CIBA2008, an experimental campaign on the atmospheric boundary layer: preliminary nocturnal results, Física de la Tierra, 21, 13–26, available at: http://revistas.ucm.es/fis/02144557/articulos/FITE0909110013A.PDF, 2009.

Zhou, B. and Ferrier, B. S.: Asymptotic Analysis of Equilibrium in Radiation fog, J. Appl. Meteor. Clim., 47, 1704–1722, 2008.

# 24

# Evaluating meteorological climate model inputs to improve coastal hydrodynamic studies

**D. Bellafiore[1], E. Bucchignani[2,3], S. Gualdi[3], S. Carniel[1], V. Djurdjeviæ[4], and G. Umgiesser[1,5]**

[1]ISMAR-CNR, Institute of Marine Science – National Research Council, Venice, Italy
[2]CIRA Italian Aerospace Research Centre, Capua, Italy
[3]CMCC Centro Euro-Mediterraneo per i Cambiamenti Climatici, Bologna, Italy
[4]SEE/VCCC – Institute of Meteorology, Faculty of Physics, University of Belgrade, Serbia
[5]Coastal Research and Planning Institute (CORPI), Klaipeda University, Klaipeda, Lithuania

**Abstract.** This work compares meteorological results from different regional climate model (RCM) implementations in the Mediterranean area, with a focus on the northern Adriatic Sea. The need to use these datasets as atmospheric forcings (wind and atmospheric pressure fields) for coastal hydrodynamic models to assess future changes in the coastal hydrodynamics, is the basis of the presented analysis. It would allow the assessment of uncertainties due to atmospheric forcings in providing coastal current, surge and wave climate changes from future implementations of hydrodynamic models.

Two regional climate models, with different spatial resolutions, downscaled from two different global climate models (whose atmospheric components are, respectively, ECHAM4 and ECHAM5), were considered. In particular, the RCM delivered wind and atmospheric pressure fields were compared with measurements at four stations along the Italian Adriatic coast. The analyses were conducted using a past control period, 1960–1990, and the A1B IPCC future scenario (2070–2100). The chosen scenario corresponds to a world of very rapid economic and demographic growth that peaks in mid-century, with a rapid introduction of new efficient technologies, which balance fossil and non-fossil resources (IPCC, 2007). Consideration is given to the accuracy of each model at reproducing the basic statistics and the trends. The role of models' spatial resolution in reproducing global and local scale meteorological processes is also discussed. The Adriatic Sea climate is affected by the orography that produces a strengthening of north-eastern katabatic winds like bora. Therefore, spatial model resolution, both for orography and for a better resolution of coastline (Cavaleri et al., 2010), is one of the important factors in providing more realistic wind forcings for future hydrodynamic models implementations. However, also the characteristics in RCM setup and parameterization can explain differences between the datasets. The analysis from an ensemble of model implementation would provide more robust indications on climatic wind and atmospheric pressure variations. The scenario-control comparison shows a general increase in the mean atmospheric pressure values while a decrease in mean wind speed and in extreme wind events is seen, particularly for the datasets with higher spatial resolution.

## 1 Introduction

One of the major aspects discussed in recent works on climate changes is how to provide information from the global scale to the local one. In fact, the impacts that sea level rise and changes in the meteorological conditions due to climate changes might have on the coastal zone are relevant for mitigation and risk assessment plans. The investigation of the coastal zone hydrodynamics, from a modelling point of view, is a challenging aspect that requires a preliminary analysis of the main forcings acting in the area, specifically the meteorological components. This work is mainly focused on the local variations in the wind and atmospheric pressure fields reproduced by climate models, so far not thoroughly investigated. Indeed, even if it is possible to find in the literature analyses of global and regional air temperature and precipitation fields (Giorgi et al., 2004; Zampieri et al., 2010), to our knowledge

**Table 1.** Summary table of the five analyzed RCM datasets.

| AOGCM | SINTEX-G | SINTEX-G | CMCC-MED | CMCC-MED | ECHAM5-MPIOM |
|---|---|---|---|---|---|
| RCM | EBU-POM | Cosmo-CLM | Cosmo-CLM | Cosmo-CLM | Cosmo-CLM |
| Experiment | E | C14E4 | C14E5 | C8E5 | S18E5 |
| Spatial Domain | 2–20° E; 40–52° N | 2–20° E; 40–52° N | 2—20° E; r 40–52° N | 2–20° E; 40–52° N | 10.7° W–36.9° E; 34.5–69.9° N |
| Resolution | 28 km | 14 km | 14 km | 8 km | 18 km |
| Vert. Levels | 42 | 40 | 40 | 40 | 32 |
| Time output | 6 h | 6 h | 6 h | 6 h | 3 h |

there is only one attempt to verify the capability to reproduce wind fields and, therefore, storminess induced in the coastal zone (Woth et al., 2006). The wind driver, compared with other variables analyzed in climate change studies, such as temperature, has the peculiarity to be highly spatially varying and strongly characterized on small scales. The challenge is to quantify climatic effects due to wind on the local scale. The capability to reproduce storminess is fundamental for local coastal management, in order to deal with the main threats of these areas, like coastal floodings and severe erosion events.

## 2 Methods

### 2.1 Climate models description and setup

Datasets from the different RCMs were analyzed comparing wind and atmospheric pressure fields both for the control period (1960–1990) and the IPCC A1B scenario (2070–2100).

Regional models consider limited domains, therefore boundary conditions are obtained from global climate simulations. Results are provided for the control period (1960–1990) and for the A1B IPCC scenario (2070–2100). The considered datasets are summarized in Table 1, defining the atmospheric-ocean global climate models (AOGCM) that provide initial and boundary conditions, besides time step, spatial domain and resolution of the RCMs.

The AOGCM SINTEX-G is a coupled atmosphere-ocean general circulation model, whose atmospheric component is ECHAM4 (Roeckner et al., 1996). The ocean model component is the reference version 8.2 of the Ocean Parallelise (OPA; Madec et al., 1999) with the ORCA2 configuration. The AOGCM CMCC-MED is formed by the atmospheric model ECHAM5 (Roeckner et al., 2003) and the global ocean component OPA in the ORCA2 configuration. In the Mediterranean region, CMCC-MED uses a high-resolution model of the Mediterranean Sea (NEMO-MFS, Oddo et al., 2009) able to resolve the small scale dynamics of the basin.

ECHAM4 has ~120 km (T106) spatial resolution, 12 h time resolution and runs on 360 days yearly, while ECHAM5 spatial resolution is ~80 km (T159), 6 h time resolution and runs on 365 days yearly.

The coupled global model where the S18E5 RCM is nested to is ECHAM5-MPIOM (MPIOM is the ocean model of the Max Planck Inst. Germany).

The first RCM is EBU-POM (Eta Belgrade University – Princeton Ocean Model), a coupled regional climate model that is the combination of two limited area models one for the atmosphere and the other for the ocean (Djurdjevic and Rajkovic, 2008, 2010; Gualdi et al., 2008). The atmospheric component is the Eta/NCEP (National Centers for Environmental Prediction) limited area model, which is a hydrostatic primitive equation, grid-point model (Janjic, 1984; Mesinger et al., 1988).

The second RCM is COSMO-CLM (Rockel et al., 2008), the climate version of the COSMO model, which is the operational non-hydrostatic mesoscale weather forecast model developed by the German Weather Service (DWD).

EBU-POM (hereafter called dataset E) has initial and boundary conditions taken from SINTEX-G. Simulated years are of 360 days, as in the global model.

The COSMO-CLM model was used to generate four different datasets. Three of them were produced at the Italian Aerospace Research Centre (CIRA), one forced by the AOGCM SINTEX-G (C14E4), and two (C14E5 and C8E5) by the AOGCM CMCC-MED. The fourth (S18E5) was generated at the DWD and forced by the AOGCM ECHAM5-MPIOM. The latter model implementation introduces a substantial smoothing of the Caucasus Mountains orography to the constant height of 150 m, which is the mean height in the ECHAM5-MPIOM model for that region.

The meteorological fields from models are bilinearly interpolated on a finite element grid to extract wind and atmospheric pressure time series in the meteorological stations presented in the next section.

### 2.2 Measurements

To evaluate the quality of the climate models in reproducing wind statistics, modeled winds will be compared with some of the longest measured datasets in the North Adriatic. The chosen stations are Trieste, Venice Tessera and Ravenna, in the northern part of the basin, and Bari station, in the South Adriatic (Fig. 1, left panel), from SYNOP National Aeronautics dataset that covers the period 1958–2004. The

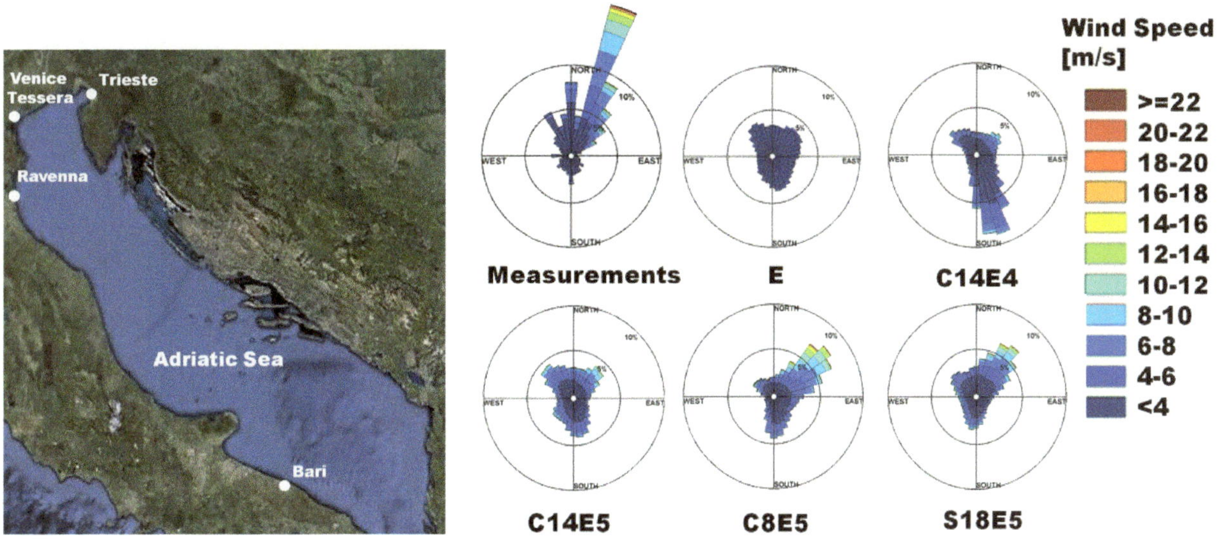


**Figure 1.** Location of wind stations (left panel). In the right panels, the wind direction distribution (direction of provenience) for E, C14E4, C14E5, C8E5, S18E5 and the measured datasets for the Trieste Station are shown for the reference period (1960–1990).

observational data sets are compared with the model simulations for the period 1960–1990, that we will refer to as the reference period. Data are provided every 3 h.

## 3 Results and discussion

### 3.1 Reference period

A preliminary analysis has been carried out to evaluate the effects related to the increase in the spatial resolution of RCMs, and the influence of the boundary conditions obtained from the different global models in reproducing wind and atmospheric pressure statistics for the reference period.

The modeled and measured mean wind speed and atmospheric pressure values are compared for the selected stations (Fig. 1, left panel). The mean wind speed in Venice Tessera is better reproduced by dataset E than by the others, though underestimated (−11 %). On the other hand, the high resolution COSMO-CLM datasets (C14E4, C8E5, S18E5) are not able to reproduce the wind speed statistics in the northwestern Adriatic. In particular, they appear to overestimate the wind speed by about 40 %. Interestingly, these models appear to perform better when compared with the other stations (e.g., Ravenna and Bari). This aspect can be explained observing that these latter stations are less characterized by specific wind regimes, also due to the lack of orographic effects. Therefore the models encounter less criticities linked with spatial resolution. Considering the two datasets C14E5 and C8E5, both laterally forced by ECHAM5 global climate model but running with different spatial resolutions, the lower resolution (C14E5) seems to be in better agreement with observations, especially in terms of mean wind speed. The C14E5 results appear to be remarkably good for the Ravenna station case, where the C8E5 dataset exhibits an overestimation of the wind speed of about 29 %. This suggests that a simple increase of the resolution does not automatically improve the results, at least in terms of mean wind field. Resolution, in any case, plays an important role in defining the correct wind regimes and, particularly in Trieste, the wind directions are fairly matched by C8E5 than C14E5 (Fig. 1, right panel) . On the other hand, resolution is not the only changing aspect between the different datasets, since different setups are used for the COSMO-CLM simulations performed at the CIRA Institute (C14E4, C14E5, C8E5) and for the one provided by the World Data Center for Climate (S18E5). In the latter case, a smoothing of the Balkan orography was performed to avoid numerical errors, even if this has led to a less realistic representation of the mountains in the region. However, despite the approximation, the model appears to perform better than the others in reproducing katabatic winds as bora in the Adriatic Sea area. A comparison with the Trieste station, in fact, indicates that S18E5 provides better results both in terms of extreme events (−0.7 % compared to higher underestimation obtained with the other models) and in the wind direction (Fig. 1, right panel). Generally, the majority of models miss the dominant NNE wind direction, probably due to the need of increased resolution and a proper dealing of orography reproduction to simulate this wind regime.

Atmospheric pressure is reproduced better than wind by all models in all stations, scoring relative differences lower than 0.4 % for the E dataset (the one that has the worst behavior). Once again the best performance is given by the C14E5 dataset, which shows a remarkably good agreement with observations for the Venice Tessera station case.

**Table 2.** Differences between A1B scenario (2070–2100) and reference period (1960–1990) – 100 (A1B – reference)/reference – for mean and 99th percentile values of atmospheric pressure (left panel) and wind speed (right panel).

| | Mean Atmospheric Pressure [%] | | | | | Mean Wind Speed [%] | | | | |
|---|---|---|---|---|---|---|---|---|---|---|
| Stations | E | C14E4 | C14E5 | C8E5 | S18E5 | E | C14E4 | C14E5 | C8E5 | S18E5 |
| VeT | 0.00 | 0.02 | 0.06 | 0.06 | 0.23 | −7.91 | −6.26 | −4.32 | −5.31 | −4.68 |
| Ts | 0.01 | 0.03 | 0.07 | 0.06 | 0.24 | −5.33 | −6.71 | −0.52 | −4.49 | −4.57 |
| Ra | 0.01 | 0.03 | 0.07 | 0.06 | 0.24 | −5.52 | −5.01 | −0.43 | −4.00 | −3.19 |
| Ba | 0.01 | 0.03 | 0.07 | 0.06 | 0.24 | −5.24 | −7.65 | −0.91 | −4.19 | −5.08 |
| | Atmospheric Pressure 1st percentile [%] | | | | | Wind Speed 99th percentile [%] | | | | |
| Stations | E | C14E4 | C14E5 | C8E5 | S18E5 | E | C14E4 | C14E5 | C8E5 | S18E5 |
| VeT | 0.07 | 0.11 | 0.16 | 0.16 | 0.30 | −5.01 | −3.93 | −4.43 | −4.55 | −0.11 |
| Ts | 0.11 | 0.06 | 0.07 | 0.02 | 0.27 | −9.26 | −6.07 | −2.69 | −3.65 | −2.18 |
| Ra | 0.11 | 0.05 | 0.07 | 0.03 | 0.29 | −9.69 | −4.89 | −1.83 | −3.97 | −3.41 |
| Ba | 0.11 | 0.06 | 0.07 | 0.03 | 0.27 | −9.70 | −6.09 | −2.64 | −1.74 | −1.64 |

The computation of extremes (99th percentiles for wind speed and 1st percentile for atmospheric pressure outputs), allows the evaluation of the model performance under extreme wind events. The best performances for atmospheric pressure are given by the high resolution models downscaled from ECHAM5 (C14E5, C8E5, overestimation ∼+0.5 % as an average). Wind speed extremes are better reproduced by C14E5 (underestimation ∼+5 % as an average). This would suggest an influence of AOGCM, where the ECHAM5 that seems to have more realistic characteristics compared to ECHAM4 performs better (increased resolution, different setups, etc.). On the other hand, the reduced Balkan orography introduced in the S18E5 can explain the underestimation of 99th percentile values of wind speed in the North Adriatic stations by this dataset.

### 3.2 A1B scenario comparison

The study of meteorological outputs for each model allows the definition of variations in the future scenarios. Analyzing the mean wind speed variations in the future A1B IPCC scenario, compared with the reference period, it is evident that the majority of models simulate a general decrease in the four coastal stations, higher for models downscaled from ECHAM4 (E and C14E4 dataset, Table 2). Considering the mean atmospheric pressure variations, the strongest increase is registered by the S18E5 (see again Table 2). A strong decrease in the wind extreme values is depicted by the E and C14E4 datasets (∼−9 % and ∼−6 %, Table 2), while the other datasets register smaller decreases. However all stations identify the same tendency. On the other hand, the analysis of the 1st percentile for the atmospheric pressure shows a slight increase in the extreme low pressure values, tendency confirmed by all models. Performing a Kolmogorov-Smirnov hypothesis test, for each station on the control period and A1B scenario wind and atmospheric pressure time series, where p-values are almost 0 everywhere, what arises is that differences in mean and distribution between them are significant.

## 4 Conclusions

The outcomes of this preliminary work on the Adriatic Sea highlight uncertainties intrinsic in the dynamical downscaling approach from GCM simulations to spatial scales suitable to study climate change impacts on coastal dynamics at the local scale. Despite the limitation and the still preliminary phase of the work, we think this analysis has pointed out the main problems in bridging the gap between the coarse information provided by the climate simulations and the detailed information required to investigate the climate change impacts and risks at the local level: increased resolution may not improve the models' reproduction of certain variables (i.e. atmospheric pressure) and the global model choice and setup could impact models' outputs. The different skills shown by the models in reproducing wind and atmospheric pressure fields suggest that ensemble of simulations would probably provide more robust climatic forcings for coastal hydrodynamic modeling implementations.

**Acknowledgements.** The support of the European Commission through FP7.2009-1, Contract 244104-THESEUS ("Innovative technologies for safer European coasts in a changing climate") is gratefully acknowledged. The authors want to thank the World Data Center for Climate for the provision of the S18E5 dataset. Thanks also to B. Rajkovic for producing the E dataset and E. Scoccimarro for the help in handling the E dataset.

Edited by: W. May

## References

Cavaleri, L., Bertotti, L., Buizza, R., Buzzi, A., Masato, V., Umgiesser, G., and Zampieri, M.: Predictability of extreme meteo-oceanographic events in the Adriatic Sea, Q. J. Roy. Meteor. Soc., 136, 400–413, 2010.

Djurdjevic, V. and Rajkovic, B.: Verification of a coupled atmosphere-ocean model using satellite observations over the Adriatic Sea, Ann. Geophys., 26, 1935–1954, doi:10.5194/angeo-26-1935-2008, 2008.

Djurdjevic, V. and Rajkovic, B.: Development of the EBU-POM coupled regional climate model and results from climate change experiments, in: Advances in Environmental Modeling and Measurements, edited by: Mihajlovic, T. D. and Lalic, B., Nova Publishers, 2010.

Giorgi, F., Bi, X., and Pal, J. S.: Mean. interannual variability and trends in a regional climate change experiment over Europe, Clim. Dynam., 22, 733–756, doi:10.1007/s00382-004-0409-x, 2004.

Gualdi, S., Rajkovic, B., Djudjevic, V., Castellari, S., Scoccimarro, E., Navarra, A., and Dadic, M.: Simulation of climate change in the mediterranean area, Web, Final Scientific, http://www.earth-prints.org/bitstream/2122/4675/1/SINTA_FInal%20Science%20Report%20_October%202008.pdf, 2008.

IPCC: Climate Change 2007, Synthesis Report, Technical report, IPCC-Intergovernamental Panel for Climate Change, 2007.

Janjic, Z.: Non-linear advection schemes and energy cascade on semi staggered grids, Mon. Weather Rev., 112, 1234–1245, 1984.

Madec, G., Delecluse, P., Imbard, M., and Levy, C.: OPA 8.1 Ocean General Circulation Model reference manual, Internal Rep. 11, Inst. Pierre-Simon Laplace, Paris, France, 1999.

Mesinger, F., Janjic, Z., Nicovic, S., Gavrilov, D., and Daven, D.: The step mountain coordinate: model description and performance for cases of alpine lee cyclogenesis and for a case of an Appalachian redevelopment, Mon. Weather Rev., 116, 1493–1518, 1988.

Oddo, P., Adani, M., Pinardi, N., Fratianni, C., Tonani, M., and Pettenuzzo, D.: A nested Atlantic-Mediterranean Sea general circulation model for operational forecasting, Ocean Sci., 5, 461–473, doi:10.5194/os-5-461-2009, 2009.

Rockel, B., Will, A., and Hense, A.: The regional Climate Model COSMO-CLM (CCLM), Meteorol. Z., 17(4), 347–348, 2008.

Roeckner, E., Arpe, K., Bengtsson, L., Christoph, M., Claussen, M., Dümenil, L., Esch, M., Giorgetta, M., Schlese, U., and Schulzweida U.: The atmospheric general circulation model ECHAM-4: Model description and simulation of present-day climate, Max Planck Institut fur Meteorologie, report 218, 90 pp., 1996.

Roeckner, E., Bäuml, G., Bonaventura, L., Brokopf, R., Esch, M., Giorgetta, M., Hagemann, S., Kirchner, I., Kornblueh, L., Manzini, E., Rhodin, A., Schlese, U., Schulzweida, U., and Tompkins, A.: The atmospheric general circulation model ECHAM 5, Max-Planck-Institut fuer Meteorologie, report 349, Hamburg, 140 pp., ISSN 0937-1060, 2003.

Woth, K., Weisse, R., and Von Storch, H.: Climate change and North Sea storm surge extremes: an ensemble study of storm surge extremes expected in a changed climate projected by four different regional climate models, Ocean Dynam., 56, 3–15, doi:10.1007/s10236-005-0024-3, 2006.

Zampieri, M., Giorgi, F., Lionello, P., and Nikulin, G.: Regional climate change in the Northern Adriatic, J. Phys. Chem. Earth, in press, doi:10.1016/j.pce.2010.02.003, 2010.

# 25

# Statistical analysis of inter-arrival times of rainfall events for Italian Sub-Alpine and Mediterranean areas

**C. Agnese[1], G. Baiamonte[1], C. Cammalleri[2], D. Cat Berro[3], S. Ferraris[4], and L. Mercalli[3]**

[1]Università degli Studi di Palermo, Dip. SAGA, Palermo, Italy
[2]Università degli Studi di Palermo, DICAM, Palermo, Italy
[3]Società Meteorologica Italiana, Castello Borello, Bussoleno (TO), Italy
[4]Università degli Studi di Torino, DIST, Torino, Italy

*Correspondence to:* C. Cammalleri (carmelo.cammalleri@unipa.it)

**Abstract.** In this work a set of time-series of inter-arrival times of rainfall events, at daily scale, was analysed, with the aim to verify the issue of increasing duration of dry periods. The set consists of 12 time-series recorded at rain gauges in 1926–2005, six of them belong to an Italian Sub-Alpine area (Piedmont) and six to a Mediterranean one (Sicily). In order to overcome the problem related to limited sample size for high values of inter-arrival times, the discrete probability polylog-series distribution was used to fit the empirical data from partial (20 yr) time-series. Moreover, a simple qualitative trend analysis was applied to some high quantiles of inter-arrival times as well as to the average extent of rain clusters. The preliminary analysis seems to confirm the issue of increasing duration of dry periods for both environments, which is limited to the "cold" season.

## 1 Introduction

Rainfall modelling represents an important issue in several research areas, such as hydrology, meteorology and climatology, because of both theoretical and practical implications. In particular, the assessment of the probabilistic structure of rainy days occurrence allows estimating the probability of both rainy and dry periods.

One interesting way to characterize the temporal structure of rainfall is the use of inter-arrival times series, which is constituted by the succession of the times, $T$, elapsed between a rainy day and the immediately preceding one (Waymire and Gupta, 1981). Because of some typical properties of rainfall, as rainy days clustering and dry periods persistence, the observed T-samples usually present some statistical peculiarities: very-high variability and skewness, relative high frequency associate to $T = 1$, "monotonically" decreasing frequencies with sporadic sampling for high values of $T$.

The main purpose of this work is to find a sign of an existing trend for some high T-quantiles (long dry periods). While a large sample is already available for some yearly-averaged features (i.e., the mean inter-arrival time), a consistent sample for high T-quantiles can be obtained only by considering a rainfall record of several years of observation; a sampling window of 20-yr has been judged large enough to fulfil this purpose. However, a time-series of any high T-quantile obtained following such procedure contains few elements only, because of the limited length of available rainfall records ($\approx$ 100 yr of observations). In order to overcome this problem, a partial time-overlap between two successive 20-yr sampling windows has been assumed. The introduction of this overlap allows obtaining more reliable time-series, but it makes tricky the detecting of possible trends using the procedures commonly adopted for yearly-averaged features.

Sample T's-distributions have been fitted by (discrete) polylog-series probability distributions; that allowed defining un-biased high T-quantiles, removing the difficulty caused by "sparse sampling". Besides, the use of theoretical distributions instead of empirical ones provides a more robust analysis tool. Furthermore, in the context of eco-hydrology, modelling of rainfall process at daily scale enables to infer statistics for some derived processes, as the mean soil moisture and/or the actual evapotranspiration, allowing us to analyse possible trends.

A set of 12 long time-series of inter-arrival times was analysed, six of them belong to an Italian Sub-Alpine area (Piedmont) and six belong to a Mediterranean one (Sicily); these two areas are characterized by different rainfall regimes and seasonality. Time-series were obtained by daily rainfall data recorded in the period 1926–2005, subdivided in 13 sub-sets of 20-yr duration progressively shifted of 5-yr. Time-series of high quantiles of associate probability distribution and theoretically-derived average cluster size were analysed for detecting possible temporal trend.

## 2 Fitting sample distribution of inter-arrival times through polylog-series probability distribution

Assuming that T-values are independent and identically distributed (i.i.d.), them can be studied as random variables that generate a renewal process (Feller, 1968), which can be seen as a generalisation of the Poisson process (implying an exponential inter-arrival times distribution). Several studies already suggested to interpret the rainfall occurrence process as a renewal process (see for example: Buishand, 1978; Foufoula-Georgiou and Lettenmaier, 1987); however, because seasonality character is frequently present in T-series, the renewal property was generally applied to each homogeneous period, distinctly. In order to describe statistical peculiarities of observed T-samples, some discrete probability distributions have been tested: the log-series distribution and the 2-parameter Extended-Yule (Martínez-Rodríguez et al., 2011) and polylog-series distributions (Gupta et al., 2008). The 2-parameter polylog-series distribution allowed to capture the main characters of T's-samples (clustering, very high variability and skewness, persistence) with better accuracy than the other distributions; accordingly, following analyses referred to this distribution only, which is defined as:

$$P\{T=k\}=\frac{w^k}{k^s Li_s(w)} \tag{1}$$

where $w$ and $s$ are the parameters of the polylog-series distribution, $Li_s(\cdot)$ is the polylogarithm (or Jonquière's) function:

$$Li_s(w)=\sum_{k=1}^{\infty}\frac{w^k}{k^s} \tag{2}$$

It can be observed that for $s=1$, the Jonquière's function (2) reduces to $-\ln(1-w)$, and the polylog-series distribution reduces to the logarithm-series distribution (Feller, 1968), often used in the past in order to regularise dry spell data (Chatfield, 1966; Kottegoda and Rosso, 1997).

The hypothesis that $T$s are i.i.d. also implies that the expected cluster size, $\langle cl\rangle$, can be derived by the following simple relationship:

$$\begin{aligned}\langle cl\rangle &= \sum_{k=1}^{\infty} kP_1^{k-1}(1-P_1)=(1-P_1)\left(1+2P_1+...+kP_1^{k-1}+...\right)\\ &= \sum_{k=0}^{\infty} P_1^k=\frac{1}{1-P_1}\end{aligned} \tag{3}$$

where $P_1=P\{T=1\}$. Thus, a good estimate of the observed average cluster size mainly depends on the capability of the distribution to well reproduce $P\{T=1\}$.

The estimate of the two parameters was performed by means of the maximum likelihood method, which consists of the solution of a system of two equations in the unknown $w$ and $s$ (for the details see Gupta et al., 2008). Goodness of fit was tested according to Martinez-Rodriguez et al. (2011); these authors highlight that the classical chi-square test is not appropriate to evaluate the goodness-of-fit for distributions characterized by long tail (as polylog-series one), because expected frequencies less than 5 are common. Also, it has to be pointed out that grouping the classes with expected frequencies less than 5 could lead to some problems (Spierdijk and Voorneveld, 2007). So, an alternative procedure to test the goodness of fit was adopted by numerically reconstructing the null hypothesis of chi-square (according to polylog-series) on the base of a Monte Carlo simulation. A set of 2000 replicates of the observation was produced by randomly sampling from the theoretical discrete distribution (using Monte Carlo approach), and the p-value associated to the actual observation was computed from the numerically reconstructed null hypothesis (Martinez-Rodriguez et al., 2011).

## 3 Dataset

A set of 12 time-series of inter-arrival rainfall times was obtained by daily rainfall data recorded in the period 1926–2005 by 12 rain gauges. As it is shown in Fig. 1, 6 stations are located in the Sicily Island (hereafter referred as SIC), which is characterized by a typical Mediterranean climate, whereas the other 6 stations are placed in a sub-Alpine region including Piedmont and Aosta Valley Italian regions (from here on referred to as PMT). The 12 series were selected in order to satisfy the following criteria: (i) minimum gaps in the 80-year recorded data (no gap filling was performed); (ii) homogeneity of the time series. Relatively to the latter point, both control of data quality and specific homogeneity tests have been carried out; in particular, the procedure suggested by Wijngaard et al. (2003) has been followed. Each time-series has been subdivided in 13 sub-sets of 20-yr duration; the first subset refers to the period 1926–1945, the other subsets were obtained by progressively shifting of 5-yr the sampling, then obtaining an overlap of 15 yr between two successive series (1931–1950, 1936–1955, ..., 1986–2005). Moreover, the analysis was performed for three different datasets separately: (1) whole year (Y); (2) six-months relatively "warm" season (WS – from April to September) and (3) six-months relatively "cold" season (CS – from October to March). As it

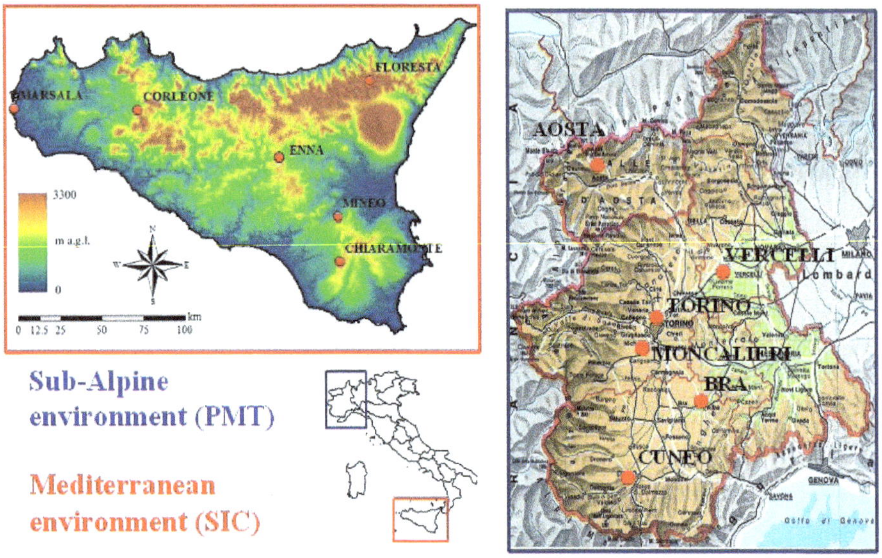


**Figure 1.** Spatial location of the rain gauges that provided rainfall time series used for the analysis. Left panel reports the 6 stations located in the Mediterranean area (SIC), whereas right panel reports the 6 stations located in the sub-Alpine (PMT) area.

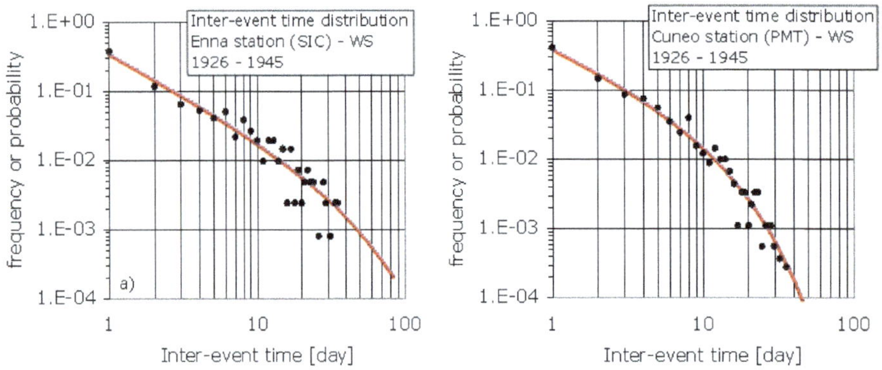


**Figure 2.** Fitting frequency distribution of inter-event times, $T$, with the polylog-series distribution, for the warm season and for Enna (SIC – left panel) and Cuneo stations (PMT – right panel). Sample data refer to the period 1926–1945.

is known, Sicily's rainfall regime is characterised by a strong seasonality, with a "warm" (dry) and a "cold" (wet) season. During the WS, rainfall is usually lower than atmospheric demand, whereas the converse is true during the CS. Rainfall regime in Piedmont and Aosta Valley is almost virtually opposite to the one in Sicily; indeed, WS is generally wetter than CS. In this phase, no additional disaggregation has been considered, mainly to not further reduce the sample size for high T-values (or for not increase the length of sampling window); therefore, similar analyses for three-month seasons have not been performed.

## 4 Results

### 4.1 Analysis of inter-arrival times series distribution

The polylog-series distribution was fitted to the 468 available datasets (12 stations × 3 periods × 13 sub-series). Test of goodness of fit was performed separately for each data set, rejecting the null hypothesis when the p-value associated to the fitted distribution was $< 0.05$.

Synthetically, the chi-square test results indicate that the null hypothesis is frequently rejected for the yearly datasets for both SIC and PMT stations (74.4 % for SIC and 78.2 % for PMT), while only in few cases it is rejected for WS and CS data sets (14.1 % and 26.9 % for WS-SIC and WS-PMT, respectively; 18.2 % and 27.1 % for CS-SIC and CS-PMT, respectively). These results generally show how polylog distribution better agrees with SIC seasonal datasets than with PMT ones.

Figure 2 shows two examples of data fitting with polylog-series distribution; for both cases the null hypothesis has not be rejected and it can be seen how the visual agreement results satisfactory.

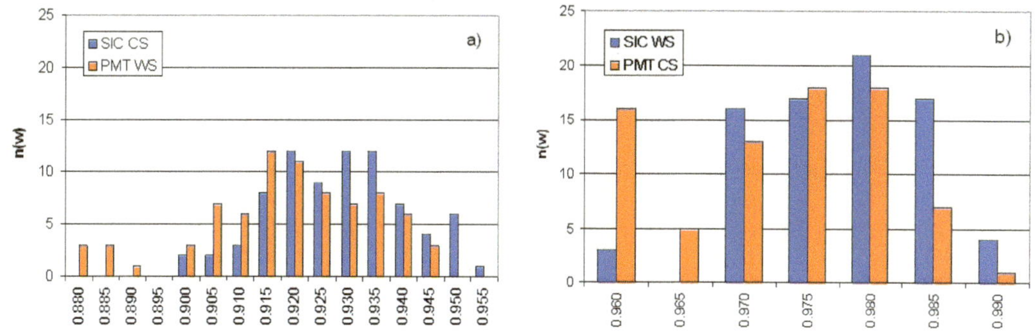


**Figure 3.** Frequency distribution of $w$ parameter for the two studied areas (SIC and PMT), for warm season (WS) and cold season (CS).

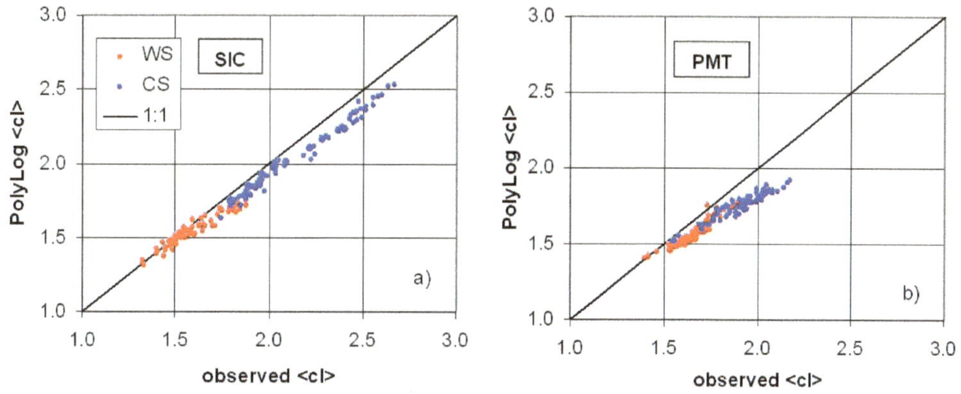


**Figure 4.** Comparison between observed and estimated average cluster size $\langle cl \rangle$ for the Mediterranean (SIC, panel **a**) and sub-Alpine (PMT, panel **b**) areas in both cold (CS) and warm (WS) season.

A preliminary analysis of range of variability of distribution parameters $w$ and $s$ has been carried out. In particular, Fig. 3 reports the frequency distributions of $w$ parameter, which is the one that mainly influences tail probabilities. Interestingly, $w$ distributions of Sicily CS and Piedmont WS (Fig. 3a), as well as of Sicily WS and Piedmont CS (Fig. 3b), are rather similar in both range of variability and frequency distribution; that is likely due to the typical features of the two studies areas, which have somewhat opposite rainfall regimes.

### 4.2 Average cluster size

As previously introduced, the use of inter-arrival times allows characterizing the temporal structure of both rainy days and dry periods. A further test of the polylog fitting was performed in terms of average cluster size, computed with Eq. (3). In Fig. 4 observed vs. estimated $\langle cl \rangle$ are represented for the stations located in the two areas (SIC, Fig. 4a and PMT, Fig. 4b) and for the two periods (WS and CS). These plots highlight in general a good performance of the model, with a slightly systematic underestimation ($<4\,\%$ on average), that becomes significant only for some recorded high values (occurred during CS for both SIC and PMT). Moreover, a significant seasonality for $\langle cl \rangle$, as well as a quite wide range of variability (from 1.2 to 2.7 days), was observed for SIC region. On the contrary, in PMT sites the observed range of variability of $\langle cl \rangle$ is quite similar for the two periods, with a wide overlap range from 1.5 to 2.0 days, and with values exceeding 2 only for CS period. This further confirm how the seasonality is more evident in SIC area than in PMT one in term of clustering of rainy days.

### 4.3 Modelling of high quantiles

Table 1 reports root mean square difference (RMSD) statistics computed for some high T-quantiles (0.80, 0.85, 0.90 and 0.95) by comparing data observed and polylog-series estimates; for both investigated area, available sample size is equal to 78 (13 periods $\times$ 6 stations). As can be noticed, RMSD values are generally very small, less than three days, with just few exceptions (SIC WS for $P \leq 0.95$; PMT CS for $P \leq 0.90$). The greatest discrepancies observed for SIC-WS could be likely explained by the reduction of the sample size with the increase of the quantiles considered. Similar considerations arise from the analysis of Fig. 5, where observed and estimated quantiles are compared for two stations (Corleone – WS and Torino – CS).

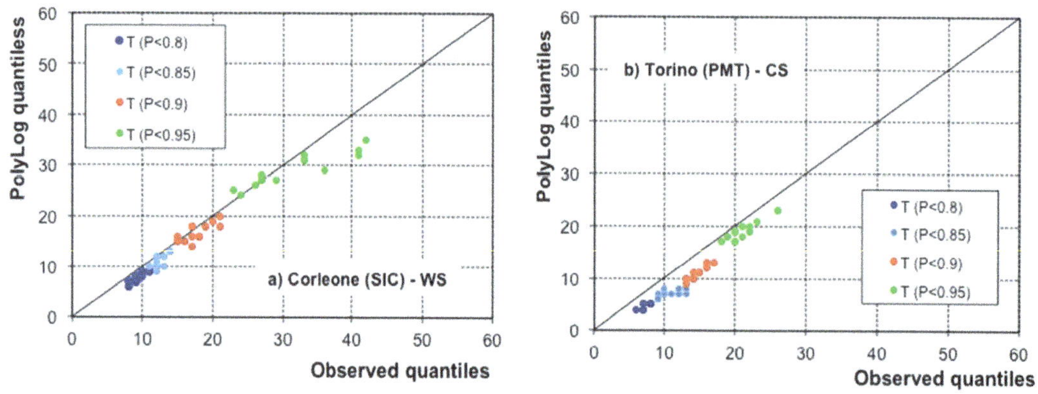


**Figure 5.** Comparison between some observed and estimated high T-quantiles for two sites: **(a)** Corleone station in SIC during WS; **(b)** Torino station in PMT during CS.

**Table 1.** Mean (M) and root mean square difference (RSMD) for some quantiles, for both seasons and both investigated areas.

| M (RMSD) in days | | $T(P < 0.80)$ | $T(P < 0.85)$ | $T(P < 0.90)$ | $T(P < 0.95)$ |
|---|---|---|---|---|---|
| PMT | WS | 6.9 (1.65) | 8.6 (1.58) | 11.1 (1.66) | 15.1 (1.02) |
| | CS | 7.5 (2.34) | 10.3 (2.84) | 14.2 (3.12) | 21.4 (2.20) |
| SIC | WS | 10.7 (1.83) | 14.3 (1.79) | 19.6 (1.72) | 33.7 (4.41) |
| | CS | 4.5 (1.47) | 5.9 (1.57) | 7.9 (1.66) | 11.5 (1.47) |

**Table 2.** Summary of the qualitative trend analysis carried out for a T-quantile ($P < 0.975$) and for the average cluster size. The symbol (+) indicates positive trend, (–) indicates negative trend and (n) no-trend.

| Area | Station | $P < 0.975$ | | $\langle cl \rangle$ | |
|---|---|---|---|---|---|
| | | WS | CS | WS | CS |
| SIC | Chiaramonte | – | n | + | + |
| | Corleone | – | + | n | – |
| | Enna | n | + | – | – |
| | Floresta | n | n | + | n |
| | Marsala | n | + | n | – |
| | Mineo | n | + | – | n |
| PMT | Aosta | n | + | n | n |
| | Bra | n | + | n | n |
| | Cuneo | + | + | – | – |
| | Moncalieri | n | + | n | n |
| | Torino | n | + | n | n |
| | Vercelli | + | n | n | n |

## 4.4 Preliminary analysis of the trends

The results reported in the previous sub-sections synthetically showed: (i) the polylog distribution usually provides a good fitting of sample $T$ distribution for seasonal (WS and CS) data; (ii) the polylog distribution is able to well estimate the average cluster sizes as well; (iii) RMSD for some high T-quantiles (0.80, 0.85, 0.90 and 0.95) are generally low (with the above-mentioned exceptions). Accordingly, both estimated high T-quantiles and average cluster sizes were considered a more robust mean to analyse the trend of such quantities than empirical ones. Moreover, the analysis can be also extended to very high quantiles, for which sample size is small.

A simple qualitative trend analysis of the quantiles was performed in order to investigate the issue of increasing duration of dry periods, of which previous studies have been dealt with yet (Alpert et al., 2002; Dai et al., 1998); due to the overlapping window used in the reconstruction of samples, the analysis of trend cannot be performed with the usual methods, which are based on the hypothesis of independence among the single values of the series. For this reason, the analysis of trend is only qualitative at this point.

As synthetically shown in Table 2, a slightly increasing trend of very high T-quantile ($P < 0.975$) has been observed for almost all the PMT and SIC stations during CS periods. In general, no trend was observed for the WS period. It is noteworthy that no significant differences between the two

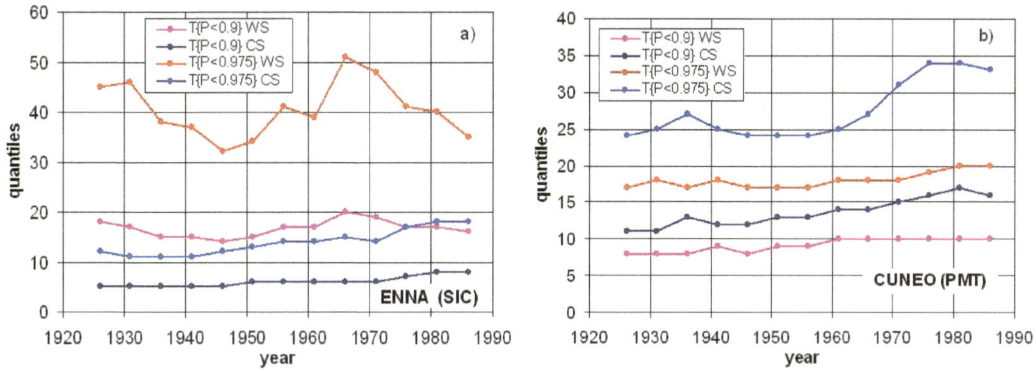


**Figure 6.** Time-series of two high quantiles $T\{P<0.9\}$ and $T\{P<0.975\}$, estimated by polylog-series distribution, for both season and for two sites: **(a)** Enna station (SIC) and **(b)** Cuneo station (PMT).

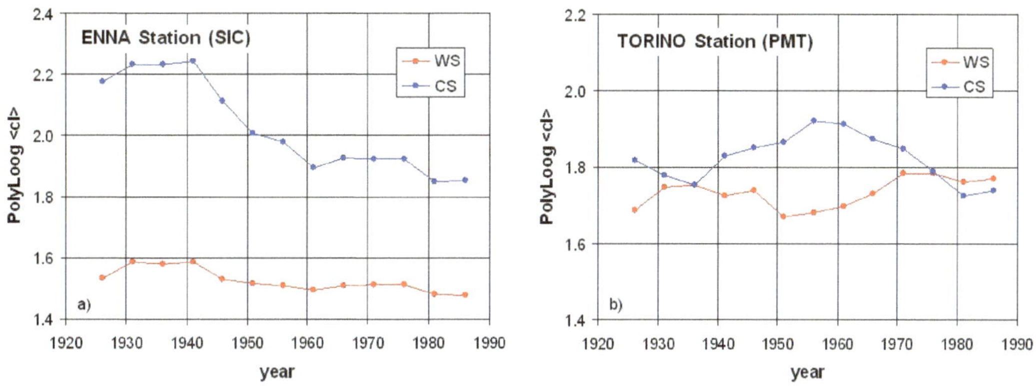


**Figure 7.** Time-series of expected cluster size estimated by polylog-series distribution, for both season and for two sites: **(a)** for Enna station (SIC); **(b)** for Torino station (PMT).

study areas were appreciated. As an example, in Fig. 6, time series of two estimated T-quantiles ($P \leq 0.9$, $P < 0.975$), for two stations, Enna (SIC) and Cuneo (PMT), are reported.

A similar qualitative trend analysis has also been performed for the average cluster size (Table 2). Generally, no trend was observed for Piedmont, while contrasting results were observed for Sicily. As an example, Fig. 7 shows the time-series of the expected cluster size for two sites, Enna (SIC) and Torino (PMT); a noticeable decreasing trend was observed for Enna, especially for CS.

## 5 Conclusions

The capability to reproduce the inter-arrival times statistical structure through a 2-parameter discrete distribution (polylog-series) was tested in two Italian regions: a Mediterranean area and a Sub-Alpine one.

A chi-square test based on the Montecarlo procedure suggests that polylog-series distribution seems to well reproduce the observed data in the whole range of variability of times and associated frequencies when the dataset is split in two sub-period: a relatively "warm" season (April–September) and a relatively "cold" one (October–March). Poor performances were instead obtained when the whole year is analysed, mainly due to the strong seasonality characterising the two study areas.

The theoretical distribution allows also reproducing well some characteristics of rainfall as: average cluster sizes and high T-quantiles. In particular, the average cluster sizes results only slightly underestimated ($\cong 5\,\%$), whereas high quantiles ($P \leq 0.8$ to 0.95) are generally well reproduced with the only exception for $P \leq 0.95$ during WS (in SIC) and CS (in PMT). This latter result can be justified by the small sample size of empirical data for high inter-arrival times.

A preliminary qualitative analysis of the trends on high quantiles and average cluster size seems to confirm the issue of increasing duration of dry periods for both environments, which is limited to the "cold" season.

No trend was instead observed for the average cluster size in Sub-Alpine areas, while discordant trends were observed in Mediterranean stations.

The polylog-series seems to account for peculiarities of inter-arrival times of the two areas by simple dividing the year in two six-month periods (a relatively "warm" season from April to September and a relatively "cold" season from October to March). This result is promising for the use of polylog-series in ecohydrological models in those areas where classical exponential relationship fails to capture the characteristics of rainfall. However, in order to better catch the differences between the characters of seasonality in different climatic regions, the same procedure followed in this work could be applied to shorter periods (three months). Additionally, the requirement of several years to sufficiently sample high inter-arrival times (and the necessary overlapping between consecutive sampling windows) requires ad-hoc methodologies for the quantitative analysis of the trends that will be investigated in the next future. Finally, the possibility to address the observed limits of the adopted distribution by using a three-parameter Lerch distribution (which generalises the polylog-series distribution) have to be tested in future researches.

**Acknowledgements.** The authors thank Osservatorio delle Acque of the Assessorato Energia e Servizi di Pubblica Utilità della Regione Siciliana and ARPA (Agenzia Regionale per la Protezione dell'Ambiente) Piemonte for providing the meteorological dataset.

Edited by: I. Auer

## References

Alpert, P., Ben-Gai, T., Baharad, A., Benjamin, J., Yekutieli, D., Colacino, M., Diodato, L., Ramis, C., Homar, V., Romero, R., Michaelides, S., and Manes, A.: The paradoxical increase of Mediterranean extreme daily rainfall in spite of decrease in total values, Geophys. Res. Lett., 29, 1536, doi:10.1029/2001GL013554, 2002.

Buishand, T. A.: Some remarks on the use of daily rainfall models, J. Hydrol., 36, 295–308, doi:10.1016/0022-1694(78)90150-6, 1978.

Chatfield, C.: Wet and dry spells, Weather, 21, 308–310, 1966.

Dai, A., Trenberth, K. E., and Karl, T. R.: Global variations in droughts and wet spells: 1900–1995, Geophys. Res. Lett., 25, 3367–3370, doi:10.1029/98GL52511, 1998.

Feller, W.: An Introduction to probability theory and its applications, Vol. 1, J. W. A. Sons, New York, USA, 1968.

Foufoula-Georgiou, E. and Lettenmaier, D. P.: A Markov renewal model for rainfall occurrences, Water Resour. Res., 23, 875–884, doi:10.1029/WR023i005p00875, 1987.

Gupta, P. L., Gupta, R. C., S.-huat Ong, S.-H., and Srivastava, H. M.: A class of Hurwitz-Lerch Zeta distributions and their applications in reliability, App. Math. Comp., 196, 521–531, doi:10.1016/j.amc.2007.06.012, 2008.

Kottegoda, N. T. and Rosso, R.: Statistics, probability, and reliability for civil and environmental engineers, McGraw-Hill, 1997.

Martínez-Rodríguez, A. M., Sáez-Castillo, A. J., and Conde-Sánchez, A.: Modelling using an extended Yule distribution, Comput. Stat. Data An., 55, 863–873, doi:10.1016/j.csda.2010.07.014, 2011.

Spierdijk, L. and Voorneveld, M.: Superstars without talent? The Yule distribution controversy, SSE/EFI Working Paper Series in Economics and Finance, No. 658, http://hdl.handle.net/10419/56199, 2007.

Waymire, E. and Gupta, V. K.: The mathematical structure of rainfall representations 1. A review of the stochastic rainfall models, Water Resour. Res., 17, 1261–1272, 1981.

Wijngaard, J. B., Klein Tank, A. M. G., and Konnen, G. P.: Homogeneity of 20th century European daily temperature and precipitation series, Int. J. Climatol., 23, 679–692, 2003.

# 26

# The influence of the new ECMWF Ensemble Prediction System resolution on wind power forecast accuracy and uncertainty estimation

**S. Alessandrini**[1], **S. Sperati**[1], **and P. Pinson**[2]

[1]RSE, Ricerca Sistema Energetico, Milano, Italy
[2]Technical University of Denmark (DTU), DTU Informatics, Lyngby, Denmark

*Correspondence to:* S. Alessandrini (stefano.alessandrini@rse-web.it)

**Abstract.** The importance of wind power forecasting (WPF) is nowadays commonly recognized because it represents a useful tool to reduce problems of grid integration and to facilitate energy trading. If on one side the prediction accuracy is fundamental to these scopes, on the other it has become also clear that a reliable estimation about their uncertainty is paramount. In fact prediction accuracy is unfortunately not constant and can depend on the location of a particular wind farm, on the forecast time and on the atmospheric situation. Previous studies indicated that the spread of power forecasts derived from the Ensemble Prediction System (EPS) in use at the European Centre for Medium-Range Weather Forecast (ECMWF) could be used as indicator of a three-hourly, three days ahead, wind power forecast's accuracy. In this paper a new application of the EPS, whose horizontal resolution was increased on January 2010 from T399/T255 (60 km) to T639/T319 (32 km), shows an improvement in the results implying that the power spread has actually enough correlation with the error calculated on the deterministic forecast in order to be used as an accuracy predictor. The periods for this comparison are from January 2008 until October 2008 (T399/T255) and from January 2011 until October 2011 (T639/T319). Moreover we have focused our attention on the influence of the new EPS configuration on the performance of a deterministic WPF conducted with the ensemble mean: the results show that increasing the EPS resolution yields a single-valued WPF whose performance is comparable with that of the new ECMWF deterministic high-resolution meteorological model, whose spatial resolution increased from T799 (25 km) to T1279 (15 km).

## 1 Introduction

Short-term forecasting of wind energy production is becoming increasingly important due to the constant growth of this renewable source, whose uncertainty require a significant commitment to meet the needs of the transmission networks operators and of other agents of the power systems and electricity markets. Wind is in fact a variable energy source, which does not allow a management similar to that of other traditional sources: in order to be able to better control its integration into the electric grids and its dispatching priority, an adequate insight of its dynamics is necessary, in particular for the short-term time horizon (0–48 h) in which trading operations are carried out in the electricity markets. The accuracy of a short-term wind power forecast is consequently a raising concern (Pinson et al., 2007; Holmgren et al., 2010). On the basis of these considerations, our application investigated the performances of the new resolution ECMWF deterministic and EPS models applied to the same wind farm as in previous work (Alessandrini et al., 2011), comparing them with the results obtained with the previous lower resolution versions of these two models. The new versions have been applied to understand if it is possible to increase the performance in both wind power prediction accuracy and uncertainty estimation.

**Table 1.** Mean and standard deviation of measured wind speed ($m\,s^{-1}$) and power (kW) for the two periods.

| | | 2008 | 2011 |
|---|---|---|---|
| Wind speed | Mean | 6.69 | 6.71 |
| | Standard deviation | 3.09 | 3.16 |
| Power | Mean | 210.02 | 204.37 |
| | Standard deviation | 225.21 | 222.71 |

## 2 Site and wind data description

The wind farm used in the application is located in a complex-terrain mountain area of Northern Sicily. It has 9 equal turbines, each one of 850 kW Nominal Power (NP). Time series power data for the single turbines are available for the years 2008 and 2011, in each case covering the period January–October. To consider valid power data we computed an average over the active turbines for each time step obtaining a single power series. Wind data has been measured at 50 m a.g.l. (same height of the hubs) by an anemometer located inside the park. Measured data have been averaged from 10-minutes timesteps to hourly periods.

The total fraction of missing hourly data is similar and lower than 5 % for both periods of the comparison. Some concerns may rise regarding the fact that we compared two models with different resolution on two different years instead of the same period. However, the mean and standard deviation of measured wind speed and power reported in Table 1 are quite similar. This allowed us to consider the two periods suitable enough for the comparison, using the same forecast system and post-processing configuration for both cases.

## 3 The wind power forecast system

The wind power forecast system used consists in the application of the ECMWF meteorological model, which produces three-hourly wind forecasts at 10 m a.g.l., for the three days ahead period and for each year using respectively the two different model resolutions. In order to reduce systematic errors and, at the same time, to convert wind to power, a single-layer, feedforward Neural Network (NN) has been applied (Cali et al., 2008). The NN has been configured linking directly the forecasted meteorological data with the measured power on the training period, in order to obtain a three-hourly forecast power series. This WPF structure has been applied also in Alessandrini et al. (2010) showing similar or better performances to those of other WPF topology (i.e. Numerical Weather Prediction + Model Output Statistic (MOS) correction + power curve).

## 4 The deterministic application

In both cases the first 4 months have been used as training period for the NN described above, while the remaining 6 months have been considered as test period. However, the NN application is recursive (i.e. self-adapting algorithm that uses recent errors to correct its weights at regular intervals), therefore the training period is continuously increased including the part of the test period previous to the forecast window. The performances have been evaluated with some statistical indices reported in Table 2, in which RMSE, MAE and bias are normalized by NP. The indices are calculated averaging over all the three hours forecast horizon between time 0 h to +24 h (Day 1), +24 h to +48 h (Day 2) and +48 h to +72 h (Day 3) where 0 h refers to 00:00 UTC, the beginning of each 3 days forecast run updated every 24 h. On the first day prediction the new resolution shows a small improvement in terms of RMSE apart from that the other indices look comparable, suggesting that the increase in resolution does not necessary assure better performances. Looking at the MAE indices, the results are in both cases in line with that of other state of the art models applied on complex terrain cases (Giebel et al., 2011).

## 5 The probabilistic application

The ECMWF EPS members are generated with perturbation of the initial conditions (ECMWF report, 2010) using singular vectors with maximum total energy growth, obtaining 50 pair of wind components plus a non-perturbed member (the control run). The 50+1 wind forecasts have been first processed by a MOS (correction), training a NN on the ensemble mean and then applying it recursively on the test period. In this case the NN (with the same structure as indicated in Sect. 4) has been configured to link the wind forecasted data with measured wind speed instead of wind power, as done in the deterministic application, in order to obtain 50+1 corrected wind speed ensemble members, on which performing a statistical calibration procedure (Pinson, 2012).

The procedure employed is based on logit-transformation of the wind speed forecasted and measured data in order to reproduce a Gaussian distribution; the ratio between the RMSE, calculated on the training period between measured wind speed and the MOS-corrected ensemble mean, and the mean of the standard deviation of the ensemble members has been used as a variance deficit to adaptively correct the mean and the variance of the transformed wind speed forecasts on the test period. The time series have been finally transformed back with the inverse logit-transformation.

The rank histograms before calibration show in both cases an overconfident model, with the measurements falling quite often outside the ensemble distribution. The situation is improved by the calibration procedure, showing a more uniform distribution between the various bins. However, the extreme bins still remain more populated than the others, as

**Table 2.** Statistical indices for three-hourly power forecast evaluation computed using measured data for the two case periods. RMSE, MAE and bias are normalized by the nominal power (NP). The indices are calculated averaging over all the three hours forecast horizon between time 0 h to +24 h (Day 1), +24 h to +48 h (Day 2) and +48 h to +72 h (Day 3) where 0 h refers to 00:00 UTC, the beginning of each 3 days forecast run updated every 24 h.

| | | Day 1 | Day 2 | Day 3 |
|---|---|---|---|---|
| Deterministic T799 (25 km) | RMSE/NP | 14.5 % | 15.2 % | 16.0 % |
| | MAE/NP | 10.3 % | 10.6 % | 11.7 % |
| | Pearson Correlation | 0.77 | 0.75 | 0.71 |
| | bias/NP | −0.02 | −0.01 | 0.0 |
| Deterministic T1279 (15 km) | RMSE/NP | 13.8 % | 15.5 % | 16.0 % |
| | MAE/NP | 10.1 % | 11.2 % | 12.1 % |
| | Pearson Correlation | 0.78 | 0.71 | 0.69 |
| | bias/NP | 0.01 | 0.01 | 0.02 |
| EPS mean T399/T255 (60 km) | RMSE/NP | 15.3 % | 15.8 % | 16.7 % |
| | MAE/NP | 11.1 % | 11.7 % | 12.6 % |
| | Pearson Correlation | 0.74 | 0.71 | 0.67 |
| | bias/NP | −0.02 | −0.01 | 0.0 |
| EPS mean T639/T319 (32 km) | RMSE/NP | 14.3 % | 15.6 % | 15.9 % |
| | MAE/NP | 10.2 % | 11.2 % | 11.7 % |
| | Pearson Correlation | 0.76 | 0.70 | 0.69 |
| | bias/NP | 0.0 | 0.0 | 0.0 |

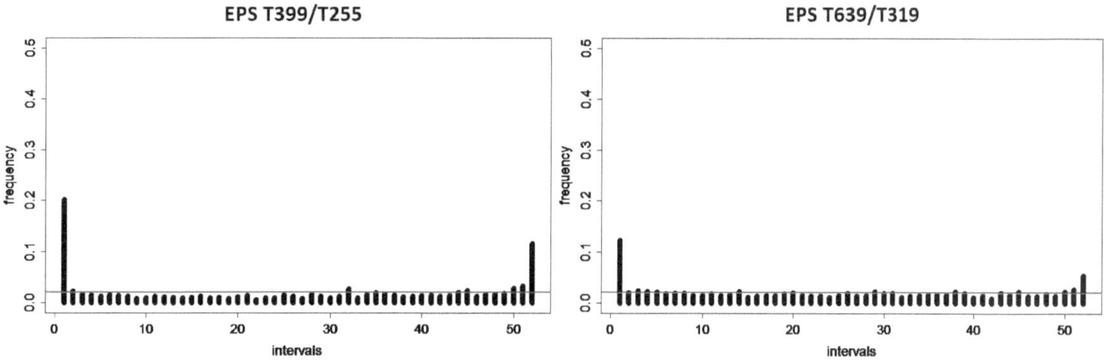


**Figure 1.** Rank histograms, calibrated wind speed ensemble members; EPS T399/T255 and EPS T639/T319 resolution.

**Table 3.** Correlation between power spread averaged on daily basis and daily RMSE, and ratio between cases falling simultaneously above or below their respective median value and total number of cases (Day 1, 2 and 3 interval are explained in Table 2 caption).

| | spread/error correlation | | | diagonal ratio | | |
|---|---|---|---|---|---|---|
| | Day 1 | Day 2 | Day 3 | Day 1 | Day 2 | Day 3 |
| 2008 | 0.57 | 0.55 | 0.57 | 0.66 | 0.66 | 0.66 |
| 2011 | 0.5 | 0.68 | 0.63 | 0.71 | 0.73 | 0.73 |

shown in Fig. 1. It is noticeable that for 2011 the MOS (system) + calibration scheme seems to work slightly better than for the 2008 case. There are probably two reasons for that. First the increase in spatial resolution slightly improves forecast accuracy. Secondarily the initial ensemble spread of the higher resolution model is in general greater than in the low resolution model requiring a lower job from the calibration process. The calibrated ensemble members are then converted to wind power using a theoretical power curve provided by the manufacturer. In Figs. 2 and 3 we show some contingency diagrams (Von Bremen, 2007): the graphs plot the daily RMSE/NP of the deterministic WPF versus the ensemble power spread, expressed as standard deviation of the ensemble power members, for the three-days period. The 2011 application shows an increase in the performances, the diagonal cases (those when the spread and forecast errore are simultaneously above or below their respective median values) are in fact more populated than the off-diagonal cases meaning that at a low ensemble spread is more often related to low forecast errors. Table 3 shows correlation

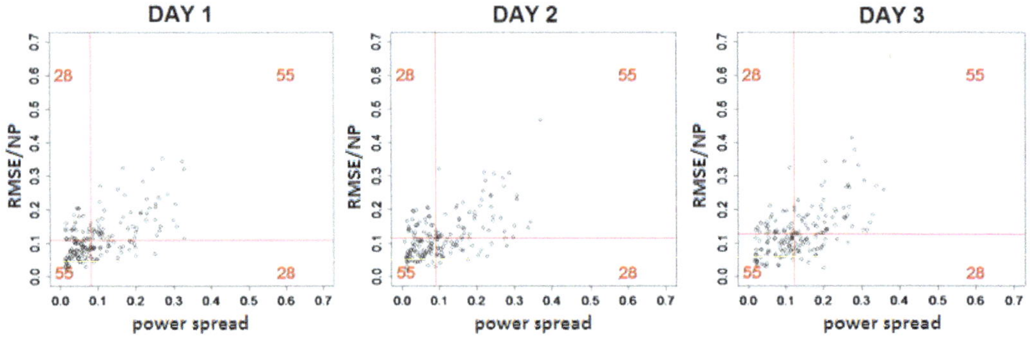


**Figure 2.** ECMWF deterministic power forecast error (RMSE/NP) vs. ensemble power spread computed as a daily average for three lead times. EPS T399/T255 (60 km). The lines are plotted at the statistical median value of the data reported in x- and y-axis. The red numbers indicate the points inside each quadrant (Day 1, 2 and 3 interval are explained in Table 2 caption).

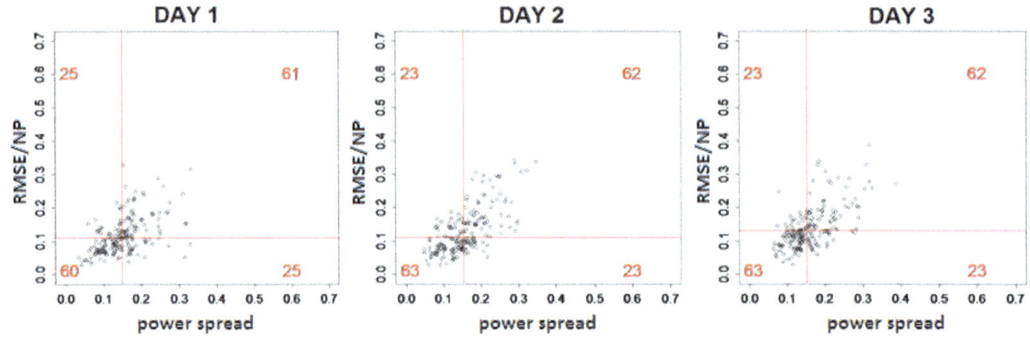


**Figure 3.** ECMWF deterministic power forecast error (RMSE/NP) vs. ensemble power spread computed as a daily average for three lead times. EPS T639/T319 (32 km). The lines are plotted on the statistical median. The red numbers indicate the points inside each quadrant (Day 1, 2 and 3 interval are explained in Table 2 caption).

indices (Pearson) and ratio between cases falling simultaneously above or below their respective median value and total number of cases. Note that contingency tables only comprise one approach to assess the potentially existing spread-skill relationship for ensemble forecasts and that other approaches may be used in order to compare several ensemble forecasting systems for the wind power application. As an example, Pinson et al. (2009) compared ensemble forecasts of wind power based on lagged-averaged poor man's ensembles, as well as NCEP and ECMWF ensemble predictions. The results presented here are not directly comparable, even though our analysis and conclusions are qualitatively similar. More work is required in the future for the thorough comparison of rival approaches to ensemble forecasting of wind power generation based on different meteorological ensemble forecasts used as input.

## 6 Deterministic approach using the ensemble mean

We conducted a WPF in a deterministic approach using the calibrated ensemble mean and comparing the obtainable performances with those of the ECMWF deterministic model; Table 2 shows that for the low resolution case the results are slightly worse, showing an increase in RMSE/NP and MAE/NP between 0.5 and 1 %. For 2011, however, the increase in resolution allows to obtain comparable performances between the power forecast based on high resolution deterministic meteorological model and on the ensemble mean.

## 7 Conclusions

In this work we showed that the increase in resolution of the ECMWF deterministic model does not necessary assure better performances of corresponding power forecasts in this test case. However, higher resolution of the EPS allowed us to obtain better performances in a WPF deterministic approach (using the ensemble mean as a deterministic model) than with lower resolution. Furthermore, the new ensemble spread seems to be usable as an indicator of a WPF accuracy with slightly better results than with the lower resolution, at least for the time horizon until 72 h ahead.

**Acknowledgements.** This work has been financed by the Research Fund for the Italian Electrical System under the Contract Agreement between RSE (formerly known as ERSE) and the

Ministry of Economic Development – General Directorate for Nuclear Energy, Renewable Energy and Energy Efficiency stipulated on 29 July 2009 in compliance with the Decree of 19 March 2009.

The authors also acknowledge Federico Fioretti and Umberto Deangelis of Enel Spa for providing meteorological and power data for the test case.

Edited by: L. Wald

## References

Alessandrini, S., Decimi, G., and Ronzio, D.: Studio della relazione energia producibile/prodotta per un campo eolico su terreno complesso con modelli meteorologici dinamici, Rapporto di Ricerca RSE no. 10000262, 2010.

Alessandrini, S., Pinson, P., Hagedorn, R., Decimi, G., and Sperati, S.: An application of ensemble/multi model approach for wind power production forecasting, Adv. Sci. Res., 6, 35–37, doi:10.5194/asr-6-35-2011, 2011.

Cali, Ue., Lange, B., Dobschinski, J., Kurt, M., Moehrlen, C., and Ernst, B.: Artificial neural network based wind power forecasting using a multi-model approach, 7th International Workshop on Large Scale Integration of Wind Power and on Transmission Networks for Offshore Wind Farms, Madrid, 2008.

ECMWF internal report: Horizontal resolution increase (www.ecmwf.int), 2010.

Giebel, G., Brownsword, R., Kariniotakis, G., Denhard, M., and Draxl, C.: The State of the Art in Short-Term Prediction of Wind Power, Anemos.plus deliverable report, 2011.

Holmgren, E., Siebert, N., and Kariniotakis, G.: Wind Power Prediction Risk Indices Based on Numerical Weather Prediction Ensembles, Proceedings of the EWEC 2010 Conference, Warsaw, Poland, 20–23 April, 2010.

Pinson, P.: Adaptive recalibration of ensemble forecasts of $(u, v)$-winds, Q. J. Roy. Meteorol. Soc., doi:10.1002/qj.1873, in press, 2012.

Pinson, P., Nielsen, H. A., Madsen, H., and Kariniotakis, G.: Skill forecasting from different wind power ensemble prediction methods, J. Phys. Conf. Ser., 75, doi:10.1088/1742-6596/75/1/012046, 2007.

Pinson, P., Nielsen, H. A., Madsen, H., and Kariniotakis, G.: Skill forecasting from ensemble predictions of wind power, Appl. Energ., 86, 1326–1334, 2009.

Von Bremen, L.: Combination of deterministic and probabilistic meteorological models to enhance wind farm forecast, J. Phys. Conf. Ser., 75, 012050, doi:10.1088/1742-6596/75/1/012050, 2007.

# 27

# Surface high-resolution temperature forecast in southern Italy

**S. Federico[1,2], E. Avolio[1], F. Fusto[3], R. Niccoli[3], and C. Bellecci[2,4]**

[1]ISAC-CNR, UOS of Lamezia Terme, Lamezia Terme (CZ), Italy
[2]CRATI, Rende (CS), Italy
[3]ARPACal, Catanzaro, Italy
[4]Engeenering Department, University of "Tor Vergata", Rome, Italy

**Abstract.** Since June 2008, 1-h temperature forecasts for the Calabria region (Southern Italy) are issued at 2.5 km horizontal resolution at CRATI/ISAC-CNR. Forecasts are available online at http://meteo.crati.it/previsioni.html (every 6-h). This paper shows the forecast performance out to three days for one climatological year (from 1 December 2008 to 30 November 2009, 365 run) for minimum, mean and maximum temperature. The forecast is evaluated against gridded analyses at the same horizontal resolution.

Gridded analysis is based on Optimal Interpolation (OI) and uses a de-trending technique for computing the background field. Observations from 87 thermometers are used in the analysis system.

In this paper cumulative statistics are shown to quantify forecast errors out to three days.

## 1 Introduction

This paper investigates the performance of the high-resolution (2.5 km horizontal resolution) operational forecast of minimum, mean, and maximum temperatures issued at CRATI/ISAC-CNR for the Calabria peninsula (Southern Italy, Fig. 1). Temperature forecasts are produced daily by the Regional Atmospheric Modeling System model (RAMS; Cotton et al., 2003), and this paper shows the forecast performance from 1 December 2008 to 30 November 2009 (365 runs).

Verification of surface temperature forecast is a complicated task because forecast and observation contain different information about the same physical quantities, and this is due to the filtering properties of the numerical weather prediction system in use. In particular, the forecast gives the average of the meteorological parameter over a grid box, and it is representative of the conditions inside the grid box, while surface observations are point measurement.

The difference between the forecast and observation temperatures is particularly important in complex orography and can be accounted by the observation representativeness (or representative) error (Lussana et al., 2010). Representative errors, which are not Gaussian in their distribution and flow dependent, are not instrumental errors but are a component of the observed value due to small scale features that cannot be resolved by the representation of the real world given by the model. They are always defined respect to a model, which in our case is the analyzed field[1].

The issues related to complex orography and representative errors are certainly important for Calabria, where there is considerable topography variability inside a grid box and where relatively high mountains peaks (1000 m high) are located few kilometres from the shoreline. For example, in presence of the sea-land contrast a sharp temperature gradient develops across the shoreline and differences of several degrees can be reported between two stations located near the coast and few kilometres inland (Simpson, 1994); or, when a cold pool develops, differences of several degrees can be reported by two close stations located, respectively, in a mountain gap and along an adjacent ridge (Myrick et al., 2005). Although the horizontal resolution of modern forecasting system is of the order of ~1–5 km, these differences can occur inside the same grid-box, and the forecast is not able to resolve these small-scale features (Myrick and Horel, 2006, 2008; Mass et al., 2003).

[1]The analyzed field depends, in turn, on the error covariances, which determine the scale of the analyzed field.

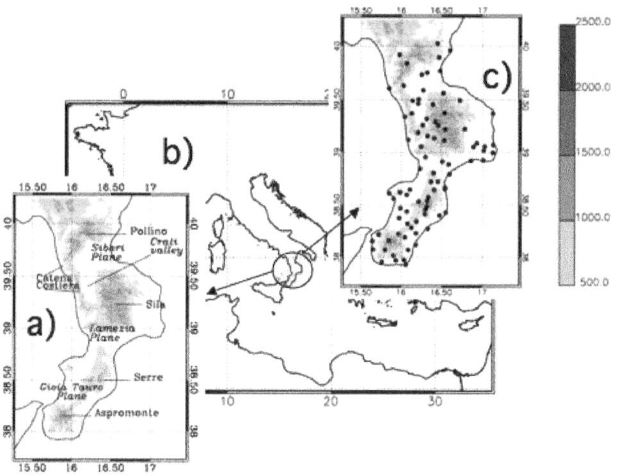


**Figure 1.** (**a**) Calabria features cited into the text; (**b**) Calabria region in the central Mediterranean; (**c**) Black-filled circles are the stations of the regional network. Grey shading shows the orographic height (m).

Because of the filtering properties of the numerical prediction model, representative errors are here treated as an additional component of the observation error and are filtered out by comparing the forecast with gridded analyses.

The analysis system, its performance in different conditions, and preliminary statistics on the forecast performance were already reported in Federico (2011). This paper, however, presents some new elements, namely: (a) it shows the model performance for a whole year, using a wider dataset, and statistics are more robust; (b) the forecast performance is presented as a function of the season; (c) it uses a different technique to compute the background field.

The paper is divided as follows: Sect. 2 introduces the objective analysis and datasets involved; Sect. 3 presents the forecast statistics, and; Sect. 4 provides conclusions.

## 2 The objective analysis and datasets involved

The observation dataset consists of minimum, mean, and maximum temperature from 87 thermometers (Fig. 1) provided by the Regional Agency for Environmental Protection of Calabria (ARPACAL). The average of the minimum distance between two thermometers is 10 km. The spatial distribution of the 87 thermometers is rather homogeneous and covers both mountains and lowlands[2].

Data are quality controlled by ARPACAL and it is here assumed that they are not affected by gross measurements errors.

The forecast is issued by the RAMS model (non-hydrostatic). A detailed description of the RAMS model is given in Cotton et al. (2003) while the model settings in the operational context are described in Federico (2011).

Three two-way nested domains with horizontal resolutions of 30 km, 7.5 km and 2.5 km are used. First domain covers the central part of the Mediterranean basin; the second domain extends over southern Italy; the third domain covers Calabria. For all domains we use thirty vertical levels, up to 16 300 m in the terrain following coordinate system. Levels are not equally spaced: layers within the Planetary Boundary Layer (PBL) are between 50 and 200 m thick, whereas layers in the middle and upper troposphere are 1000 m thick.

Atmospheric initial and dynamic boundary conditions, available every 6 h at $1 \times 1$ degree horizontal resolution, are derived from the 12:00 UTC Global Forecasting System (GFS) run of the National Centres for Environmental Prediction (NCEP; Sela, 1982). A four-dimensional data assimilation technique is used to define the forcing at the lateral boundaries of the five outermost grid cells of the largest domain.

To consider a completely independent dataset for verifying the forecast, the background field is obtained by detrending the observations in the vertical direction, similarly to the methodology of Uboldi et al. (2008). In particular, the vertical temperature gradient is computed for each day and meteorological parameter by performing a least-square minimization of the observations. For example, in the case of mean temperature for the day $i$-th we compute:

$$T_{i,\mathrm{mean}}(z) = T_{i,0\_\mathrm{mean}} + \gamma_{i,\mathrm{mean}} z \quad (1)$$

where $\gamma_{i,\mathrm{mean}}$ and $T_{i,0\_\mathrm{mean}}$ are the parameters computed from the linear fitting and $z$ is the height.

The background field is then reconstructed for each day and meteorological parameter by using the same 2.5 km digital elevation model (DEM) used by the RAMS, and applying to this DEM the result of the linear fitting. Again, considering the mean temperature for the day $i$-th we have:

$$T_{i,\mathrm{mean}}(x,y) = T_{i,0\_\mathrm{mean}} + \gamma_{i,\mathrm{mean}}\, z_R(x,y) \quad (2)$$

where $T_{i,\mathrm{mean}}(x,y)$ is the background field, $z_R(x,y)$ is the DEM used by the RAMS grid at 2.5 km horizontal resolution, and $\gamma_{i,\mathrm{mean}}$ and $T_{i,0\_\mathrm{mean}}$ are those of Eq. (1).

The objective analysis implemented, which relies on OI, is described in Federico (2011). In this section we give some details for the paper readability.

The analyzed field, that is the two-dimensional field of minimum, mean or maximum temperature, is given by the equation:

$$\boldsymbol{x}_\mathrm{a} = \boldsymbol{x}_\mathrm{b} + \mathbf{W}[\boldsymbol{y}_\mathrm{o} - H(\boldsymbol{x}_\mathrm{b})] \quad (3)$$

where $\boldsymbol{x}_\mathrm{a}$ is the analyzed vector (i.e., the best estimate of the unknown "truth"), $\boldsymbol{x}_\mathrm{b}$ is the background (or first guess) field, $\boldsymbol{y}_\mathrm{o}$ is the observation vector, whose elements are the measurements of the 87 thermometers, $H$ is the observation operator that transforms the background field into first guesses

[2]More information about the automated weather stations of the regional network can be found at www.cfcalabria.it.

**Table 1.** Matrices dimensions in the OI algorithm: $n$ (7810) is the grid-points number of the third RAMS domain and $p$ (87) is the number of thermometers.

| Matrix/vector | rows × columns |
|---|---|
| $\mathbf{x}$ | $n\times 1$ |
| $\mathbf{y}$ | $p\times 1$ |
| $\mathbf{H}$ | $p\times n$ |
| $\mathbf{W}$ | $n\times p$ |
| $\mathbf{P}$ | $n\times n$ |
| $\mathbf{B}$ | $n\times n$ |

of the observations, and **W** is the optimal weight (or gain) matrix. Matrices dimensions are given in Table 1, where $n$ is the number of grid points of the third RAMS domain (7810) and $p$ is the thermometers number (87).

The gain matrix **W** is given by:

$$\mathbf{W}=\mathbf{BH}^{T}(\mathbf{R}+\mathbf{HBH}^{T})^{-1} \tag{4}$$

where **B** and **R** are the background and observational error covariance matrices, respectively, and $\mathbf{H}^{T}$ is the transpose of the Jacobian of the forward observation operator (which transforms observation points back to grid points). **H** is a bilinear interpolation operator, which accounts for the altitude differences between grid points and stations, through the vertical gradient of the minimum, mean, and maximum temperature. Its formulation is reported in Federico (2011).

The **R** and **B** matrices depend on the observation ($\sigma_{\mathrm{o}}^{2}$) and background ($\sigma_{\mathrm{b}}^{2}$) error covariances, respectively, whose magnitudes are estimated by fitting, as a function of the distance $r$, the covariance between observational innovations computed from all background field-observation pairs (Lönnberg and Hollingsworth, 1986; Myrick and Horel, 2006). Assuming that: (a) the observational errors are uncorrelated with one another; and (b) the background and the observational errors are uncorrelated, we obtain the following two equations for the innovations' covariance:

$$\mathrm{cov}\ (r{=}0)=\sigma_{\mathrm{o}}^{2}+\sigma_{\mathrm{b}}^{2} \tag{5a}$$

and:

$$\mathrm{cov}\ (r\neq 0)=\sigma_{\mathrm{b}}^{2}\,\rho(r) \tag{5b}$$

where $\rho(r)$ is the background error correlation function, which is assumed as an isotropic function of the distance From the data fitting we obtain: $\sigma_{\mathrm{o}}^{2}=0.7,\ 0.4,\ 1.2\,°\mathrm{C}^{2}$ for minimum, mean, and maximum temperature, respectively; $\sigma_{\mathrm{b}}^{2}=1.0,\ 0.6,\ 1.2\,°\mathrm{C}^{2}$ for minimum, mean, and maximum temperature, respectively. The background error correlation function, $\rho(r)$, is a sixth-order polynomial, which minimizes the $\chi^{2}$ among interpolating polynomials[3].

Once observation and background error covariances are determined, the matrices **R** and **B** are easily formed for each parameter. **R** is a $p\times p$ diagonal matrix whose elements are all equal to $\sigma_{\mathrm{o}}^{2}$. **B** is an $n\times n$ matrix whose element $ij$ is the value of the background error correlation function computed for the distance between grid points $i$ and $j$ multiplied by the value $\sigma_{\mathrm{b}}^{2}$.

The analysis error, which by definition quantifies the difference between the analysis and the unknown truth (Kalnay, 2003), is given by the square root of the diagonal elements of the precision matrix $\mathbf{P}_{\mathrm{a}}$. The precision matrix is a $n\times n$ matrix given by:

$$\mathbf{P}_{\mathrm{a}}=(\mathbf{I}_{n}-\mathbf{WH})\mathbf{B} \tag{6}$$

Where $\mathbf{I}_{n}$ is the $n$-dimension identity matrix. It is worth noting that the estimate of the analysis precision is dependent on the statistical estimates/assumptions of the errors being accurate. If the observations and/or background error covariances are poorly known, if there are biases, or if the observations and background errors are correlated, the analysis precision can be considerably worse than implied by Eq. (6).

Before concluding this section it is useful to remark that the grid resolution (2.5 km) does not correspond to the scale resolution of the analyses, which depends on several factors as the station density and distribution, the parameter considered, and the choices made in specifying the error covariance matrices. In this paper, the distance of the background correlation error is much larger than 2.5 km[4], which spreads the observational innovations at distances larger than the grid resolution, smearing small scale features.

While the scale resolution of the analysis is not easy to quantify, it is here stressed that, given the initial assumption on error covariance matrix, the precision matrix returns the bulk estimate of the analysis error to the unknown truth for each grid point, taking into account for the different factors determining the analysis resolution scale.

## 3 Results

Cumulative statistics are reported in Table 2 and show the bias, the Mean Absolute Error (MAE) and the RMSE for the whole period averaged for the land grid points of the third model domain[5]. For example the RMSE reported in Table 2 for each parameter and forecast day is given by:

$$\mathrm{RMSE}=\frac{1}{N}\sum_{i=1}^{N}\sqrt{\frac{1}{N_{\mathrm{d}}}\sum_{k=1}^{N_{\mathrm{d}}}(T_{k}-A_{k})^{2}} \tag{7}$$

[3]Polynomials from first to eighth order were tried. Different coefficients of the sixth-order polynomial are used for minimum, mean, and maximum temperature.

[4]In this paper we do not show the detail of the Lönnberg and Hollingsworth method, and the reader is referred to Federico (2011). However, the data show that innovations are correlated for distances of the order of 30 km.

[5]We consider land grid-points only because measurements are available over the land.

**Table 2.** Bias (°C), MAE (°C), and RMSE (°C) averaged for land grid points respect to the analyses. Values for each grid point are computed for the whole period. The rows show the results of each forecast day. Min, med, and max refers to minimum, mean and maximum temperature, respectively. The analysis error is shown only for the first-day and is equal for all days. The values in parentheses are the statistics for the analyses grid-points nearest to the observations. Shaded rows: statistics computed respect to ROI-analyses. The first-day ROI row (in italic) is not intended for verification purposes but shows the performance of the OI scheme using the RAMS first-day forecast as background field.

| | Bias min | Bias med | Bias max | MAE min | MAE med | MAE max | RMSE min | RMSE med | RMSE max |
|---|---|---|---|---|---|---|---|---|---|
| 1D | −1.0 ± 0.8 (−0.8) | −0.6 ± 0.6 (−0.5) | 0.0 ± 0.9 (−0.1) | 1.6 ± 0.8 (1.5) | 1.2 ± 0.6 (1.2) | 1.6 ± 0.9 (1.7) | 1.9 ± 0.8 (1.9) | 1.5 ± 0.6 (1.6) | 2.1 ± 0.9 (2.2) |
| 2D | −1.0 (−0.8) | −0.7 (−0.7) | −0.2 (−0.3) | 1.6 (1.6) | 1.4 (1.4) | 1.8 (1.8) | 2.0 (2.0) | 1.7 (1.7) | 2.3 (2.4) |
| 3D | −1.0 (−0.8) | −0.7 (−0.7) | −0.2 (−0.3) | 1.7 (1.6) | 1.4 (1.4) | 1.8 (1.9) | 2.1 (2.1) | 1.8 (1.7) | 2.4 (2.4) |
| *1D-ROI* | *−0.6 ± 0.9 (−0.5)* | *−0.4 ± 0.8 (−0.4)* | *−0.1 ± 1.1 (−0.1)* | *1.0 ± 0.9 (1.0)* | *0.8 ± 0.8 (0.8)* | *1.0 ± 1.1 (1.0)* | *1.2 ± 0.9 (1.3)* | *1.0 ± 0.8 (1.1)* | *1.3 ± 1.1 (1.4)* |
| 2D-ROI | −0.6 (−0.5) | −0.5 (−0.5) | −0.3 (−0.4) | 1.2 (1.2) | 1.0 (1.0) | 1.4 (1.3) | 1.6 (1.6) | 1.3 (1.4) | 1.9 (1.8) |
| 3D-ROI | −0.6 (−0.6) | −0.5 (−0.5) | −0.3 (−0.3) | 1.4 (1.3) | 1.1 (1.1) | 1.6 (1.4) | 1.8 (1.8) | 1.5 (1.5) | 2.1 (2.0) |

where $T_k$ is the temperature forecast for the day $k$-th, $A_k$ is the corresponding analysis, $N_\mathrm{d}$ is the number of days (365), $N$ is the number of land grid points of the third RAMS domain.

The bias is negative for all parameters and forecast days showing that the forecast is colder than analysis on average. The MAE shows a good performance because the error it is below 2 °C for all parameters and forecast days. The RMSE shows also good performance being about 2 °C for minimum and mean temperature for the third forecast day, which scores the worst. For maximum temperature, RMSE varies from 2.1 to 2.4 °C from the first to the third day.

As noticed in the introduction, the use of gridded analyses is motivated by the filtering of small scales features that cannot be represented by the numerical weather prediction system in use. The filtering properties of the OI scheme are determined by the error covariances, which were computed using an observation-based field as background (hereafter we will refer to these analyses as OI-analyses). These filtering properties may not match those of the numerical model and it is interesting to show the cumulative statistics respect to OI analyses using the RAMS first day forecast as background field (hereafter ROI-analyses)[6]. These statistics cannot quantify the forecast error for the first-day forecast because analyses and forecasts are statistically dependent.

[6]The parameters of the OI scheme were determined by the Lönnberg and Hollingsworth method giving: $\sigma_\mathrm{o}^2 = 1.1, 0.5, 1.2\,°\mathrm{C}^2$ for minimum, mean, and maximum temperature, respectively; and $\sigma_\mathrm{b}^2 = 1.5, 1.5, 2.7\,°\mathrm{C}^2$ for minimum, mean, and maximum temperature, respectively.

Table 2 shows that forecast errors decrease when compared to ROI-analyses. This is expected because: (a) the filtering properties of the ROI-analyses match better those of the numerical model; (b) as we didn't attempt to remove the forecast bias, the ROI-analyses have a small bias respect to OI-analyses (absolute value less than 0.5 °C for all parameters when averaged over the entire domain), which improves the forecast error.

It is also interesting to note that the ROI-analysis error is larger than OI-analysis using de-trended fields because the innovations, i.e. the differences between the background at the observation locations and the observations, are larger for ROI-analyses. So, the better forecast skill respect to ROI-analyses is partially compensated by the larger error of the ROI-analyses respect to the unknown truth, given the initial assumption on error covariances.

The statistics of Table 2 refer to all land grid points of the third domain. Because the grid cells containing at least one station correspond to less than 3 % of the land grid points, it is interesting to consider the verification over the small fraction of the grid-points adjacent to observation locations. These statistics are reported in Table 2 and show small differences compared to the statistics over the whole domain. This result is not surprising because local features are smoothed by the OI scheme and stations are rather homogeneously distributed over the country.

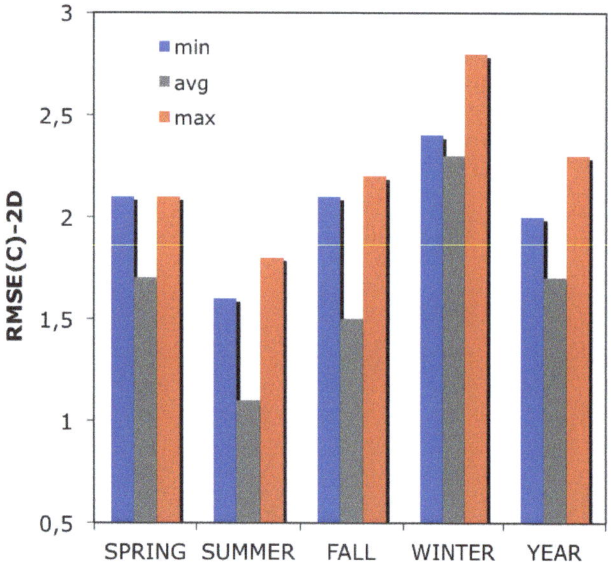


**Figure 2.** RMSE (°C) seasonal distribution for minimum (min), mean (avg), and maximum (max) temperature.

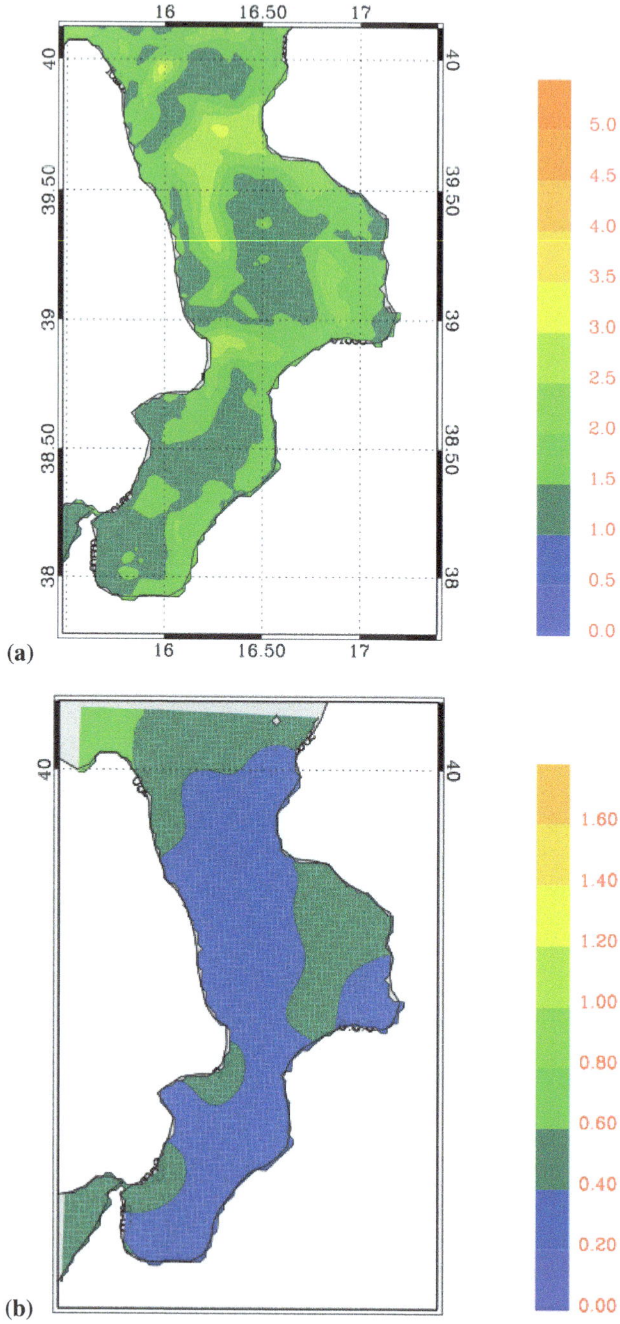


**Figure 3.** **(a)** Mean temperature RMSE (°C) for the second-day forecast; **(b)** averaged precision (°C$^2$) for mean temperature.

Thereafter, we will consider the statistics for the second-day forecast and for the whole (land) domain because: (a) the second day scores between the first and third day and it is representative of the whole forecast behaviour; (b) the results for the subset of grid points nearest to the observations and for the whole domain are similar, as shown by the results of Table 2. Statistics are shown against OI-analyses because they give a conservative estimate of the forecast error.

To analyze the model performance with season, Fig. 2 shows the land averaged RMSE for minimum, mean, and maximum temperature, for different seasons and for the second-day forecast. Spring contains statistics for MAM, summer for JJA, fall for SON and winter for DJF. The December month is taken from 2008.

Summer has the best performance while winter scores the worst. For example, mean temperature RMSE is 1.1 °C for summer and 2.3 °C for winter. Spring and fall errors are between those of summer and winter. Comparing spring and fall it follows that the former scores better for maximum temperature while the latter has a lower error for mean temperature.

Previous result is expected considering the Mediterranean characteristics of the Calabria climate (Colacino, 1990; Federico et al., 2009). In summer the weather is stable and fair and its predictability is expected longer; in winter the incidence of synoptic scale cyclones is the highest and forecast errors are larger in these conditions.

The spatial distribution of the RMSE is shown in Fig. 3a for mean temperature and for the second day forecast. First we remark the rather good performance of the forecast because RMSE is less than 1.5 °C for more than half of the domain area. The largest error occurs in the Crati valley and in the Sibari Plain where RMSE is larger than 2.0–2.5 °C with a maximum of more than 3.0 °C in the plain centre. Errors are lower than 1.5 °C in southern Calabria (roughly south of 39° N), over the Sila plateau, and for northwest Calabria.

It is interesting to note that areas with larger RMSE are co-located with well-defined topographical features: the Crati valley and Sibari Plain, the Lamezia Terme Plain, and the deep small valley in northwest Calabria (Lao river valley).

The reason for this issue is that the forecast smears the meteorological features in these valleys, and the 2.5 km horizontal resolution is still not enough to represent several local phenomena induced by the physiographical features of Calabria, as the sharp temperature gradient across the shoreline or along the slopes of the ridges surrounding the plains. It is also interesting to point out the two local error maxima over the Sila plateau. They correspond to the two main lakes of Calabria (Cecita and Arvo) and are likely caused by an overestimation of the water surface temperature in the initial field used by the meteorological model. Work is in progress to assess this point.

To gain a complete knowledge of the RMSE spatial patterns, Fig. 3b shows the precision of the analysis for mean temperature averaged for the whole period. The square root of Fig. 3b gives the analysis error.

The analysis error is larger in data void areas, namely northern Calabria and the eastern flank of the Sila, and its value is modulated by the gain matrix and the background error by Eq. (6). Comparing Fig. 3a and b we note that the model performance is particularly good for southern Calabria and for the western Sila plateau because they are data rich regions and the analysis and forecast RMSE are low.

Root mean square error patterns for other days and for other parameters are similar to Fig. 3a (not shown) and errors are lower/larger than Fig. 3a depending on the forecast day and on the parameter considered, according to the results of Table 2.

## 4 Conclusions

This paper shows the verification of a high-resolution surface temperature forecast for one climatological year (1 December 2008–30 November 2009) in Calabria.

There is a main issue in verifying the forecast in complex terrain because forecast and observation contain different information about the same physical quantities and this is due to the filtering properties of the numerical weather prediction system in use. To cope with this issue, we compare the forecast with analyses of surface temperature, which are representative of the average conditions inside the grid box and are ideally equal to the average of an infinite number of unbiased observations inside the grid box. Analyses are generated by optimal interpolation. The background field is obtained by de-trending the observations in the vertical direction, and reconstructing the field by the digital elevation model of the numerical weather prediction model.

The analysis algorithm and some example of analysis were already reported in the bibliography (Federico, 2011), and the aim of this paper is to quantify model errors using a wider and different dataset, which is more relevant for statistical and climatic considerations, as well as to present new statistics.

Verification of surface temperature forecast is a high dimensional problem and cumulative statistics are used to quantify the forecast performance. Mean absolute error and RMSE show good results. The largest error is for maximum temperature, whose RMSE varies from 2.1 °C to 2.4 °C from the first- to the third-day forecast.

The seasonal dependence of the error is modulated by the Mediterranean characteristics of the Calabria climate. Errors are the smallest in summer, when the weather is fair and stable, while they are the largest in winter, which has the maximum incidence of synoptic-scale cyclones.

The RMSE spatial pattern highlights areas with larger (lower) errors. Southern Calabria has the smallest errors, while the largest errors are co-located with three main topographical features: the Crati valley and Sibari Plain, the Lamezia Terme Plain, and Lao river valley. The forecast inspection shows that errors are caused by the smearing of the meteorological features induced by local physiographical features of Calabria and that the 2.5 km horizontal resolution is still not enough to represent the sharp temperature gradients generated by those features. Errors patterns are similar for all forecast days and were also found in Federico (2011) for a different dataset, thus revealing areas in which the model forecast need to be improved the more.

The analysis error, which quantifies the difference between the analyzed field and the unknown truth, given the initial assumption on error covariance matrix, increases in data void areas and takes into account for the several factors determining the analysis scale.

**Acknowledgements.** This work was partially funded by the project "MAPVIC - Modellistica Agrometeorologia di Precisione per la Viticoltura Calabrese" funded by the Calabria Region. The Regional Agency for Environmental Protection (ARPACAL) is acknowledged for the regional network data. We are grateful to the reviewers who helped to improve the quality of the paper.

Edited by: O. E. Tveito

## References

Colacino, M.: Mediterranean Meteorology, in: Proceedings NATO-ASI, edited by: Charnock, H., Reports in Meteorology and Oceanography, 40, 1–38, 1990.

Cotton, W. R., Pielke Sr., R. A., Walko, R. L., Liston, G. E., Tremback, C. J., Jiang, H., McAnelly, R. L., Harrington, J. Y., Nicholls, M. E., Carrio, G. G., and McFadden, J. P.: RAMS 2001: Current status and future directions, Meteor. Atmos., Phys., 82, 5–29, 2003.

Federico, S.: Verification of surface minimum, mean, and maximum temperature forecasts in Calabria for summer 2008, Nat. Hazards Earth Syst. Sci., 11, 487–500, doi:10.5194/nhess-11-487-2011, 2011.

Federico, S., Avolio, E., Pasqualoni, L., De Leo, L., Sempreviva, A. M., and Bellecci, C.: Preliminary results of a 30-year daily rainfall data base in southern Italy, Atmos. Research, 94, 641–651, doi:10.1016/j.atmosres.2009.03.008, 2009.

Kalnay, E.: Atmospheric Modeling, Data Assimilation and Predictability, Cambridge University Press, 2003.

Lönnberg, P. and Hollingsworth, A.: The statistical structure of short-range forecast errors as determined from radiosonde data. Part II: The covariance of height and wind errors, Tellus, 38A, 137–161, 1986.

Lussana, C., Uboldi, F., and Salvati, M. R.: A spatial consistency test for surface observations from mesoscale meteorological networks, Q. J. Roy. Meteor. Soc., 136, 1075–1088, 2010.

Mass, C. F., Albright, M., Ovens, D., and coauthors: Regional Environmental Prediction Over the Pacific Northwest, B. Am. Meteorol. Soc., 84(10), 1353–1366, 2003.

Myrick, D. T. and Horel, J. H.: Verification of surface temperature from the National Digital Forecast Database over the western United States, Weather Forecast., 21, 869–892, 2006.

Myrick, D. T. and Horel, J. H.: Sensitivity of Surface Analyses over the Western United States to RAWS Observations, Weather Forecast., 23, 145–158, 2008.

Myrick, D. T., Horel, J. D., and Lazarus, S. M.: Local adjustment of the background error correlation for surface analyses over complex terrain, Weather Forecast., 20, 149–160, 2005.

Simpson, J. E.: Sea breeze and local winds, Cambridge University Press, Cambridge, 1994.

Sela, J. G.: The NMC Spectral Model, NOAA Tech. Rep. NWS-30, 36 pp., 1982.

Uboldi, F., Lussana, C., and Salvati, M: Three-dimensional spatial interpolation of surface meteorological observations from high-resolution local networks, Meteorol. Appl., 15, 331–345, 2008.

# Quality control of 10-min soil temperatures data at RMI

**C. Bertrand, L. González Sotelino, and M. Journée**

Royal Meteorological Institute of Belgium, Brussels, Belgium

*Correspondence to:* C. Bertrand (cedric.bertrand@meteo.be)

**Abstract.** Soil temperatures at various depths are unique parameters useful to describe both the surface energy processes and regional environmental and climate conditions. To provide soil temperature observation in different regions across Belgium for agricultural management as well as for climate research, soil temperatures are recorded in 13 of the 20 automated weather stations operated by the Royal Meteorological Institute (RMI) of Belgium. At each station, soil temperature can be measured at up to 5 different depths (from 5 to 100 cm) in addition to the bare soil and grass temperature records. Although many methods have been developed to identify erroneous air temperatures, little attention has been paid to quality control of soil temperature data. This contribution describes the newly developed semi-automatic quality control of 10-min soil temperatures data at RMI.

## 1 Introduction

Of great importance in agriculture, soil temperature affects plant growth directly, e.g. in seed germination, root growth and nutrient uptake as well as indirectly in soil water and gas flow, soil structure and nutrient availability (Hillel, 1998). Soil temperature is an important parameter in energy balance applications such as land surface modeling, numerical weather forecasting, and climate prediction (e.g. Best et al., 2005). It is also important in radiative transfer applications, such as in the retrieval of land surface properties with satellite sensors, and especially in the retrieval of surface moisture with microwave sensors (e.g. de Jeu et al., 2008). Soil temperature varies in response to exchange processes that take place primarily through the soil surface. These effects are propagated into the soil profile by a complex series of transport processes, the rates of which are affected by time-variable and space-variable soil properties. At each succeeding depth, the peak temperature is dampened and shifted progressively in time (see Fig. 1). The degree of damping increases with depth and is related to the thermal properties of the soil and the frequency of the temperature fluctuation. Due to the much higher heat capacity of soil relative to air and the thermal insulation provided by vegetation and surface soil layers, soil heat anomalies of daily or weekly timescales in shallow layers near the surface do not propagate to the deeper layers. Only persistent long-term anomalies (e.g. at the inter annual and decadal scale) affect temperature variations in those layers (e.g. Lachenbruch and Marshall, 1986).

While many methods have been proposed to identify erroneous air temperatures, only little attention has been paid to quality control of soil temperature data. This situation imposes a significant effort toward providing quality control and assurance (QC/QA) especially devoted to soil temperature measurements. Here we propose a semi-automatic quality control method to check 10-min grass and soil temperature records. In developing quality control tests for the automatic weather stations (AWSs) operated by the Royal Meteorological Institute of Belgium (RMI), we adapted some of the tools developed in Hu et al. (2002) and Hu and Feng (2003) to examine daily and hourly soil temperature data and expanded their approach to include additional tests introduced for some of them in quality control of air temperatures at RMI (Bertrand et al., 2013). The framework is similar to Gandin's concept of complex QA (Gandin, 1988), in that it approaches the question of the validity of a given datum from several different angles and considers errors of different types. However, in the present approach, the automated QA functions are included in a larger QA protocol involving manual inspections similarly to the method implemented at RMI for the quality control of 10-min air temperature data (Bertrand et al., 2013).

**Table 1.** List of the 13 RMI's Automatic Weather Stations performing at least one soil temperature record and the associated measurements.

| AWS | | | Air temperature measurement (s) | | | | | | | QC | Coordinates | |
|---|---|---|---|---|---|---|---|---|---|---|---|---|
| Code | Name | Climate zone | Grass | Bare soil 0 cm | Soil −5 cm | Soil −10 cm | Soil −20 cm | Soil −50 cm | Soil −1 m | group | Lat. (° N) | Lon. (° E) |
| 6414 | Beitem | 1 | | X | X | X | X | X | | 3 | 50.91 | 3.12 |
| 6431 | Zelzate | 1 | | X | | | | | | 5 | 51.18 | 3.81 |
| 6434 | Melle | 1 | X | X | X | X | X | X | | 2 | 50.98 | 3.83 |
| 6438 | Stabroek | 1 | X | X | X | X | X | X | | 2 | 51.33 | 4.37 |
| 6439 | Sint Katelijn Waver | 2 | X | X | | X | X | X | | 4 | 51.08 | 4.53 |
| 6447 | Uccle | 1 | X | X | X | X | X | X | | 2 | 50.80 | 4.36 |
| 6455 | Dourbes | 3 | X | X | X | X | X | X | | 2 | 50.10 | 4.60 |
| 6459 | Ernage | 3 | X | X | X | X | X | X | | 2 | 50.58 | 4.69 |
| 6464 | Retie | 2 | X | X | X | X | X | X | | 2 | 51.22 | 5.03 |
| 6472 | Humain | 4 | X | X | X | X | X | X | X | 1 | 50.19 | 5.26 |
| 6477 | Diepenbeek | 2 | X | X | X | X | X | X | | 2 | 50.92 | 5.45 |
| 6484 | Buzenol | 4 | | X | X | X | X | X | | 3 | 49.62 | 5.59 |
| 6494 | Mont Rigi | 5 | | X | X | X | X | X | | 3 | 50.51 | 6.07 |

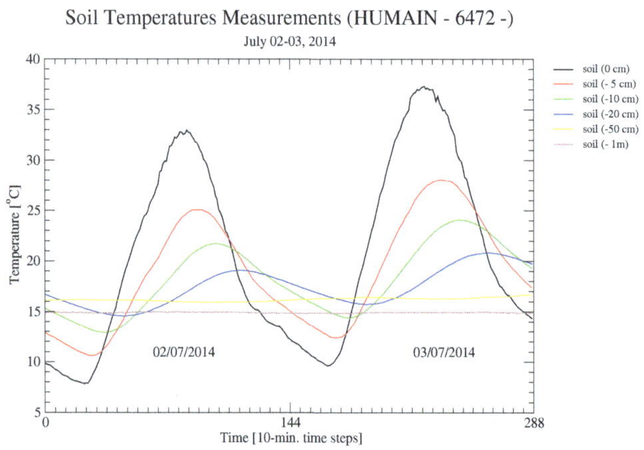


**Figure 1.** 10-min soil temperature at various depths vs time of the day for 2–3 July 2014, at the Humain station (RMI's AWS 6472, see Table 1).

## 2 Soil temperature records at RMI

Automatic weather stations operated by RMI range from basic climatological stations to fully equipped synoptic stations performing a complete set of meteorological observations. Soil temperature measurements are performed in 13 of them to provide soil temperature profile in different regions across Belgium for agricultural management as well as for climate research. At each station, soil temperature can be measured at up to 5 different depths (from 5 to 100 cm) in addition to the bare soil and grass temperatures records (see Table 1). Due to the large heterogeneity within the RMI's AWSs, five groups based on the recorded temperature parameters are distinguished for the automated data QC and the stations are gathered in five climatic zones.

## 3 Complex automatic QA

RMI's AWSs are built around a programmable data logger that measures the sensors, then processes, stores and transmits the data to the central database (DB) in Uccle, Brussels. Once converted to digital values a first processing is performed on the raw data at the data logger level allowing calculation of 10-min temperatures values from the 5-s measurements. To ensure that gross errors are trapped before being further transmitted in the central DB a first basic QC is performed on all temperature values once acquired centrally. Automated procedures monitor the data to make sure they are collected and that the system performance is acceptable. After an existence test, a module checks for physical limits and flags the data violating these limits (erroneous when data lie outside physical limits and suspect when lying outside basic long-term climatological extremes that do not take into account the time of year and location). A list of missing and flagged data is automatically produced after each control cycle and transmitted to the AWS network maintenance team for further intervention. Note that values flagged as erroneous fail immediately and do not require further testing.

Second, each night automated QA procedures check the previous day 10-min temperatures values for more subtle errors. Implementation of the complex automated QA is diagrammed in Fig. 2. Daily temperature values are first checked using annual variation expectancies envelopes and the shape of the diurnal variation of the soil temperature at the different depths are compared to ensure consistency between them (see Sect. 3.1 for details). Individual 10-min records are further scrutinized for plausibility (using adjusted limits to reflect climatic conditions more precisely than in the first near real time range test), internal consistency, temporal consistency and spatial consistency (see Sect. 3.2 for details).

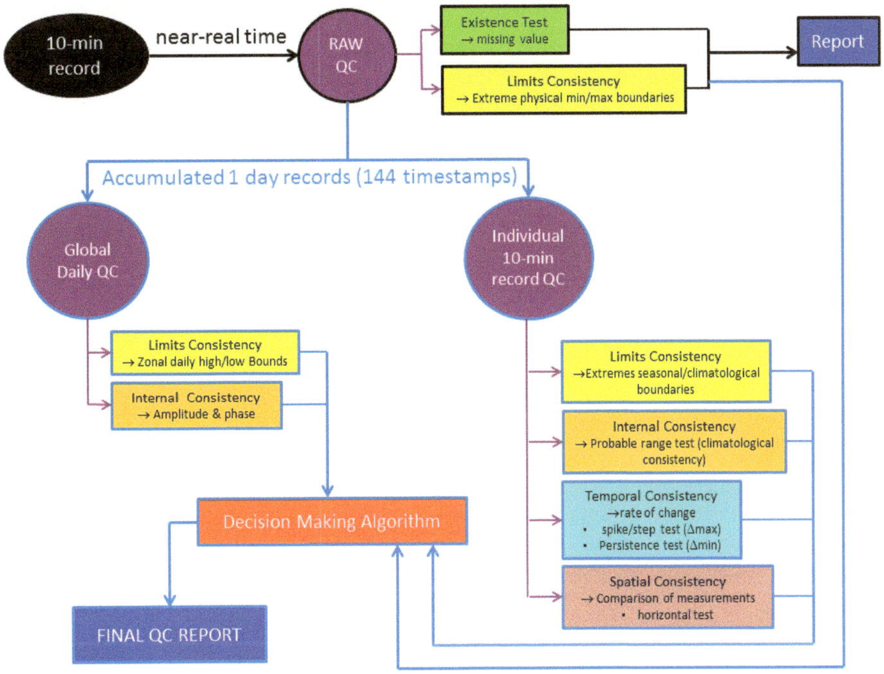


**Figure 2.** Complex QA of 10-min grass/soil temperatures records.

At the end of the checks, when all 10-min daily temperature time series recorded in a given station have been analyzed, a decision algorithm (that is applicable to all variables and all sites) interprets the scores obtained at each of the individual tests and attributes a final flag (i.e. erroneous, suspect or valid) to each particular data given the weight of evidence. At the end of the process, a report is automatically generated for each AWS and sent to the QC staff.

## 3.1 Global daily QC

In nature, soil temperature varies continuously in response to the ever changing meteorological regime acting upon the soil-atmosphere interface. That regime is characterized by a regular periodic succession of days and nights, and of summers and winters. The regular diurnal and annual cycles are perturbed by irregular episodic phenomena as cloudiness, cold wave, warm waves, rain storms or snow storms, and periods of drought. In addition to these external influences, there are the soil's own changing properties (i.e. temporal changes in reflectivity, heat capacity, and thermal conductivity as the soil alternately wets and dries, and the variation of all these properties with depth), as well as the influences of geographic location and vegetation. While the thermal regime of soil profiles is very complex, a simple mathematical representation of the fluctuating thermal regime in a soil profile is obtained by assuming that at all depths in the soil the temperature oscillates as a pure harmonic (sinusoidal) function of time around an average value (van Wijk and de Vries, 1963).

To rapidly identify data outside the variation range of soil temperatures at each depth, daily values (i.e. computed from the 10-min measurements) are compared to lower and upper bounds for soil temperature at given depths. Similarly to the LIM test in Hu and Feng (2003), lower and upper bounds for a given temperature data series (i.e. grass and soil temperatures) are constructed for each of the five climate zones by retrieving the highest and lowest daily values on each calendar day of the year from 9 years (2005–2013) of manually quality controlled historical data. Assuming that annual soil/grass temperature variations follow a sinusoidal curve, envelopes of annual variation of these extreme temperatures were then defined using wave functions of the form:

$$T_{L/U}(z,d) = T_{Lo/Uo}(z) + A_{Lo/Uo}(z) \cdot \sin(\omega_o d) \tag{1}$$

where, $T_L$ and $T_U$ are the lower and upper bounds of soil temperature variations, respectively, $d$ is the day of year, $\omega_o$ is the angular frequency, which is $2\pi$ times the actual frequency (i.e. $2\pi/365$ in case of an annual forcing), $z$ is depth, and $T_{Lo/Uo}(z)$ and $A_{Lo/Uo}(z)$ are the annual mean of these extreme soil temperatures and their amplitude of variations, respectively. To include the extreme values within the derived expectancy envelopes from Eq. (1), the boundaries are adjusted as follows (see Fig. 3 for an illustration): data satisfying

$$T_L(z,d) - 2 \leq T(z,d) \leq T_U(z,d) + 2$$

succeed the limits consistency test, and data failing it are soft flagged if they are less than 10 % outside the range delimited by the adjusted boundaries and hard flagged otherwise.

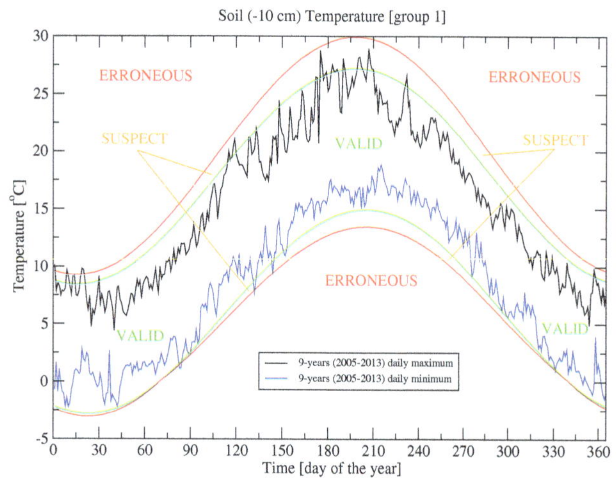


Figure 3. Example of annual variation expectancy envelopes used in the Global Daily QC Limits Consistency test. Illustrated limits apply to the daily mean soil temperature recorded at −10 cm for stations of climate zone 1 (see Table 1)

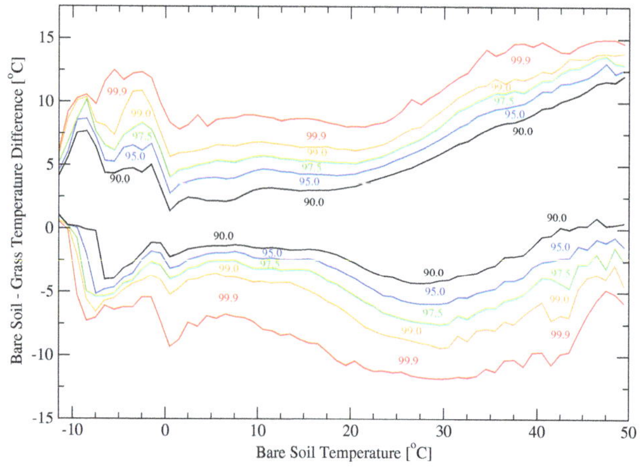


Figure 4. Probable bare soil–grass temperature differences as a function of the bare soil temperature.

To check the reduction in amplitude of the diurnal temperature cycle and the phase shift of the temperature maximum and minimum with depth (see Fig. 1), the recorded diurnal temperature cycle at any depth $z$, and time $t$, is modeled as follows:

$$T(z,t) = \overline{T}(z) + A(z) \cdot \sin(\omega t - \phi(z)) \quad (2)$$

where, $\overline{T}(z)$ is the daily mean temperature at depth $z$, $\omega$ is the angular frequency (diurnal forcing), $A(z)$ is the amplitude of the temperature wave at depth $z$, and $\phi(z)$ is the phase constant at depth $z$ (aligns soil temperature variation with the forcing). The different $A(z)$ and $\phi(z)$ which are functions of $z$ but not of $t$ are then compared and temperature data series which do not respect the damping and retarding of the temperature waves with depth fail the internal consistency test.

Note that these two tests are denoted as global because they cannot discern which observations within the daily time series are responsible for the offense.

## 3.2 Individual 10-min record QC

The question of the validity of a given datum is approached using a number of additional tests applied to each particular 10-min temperature record. To verify whether the values are within acceptable range limits depending on the climatic conditions of the measurement site, individual data values are compared with upper and lower seasonal bounds derived for each temperature parameters and zones from several years of previous manually controlled data. The check provides information as to whether the values are erroneous or suspect. To minimize the possibility of a false positive identification, the algorithm does not report an anomaly in case where the majority of the recorded parameters in a given site at time $t$ are flagged as being either to cool or to warm. Grass temperature and bare soil temperature are further examined for climatological consistency.

Bare soil temperature is measured by a PT100 sensor in contact with the ground on a horizontal surface fully exposed to the open sky. Similarly, grass temperature is measured by a PT100 sensor, fully exposed to the open sky, suspended horizontally over an area covered with short cropped turf and in contact with the tips of grass blades. With the advent of automation and the lack of daily attention by an observer or caretaker, this set up has proved limitations (e.g. lack/lost of contact between the temperature probe and the ground surface or the grass blades, probe fully covered by grass, …). The probable range test provides a more stringent constraint than simple valid maximum/minimum limit test by requiring consistency among temperature parameters as well as consistency with historical data. Basically, the differences between bare soil and grass temperatures as a function of the bare soil temperature are compared to probable difference determined from several years of previous data. Contours in Fig. 4 indicate which combinations of grass and bare soil temperatures fall within a given percentile of joint probability density. Following a review of the values that fall outside the 99.9 % boundary, the 99.9th percentile was selected as the boundary of acceptability. Combinations are hard flagged when falling outside the 99.9 % boundary and soft flagged when falling between the 99.0 and 99.9 % boundaries. Because two comparisons (involving three parameters) are necessary to unambiguously identify which parameter is problematic, similar joint probability densities were established involving the soil temperature at −10 cm. This last parameter has been chosen for the probable range test as it is systematically recorded in stations where both grass and bare soil temperatures are measured (see Table 1).

To examine the temporal consistency of the data, two tests involving the rate of change of the variables from a preceding acceptable level are applied: the spike/step (ΔMax) test and the persistence (ΔMin) test. Both, the maximum and minimum probable changes for each analyzed parameters (i.e. grass and soil temperatures) are based on the 99.9th percentile change for several years of previous data. Because the rate of change for deep soil temperatures can be very small, the persistence test does not apply to soil temperature below −20 cm. Values are checked for 10-min, 1, 2, 3 and 6 h time steps. To minimize the possibility of a false positive identification, the data must fail in at least 3 of the 5 tested time steps prior to be flagged as suspect or erroneous. Moreover, because in case of extreme meteorological conditions, unusual variability in the air temperature may occur, grass and bare soil temperatures data may be flagged as suspect, although correct. To prevent from this, the algorithm does not report a spike/step anomaly for grass and bare soil temperatures if both temperatures at the same site fail the spike/step test. Similarly, the algorithm does not report a persistence anomaly for the grass and bare soil temperatures in case of snow cover. Below 0 °C in snow free condition, a persistence anomaly is reported based on the assumption that freezing conditions can affect multiple sensors. Note that for stations where only one of the grass or bare soil temperatures is recorded, the −5 cm soil temperature value is used (when applicable) in place of the lacking parameter to adjust the ΔMax and ΔMin tests.

Finally, horizontal comparisons of the same measurement at different stations are performed for all recorded temperature parameters. As for the quality control of 10-min air temperature data implemented at RMI (Bertrand et al., 2013), the horizontal check works in two steps. First, an outlier detection is performed on both the station data being quality controlled and the data of the surrounding stations using the daily 10-min temperature time series of each stations.

Let $T_{i,t}$ be a 10-min temperature record at station $i$ ($i = 1, M$ with $M \leq 13$) at time $t$ ($t = 1, N$ with $N \leq 144$) in a given day. $Y_{i,t} = T_{i,t} - \overline{T_i}$ with $\overline{T_i} = \frac{1}{N}\sum_{t=1}^{N} T_{i,t}$ the daily mean temperature at station $i$. $Z_{i,t} = Y_{i,t} - \overline{Y_t}$ with $\overline{Y_t} = \frac{1}{M}\sum_{i=1}^{M} Y_{i,t}$ the stations' mean at time $t$. We test whether or not the $Z_{i,t}$ values fall within the confidence interval defined by:

$$\overline{Z_t} - C\sigma_{Z_t} \leq Z_{i,t} \leq \overline{Z_t} + C\sigma_{Z_t} \quad (3)$$

where $\overline{Z_t} = \frac{1}{M}\sum_{i=1}^{M} Z_{i,t} = 0$, $\sigma_{Z_t} = \sqrt{\frac{1}{M-1}\sum_{i=1}^{M}(Z_{i,t} - \overline{Z_t})^2}$ is the estimated standard deviation at time $t$, and $C$ is an adjustment parameter function of the considered soil/grass temperature parameter. Values $T_{i,t}$ that do not satisfy the relation in Eq. (3) are considered as outliers.

If an outlier is detected for the station being quality checked, then the data fails the horizontal consistency test. Otherwise, the algorithm tests on a 10-min basis whether the analyzed station value, $T_{i,t}$, falls inside a confidence interval formed from surrounding stations data that were not classified as outliers. Measurements that fail the test are soft or hard flagged depending upon the departure of the data from the confidence interval. Note that the outlier check in Eq. (3) can lead to false positives if one or several of the comparison measurements are spatial outliers able to influence the stations mean, $\overline{Y_t}$, in such a way that the measurement under the test is erroneously flagged as an outlier while being valid. In such cases either the decision algorithm at the end of the checking process identifies the false positives as valid based on the scores obtained at the other tests involved in the automated QA system or they will be reviewed during the manual follow up (see Sect. 4) and set to valid if justified.

In developing quality control methods for the US Department of Agriculture (USDA) Natural Resources Conservation Service (NRCS) Soil Moisture-Soil Temperature (SM-ST) network, Hu et al. (2002) established a soil heat diffusion model to screen and identify erroneous soil temperature data. Because such kind of model was found to perform well only in sunny and clear days situations, in our case, modeled data are not used to examine the soil temperatures records. Instead a soil model is used to assist the QC staff in their corrections and estimations (see Sect. 4).

**Table 2.** Overall performance of the automatic QC. The evaluation is performed over the full month of November 2014. A total of 298 512 (100 %) 10-min records including all soil/grass temperature parameters recorded within the RMI AWS network have been analyzed.

| 10-min records | | | |
|---|---|---|---|
| $QC_{AUTO}$ | Manual QC TRUE | Manual QC FALSE | Total |
| Valid | 97.71 % | 1.60 % | 99.31 % |
| Suspect | 0.17 % | 0.09 % | 0.26 % |
| Erroneous | < 0.01 % | 0.43 % | 0.43 % |
| Total | 97.88 % | 2.12 % | 100 % |

### 3.3 Automated QC performance

Quality assurance consists of procedures or rules against which data are tested. Each procedure will either detect the data as being valid, suspect or erroneous. False positives (i.e. type I error) increase the burden on the manual QC, and false negatives (i.e. type II error) reduce the quality of the data. One month of data (e.g. November 2014) has been used to determine the overall performance of the automated QA system. Independent manual QC applied on the recorded 10-min soil temperatures during the same month has been considered as reference for the evaluation. Table 2 presents a general overview of the performance of the newly developed

**Table 3.** Quantitative evaluation of the different tests involved in the automated QC of the 10-min soil/grass temperatures records (QC1 = physical limits test, QC2 = Min–Max range test, QC3 = spike/step – persistence test, QC4 = spatial horizontal test and QC5 = probable range test). The evaluation is performed over the full month of November 2014. A total of 298 512 (100 %) 10-min records including all soil/grass temperature parameters recorded within the RMI AWS network have been analyzed (v = valid, s = suspect, e = erroneous, and nt=no check).

| | | 10-min records | |
|---|---|---|---|
| QC TESTS | | Manual TRUE 292 187 = 100 % | Manual FALSE 6325 = 100 % |
| QC1 | v | 100 % | 100% |
| | s | | |
| | e | | |
| | nt | | |
| QC2 | v | 98.15 % | 71.40 % |
| | s | 1.81 % | 26.12 % |
| | e | 0.04 % | 2.34 % |
| | nt | | 0.14% |
| QC3 | v | 99.36 % | 89.38 % |
| | s | 0.51 % | 7.67 % |
| | e | 0.13 % | 2.81 % |
| | nt | | 0.14 % |
| QC4 | v | 91.45 % | 96.13 % |
| | s | 7.07 % | 1.44 % |
| | e | < 0.01 % | 2.43 % |
| | nt | 1.48 % | |
| QC5 | v | 99.96 % | 78.95 % |
| | s | 0.04 % | 1.56 % |
| | e | < 0.01 % | 19.34 % |
| | nt | | 0.15 % |
| $QC_{AUTO}$ | v | 99.83 % | 75.38 % |
| | s | 0.17 % | 4.49 % |
| | e | < 0.01 % | 20.13 % |
| | nt | | |

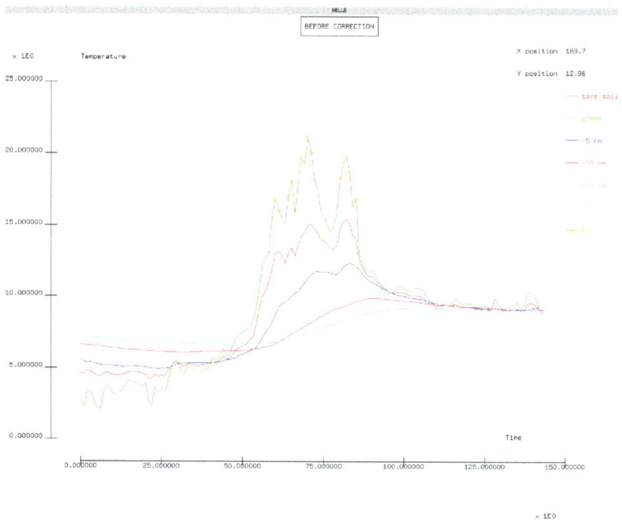


**Figure 5.** Visualization of the automated QC applied on the 10-min soil/grass temperatures records performed on 22 November 2014 at the Melle station (AWS 6434, see Table 1). Erroneous grass temperature data are indicated by a orange circle on the green curve.

complex QA system while Table 3 provides a quantitative evaluation of the various tests involved in the data checking. It is worth pointing out that both tables refer to the 10-min data tests (daily tests cannot identify which observations are responsible for the offense) and that the results could differ for a given temperature parameter or a station type (i.e. QC group in Table 1). Table 2 indicates that type I errors generated by the automated QA system are very low (less than 0.01 % of the true 10-min records were detected as erroneous by the algorithm). By contrast the percentage of type II errors is very large (more than 75 % of the false 10-min records were found as valid by the automated QC as indicated in Table 3). However, this apparent very bad performance of the automated QA system in term of type II errors has to be handle with caution. First, it often occurred that while the algorithm effectively detected 10-min erroneous/suspect measurements in a daily parameter time series, the operators corrected more records than the ones found problematic by the system. As an example, Fig. 5 indicates that the 10-min grass temperature records were found erroneous by the automated QA system 14 times (orange circles on the green curve) on 22 November 2014 at the Melle station. After visualization of the station grass temperatures time series on 22 November 2014, 33 corrections were performed by the operator on the 10-min records (i.e. the full time segment where problematic measurements were detected by the algorithm was manually corrected). Second, Tables 2 and 3 only deal with the 10-min tests and do not account for the daily tests. It is worth pointing out that during the month of November 2014, the grass temperature measurements performed in the Stabroek station were found systematically wrong by the QC staff during 28 days. Over this time period, the grass temperature parameter in this station was detected as erroneous (suspect) 23 (4) times on a daily basis while the 10-min tests did not necessarily reported any erroneous/suspect measurements for this parameter as illustrated in Fig. 6.

When accounting for both daily and 10-min automated QC results, the algorithm succeeded to identify the stations, days and parameters on which corrections were made by the QC staff. A detailed analysis of the type II errors revealed that they mainly concern grass temperature measurements. We strongly suspect that the grass temperature database (and to a lesser extend the bare soil temperature database) used to derive the tests was not validated as it should have been. Typically, probable range test aims at detecting problematic situations as the one illustrated in Fig. 6 for the grass tem-

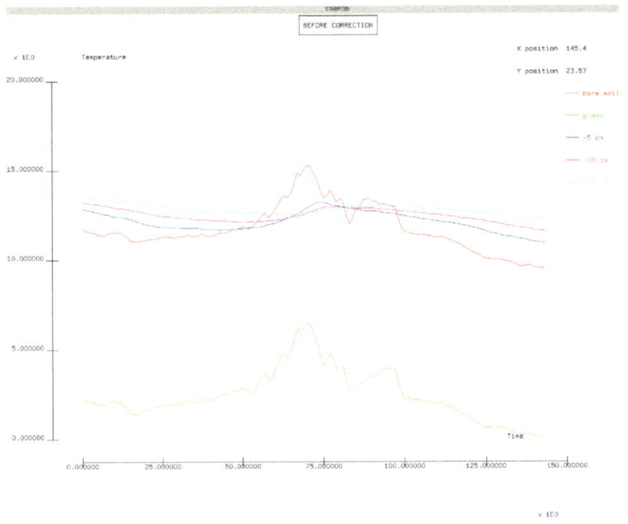


**Figure 6.** Visualization of the automated QC applied on the 10-min soil/grass temperatures records performed on 3 November 2014 at the Stabroek station (AWS 6438, see Table 1). No erroneous or suspect data were found in the daily 10-min temperatures time series.

perature 10-min daily time series. Because erroneous data have been involved when defining the boundaries used in the automated tests, probable differences between bare soil and grass temperatures given in Fig. 4 are certainly too permissive. This drawback has a direct impact on the detection performance of problematic grass, bare soil and $-10$ cm soil temperatures records by the probable range test. As an example, Table 3 indicates that this test (i.e. QC5) produced type II errors in about 79 % of the cases while this category of test has proven to be one of the most efficient in the detection of erroneous 10-min air temperature records performed by the RMI's AWSs.

## 4 Manual QA

Each day, the QC staff analyses the preceding day 10-min temperature records in the light of the assigned quality flags from the automated QA system. Results of the automated QA system can be graphically plotted on the operator terminal screen as illustrated in Figs. 5 and 6. In that case, all the analyzed station 10-min soil/grass temperatures records of the inspected day are illustrated in a graphic window and erroneous or suspect data are indicated in the corresponding parameter daily time series (e.g. orange circles on the green line in Fig. 5). Visual inspection of all records flagged by the automated decision making algorithm is done to distinguish instrumental problems from plausible behaviors. It is the human decision whether or not a value is accepted. When errors are verified or visually detected, faulty records are eliminated and "trouble tickets" are issued where needed to the maintenance team so that sensors can be replaced or repaired. More than simply deleting erroneous measurements, human operators supply corrections and estimations (i.e. when values are missing) where possible. They are supported in this task by automated procedures. As an example, assuming that the thermal properties are constant with depth, soil temperature at any depth below the ground surface (i.e. $0 < z < \infty$) can be estimated using a soil heat diffusion model (van Wijk and de Vries, 1963).

The correction/estimation process is fully interactive, operators directly visualize on screen the corrections they applied on the parameters time series (the graphic window displaying the station temperatures time series being automatically updated after each modification). They have the opportunity to visualize different corrections on the problematic time series in order to determine the most appropriate in their specific case. When the correction/estimation process is completed, all modifications introduced by the operator are automatically implemented in the central RMI database. Note that the original parameters values are kept in the database and still accessible by the QC staff if required.

## 5 Conclusions

Automation of the RMI's AWSs data quality control is ongoing. After the automated quality control of 10-min air temperature data (Bertrand et al., 2013), automated quality assurance procedures devoted to 10-min grass/soils temperature records have been operationally implemented to support the QC staff in their work. The purpose of this automated data screening is to objectively identify abnormal data values for subsequent review by an experienced data analyst. The review is necessary to determine whether an anomaly results from a problem with instrumentation or whether it accurately reflects unusual meteorological conditions. Validation exercises have revealed that the complex automatic QA system is able to correctly identify problematic parameters in a particular station on a given day. However, automated tests applied to 10-min temperature records produce a very high percentage of type II error. In depth analysis of type II errors indicates that because the database of grass temperature records (and to a lesser extent bare soil temperature records) used to derive the boundaries involved in the automated tests was not validated as it should be, the probable range test fails to perform correctly. To overcome such a limitation an extensive validation of our historical records of 10-min grass and bare soil temperatures will be undergone as soon as possible. Once available, the new validated database will be used to refine the automated tests in general and in particular the probable range test involving the grass, bare soil and -10 cm soil temperatures. This forthcoming version of the algorithm will be evaluated using test data days from a whole year as the use of one single month of data could have masked sensitivities of the automated QA system to seasonal variations.

Edited by: I. Auer

## References

Bertrand, C., Gonzalez Sotelino, L., and Journée, M.: Quality control of 10-min air temperature data at RMI, Adv. Sci. Res., 10, 1–5, doi:10.5194/asr-10-1-2013, 2013.

Best, M. J., Cox, P. M., and Warrilow, D.: determining the optimal soil temperature scheme for atmospheric modeling applications, Bound.-Lay. Meteorol., 114, 111–142, 2005.

de Jeu, R. A. M., Wagner, W., Holmes, T. R. H., Dolman, A. J., van de Giesen, N. C., and Friesen, J.: Global soil moisture patterns observed by space borne microwave radiometers and scatterometers, Surv. Geophys., 29, 399–420, 2008.

Gandin, L. S.: Complex quality control of meteorological observations, Mon. Weather Rev., 116, 1137–1156, 1988.

Hillel, D.: Environmental soil physics. Academic press, London, 771 pp., 1998.

Hu, Q. S. and Feng, S.: A daily soil temperature dataset and soil temperature climatology of the contiguous United states, J. Appl. Meteorol., 42, 1139–1156, 2003.

Hu, Q. S., Feng, S., and Schaefer, G.: Quality control for USDA NRCS SM-ST Network soil temperatures: a method and a dataset, J. Appl. Meteorol., 41, 607–619, 2002.

Lachenbruch, A. H. and Marshall, B. V.: Changing climate: geothermal evidence from permafrost in the Alaskan Arctic, Science, 234, 689–696, 1986.

van Wijk, W. R. and de Vries, D. A.: Periodic temperature variations in homogeneous soil, in: Physics of plant environment, edited by: van Wijk, W. R., North-Holland Publ. Co., Amsterdam, 102–143, 1963.

# Satellite-based climate information within the WMO RA VI Regional Climate Centre on Climate Monitoring

**A. Obregón, H. Nitsche, M. Körber, A. Kreis, P. Bissolli, K. Friedrich, and S. Rösner**

Deutscher Wetterdienst, WMO RA VI Regional Climate Centre Network Offenbach Node on Climate Monitoring, Frankfurter Str. 135, 63067 Offenbach, Germany

*Correspondence to:* A. Obregón (andre.obregon@dwd.de)

**Abstract.** The World Meteorological Organization (WMO) established Regional Climate Centres (RCCs) around the world to create science-based climate information on a regional scale within the Global Framework for Climate Services (GFCS). The paper introduces the satellite component of the WMO Regional Climate Centre on Climate Monitoring (RCC-CM) for Europe and the Middle East. The RCC-CM product portfolio is based on essential climate variables (ECVs) as defined by the Global Climate Observing System (GCOS), spanning the atmospheric (radiation, clouds, water vapour) and terrestrial domains (snow cover, soil moisture). In the first part, the input data sets are briefly described, which are provided by the EUMETSAT (European Organisation for the Exploitation of Meteorological Satellites) Satellite Application Facilities (SAF), in particular CM SAF, and by the ESA (European Space Agency) Climate Change Initiative (CCI). In the second part, the derived RCC-CM products are presented, which are divided into two groups: (i) operational monitoring products (e.g. monthly means and anomalies) based on near-real-time environmental data records (EDRs) and (ii) climate information records (e.g. climatologies, time series, trend maps) based on long-term thematic climate data records (TCDRs) with adequate stability, accuracy and homogeneity. The products are provided as maps, statistical plots and gridded data, which are made available through the RCC-CM website (www.dwd.de/rcc-cm).

## 1 Introduction

Since the launch of the first satellites observing Earth's atmosphere in the 1970s (e.g. Meteosat-1 in 1977), more than 30 years of satellite imagery have become available. Satellite data with adequate quality for the generation of long-term climate data sets exist since the beginning of the 1980s. Although meteorological satellites have been primarily designed for weather forecasting, substantial international efforts have been made in recent years highlighting the importance of climate monitoring and the detection of subtle trends and change in climate. In 2008, the WMO (World Meteorological Organization) network for Sustained and Coordinated Processing of Environmental Satellite Data for Climate Monitoring (SCOPE-CM) was established for the coordination of the sustainable generation of long-term climate data records (CDR), building upon essential climate variables (ECV) as defined by the Global Climate Observing System (GCOS). SCOPE-CM interfaces with the WMO Integrated Global Observing System (WIGOS) and the Global Space-based Inter Calibration System (GSICS), building an end-to-end system for climate monitoring (Lattanzio et al., 2013). CDRs are time series of measurements of sufficient length, consistency, and continuity to determine climate variability and change (NRC, 2004). They comprise homogeneous long-term data records of radiances or brightness temperatures (fundamental climate data records, FCDR) and their derived geophysical variables (thematic climate data records, TCDR). The latency of CDR product releases or updates is usually several years. A better timeliness of climate data is enabled by the provision of so-called environmental data records (EDR), which are constructed as instantaneous estimates of geophysical variables (Schulz et al., 2009). The term "interim climate data records" (ICDR) may be used if certain automated corrections are established in the processing system. EDRs and ICDRs are generally provided in operational mode in near-real time (NRT) and do not meet the strict requirements made for CDRs. They generally lack stability, accuracy and sufficient length for climate studies.

Since climate change and its potential impacts have boosted social demands for tailored climate services, the WMO is making substantial efforts to implement Regional Climate Centres (RCC) around the world in order to assist national meteorological and hydrological services (NMHS) to deliver better climate services and products within the Global Framework for Climate Services (GFCS). There is increasing demand for climate monitoring information to enable countries to cope with short-term climate variability (Peterson and Baddour, 2011). The WMO Regional Climate Centre Network (RA – Regional Association – VI) Offenbach Node on Climate Monitoring (www.dwd.de/rcc-cm) provides reliable services based on scientific standards and made available in an operational mode. Figure 1 illustrates where RCC-CM interfaces in the concept of a sustained climate information flow. Formally, the primary users of RCCs are the NMHS, which provide their national end users with the needed information. However, most RCC-CM products are freely available via the web portal (refer to Sect. 4). The acquisition of users is an ongoing effort, e.g. through participation in international projects (e.g. EU FP7 project EUPORIAS, www.euporias.eu). The RCC-CM product portfolio is based on an implementation plan agreed at the 15th session of the WMO RA VI and comprises in situ-based and satellite-based products. The satellite component of RCC-CM offers various products spanning the atmospheric (radiation, clouds, water vapour) and terrestrial domain (snow cover, soil moisture).

Major challenges remain in the understanding of interaction processes between water vapour and clouds as well as their radiative impact (Schulz et al., 2009). Cloud properties, surface radiation budget and water vapour are defined by GCOS as ECVs. Clouds play a crucial role in the climate system by altering Earth's radiation and water balance. Clouds have a large influence on climate but, in turn, a changing climate will also affect the formation of clouds. Monitoring of the shortwave and longwave fluxes at Earth's surface is of crucial importance for the understanding of the climate system and for the detection of global and regional changes. Water vapour in the atmosphere is the most important greenhouse gas and the main contributor to the natural greenhouse effect. Since warmer air can hold more moisture, the water vapour in the atmosphere is directly related to temperature. Monitoring the spatial distribution of water vapour is therefore of great interest.

Sunshine duration (SD), together with surface temperature and precipitation, is one of the most important and widely used parameters in climate monitoring and a key variable for various sectors, including tourism, public health, agriculture, vegetation modelling, and solar energy applications. SD is strongly related to the ECVs cloud properties and surface radiation budget (Kothe et al., 2013).

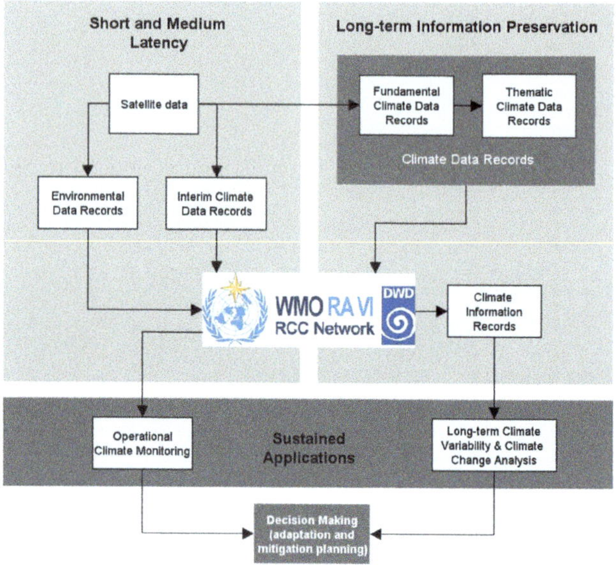


**Figure 1.** Sustained climate information flow and the interfaces with RCC-CM. Figure adapted and modified from Dowell et al. (2013, Fig. 6.2) and WMO (2012, Fig. 2).

Soil moisture is a key parameter in the water, energy and carbon cycles and a critical variable in monitoring hazards such as floods and droughts. It is of high relevance for a variety of agricultural and environmental applications. Therefore, soil moisture was recognized as an ECV by GCOS and as a second top-priority parameter by the Group on Earth Observation (GEO). Within the RA VI RCC Network (www.rccra6.org) it was recommended to consider adding soil moisture products to the product portfolio.

Snow properties are highly sensitive to changes in temperature and precipitation and are therefore a strong indicator for climate variability and change. A global decline of snow would affect planetary albedo, soil moisture, vegetation growth and the hydrological balance (GCOS, 2011). The monitoring of snow properties is therefore of great importance.

## 2 Variables and data sets

The atmospheric data used are provided by the EUMETSAT (European Organisation for the Exploitation of Meteorological Satellites) Satellite Application Facility on Climate Monitoring (CM SAF) (Schulz et al., 2009), while the terrestrial data are mainly provided by other EUMETSAT SAFs and the ESA (European Space Agency) Climate Change Initiative (CCI) (Hollmann et al., 2013). Table 1 gives an overview on all data sets used for the generation of the RCC-CM climate monitoring products. Input data are near-real-time data (for the generation of operational climate monitoring products)

**Table 1.** RCC-CM satellite-based climate variables and corresponding input data sets (EDR – environmental data record, ICDR – interim climate data record, TCDR – thematic climate data record).

| Variable | EDR/ICDR | TCDR |
|---|---|---|
| Global radiation | CM SAF SIS SEVIRI | CM SAF SIS MVIRI 1983–2005 |
| Direct radiation | CM SAF SID SEVIRI | CM SAF SID MVIRI 1983–2005 |
| Cloud fractional cover | CM SAF CFC AVHRR-LAC | CM SAF CFC CLARA-A1 1982–2009 |
| Cloudy/sunny days | CM SAF CFC AVHRR-LAC | CM SAF CFC CLARA-A1 1982–2009 |
| Total water vapour | CM SAF TPW ATOVS | CM SAF TPW ATOVS 1999–2001 |
| Surface albedo | CM SAF SAL AVHRR-LAC | CM SAF SAL CLARA-A1 1982–2009 |
| Sunshine duration | CM SAF SID SEVIRI | CM SAF SID MVIRI 1983–2005 |
| Soil moisture | H-SAF SM-DAS-2/Geoland2 SWI | ESA Soil Moisture CCI 1978–2010 |
| Snow cover extent | H-SAF AVHRR SN-OBS-3 | ESA GlobSnow 1995–2012 |

and climate data records (for the generation of climate information records); see also Fig. 1.

### 2.1 Clouds, radiation and water vapour

RCC-CM uses input data provided by CM SAF for the following parameters:

- cloud fractional cover (CFC)
- global radiation (surface incoming shortwave radiation, SIS)
- direct radiation (surface incoming direct radiation, SID)
- surface albedo (SAL)
- total precipitable water (TPW).

CM SAF data are based on instruments on geostationary (GEO) and polar-orbiting (LEO) satellites. The CFC product is based on the AVHRR (Advanced Very High Resolution Radiometer) instrument, flying on the NOAA (National Oceanic and Atmospheric Administration) and MetOp satellites. CFC refers to the fraction of clouds within each grid box. Fractional cloud cover is expressed as percent. The accuracy of the near-real-time CFC product for Europe is defined as $\pm 10\,\%$ compared to MODIS (Moderate Resolution Imaging Spectroradiometer). A CDR of climate quality is provided with the AVHRR-based CLARA (cloud, albedo and radiation)-A1 data set (Karlsson et al., 2013) which has a high accuracy and stability for analysis of climate variability and change. This data set also includes SAL.

SIS and SID refer to global and direct radiation (global minus diffuse radiation), respectively. RCC-CM SIS and SID products are based on CM SAF data sets using data from geostationary Meteosat satellites. The near-real-time products are based on the SEVIRI (Spinning Enhanced Visible and Infrared Imager) and GERB (Geostationary Earth Radiation Budget) instruments. The accuracy of the CM SAF input data is defined as 90 % of absolute bias values below $10\,\mathrm{W\,m^{-2}}$ compared with in situ data (plus the uncertainty of ground based measurements) for monthly means. Higher bias values occur in the Alpine and other mountainous regions, e.g. due to uncertainties in area to point comparison. An MVIRI (Meteosat Visible and Infrared Imager)-based climate data record (1983–2005) is provided for SIS and SID (Posselt et al., 2012).

CM SAF provides several water vapour products, including the TPW, which represents the vertically integrated water vapour of an atmospheric unit's vertical column. The accuracy of the near-real-time TPW product is defined as a bias of 1 mm and a RMSE (root mean square error) of 4.5 mm using GUAN (GCOS Upper Air Network) radiosondes as a reference. A CDR for TPW exists, but only for a relatively short period covering the years 1999–2011 (ATOVS edition 1). Due to missing intercalibration for the first years, only data between 2001 and 2011 are used within RCC-CM.

### 2.2 Sunshine duration

The RCC-CM sunshine duration products are derived by using a simple method proposed by Kothe et al. (2013) using hourly CM SAF SID data. The method is based on the WMO definition for bright sunshine, occurring when direct solar radiation exceeds a threshold of $120\,\mathrm{W\,m^{-2}}$ (WMO, 2008). An evaluation by product intercomparison and validation against station measurements showed a good agreement with accuracy within the range of $\pm 1\,\mathrm{h\,day^{-1}}$. The uncertainty of the satellite products is quite homogeneous over land in comparison to products based on station interpolation, whose uncertainty increase in areas with low station density or in mountainous regions. The satellite-derived SD product gives valuable additional spatial information over the European continent (Kothe et al., 2013). The SEVIRI-based SID near-real-time products are used for the generation of operational SD-monitoring products, while the MVIRI-based SID climate data record is used for the generation of a long-term data set, which can serve as reference climatology and for climate-related studies.

### 2.3 Soil moisture

A 32 year long surface soil moisture data set (1978–2010) was derived within the ESA Soil Moisture CCI by combining active and passive satellite instruments, with emphasis on preserving long-term trends (Liu et al., 2012; Wagner et al., 2012). RCC-CM long-term climate information records are based on this data set. RCC-CM near-real-time soil moisture monitoring products are currently in development and will be based on the soil moisture profile (also: root-zone soil moisture), which also characterizes the soil wetness in deeper soil layers. One candidate for the operational monitoring product is the ASCAT (Advanced Scatterometer)-based product provided by the Satellite Application Facility on Hydrology (H-SAF; SM-DAS-2 product), which uses a soil moisture assimilation system, propagating the surface observation from ASCAT towards the roots region. A second candidate is the soil water index (SWI), provided within the Copernicus Geoland2 project. SWI is derived by an exponential filter simulating the soil moisture value of deeper soil layers (Brocca et al., 2010).

### 2.4 Snow cover

Recently, a long-term data set of snow extent (SE) for the years 1995–2012 has been developed within the ESA GlobSnow project, which relies on a method by Metsämäki et al. (2012). The ESA GlobProjects form a series of precursor projects for ESA CCI. SE is given as fractional snow cover in percentage per satellite grid cell. The GlobSnow SE data set is based on optical data from Envisat's AATSR (Advanced Along Track Scanning Radiometer) and ERS-2's (European Remote-Sensing satellite) ATSR-2 sensors covering the Northern Hemisphere. According to GlobSnow this is the longest historical time series of SE with a moderate spatial resolution and potentially the highest thematic accuracy available. RCC-CM uses the monthly SE composites. In addition to the long-term series, GlobSnow initiated an operational near-real time service in October 2010 based on AVHRR and NPP (National Polar-orbiting Partnership) Suomi VIIRS (Visible Infrared Imaging Radiometer Suite) instruments. The GlobSnow NRT product was selected as a candidate for an operational monitoring within RCC-CM. A second candidate is the H-SAF effective snow cover product (SN-OBS-3) which is currently in a pre-operational stage. Monitoring of snow water equivalent will be considered in the future.

## 3 Products

The RCC-CM products fall into two main categories:

1. Operational climate-monitoring products based on near-real-time data (with adequate timeliness).

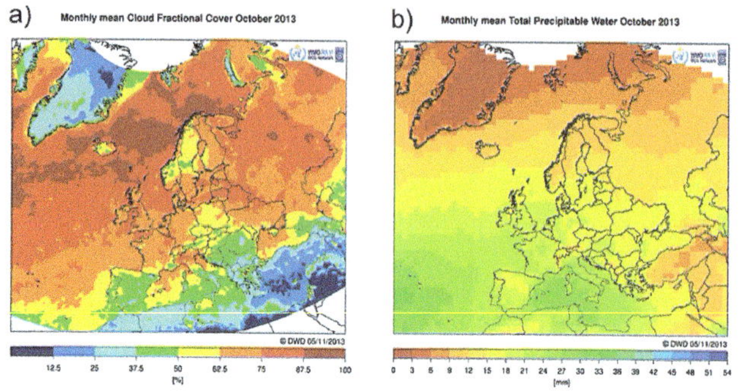


**Figure 2.** Monthly means of cloud fractional cover **(a)** and total precipitable water **(b)** for October 2013.

2. Climate information records based on long-term climate data records (with adequate accuracy, homogeneity and stability).

RCC-CM products are provided as maps, statistical plots and gridded data. Monthly, seasonal and annual products are generated, which are available for Europe (60° W–60° E, 30°–80° N) and the following sub-regions:

- northern Europe (25° W–40° E, 55°–70° N)
- central Europe (15° W–25° E, 45°–60° N)
- Mediterranean (10° W–40° E, 30°–45° N)
- eastern Europe (15°–45° E, 40°–60° N).

The geographic coordinates of the given regions refer to the statistical products "annual course" (Sect. 3.1.4), "time series" (Sect. 3.2.3) and "spatio-temporal diagrams" (Sect. 3.2.4). The maps show a slightly larger coverage. All analyses were carried out using climate data operators (CDO) and the statistical software R.

### 3.1 Operational climate monitoring

#### 3.1.1 Means

The input data sets presented in Table 1 are mainly provided as monthly means, except for the cloud indices for which daily data are used and for sunshine duration for which hourly data are used. For the RCC-CM mean products several pre-processing steps are made including format conversion, geographic subsetting and remapping onto a regular latitude–longitude grid. The spatial resolution of the target grid depends on the original input data set and ranges between 0.05° × 0.05° and 1° × 1°. Figure 2 shows monthly mean products of cloud fractional cover (0.25° × 0.25° resolution) and total precipitable water (1° × 1° resolution).

Seasonal and annual means are calculated by averaging of monthly means. If missing values occur, no seasonal or annual means are derived for the respective grid box. For the production of maps, data have been transformed to a rotated

0.1° × 0.1° lat–long grid using bilinear resampling. TPW data have been resampled to a rotated 1° × 1° lat–long grid. Maps for all other product types are produced in the same way.

### 3.1.2 Anomalies

Monthly, seasonal and annual anomalies are calculated using the corresponding means (Sect. 3.1.1) and long-term averages (Sect. 3.2.1). The reference climatology for each variable is built from the corresponding climate data records (refer to Table 1).

The following types of climate anomalies are calculated:

- absolute anomalies (absolute deviations from the long-term average)
- relative anomalies (deviations from the long-term average expressed as percentage)
- standardized anomaly (absolute deviation expressed as times of standard deviation)
- percentiles (values below or above the lower or upper tercile and decile of the probability density function of the climate element).

Absolute and relative anomalies are common standards in climate monitoring. They are easy to interpret and reveal the simple departures from the long-term average. However, they do not take into account the natural variability of a parameter at a given location. Hence, no proper statements can be made on the unusualness of an anomaly. Therefore, two different measures are introduced in RCC-CM which are able to characterize the rareness of an event by taking into account the natural variability of a parameter for every grid box. The so-called standardized anomaly uses the standard deviation as a reference while the percentile product relies on deciles and terciles derived through probability density functions. As an example, Fig. 3c shows the absolute anomaly for global radiation in July 2013. High positive anomalies occur over the Atlantic Ocean and central Europe. Figure 3d shows the standardized anomaly, revealing discrepancies between these two regions. The standardized anomalies over central Europe are lower compared to those over the Atlantic Ocean, due to a general higher natural variability of global radiation in July over central Europe (see standard deviation in Fig. 3b).

### 3.1.3 Indices

CDC daily data are used for the calculation of the indices "cloudy days" and "sunny days" according to the definition of the CCl/CLIVAR/JCOMM Expert Team on Climate Change Detection and Indices (ETCCDI). Cloudy days are defined as days where average cloud cover exceeds 6 octas, while sunny days refer to a mean cloudiness of 2 octas or below. For the monthly, seasonal and annual RCC-CM products sunny and cloudy days, respectively, are counted for the according period. Anomalies are calculated as described in Sect. 3.1.2.

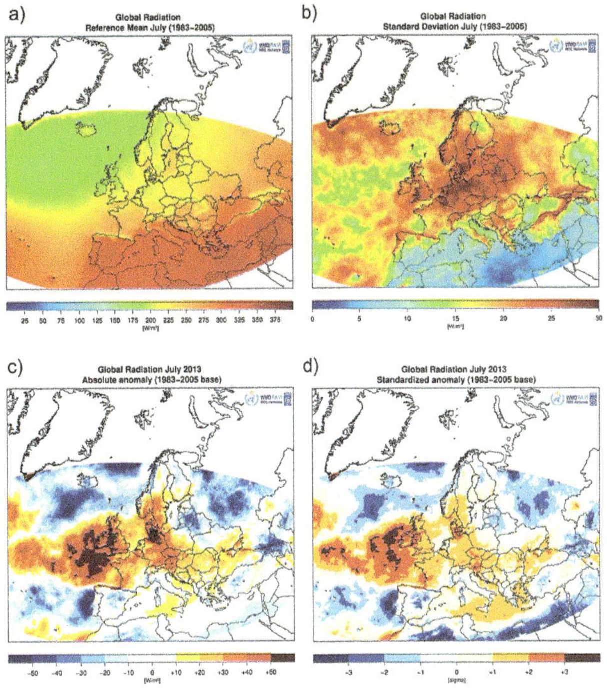

**Figure 3.** Analysis of global radiation for July 2013 compared to the long-term climate average of 1983–2005 **(a)** and the standard deviation **(b)**. Absolute anomalies **(c)** and standardized anomalies **(d)** are shown. The upper maps fall into the category of "climate information records" while the bottom maps are "operational climate monitoring" products, which are provided on a monthly basis.

### 3.1.4 Annual course

In operational climate monitoring the annual cycle of a parameter compared to its long-term average and variability is of great interest. The plots for the annual course show the mean of the current year compared to the reference climatology, based on calculations of area-weighted means. For reference, long-term means and standard deviations are shown, as well as the 10th and 90th percentiles (Fig. 4).

Figure 4 illustrates the annual course of cloud fractional cover and direct radiation for central Europe in 2013. For the beginning of the year, the products reveal well-above average cloudiness and hence negative anomalies of direct radiation in the range of the lower decile. While most of 2013 shows a normal course in terms of CFC and SID, July 2013 is a distinct exception. Cloudiness was in the range of the lower 10th percentile while direct radiation clearly exceeded the 90th percentile. This is consistent with other related parameters like sunshine duration, which was exceptionally high in central Europe in July 2013. In Germany, it was the second sunniest July on record.

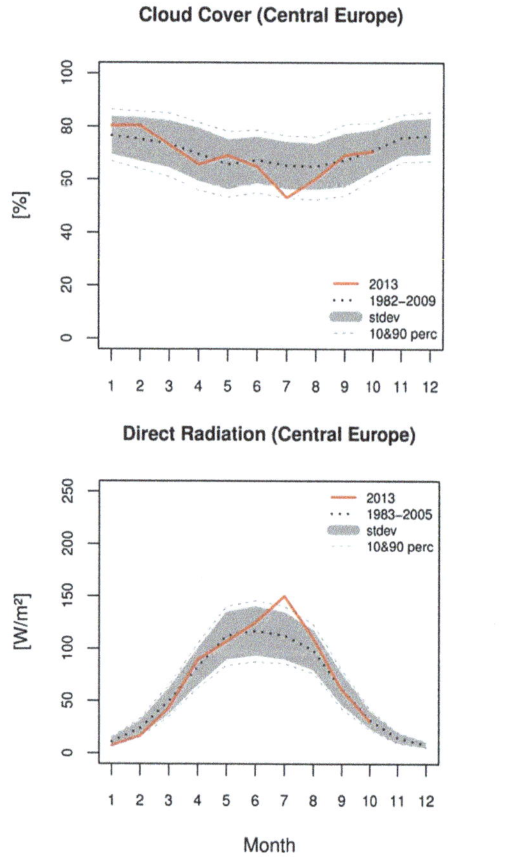


**Figure 4.** Annual course of monthly means of cloud cover and direct radiation for the central European subregion. Long-term averages, standard deviations and extreme percentiles are provided as a reference. The products for October 2013 are shown (provided in November 2013).

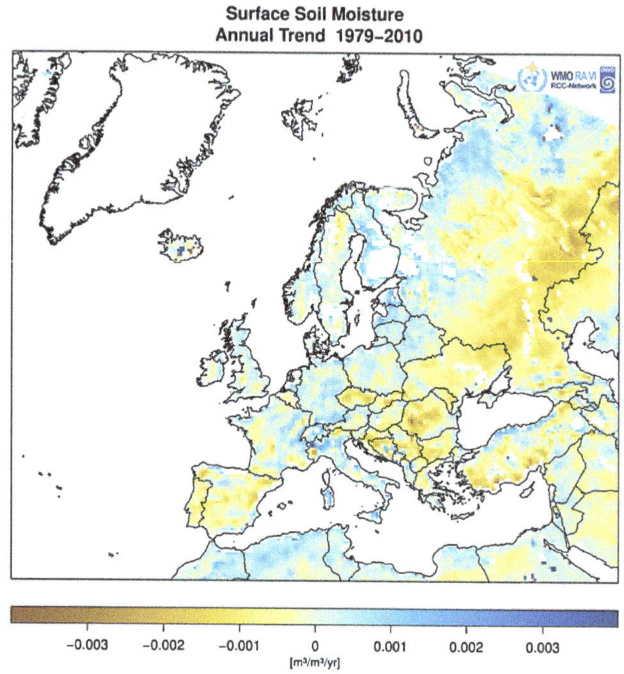


**Figure 5.** Annual trend in surface soil moisture for the period 1979–2010, based on a harmonized data set of active and passive microwave observations provided by the ESA Soil Moisture CCI.

## 3.2 Climate information records

### 3.2.1 Climatologies

The climate data records in Table 1 are used for the calculation of long-term averages and standard deviations for single months, seasons and for the whole time period. The long-term averages are used as a reference for the calculation of anomalies (Sect. 3.1.2) and for the climatological interpretation of the current annual cycle (Sect. 3.1.4). In addition, long-term means and standard deviations are also provided as maps (Fig. 3a, b).

### 3.2.2 Trend maps

Monthly, seasonal and annual trends are calculated using a simple linear regression for all climate data records and provided as maps for each parameter. As an example, Fig. 5 shows the annual trend in surface soil moisture (in $m^3\,m^3\,yr^{-1}$) for the period 1979–2010, revealing a trend towards drier soils across large parts of eastern Europe.

Generally, the generation of satellite-based data sets of true climate quality which are suitable for the detection of climate trends is a major challenge. CM SAF focuses its efforts on the development of high-quality climate data records suitable for the analysis of climate variability and potentially for the detection of trends. However, since most CM SAF CDRs used within RCC-CM are version 1 data sets, there are still some issues remaining. Based on own investigations and literature review, some data sets are not used for the whole time period due to uncertain homogeneity. RCC-CM trend maps for CFC are only calculated for the period 1994–2009, since some inhomogeneities might be present before. Trend maps for SIS and SID are calculated for the period 1994–2005, since this period was found to be homogeneous in Europe. Before 1994, some shifts and artificial trends were detected (Krähenmann et al., 2013). Sanchez-Lorenzo et al. (2013) found inhomogeneities over Europe in the late 1980s and early 1990s using homogenized station data sets as the reference. Generally, the trend maps should be interpreted with caution. Also, no significance tests have been applied.

### 3.2.3 Time series

Time series are derived by calculating area-weighted means for Europe and the defined subregions using the climate data records in Table 1. The RCC-CM products show monthly, seasonal, and annual means for the whole period of the long-term record and include a 3 year running mean (Fig. 6 top).

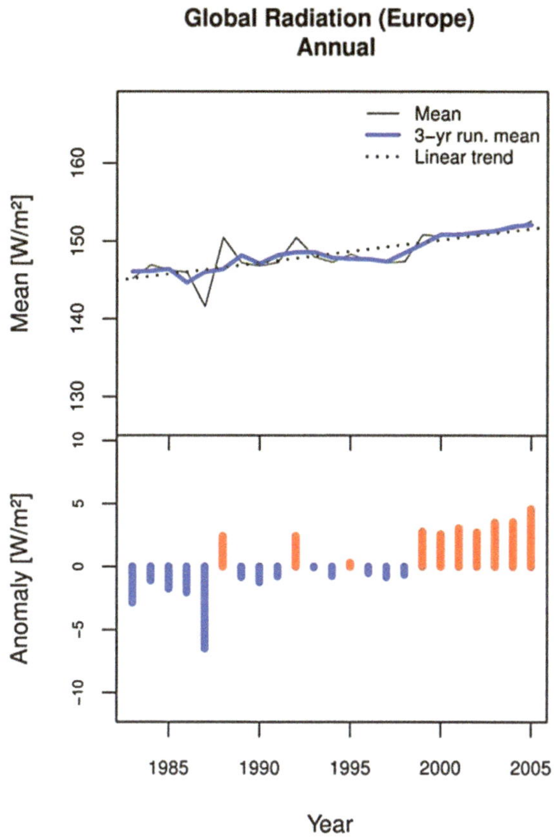


**Figure 6.** Time series of average annual global radiation for Europe (top) and time series of anomalies (bottom) based on the CM SAF MVIRI data set.

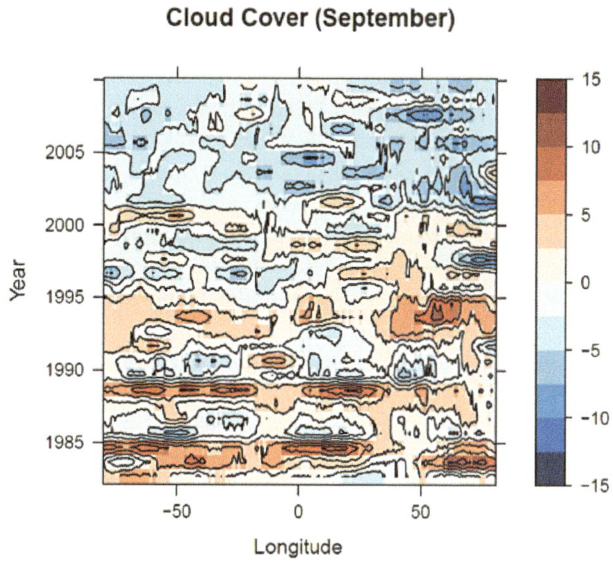


**Figure 7.** Hovmöller diagram of cloud cover anomalies (in %) for Europe from 1982 till 2009 based on CM SAF CLARA-A1.

Time series of corresponding anomalies (absolute departures from the long-term average) are also provided (Fig. 6 bottom). The time series for mean annual global radiation is shown in Fig. 6. The data show a positive trend of global radiation in recent years, which is consistent with a decrease of cloudiness (CFC time series, not shown) and the observed global brightening (e.g. Wild, 2012). The trend in the satellite data could also be caused by inhomogeneities in the time series, though. It is known that due to possible inhomogeneities before 1994 in the CFC, SIS and SID data sets (refer to Sect. 3.2.2) time series before 1994 should be interpreted with caution.

### 3.2.4 Spatio-temporal diagrams

The visualization of climate data by means of time–latitude or time–longitude cross sections provides information about temporal and spatial variability and changes in the climate system on yearly to decadal scales. These spatio-temporal diagrams are referred to as Hovmöller plots and ease the understanding of climate dynamics (Hocke and Kämpfer, 2009).

RCC-CM provides Hovmöller diagrams for all satellite CDRs in Table 1, visualizing the temporal propagation of zonal and meridional anomalies. Figure 7 provides an example for the annual meridional change in cloud cover for September, revealing a tendency towards higher values in recent years, in particular over continental Europe. The cause of the trend – whether a real phenomenon or caused by instrumental inhomogeneities – should be studied thoroughly.

## 4 Access, documentation and future perspectives

The RCC-CM maps, statistical plots and gridded data are made available through the RCC-CM website (www.dwd.de/rcc-cm). User documentation is available via so-called "product description sheets" for each climate variable. These documents provide product specifications such as spatial and temporal resolution and coverage, timeliness, data format, units and information on input data sets. The scientific methodology and data quality are briefly described and implications for usage are given. The provision of information on uncertainties and accuracies is generally of great importance. Information regarding the accuracy of the input data is provided within the product description sheets, where the corresponding provider documents are referenced. For example, the target accuracy of the CM SAF data is defined within the CM SAF service specifications found at the CM SAF website (www.cmsaf.eu). The quality of the CM SAF input data is documented on a regular basis within the CM SAF validation reports. Limitations of the CM SAF data are described in the CM SAF product user manuals. Similar documentation is available for data sets of other providers, which is referenced in the corresponding RCC-CM product description sheets.

For some satellite-based products, comparisons with in situ measurements are done (e.g. for clouds, water vapour and sunshine duration) and validation results are provided. A detailed comparison of gridded station-based and satellite-based sunshine duration showed rather good consistencies between the products (for a broader discussion refer to Kothe et al., 2013). For comparison purposes, also reanalyses data are used for some parameters (e.g. global radiation; refer to Krähenmann et al., 2013).

As an outlook, the RCC-CM satellite monitoring aims on updating the climate information records as soon as new climate data records become available. Generally, there is a need for more regularly updated CDRs. Trenberth et al. (2013) point out the increasing demands for short-term assessments of the current climate, especially when major climate events occur. Within an RCC-CM-led study, Krähenmann et al. (2013) proposed a method for the adjustment of CDRs for the integrated use with near-real time data, based on an example using data sets for global radiation. The method consists of homogeneity tests and shift-size adjustments and motivates for the operational generation of near-real-time climate data records in order to facilitate a better analysis of the current state of the climate and to provide better climate services. Although the method has been developed for SIS data sets it is, in principle, applicable for the adjustment of CDRs addressing other ECVs as well.

Future perspectives of the RCC-CM activities are the provision of other climate indices, the exploitation of daily data for the monitoring of more flexible timescales (e.g. for extreme events) and the ongoing improvement of products.

**Acknowledgements.** Cloud, albedo, radiation and water vapour data were provided by the EUMETSAT CM-SAF hosted by Deutscher Wetterdienst (www.cmsaf.eu). Long-term records of soil moisture and snow cover were provided by ESA Soil Moisture CCI (www.esa-soilmoisture-cci.org) and ESA GlobSnow (www.globsnow.info), respectively. EUMETSAT H-SAF (hsaf.meteoam.it) is acknowledged for the provision of near-real-time data of soil moisture and snow cover. CDO was used in this work, which is a development by Uwe Schulzweida (2014) at the Max Planck Institute for Meteorology. The R project and many volunteers providing additional packages are also acknowledged. The *rasterVis* package was used for the generation of Hovmöller plots and the *solaR* package for the sunshine duration retrieval. These packages were developed by Oscar Perpiñán Lamigueiro, who is gratefully acknowledged for providing support. We thank Andrea Kaiser-Weiss and two anonymous reviewers for constructive suggestions for improving the manuscript.

Edited by: C. Buontempo

## References

Brocca, L., Melone, F., Moramarco, T., Wagner, W., and Hasenauer, S.: ASCAT SoilWetness Index validation through in-situ and modeled soil moisture data in Central Italy, Remote Sens. Environ., 114, 2745–2755, 2010.

Dowell, M., Lecomte, P., Husband, R., Schulz, J., Mohr, T., Tahara, Y., Eckman, R., Lindstrom, E., Wooldridge, C., Hilding, S., Bates, J., Ryan, B., Lafeuille, J., and Bojinski, S.: Strategy Towards an Architecture for Climate Monitoring from Space, available at: http://www.wmo.int/pages/prog/sat/documents/ARCH_strategy-climate-architecture-space.pdf (last access: 25 January 2014), 2013.

GCOS: Systematic observation requirements for satellite-based products for climate, 2011 update, WMO GCOS Report No. 154, 2011.

Hocke, K. and Kämpfer, N.: Hovmöller diagrams of climate anomalies in NCEP/NCAR reanalysis from 1948 to 2009, Clim. Dynam., 36, 355–364, 2009.

Hollmann, R., Merchant, C., Saunders, R., Downy, C., Buchwitz, M., Cazenave, A., Chuvieco, E., Defourny, P., de Leeuw, G., Forsberg, R., Holzer-Popp, T., Paul, F., Sandven, S., Sathyendranath, S., van Roozendael, M., and Wagner, W.: The ESA Climate Change Initiative: Satellite Data Records for Essential Climate Variables, B. Am. Meteorol. Soc., 94, 1541–1552, 2013.

Karlsson, K.-G., Riihelä, A., Müller, R., Meirink, J. F., Sedlar, J., Stengel, M., Lockhoff, M., Trentmann, J., Kaspar, F., Hollmann, R., and Wolters, E.: CLARA-A1: a cloud, albedo, and radiation dataset from 28 yr of global AVHRR data, Atmos. Chem. Phys., 13, 5351–5367, doi:10.5194/acp-13-5351-2013, 2013.

Kothe, S., Good, E., Obregon, A., Ahrens, B., and Nitsche, H.: Satellite-based sunshine duration for Europe, Remote Sens., 5, 2943–2972, 2013.

Krähenmann, S., Obregon, A., Müller, R., Trentmann, J., and Ahrens, B.: A Satellite-Based Surface Radiation Climatology Derived by Combining Climate Data Records and Near-Real-Time Data, Remote Sens., 5, 4693–4718, 2013.

Lattanzio, A., Schulz, J., Matthews, J., Okuyama, A., Theodore, B., Bates, J. J., Knapp, K. R., Kosaka, Y., and Schüller, L.: Land Surface Albedo from Geostationary Satellites: A Multiagency Collaboration within SCOPE-CM, B. Am. Meteorol. Soc., 94, 205–214, 2013.

Liu, Y. Y., Dorigo, W. A., Parinussa, R. M., de Jeu, R. A. M., Wagner, W., McCabe, M. F., Evans, J. P., and van Dijk, A. I. J. M.: Trend-preserving blending of passive and active microwave soil moisture retrievals, Remote Sens. Environ., 123, 280–297, 2012.

Metsämäki, S., Mattila, O.-P., Pulliainen, J., Niemi, K., Luojus, K., and Böttcher, K.: An optical reflectance model-based method for fractional snow cover mapping applicable to continental scale, Remote Sens. Environ., 123, 508–521, 2012.

NRC: Climate Data Records from Environmental Satellites, National Academies Press, Washington D.C., 2004.

Peterson, T. C. and Baddour O.: Towards an enhanced climate system monitoring: challenges and perspectives, Clim. Res., 47, 21–28, 2011.

Posselt, R., Müller, R. W., Stöckli, R., and Trentmann, J.: Remote sensing of solar surface radiation for climate monitoring – the CM-SAF retrieval in international comparison, Remote Sens. Environ., 118, 186–198, 2012.

Sanchez-Lorenzo, A., Wild, M., and Trentmann, J.: Validation and stability assessment of the monthly mean CM SAF surface solar radiation dataset over Europe against a homogenized surface dataset (1983–2005), Remote Sens. Environ., 134, 355–366, 2013.

Schulz, J., Albert, P., Behr, H.-D., Caprion, D., Deneke, H., Dewitte, S., Dürr, B., Fuchs, P., Gratzki, A., Hechler, P., Hollmann, R., Johnston, S., Karlsson, K.-G., Manninen, T., Müller, R., Reuter, M., Riihelä, A., Roebeling, R., Selbach, N., Tetzlaff, A., Thomas, W., Werscheck, M., Wolters, E., and Zelenka, A.: Operational climate monitoring from space: the EUMETSAT Satellite Application Facility on Climate Monitoring (CM-SAF), Atmos. Chem. Phys., 9, 1687–1709, doi:10.5194/acp-9-1687-2009, 2009.

Schulzweida, U.: CDO User's Guide, version 1.6.3, MPI for Meteorology, Germany, available at: https://code.zmaw.de/projects/cdo/embedded/1.6.3/cdo.pdf (last access: 12 May 2014), 2014.

Trenberth, K. E., Anthes, R. A., Belward, A., Brown, O., Haberman, E., Karl, T. R., Running, S., Ryan, B., Tanner, M., and Wielicki, B.: Challenges of a Sustained Climate Observing System, in: Climate Science for Serving Society: Research, Modelling and Prediction Priorities, edited by: Hurrell, J. W. and Asrar, G., Springer, Berlin, Germany, 2013.

Wagner, W., Dorigo, W., de Jeu, R., Fernandez, D., Benveniste, J., Haas, E., and Ertl, M.: Fusion of active and passive microwave observations to create an Essential Climate Variable data record on soil moisture, ISPRS Annals of the Photogrammetry, Remote Sensing and Spatial Information Sciences, Volume I-7, 2012, XXII ISPRS Congress, 25 August–1 September 2012, Melbourne, Australia, 2012.

Wild, M.: Enlightening Global Dimming and Brightening, B. Am. Meteorol. Soc., 93, 27–37, 2012.

World Meteorological Organization (WMO): Guide to Meteorological Instruments and Methods of Observation, 7th Edn., WMO, Geneva, Switzerland, 2008.

World Meteorological Organization (WMO): SCOPE-CM Phase 2 Implementation Plan: Final Version, Version 1.0, 2012, available at: http://www.wmo.int/pages/prog/sat/documents/SCOPE-CM_Phase-2-Implementation-Plan (last access: 25 January 2014), 2012.

# 30

# Digital Climate Atlas of the Carpathian Region

**I. Antolović[1], V. Mihajlović[1], D. Rančić[1], D. Mihić[2], and V. Djurdjević[2]**

[1]Faculty of Electronic Engineering, Niš, Serbia
[2]Hydrometeorological Service of Serbia, Belgrade, Serbia

*Correspondence to:* D. Mihić (dragan.mihic@hidmet.gov.rs)

**Abstract.** The main goal of the CARPATCLIM (Climate of the Carpathian Region) project is to construct the gridded climatological database for the region in a daily temporal resolution for the period 1961–2010 by using 0.1° spatial resolution. The solution of this requirement as well as one of the final products of the CARPATCLIM project is a Digital Climate Atlas which is designed as the main entry point for all the gridded data and maps generated during the project, together with metadata for all data sets (original data as well as data created during the project). With respect to the INSPIRE (Infrastructure for Spatial Information in the European Community) directive, the Digital Climate Atlas is developed as a rich Web GIS (Geographic Information System) application based on modern Web standards offering all necessary tools for climate data visualization and extraction. Another important product of the CARPATCLIM project is the Metadata Catalog which is designed as a tool for searching of climate metadata by various parameters (i.e. period, variable, region etc.).

## 1 Introduction

In the past years scientific community shows a constant growing demand for online digital climate information. Systems (Geoportal ARSO, 2013; Czech Republic Climate Atlas, 2013) that provide these kind of services are commonly implemented as specialized GIS (Geographic Information Systems) applications. These systems are designed as online Digital Atlas Web applications serving as entry points for efficiently accessing and querying climatology data.

In this paper we describe the Digital Atlas Web GIS application as a part of the CARPATCLIM (CARPATCLIM, 2013) project. The Carpathian Region has gained a lot of focus lately considering that the climate change is expected to result in important changes in the Carpathian climate, with consequence on natural ecosystems and human societies in the mountains and surrounding plains (Bartholy and Pongracz, 2007; Bartholy et al., 2008, 2009). The main aim of CARPATCLIM project is to improve the climate data information basis in the Carpathian Region for applied regional climatological studies such as drought monitoring (Oszlanyi et al., 2004). The service will improve the digital data basis at national meteorological services in the Carpathian Region, and will facilitate access to derived datasets by the wider scientific community. All this was achieved by improving the availability and accessibility of homogeneous and spatially representative time series of climatological data through data rescue, quality control and data homogenization, done on national level using common software MASH (Szentimrey, 1999, 2008). Additionally it was necessary to ensure Carpathian countries data harmonization with special emphasis on across-country harmonization and production of gridded climatologies per country, implemented by common software MISH (Szentimrey and Bihari, 2007, 2011). A direct product of this project is a gridded climatological database for the Carpathian Region (between 50° N and 44° N latitude and 17° E to 27° E longitudes, with a 0.1° spatial resolution) in a daily temporal resolution for the period 1961–2010.

Considering the long period of time and the need to organize and visualize data for various climate variables and indicators it was necessary to design an efficient infrastructure and tools for proficient management of vast amount of climate data. All these requirements are solved by designing a Web based Digital Climate Atlas application which will act as the main entry point for all gridded data and maps

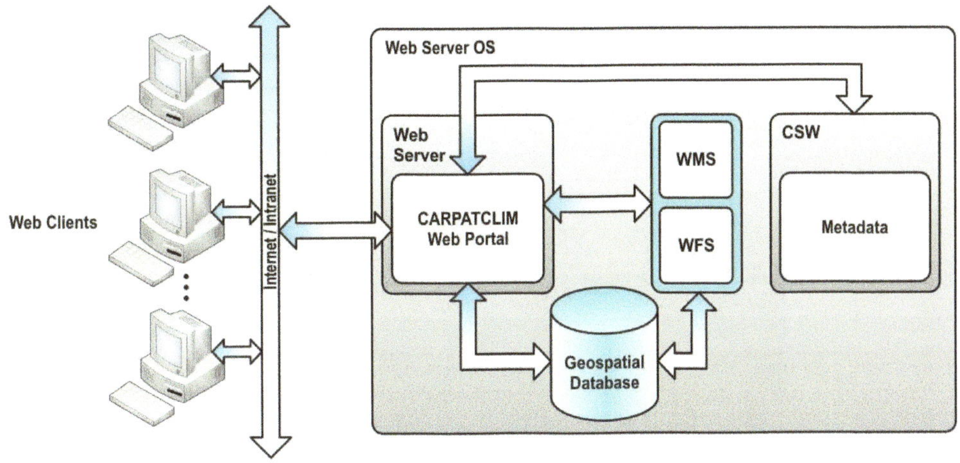


**Figure 1.** Digital Atlas Client-Server architecture.

generated during the CARPATCLIM project as well as metadata for all climate variables and indicators. With respect to the INSPIRE directive, the Digital Climate Atlas is developed as a rich Web GIS application based on modern Web standards offering all necessary tools for visualization and extraction of gridded climate data. Additionally the main gain of using modern Web standards was the compatibility with all major web browsers thus making the Digital Climate Atlas usable and accessible from wide range of platforms and users.

Fourteen essential meteorological variables (minimum, maximum and mean air temperature, precipitation, wind speed and direction, sunshine duration, cloud cover, global radiation, relative humidity, surface vapor pressure, surface air pressure, snow depth and snow water equivalent) and variety of climate indices (Palfai Drought Index, Standardized Precipitation Index, Reconnaissance Drought Index, Palmer Drought Severity Index etc.) are presented in the form of maps with possibility of downloading the gridded data as ascii files. These files are accompanied with the interpretation sheets that explain concisely the characteristics and the derivation of the data.

In this paper we further discuss and describe the architecture of the Digital Climate Atlas of the Carpathian Region. In the second part we present the components of the client-server architecture of the digital atlas, in the third part we examine the features of the Web GIS client, in part four we give a brief description of the metadata catalog and finally Chapter 5 presents the conclusions and directions for future research and development.

## 2 Digital Atlas Client-Server architecture

The Digital Atlas Client-Server architecture was designed with the goal to enable easy integration with standard OGC (Open Geospatial Consortium) Web services. The Client-Server architecture is shown in Fig. 1 and the main components of this architecture are:

- *Web Map Server* – this component integrates both WMS (Web Map Services) and WFS (Web Feature Service). The WMS service is used for map rendering while the WFS services is used for storing data relevant to meteorological stations (position, id, owner country etc.).
- *Geospatial database* – this component relies on the PostgreSQL database with PostGIS geospatial extension and it is being used for storing of climatological grid data.
- *Web server* – this component hosts the Carpatclim Digital Atlas Web GIS application which features rich user interface and geospatial data management. For the purpose of this project IIS7 (Internet Information Services) Web server was used.
- *CSW (Catalog Service for the Web) server* – for the purpose of this project Geonetwork open source Java based framework was used.

## 3 Digital Atlas Web GIS client

In order to maximize the user experience and provide rich user interface without the need to set up additional plugins the Web GIS client is based on HTML5 technology which also has the benefit of gaining compatibility with all major Web browsers.

The Digital Atlas Web GIS is a Web application which enables users to visualize and query geospatial data (Fig. 2). It is important to emphasize that this Web GIS client belongs to a group of medium thick clients which means that it combines advantages of a rich user interface with centralized data control. This is accomplished through the modular application architecture which is shown on Fig. 3. The Digital

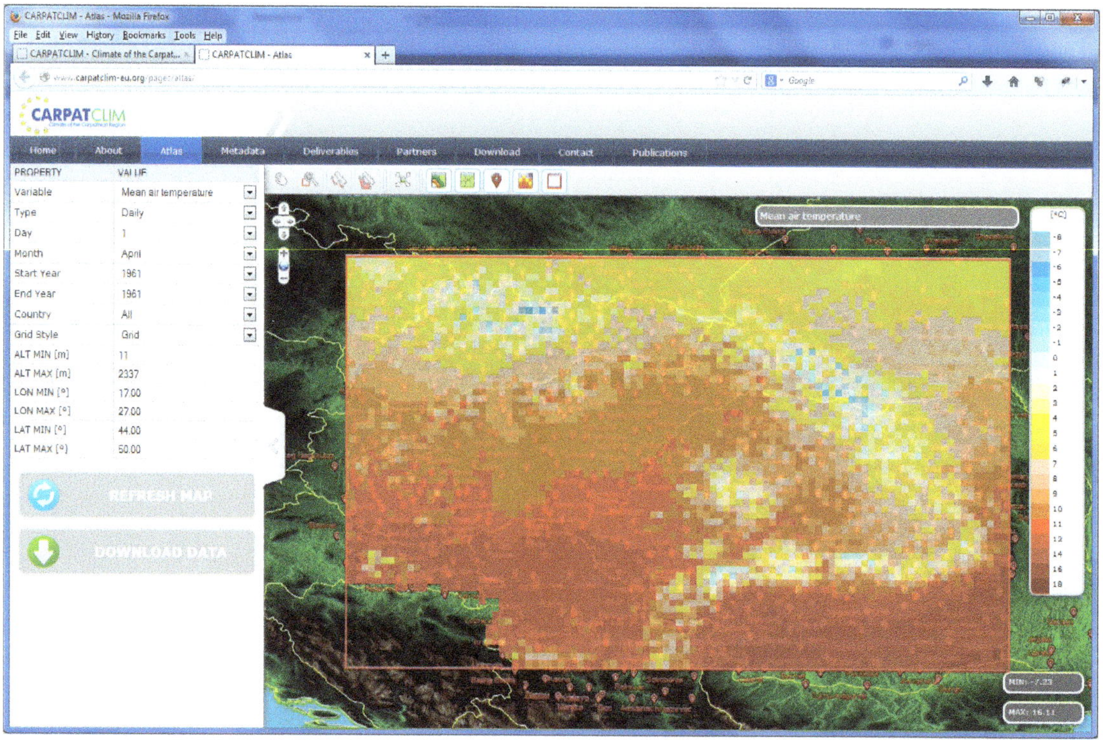


**Figure 2.** Digital Atlas Web GIS client.

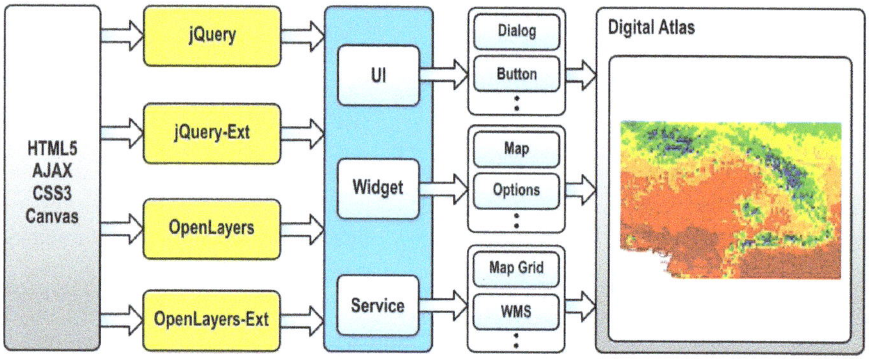


**Figure 3.** Digital Atlas Web application architecture.

Atlas Web GIS client relies on the GinisWeb AJAX framework (Bogdanović et al., 2011) which combines best qualities from several Web AJAX GIS libraries into a single high level application programming interface (API) framework.

The libraries which are included within the framework are:

- *OpenLayers* – is a framework for Web GIS application development and it is maintained by the Open Source Geospatial Foundation. This framework is used for easy prototyping of interactive Web GIS applications. Additionally it enables adding and displaying layers of various standard raster data sources (WMS, Google Maps. Bing Maps. Open Street Maps etc.) as well as vector layers based on SVG (Scalable Vector Graphics).
- *OpenLayersExt* – is a collection of custom OpenLayers extensions which, among other things, add support for Climate Grid layers (extended WMS layers with additional parameters for time periods, altitude threshold etc.)
- *jQuery* – is a library primarily designed for Document Object Model (DOM) manipulation and it is compatible with most current Web browsers. This library is an essential tool for building interactive HTML5 application therefore it is a very important part of the Digital Atlas core.

- *jQueryExt* – is a collection of jQuery extensions. These extensions include various third party addons which are not included in the official jQuery build (date/time controls, charting tools etc.).

Climate data related GIS functionalities include the following features:

- *Time span selection* – provides rendering of selected Daily/Monthly/Yearly time spans. For example the user can choose to render Tmax (Maximum temperature) Monthly grid for the time span from 1961 to 1971.
- *Per country filtering* – provides rendering of climate grid data only for selected countries,
- *Altitude filtering* – provides rendering of climate grid data only for regions which are above minimum altitude and below maximal altitude,
- *Data download* – provides easy download of grid data for selected parameters. The user downloads a compressed archive which contains three parts:
  - *Grid data values in ESRI GRID format* – the text format of the data enables easy reading and it is supported by wide range of GIS applications.
  - *Visualized grid data image in JPEG/PNG format* – in that way the image with rendered data can be uploaded to a web site.
  - *Interpretation sheet in PDF format* – this document contains all relevant information (origin, parameters etc.) which describes the downloaded grid data.

## 4 Metadata catalogue

An important step in creating the Climate Atlas of the Carpathian Region is building the metadata catalogue that describes the underlying datasets used in this project. Depending on different layers of data, four metadata groups were created:

- Metadata of the original climate observations listed in the data inventory, per country.
- Metadata of the climate datasets after data rescue, quality check and data homogenization, per country. This includes all relevant information to trace back the processing executed in order to arrive at the current datasets.
- Metadata of national gridded datasets.
- Metadata of all datasets and products of the Climate Atlas of the Carpathian Region.

The metadata catalog relies on the GeoNetwork CSW engine. This solution is chosen considering that Geonetwork is opensource and complies with various standards (e.g. ISO19115/ISO19119/ISO19110 following ISO19139, FGDC and Dublin Core). The above mentioned metadata was generated only once (encapsulating the ISO19139/19119 standard) from existing climate variable data contained in the PostgreSQL database. The generated metadata was then imported into the GeoNetwork database by using GeoNetwork embedded xml services.

## 5 Conclusions

The Digital Climate Atlas and the Metadata Catalog make a powerful online utility which offers a free, fast and easy access to Carpath Climate data. The Digital Atlas is a WebGIS based solution which relies on the HTML5 technology. As a result the Digital Atlas has rich graphic capabilities and runs on all modern Web browsers without the need for any kind of set up. The user interface of the Digital Atlas WebGIS application is designed to be highly intuitive. All necessary parameters (variable name, time frequency, time period etc.) for grid visualization can be customized on a standard Web form. The Digital Atlas is publically available thus usable not only to experts in the field of climatology, agriculture, water management, regional climate modeling etc., but to common users as well.

**Acknowledgements.** Authors thank the European Commission, Joint Research Centre (JRC), Institute for Environment and Sustainability, Ispra, Italy for Contract Notice OJEU 2010/S 110-166082 dated 9 June 2010. The authors would also like to take the opportunity to thank the following institutions and its members who participated in the CARPATCLIM project:

- Central Institute for Meteorology and Geodynamics, Austria: Ingeborg Auer, Johann Hiebl
- Meteorological and Hydrological Service of Croatia: Janja Milković
- Czech Hydrometeorological Institute: Pavel Zahradníček, Petr Štěpánek, Radim Tolasz
- Hungarian Meteorological Service (leading organisation): Tamás Szentimrey, Zita Bihari, Mónika Lakatos, Tamás Kovács, Ákos Németh, Sándor Szalai
- Institute of Meteorology and Water Management – National Research Institute, Poland: Piotr Kilar, Robert Pyrc, Danuta Limanowka
- National Institute for Research and Development in Environmental Protection of Romania: Sorin Cheval, Monica Matei
- Republic Hydrometeorological Service of Serbia: Predrag Petrović, Tatjana Savić, Aleksandra Kržič
- Slovak Hydrometeorological Institute: Peter Kajaba, Gabriela Ivanakova, Oliver Bochnicek, Pavol Nejedlik, Pavel Šastný
- Ukrainian Research Hydrometeorological Institute: Oleg Skrynyk, Yurii Nabyvanets, Natalia Gnatiuk
- Joint Research Centre: Tiberiu Antofie, Jonathan Spinoni, Jürgen Vogt

Edited by: I. Auer

## References

Bartholy, J. and Pongracz, R.: Regional analysis of extreme temperature and precipitation indices for the Carphatian Basin from 1946 to 2001, Global Planet. Change, 57, 83–95, 2007.

Bartholy, J., Pongracz, R., Gelybo, G., and Szabo, P.: Analysis of expected climate change in the Carphatian Basin using the PRUDENCE results, Idojaras, 112, 249–246, 2008.

Bartholy, J., Pongracz, R., Torma, C., Pieczka, I., Kardos, P., and Hunyady, A.: Analysis of regional climate change modeling experiments for the Carpathian Basin, International Journal of Global Warming, 1, 238–252, 2009.

Bogdanović, M., Antolović, I., Stoimenov, L., Rančić, D., and Pejanović, M.: GinisWeb – A framework for the development of web gis applications in the field of security and defense, OTEH 2011, Belgrade, Serbia, CD, 6–7 October, 2011.

CARPATCLIM – Climate of the Carpathian region: http://www.carpatclim-eu.org/, last access: 17 April 2013.

Czech Republic Climate Atlas: http://www.chmi.eu/meteo/ok/atlas/en/menu.html, last access: 17 April 2013.

Geoportal ARSO: http://gis.arso.gov.si, last access: 17 April 2013.

Oszlanyi, J., Grodyinska, K., Badea, O., and Shparyk, Y.: Nature conservation in Central and Eastern Europe with special emphasis on the Carpathian Mountains, Environ. Pollut., 130, 127–134, 2004.

Szentimrey, T.: Multiple Analysis of Series for Homogenization (MASH), Proceedings of the Second Seminar for Homogenization of Surface Climatological Data, Budapest, Hungary; WMO, WCDMP-No. 41, 27–46, 1999.

Szentimrey, T.: Development of MASH homogenization procedure for daily data, Proceedings of the Fifth Seminar for Homogenization and Quality Control in Climatological Databases, Budapest, Hungary, 2006; WCDMP-No. 68, WMO-TD NO. 1434, 2008, 116–125, 2008.

Szentimrey, T. and Bihari, Z.: Mathematical background of the spatial interpolation methods and the software MISH (Meteorological Interpolation based on Surface Homogenized Data Basis), Proceedings from the Conference on Spatial Interpolation in Climatology and Meteorology, Budapest, Hungary, COST Action 719, COST Office, 17–27, 2007.

Szentimrey, T. and Bihari, Z.: MISH (Meteorological Interpolation based on Surface Homogenized Data Basis), COST Action 719 Final Report, in: The use of GIS in climatology and meteorology, edited by: Tveito, O. E., Wegehenkel, M., van der Wel, F., and Dobesch, H., 54–56, 2011.

# Modelling static 3-D spatial background error covariances – the effect of vertical and horizontal transform order

**M. A. Wlasak and M. J. P. Cullen**

Met Office, FitzRoy Road, Exeter, Devon, EX1 3PB, UK

*Correspondence to:* M. A. Wlasak (marek.wlasak@metoffice.gov.uk)

**Abstract.** A major difference in the formulation of the univariate part of static background error covariance models for use in global operational 4DVAR arises from the order in which the horizontal and vertical transforms are applied. This is because the atmosphere is non-separable with large horizontal scales generally tied to large vertical scales and small horizontal scales tied to small vertical scales. Also horizontal length scales increase dramatically as one enters the stratosphere. A study is presented which evaluates the strengths and weaknesses of each approach with the Met Office Unified Model.

It is shown that if the vertical transform is applied as a function of horizontal wavenumber then the horizontal globally-averaged variance and the homogenous, isotropic length scale on each model level for each control variable of the training data is preserved by the covariance model. In addition the wind variance and associated length scales are preserved as the scheme preserves the variances and length scales of horizontal derivatives. If the vertical transform is applied in physical space, it is possible to make it a function of latitude at the cost of not preserving the variances and length scales of the horizontal derivatives.

Summer and winter global 4DVAR trials have been run with both background error covariance models. A clear benefit is seen in the fit to observations when the vertical transform is in spectral space and is a function of total horizontal wavenumber.

## 1 Introduction

For the last couple of decades variational data assimilation schemes have been the preferred method of determining the appropriate initial conditions for global atmospheric forecasts. The schemes need to represent approximately the error statistics of the Bayesian prior which is normally the error associated with the previous forecast. A typical forecast model can have $10^8$ or more degrees of freedom. To get a reasonably accurate full-rank background error covariance without making any underlying assumptions as an approximation to the prior, one would need to use an unworkably large number of samples. In this study the input data is taken from ensembles of forecasts and forecast differences are considered to be proxies for forecast error. Even if it were possible to get enough samples the data storage requirements of this covariance matrix would be also be impractical. To define any full-rank covariance matrix of degree $n$ one needs $n(n+1)/2$ numbers.

The majority of these schemes include a background error covariance model - a mathematical model that uses physical, dynamical and statistical insight to reduce the amount of information needed to determine the background error covariance. One form of covariance model that is getting increasingly popular approximates the background error covariance matrix by a low rank sample covariance computed from an ensemble and then removes (localises) spurious correlations caused by sampling error. Such models have variances and correlations that change for every data assimilation cycle and are outside the scope of this paper. This paper considers full-rank covariance models. Such models are designed to be full-rank by using statistics that are averaged in space and/or time by resorting to assumptions like homogeneity, isotropy and stationarity.

Most covariance models have a multivariate step, normally called the parameter transform, which typically involves a reduction of the number of atmospheric variables considered, to a smaller set called control variables that are assumed to be uncorrelated with one another. The standard control variables used at the Met Office are streamfunction $\psi$, velocity potential $\chi$, ageostrophic pressure Ap and humidity $\mu$. For each uncorrelated "control variable" a spatial decomposition is done to precondition the problem and to separate structures that hold large background error variances from those that do not. This paper focuses on this spatial decomposition.

The three dimensional aspects of forecast errors vary with horizontal and vertical position and cannot be represented completely without further approximation. The standard approximation that is used is to assume that the errors are homogenous and isotropic on either some horizontal model level in some vertical co-ordinate or some empirical vertical mode basis. This paper looks at the differences between these two approaches within the Met Office's global data assimilation system.

The rest of this paper is organised as follows. Section 2 describes the current and proposed static background error covariance model used in global 4DVAR at the Met Office and lists their corresponding strengths and weaknesses. Section 3 gives details of the forecast trials and shows the effect of the proposed static background error covariance model. Section 4 gives the conclusions.

## 2 A comparison between current and proposed static background error covariance model

The two static covariance models can be viewed in terms of their implied background error covariances $\mathbf{B}_{\mathrm{s1}}$ $\mathbf{B}_{\mathrm{s2}}$ and their respective parameter, horizontal and vertical transforms:

$$\mathbf{B}_{\mathrm{s1}} = \mathbf{U}_{\mathrm{p}}\mathbf{U}_{\mathrm{v}}\mathbf{U}_{\mathrm{h}}\mathbf{U}_{\mathrm{h}}^{T}\mathbf{U}_{\mathrm{v}}^{T}\mathbf{U}_{\mathrm{p}}^{T} \quad (1)$$

$$\mathbf{B}_{\mathrm{s2}} = \mathbf{U}_{\mathrm{p}}\mathbf{U}_{\mathrm{h}}\mathbf{U}_{\mathrm{vs}}\mathbf{U}_{\mathrm{vs}}^{T}\mathbf{U}_{\mathrm{h}}^{T}\mathbf{U}_{\mathrm{p}}^{T} \quad (2)$$

where $\mathbf{U}_{\mathrm{p}}$ is the parameter transform from control variables to model variables, $\mathbf{U}_{\mathrm{h}}$ is a spectral horizontal transform from spectral space to grid point space, $\mathbf{U}_{\mathrm{v}}$ is a vertical transform applied in grid-point space and $\mathbf{U}_{\mathrm{vs}}$ is a vertical transform applied in spectral space.

The covariance model associated with $\mathbf{B}_{\mathrm{s1}}$ is the current static background error covariance model as described in Lorenc et. al. (2000). Since the Unified Model (UM) is a grid-point based forecasting model, it is natural to consider applying the vertical transform in grid-point space. The vertical transform from model levels to empirical vertical modes is constructed by calculating the eigenvectors of a horizontally-area-averaged vertical covariance matrix which has been weighted in vertical. This vertical scaling for $\psi$ is a pressure weighting that takes account of the mass of each layer. The key feature involves a vertical mode basis defined by the eigenvectors that are the same across the whole globe, but whose variance in vertical mode space varies as a function of latitude. A latitude-dependent vertical regression is also applied between the full pressure increment and a geostrophic pressure increment derived from a variant of the linear balance relation. This regression makes $\psi$ and Ap uncorrelated with each other.

One of the advantages of the operational covariance model is that it can to a degree take into account the latitude dependence of the covariance statistics. However, since a single vertical covariance matrix is used for each control variable, the scheme is unable to preserve the variances of horizontal derivatives that are needed to get the wind variances and associated length scales right as a function of model level. Instead of using the actual calibration statistics for $\psi$ and $\chi$, the rotational kinetic energy (RKE) and divergent kinetic energy (DKE) are used. These substitutes have flatter power spectra and as a result each has a single global vertical covariance that is better at representing the vertical structure across the full range of horizontal wavenumbers. This choice also shortens the length scales of $\psi$ and $\chi$. Ingleby (2001) shows that this makes the implied pressure and temperature variances smaller than what is observed in the training data.

Strategically it makes sense to have a covariance model that preserves faithfully some key properties of the training data. It would allow a shift in emphasis towards getting the training data right instead of concentrating on tuning the background error covariance model as has been the case in the past.

To this end we have developed an alternate simpler covariance model that is represented by $\mathbf{B}_{\mathrm{s2}}$. This model has features that are similar to the old ECMWF covariance system (Courtier et al., 1998). While the old ECMWF system assumed that the correlations of the control variables were homogenous and isotropic on each model level, $\mathbf{B}_{\mathrm{s2}}$ assumes that the covariances are homogenous and isotropic. The main difference between $\mathbf{B}_{\mathrm{s2}}$ and the current operational Met Office covariance model is as follows. Instead of storing a square root of the vertical covariance matrix and a power spectrum for each vertical mode, the order of the horizontal and vertical transforms is reversed and the square root of each vertical covariance matrix for each total horizontal wavenumber is stored. Also a global vertical regression is stored between the full pressure increment and the balanced pressure increment in order to get Ap to be uncorrelated with $\psi$. By not allowing latitudinal variation in the vertical regression, the horizontal-average variance estimate of the vertical derivative of the training data is preserved. This is needed as the hydrostatic equation is used to calculate temperatures from the vertical derivative of pressure. By preserving the horizontal-average variance estimate of the vertical derivatives, the horizontal-average estimate of the temperature variance is preserved on each model level.

The advantage of this approach is that the horizontally-averaged variance estimate for each model level is preserved for the observed winds, temperatures and pressures. Also it preserves the variance for each horizontal wavenumber and thus preserves homogeneous and isotropic length scale information of the training data for each model level. By preserving the homogeneous and isotropic length scales of the derivatives of the control variables on each model level, the homogeneous and isotropic length scales of the training data for the RKE and DKE will also be preserved as well as the horizontal-averaged variance estimate of the wind.

The control variable $\psi$ determines the balanced pressures and temperatures. By preserving the length scales of $\psi$, we preserve also the balanced pressure and temperatures.

The mathematical mechanism can be seen most easily if we consider $\psi$, $\chi$ in spectral space. The streamfunction can be written in terms of its spectral coefficients $\psi_l^m$ as

$$\psi = \sum_{l=0}^{L} \sum_{m=-l}^{l} \psi_l^m Y_l^m(\lambda, \sin\varphi), \tag{3}$$

where $L$ is the order of the triangular truncation, $l$ and $m$ are the total and meridional wavenumbers respectively, and $Y_l^m(\lambda, \sin\varphi)$ is the spherical harmonic function for latitude $\varphi$ and longitude $\lambda$.

One property of the gradient of a spherical harmonic is

$$\frac{1}{4\Pi} \int_{-1}^{1} \int_{0}^{2\Pi} \nabla Y_s^t[\lambda, \sin\varphi] \cdot \nabla Y_l^{m*}[\lambda, \sin\varphi] \partial\lambda \partial \sin\varphi$$
$$= l(l+1)\delta_s^l \delta_t^m, \tag{4}$$

where $*$ denotes the adjoint operator, $s$ and $l$ are total wavenumbers, $t$ and $m$ are meridional wavenumbers, $\nabla$ is the gradient operator, $\delta$ is the Dirac delta function and $\cdot$ is the scalar product operator between two vectors.

We first consider the variance of the divergent wind, which is given by the variance of the gradient of the velocity potential. The divergent wind can be calculated by the sum of the product of the spectral coefficients of the velocity potential and the derivative of the spherical harmonics. The above Eq. (4) tells us that the gradient of the spherical harmonics has an orthogonal basis for each total wavenumber. So provided the total variance of the spectral coefficients of velocity potential is kept for each total wavenumber then we will be able to conserve the full variance of the divergent wind.

A similar argument holds for streamfunction spectral coefficients and the variance of the rotational wind, noting that $(\mathbf{k} \times \nabla Y) \cdot (\mathbf{k} \times \nabla Y) = \nabla Y \cdot \nabla Y$, where $\mathbf{k}$ is the unit vector in the vertical and $\times$ is the vector product of two vectors.

This argument is not limited to variances but extends to cross-covariances between different model levels as we do not need the product of the spectral coeffficients to be positive.

**Table 1.** Comparison of covariance models. Strengths denoted by (+), weaknesses denoted by (−).

| Current static operational covariance model | Proposed static covariance model |
|---|---|
| Latitude dependence built into vertical and parameter transform (+) | Latitude dependence built into parameter transform only (−) |
| Power spectra not conserved for model variables (−) | Global variances and power spectra conserved for model variables on model levels (+) |

If the streamfunction and velocity potential power spectra do not vary much with model level then covariance model $\mathbf{B}_{s1}$ would sufffice. This was indeed the case when the atmospheric model was predominantly limited to the troposphere. However as the atmospheric model now includes the stratosphere and the mesosphere, the characterisation of the power spectra varies with height with successively longer length scales seen in the upper regions. Also there are huge differences in the vertical correlations across horizontal wavenumbers, with much broader vertical correlations for $\psi$ at low wavenumbers.

A weakness of this approach is that the covariance statistics can only take account of latitudinal dependency through the parameter transform. Also both methods suffer from the degree to which homogeneity and isotropy assumptions are valid.

A summary of the strengths and weaknesses of the two methods is given in Table 1.

## 3 Results

### 3.1 Trial results

Two forecast trials were run to compare the strengths and weaknesses of the two covariance model formulations. 4DVAR non-hybrid trials were used for this comparison so that the behaviour of the two models could be seen clearly and not obscured by the effects of the errors of the day from the ensemble. The UM (version 7.9) was used at N320 resolution. The data assimilation VAR system (version 29.2.1) was run in a two step process with the first step running at N108 resolution, giving Hessian eigenvectors for a subsequent N216 resolution minimisation.

The background error statistics were generated using the same set of training data – 300 samples of forecast errors from a spun up early version of the ECMWF ensemble of analyses system that was propagated for 6 h by the UM. The 6 h run ensures compatibility with the UM while not changing the statistics much. The period from which the ensemble of analyses was extracted was 1–17 October 2006.

The first trial was a summer trial from 28 June to 31 July 2012. The overall measure used at the Met Office for verification is the NWP index for fit to observations and

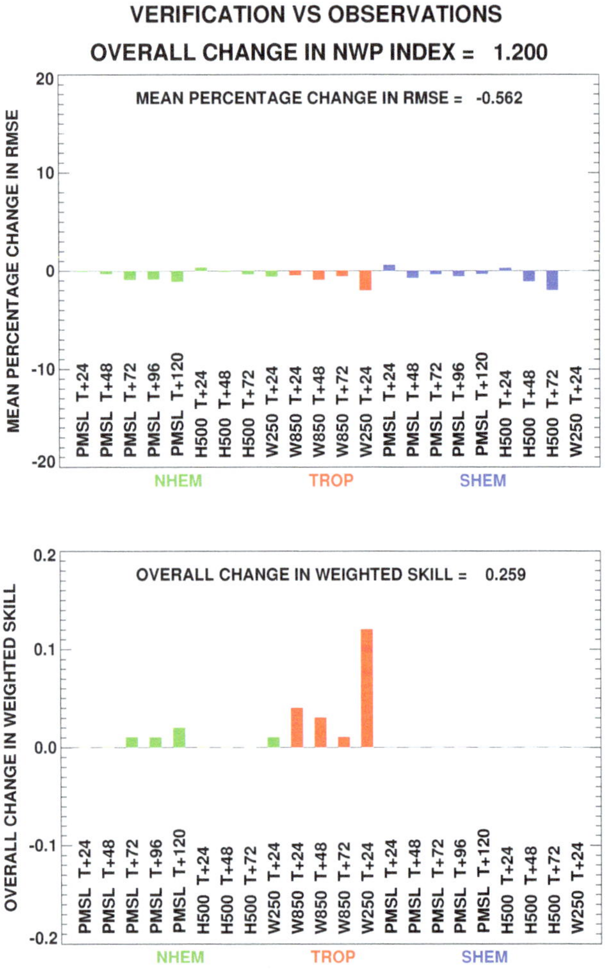


**Figure 1.** Verification results for winter trial 2012: new covariance model versus current covariance model. Upper plot shows reduction in RMSE. Lower plot shows how the change in RMSE affects NWP observation index.

own analyses. Verification against analyses is only relevant when making small changes to the nature of the analysis; this verification measure is not meaningful when changing the covariance model in a profound way. For this reason we stick to presenting only the verification scores for fit to observations.

The NWP index is a basket of scores of different forecast parameters. Detail of the NWP index is found in the Appendix of paper Rawlins et al. (2007). There was a marked improvement in skill in fit to observations of +0.7. This value far exceeds what is typically considered to be the standard threshold of +0.3 which indicates a statistically significant positive result at the 95 % confidence level (P. Jermey, personal communication, 2014). The benefit was seen mainly in the southern hemisphere in the surface pressure, 500 hPa geopotential height and 250 hPa winds. Also there was a very large improvement in the fit to 250 hPa winds in the tropics.

In addition a winter trial was run from 18th January 2012 to 28 Febuary 2012. The results for this trial were even more favourable with an improvement in the NWP index fit to observations of +1.2. As shown in Fig. 1, the benefit was seen for most measures in north, south and tropical regions.

### 3.2 Analysis increments

The biggest difference seen between the analysis increments from the two systems is in the horizontal scale and variance of the pressure and potential temperature. The new covariance model has doubled the variance of the pressure and potential temperature with much more variance at lower horizontal wave numbers due to preserving the characteristics of the training data. The larger length scale also manifests itself in the first Hessian eigenvector.

## 4 Conclusions

The paper shows that a simple covariance model designed to preserve global length scales and variances for each model level of each model variable can beat a long standing operational covariance model. Getting the global length scales and variances right on each model level matters.

This study has limited itself to two static covariance models. When including a non-static component to the covariance model as is described in Clayton et al. (2013) we see a neutral signal. Further work on tuning the non-static model is required.

It is noted that homogeneity and isotropy are quite harsh approximations. It would be good to identify an intermediate vertical grid that varies with latitude such that for each transformed model level the assumption of homogeneity of the variance holds more strongly.

**Acknowledgements.** The authors like to thank N. Bowler for helpful suggestions as well as R. Bannister and the two anonymous referees for their helpful reviewing of this paper.

Edited by: S. Dance

## References

Clayton, A. M., Lorenc, A. C., and Barker, D. M.: Operational implementation of a hybrid ensemble/4D-Var global data assimilation system at the Met Office, Q. J. Roy. Meteor. Soc., 139, 1445–1461, doi:10.1002/qj.2054, 2013.

Courtier, P., Andersson, E., Heckley, W., Vasiljevic, D., Hamrud, M., Hollingsworth, A., Rabier, F., Fisher, M., and Pailleux, J.: The ECMWF implementation of three-dimensional variational assimilation (3D-Var). I: Formulation, Q. J. Roy. Meteor. Soc., 124, 1783–1807, doi:10.1002/qj.49712455002, 1998.

Ingleby, N. B.: The statistical structure of forecast errors and its representation in The Met. Office Global 3-D Variational Data

Assimilation Scheme. Q.J.R. Meteorol. Soc., 127: 209–231. doi:10.1002/qj.49712757112, 2001.

Lorenc, A. C., Ballard, S. P., Bell, R. S., Ingleby, N. B., Andrews, P. L. F., Barker, D. M., Bray, J. R., Clayton, A. M., Dalby, T., Li, D., Payne, T. J. and Saunders, F. W.: The Met. Office global three-dimensional variational data assimilation scheme. Q.J.R. Meteorol. Soc., 126: 2991–3012. doi:10.1002/qj.49712657002, 2000.

Rawlins, F., Ballard, S. P., Bovis, K. J., Clayton, A. M., Li, D., Inverarity, G. W., Lorenc, A. C., and Payne, T. J.: The Met Ofiice Global 4-Dimensional Data Assimilation System, Q. J. Roy. Meteor. Soc., 133, 347–362,doi:10.1002/qj.32, 2007.

# 32

# WegenerNet climate station network region Feldbach, Austria: network structure, processing system, and example results

**T. Kabas, A. Leuprecht, C. Bichler, and G. Kirchengast**

Wegener Center for Climate and Global Change (WEGC), and Institute for Geophysics, Astrophysics and Meteorology/Institute of Physics (IGAM/IP), University of Graz, Graz, Austria

**Abstract.** The WegenerNet climate station network is a pioneering weather and climate observation experiment at very high resolution in southeastern Austria. The network comprises 151 meteorological stations within a limited area of approximately 20 km × 15 km centered near the town of Feldbach. Measurements include the parameters air temperature, relative humidity, precipitation amount, and others at selected stations (e.g. wind and soil parameters). The temporal sampling is 5 min except 30 min sampling of soil measurements. All data pass a Quality Control System and the provided data products include station data (~1.4 km × 1.4 km grid) and gridded data (1 km × 1 km and 0.01° × 0.01° grids) on various temporal scales (from 5 min to annual). For application purposes all data are available in near real time (data latency less than 1–2 h in standard operation) via the WegenerNet data portal (www.wegenernet.org).

## 1 Introduction

Analyses of climate variability on regional or finer scales are a great challenge for the climate change and impact research community. Investigations of local characteristics and further questions in microclimatology are limited by the availability of high-resolution temporal and spatial observations. Operationally, such in-situ observations are yet not available according to the authors' knowledge. Only temporary field campaigns are conducted, partially backed by further measuring methods. In the Convective and Orographically Induced Precipitation Study, for instance, various methods were combined including radar measurements and highly resolved station observations in a limited area in southwest Germany (Wulfmeyer et al., 2008). Such comparative studies demonstrate the technical possibilities to fullfill the users' demand for high resolution data besides the usage of in-situ measurements (also compare www.microradarnet.net).

However, in order to obtain accurate information these measurements have to be calibrated on and validated against dense observational station data. For this purpose, and lot of other research applications as well, long-term station observations are required. The current paper presents the very dense network WegenerNet located in southeast Austria which is intended to provide such long-term observational records. Since the beginning of 2007, regular measurements of 151 meteorological stations and derived gridded data are provided. In this paper, a brief overview of the study region and the network structure is given in Sect. 2. Section 3 introduces the data processing followed by the provided weather and climate data products (Sect. 4). The resulting high-resolution maps of two exemplary events are illustrated in Sect. 5 and finally, concluding remarks are presented in Sect. 6.

## 2 Network structure and measured parameters

The WegenerNet is located in southeastern Styria, Austria, centered near the town of Feldbach. Results of previous studies underline the sensitivity of southeast Austria to changes in climate conditions (e.g. Auer et al., 2001; Prettenthaler et al., 2007; Heinrich, 2008; Kabas et al., 2011) and a limited area of approximately 20 km × 15 km was selected as study area for this observational experiment. Within the complex landscape of Austria, the enclosed territory is part of the Alpine foreland in which lowlands and hilly landscapes with comparably low altitudes represent the prevailing landforms (257–520 m station altitudes). In the study region,

**Table 1.** Different types of WegenerNet stations and the measured parameters completed by their temporal resolution. Heated rain gauges are marked by *.

| Station type | Number of stations | Parameters | Temporal resolution |
|---|---|---|---|
| Base station | 127 | Air temperature, Relative humidity, Precipitation amount | 5 min |
| Special base station | 11 | Air temperature, Relative humidity, Precipitation amount | 5 min |
| | | Soil temperature, pF-value | 30 min |
| | 1 | Air temperature, Relative humidity | 5 min |
| Primary station | 11 | Air temperature, Relative humidity, Precipitation amount* | 5 min |
| | | Average wind speed and direction | 5 min |
| | | Peak gust and its direction | 5 min |
| Reference station | 1 | Air temperature, Relative humidity, Precipitation amount* | 5 min |
| | | Average wind speed and direction | 5 min |
| | | Peak gust and its direction | 5 min |
| | | Soil temperature, pF-value | 30 min |
| | | Air pressure, Net radiation | 5 min |

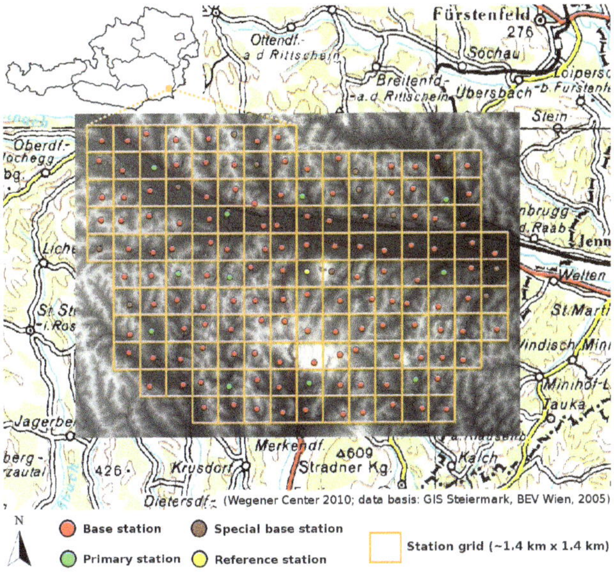


**Figure 1.** The WegenerNet study region (orographic map) in Austria and the station locations in the station grid.

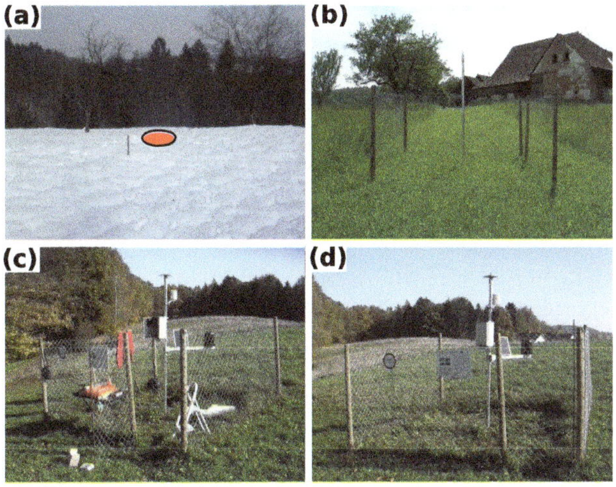


**Figure 2.** Setup of the WegenerNet climate station No. 6 during the year 2006.

which is characterized by the river Raab valley and the surrounding hilly landscape, 151 meteorological stations are arranged based on an approximately 1.4 km × 1.4 km grid (see Fig. 1). The station locations are determined in consideration of representative meteorological conditions as well as technical needs of the installed sensors depending on the station type and the measured parameters.

As shown in Table 1, the WegenerNet stations are separated in four different types. Base stations measure the main parameters air temperature, relative humidity and precipitation amount. Selected further stations include measurements of soil and wind parameters at suitable sites within the network. At the reference station, the main parameters as well as soil and wind parameters are completed by measurements of air pressure and net radiation. The temporal sampling is 5 min with the exception of soil parameters sampled every 30 min (as they react more inert at short time scales).

The network was mainly set up during the year 2006 supported by partners in the region and students of the University of Graz. Figure 2 illustrates the main steps of construction: (a) location selection, (b) infrastructure setup, (c) sensor installation and (d) test mode. Further information on the network implementation is given in Kirchengast et al.

**Table 2.** Quality layers of the Quality Control System.

| Layer No. | Quality layer | Quality flag |
|---|---|---|
| 0 | operational check | 1 |
| 1 | availability check | 2 |
| 2 | sensor check | 4 |
| 3 | climatological check | 8 |
| 4 | time variability check | 16 |
| 5 | intra-station consistency check | 32 |
| 6 | inter-station consistency check | 64 |
| 7 | external reference check | 128 |

**Table 3.** Maximum period of linearly interpolated data.

| Parameters | Maximum period |
|---|---|
| Air temperature, Relative humidity | 30 min |
| Precipitation amount | 15 min |
| Average wind speed and direction | 15 min |
| Peak gust and its direction | 15 min |
| Soil temperature, pF-value | 180 min |
| Air pressure | 60 min |
| Net raditation | 15 min |

(2008). Data started to be delivered in the fall of 2006 and regular measurements of 150 stations are provided since January 2007. Due to the implementation of additional sensors at the reference station, preparatory modifications of the technical infrastructure were needed and data are available since mid-October 2007.

## 3 Processing system

The collected data are integrated in an automatic processing system including data transfer, quality control, product generation, and presentation. Each station is equipped with an internet-attached data logger and the measurements are transmitted as binary files via GPRS to the WegenerNet server in 1 h intervals. Within this hourly schedule, all stations are split in 10 packages with a 5 min time lag in transfer starting every full hour. The incoming raw data files of measured parameters as well as several housekeeping values of the data logger (e.g. power supply level, internal temperature and humidity conditions) are stored in a relational database (PostgreSQL) in which important metadata of technical specifications and information on the station locations are recorded as well.

After the hourly cycle of data transfer, the raw data pass the Quality Control System (QCS) in which the measurements are checked for their technical and physical plausibility. The QCS consists of seven quality layers (see Table 2) holding various tests on the data. The quality control starts with specifications of the installed sensors to detect malfunctions right at the beginning. Next, the climatological plausibility is determined by climatic bounds on monthly scale which are derived from two long-term stations of the Austrian Central Institute for Meteorology and Geodynamics (ZAMG) located in the study region. The next layer indicates interfering signals and sensor errors (e.g. frozen measurement devices) by analyzing the temporal variability and constancy within parameter-specific periods. In the intra-stational consistency check, meteorological and physical relations of measured parameters at each station are compared. For example, the peak gust is needed to be higher than the average wind speed and higher amounts of precipitation require increased humidity conditions. The following quality layer of inter-stational consistency is based on the spatial density of measured main parameters. According to a candidate-reference method, neighboring stations are selected as references in consideration of defined requirements for each parameter (e.g. distance, topography, altitude difference). As a result, the small-scale variability could be included in more detail in the detection of errors like blocked rain gauges or manual water input by a person. In the final layer, WegenerNet data is compared with external references as used for the comparison of air pressure measurements at the reference station with the two ZAMG stations in the study area. In the course of the QCS processing, each data value gets marked by an appropriate quality flag. If some test fails in layer $i$ the $i$-th bit of the flag is set (see Table 2). Thus a quality flag 0 corresponds to valid data. Looking at data records within periods of active observations between 2007 and 2010, 93.9% of measurements are marked by a flag 0. Out of values with flags of higher order, only one fifth belongs to technical problems and the majority of values identified as erroneous is related to meteorological and/or climatological inconsistencies.

Subsequently to the quality control, only measurements of best quality (flag 0) are considered for the Data Product Generator (DPG) processing. In the DPG, the time series of all parameters are checked for gaps resulting in missing measurements or erroneous values detected by the QCS. For each parameter a maximum period of consecutively missing data is defined (see Table 3) and short periods are linearly interpolated. In case of longer gaps they get filled by a predefined missing value. Further processes of gap-filling and grid interpolation are realized for the main parameters and briefly described in the two following steps. First, data of neighboring stations defined in the QCS are used for inverse-distance-squared weighting interpolation (IDSWI) of still missing values if a sufficient number of neighbors is available. For air temperature, the interpolation is performed at the reference altitude of 300 m. The required lapse rate to project to this altitude is calculated by linear regression analysis from temperature data of stations and their altitudes for every 1 h time window. Next, gridded data are derived from neighboring sites by IDSWI. Regular grids are calculated in UTM

(1 km × 1 km grid) and longitude/latitude (0.01° × 0.01° grid) coordinates. For air temperature, a 10 m × 10 m digital elevation model is used to complement the gridded data at 300 m altitude by two orographic grids: one is computed for the mean altitude of each grid cell (T_zAvgTerr) and the other one for the altitude of the central point of the cell (T_zTerrain). After the grid interpolation, still missing station data of the main parameters are derived from the grid by IDSWI. Air temperature is interpolated at the reference altitude and projected to the station altitude afterwards. Out of missing input station data for the DPG processing from 2007 to 2010, 22.9% are filled by linear interpolation, 36.9% are interpolated by inverse-distance-squared weighting using data of neighboring stations and 25.8% are derived from the grid by IDSWI. The remaining 14.4 % cannot be interpolated which corresponds to more than 99% of overall data availability. Finally, the DPG uses the 5 min data and produces weather and climate data products for station and gridded data on various temporal scales (from half-hourly to annual). Both subsystems (QCS and DPG) are realized by the programming language Python.

## 4 Data products and availability

The provided weather and climate data products consist of station data and gridded data. Regular grids are realized for the main parameters in 1 km × 1 km (UTM) and 0.01° × 0.01° (longitude/latitude) resolution. The temporal resolution includes 5 min, half-hourly, hourly, daily, seasonal, and annual data. For application purposes the resulting data products are available via the bi-lingual (dt, en) WegenerNet data portal (www.wegenernet.org). Presently (December 2010), the main web interface is still online as a system in which MapServer is used to import spatial data by its database interface and to generate images out of static geographic formats. Furthermore, station data are visualized as time series by the scripting language PHP. Since February 2010, the visualization of gridded data products is a first step to a new data portal based on OpenLayers. In this GIS framework, all geographic information (e.g. OpenStreetMap) is displayed with MapServer. Maps of meteorological parameters are generated on the fly by a Python CGI script and transparently overlayed on optional base layers. In this way, station data and gridded data are visualized and furthermore prepared for download in common data formats (csv, NetCDF). In terms of availability, measured data and calculated data products are provided in near real time (data latency less than 1–2 h in standard operation).

## 5 Results on 1 km-scale resolution

The resulting WegenerNet data sets enable a detailed view on small-scale variability as briefly shown in the two following examples.

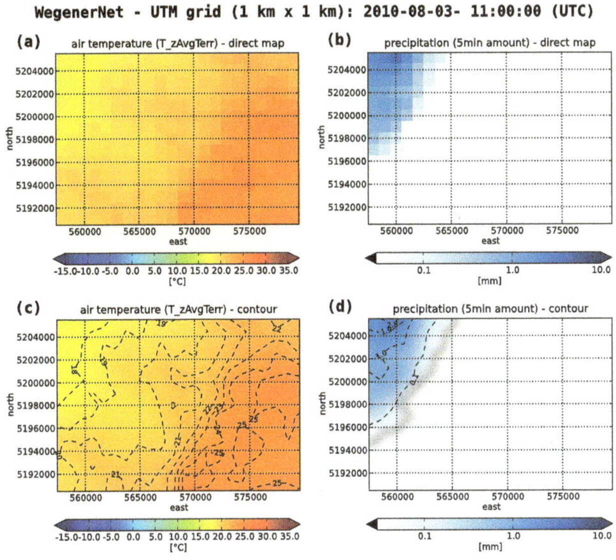


**Figure 3.** Temperature and precipitation conditions on August 3, 2010, in the WegenerNet region at 1 km resolution in UTM coordinates (dotted lines): direct map of grid cells in **(a)** and **(b)**, smoothed cell boundaries and inserted contours (dashed lines) in **(c)** and **(d)**. Precipitation amounts (5 min) are given on a logarithmic scale.

In Fig. 3 a frontal system reaches the study area from the Northwest. The prevailing temperature conditions (T_zAvgTerr) and precipitation amounts (5 min) are shown at an exemplary time step in two ways. The upper part (Fig. 3a and b) illustrates the direct map of the grid with clearly recognizable 1 km × 1 km grid cells. In Fig. 3c and d, the cell boundaries are smoothed and dashed contours are inserted. Looking at Fig. 3a, a strong temperature gradient occurs from the Northwest to the Southeast. In the forefront of areas with measureable precipitation in Fig. 3d, which ranges up to amounts of about 3 mm per 5 min, the cold air extends to the temperature difference of about 2 °C illustrated by the contours in Fig. 3c. Thus, the north-western part has already cooled down to about 18 °C, however, still warmer conditions of about 24 °C appear in the Southeast. Over the next hours, precipitation can be observed throughout the study area and lead to daily amounts up to 19 mm.

The second example demonstrates the temporal and spatial variability observed for precipitation. Figure 4 shows 5 min amounts at six timesteps of a summer precipitation event which passed the WegenerNet within one hour. As in the example before, precipitation is observed at the north-western climate stations first. In the next two timesteps (Fig. 4b and c), a second precipitation cell emerges close to the center of the study area and the precipitation amounts increase to more than 2.5 mm per 5 min. The two cells merge in Fig. 4d with up to 3.3 mm and finally leave the region southwards.

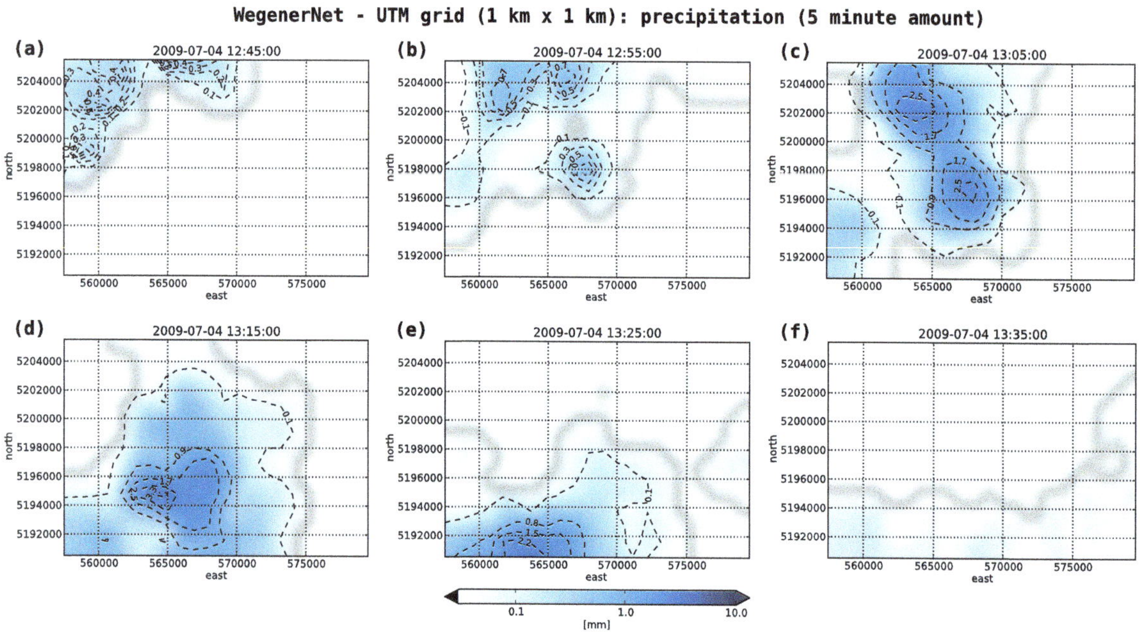


**Figure 4.** Temporal and spatial variability of precipitation on 4 July 2009, at 1 km resolution in UTM coordinates (dotted lines). Amounts with 5 min sampling are shown on a logarithmic scale at every second timestep within one hour.

## 6 Conclusions

The climate station network WegenerNet provides a new data set of very high resolution in the southeastern part of Austria. Local weather and climate conditions are observed by 151 meteorological stations with an average distance between stations of approximately 1.4 km. An automatic processing system was implemented including the data transfer in 1 h intervals and the quality control of measurements in seven quality layers. Furthermore, regular grids are computed for the main parameters air temperature, relative humidity and precipitation amount at 1 km and 0.01° resolution, respectively. Station data and gridded data are prepared on various temporal scales and available for data users in near real time at the WegenerNet data portal.

This study confirms the applicability of the WegenerNet as a long-term observation network for the climate and climate impact research community. A range of possible research applications include convective events, terrain climatology, extreme events and further studies on local scale variability. The high resolution in space and time allows the evaluation of remote sensing techniques (e.g. radar) and the validation of small-scale dynamics in a model and analysis system (e.g. Haiden et al., 2010). Furthermore, the findings can be a basis for weather- and climate-related decisions in the study region such as questions of water balance assessments, agricultural productivity and area development.

**Acknowledgements.** The climate station network WegenerNet was funded by the Austrian government, the state of Styria, the 27 municipalities in the study region, the City of Graz and the University of Graz. The authors also thank all our regional partners and further sponsors supporting the network setup and its operation. Furthermore, we thank the Austrian Central Institute for Meteorology and Geodynamics (ZAMG) for providing comparison data from their observational station network.

Edited by: I. Auer

## References

Auer, I., Böhm, R., and Schöner, W.: Austrian long-term climate 1767–2000, multiple instrumental climate time series from Central Europe, Österreichische Beiträge zu Meteorologie und Geophysik 25, Zentralanstalt für Meteorologie und Geodynamik, Wien, 2001.

Haiden, T., Kann, A., Wittmann, C., Pistotnik, G., Bica, B., and Gruber, C.: The Integrated Nowcasting through Comprehensive Analysis (INCA) system and its validation over the Eastern Alpine region, Weather Forecast., in press, doi:10.1175/2010WAF2222451.1, 2010.

Heinrich, G.: Klimawandel und Trockengefährdung in der Landwirtschaft: Eine Analyse für die Steiermark, Wiss. Ber. 20-2008, Wegener Zentrum für Klima und Globalen Wandel, Graz, 2008.

Kabas, T., Foelsche, U., and Kirchengast, G.: Seasonal and annual trends of temperature and precipitation within 1951/1971–2007 in south-eastern Styria, Austria, Meteorol. Z., under review, 2011.

Kirchengast, G., Kabas, T., Binder, S., Leuprecht, A., and Bichler, C.: Pionierexperiment WegenerNet Klimastationsnetz: Ein neuartiges Messnetz in der Region Feldbach (Steiermark/Österreich) zur Beobachtung von Wetter und Klima mit sehr hoher Auflösung, Wiss. Ber. 23-2008, Wegener Zentrum für Klima und Globalen Wandel, Graz, 2008.

Prettenthaler, F., Dalla-Via, A., Gobiet, A., Fank, J., Kurzmann, R., Oberauner, I., Töglhofer, C., Truhetz, H., and Zakarias, G.: Wasser und Wirtschaft im Klimawandel, in: Konkrete Ergebnisse am Beispiel der sensiblen Region Oststeiermark, edited by: Prettenthaler, F. and Dalla-Via, A., Vol. 1, Studien zum Klimawandel in Österreich, Verlag der Österreichischen Akademie der Wissenschaften, Wien, 2007.

Wulfmeyer, V., Behrendt, A., Bauer, H.-S., Kottmeier, C., Corsmeier, U., Blyth, A., Craig, G., Schumann, U., Hagen, M., Crewell, S., Girolamo, P. D., Flamant, C., Miller, M., Montani, A., Mobbs, S., Richard, E., Rotach, M. W., Arpagaus, M., Russchenberg, H., Schlüssel, P., König, M., Gärtner, V., Steinacker, R., Dorninger, M., Turner, D. D., Weckwerth, T., Hense, A., and Simmer, C.: The Convective and Orographically Induced Precipitation Study: A research and development project of the World Weather Research Program for improving quantitative precipitation forecasting in low-mountain regions, Bull. Am. Meteorol. Soc., 89, 1477–1486, doi:10.1175/2008BAMS2367.1, 2008.

# Diagnostic study of the influence of lateral boundary conditions for the REMO RCM simulations over the Carpathian Basin

G. Szépszó

Hungarian Meteorological Service, P.O. Box 38, 1525 Budapest, Hungary

**Abstract.** At the Hungarian Meteorological Service, two experiments were accomplished with the REMO5.0 regional climate model: (1) a simulation of a past period from 1961 to 2000, driven by the ECMWF ERA40 re-analysis data, and (2) a transient run from 1951 to 2100 driven by the ECHAM5/MPI-OM global coupled atmosphere-ocean model using SRES A1B forcing. The integration domain covers continental Europe with 25 km horizontal resolution in both experiments. Present article is dedicated to the investigation of the simulation results for the past period. The results for 1961–1990 were compared on the one hand with observations, and on the other hand, with each other and the corresponding global fields in order to assess the impact of the different lateral boundary conditions on the results focusing on the area of our interest, i.e., Hungary. The evaluation indicated that the re-analysis driven experiment provides warm and in summer dry past climate over the Carpathian Basin, whereas lower temperature and higher precipitation values are obtained when the lateral boundary information is derived from a global climate model. Based on the validation, it is concluded that the temperature characteristics in the simulation-driven case outperformed the experiment forced by quasi-perfect (i.e., re-analysis) data, however, similar apparent conclusion cannot be drawn for precipitation. This paper is undertaking to give deeper insight into the details and possible reasons for these outcomes.

## 1 Introduction

Recently the continuously improving global climate models (GCMs) are providing realistic projections for synoptic-scale characteristics of the climate, however, they are insufficient for detailed regional-scale estimations. The use of regional climate models (RCMs) ensures a dynamically based opportunity to enhance the global results for regional scale. The large-scale information for the regional simulations is coming from global models through the lateral boundary conditions (LBCs). Several experiments (Giorgi and Bates, 1989; Radu et al., 2008) provided evidence, that RCMs are able to reproduce the synoptic-scale climate features specified by the LBCs. Nevertheless, due to the limited area character and the higher resolution, they have to be capable of describing the small-scale physical processes. Moreover, the Big-Brother Experiments of Denis et al. (2002) showed, that not only the topography-induced patterns can be more realistically reflected by RCMs, but also the fine-scale processes forced by atmospheric nonlinearities and hydrodynamic instabilities due to the application of more sophisticated physical parameterization schemes.

The present study is dedicated to explore in detail the relation between the driving fields and driven regional climate models over the Carpathian Basin focusing on the period of 1961–1990 through a comparative analysis of the results of REMO regional climate model used at the Hungarian Meteorological Service (HMS) and its lateral boundary conditions as well as the observations. In Sect. 2 the most important characteristics of the accomplished experiments are introduced. Section 3 deals with thorough analysis of the results over Central Europe with special emphasis on the Hungarian territory. In Sect. 4 several open issues are discussed and the major conclusions are drawn.

## 2 Experimental design

The REMO5.0 regional climate model was adapted at the Hungarian Meteorological Service in 2004. REMO (Jacob and Podzun, 1997) was developed by the Max Planck Institute for Meteorology in Hamburg based on the Europa Model, i.e., the former numerical weather prediction model of the German Weather Service, with the inclusion of the global ECHAM4's physical parameterization package. The physics of REMO5.0 and the current cycle of the ECHAM model (ECHAM5; Roeckner et al., 2003) are built on the same basis, however, ECHAM5 includes several improvements covering the parameterization of longwave radiation, cloud microphysics, etc. At the HMS two experiments were accomplished with the REMO model: (1) a simulation of a past period from 1961 to 2000, driven by the ERA40 reanalyses (Uppala et al., 2005), and (2) a transient run for the period of 1951–2100 driven by the ECHAM5/MPI-OM (Roeckner et al., 2003; Marsland et al., 2003) global coupled atmosphere-ocean model. (Hereafter, the simulations are referred to as ERA40-driven or REMO_ERA experiment and ECHAM-driven or REMO_ECHAM one, respectively.) In the (1951–2000) control period, the observed atmospheric concentrations of greenhouse gases and aerosols were taken for the global climate model integration, while for the 21st century these concentrations were prescribed according to the SRES A1B scenario (Nakicenovic et al., 2000). The regional integration domain was the same in both experiments: it covers continental Europe with 25 km horizontal resolution and in vertical 20 levels were applied. As far as the lateral boundary conditions are concerned, the horizontal resolution of the ERA-40 dataset is 1.125 degree (near 125 km), while that of the ECHAM5 is 1.875 degree (approximately 200 km).

The evaluation is addressing on the one hand, the validation, and on the other hand, the assessment of the influence of the different lateral boundary conditions on the results with special focus on Hungary. Therefore, the regional and global results were compared with the 10-min resolution monthly CRU (Mitchell and Jones, 2005) observational dataset and with each other and the driving fields for the period of 1961–1990. The validation was achieved for the 2-m mean temperature and precipitation amount, whereas the RCM-GCM intercomparison was extended with the investigation of 850-hPa temperature and 700-hPa relative humidity. (Validation of the last two parameters was not carried out in the absence of reliable and spatially homogenous measurements, though 850-hPa temperature provided by ERA-40 can be considered also as observation.) The reason for the choice of relative humidity was that despite the fact that specific humidity gives more direct information about the precipitation potential of the atmosphere than relative humidity, the evaluation of the latter one also provides an insight into the moisture and precipitation conditions of the simulations, especially in the summer months.

## 3 Results

### 3.1 Temperature

Investigating the results of the ERA40-driven simulation, it can be generally concluded, that REMO5.0 produces too high 2-m temperature values in every season almost over the whole integration area (not shown). This overestimation is especially large over Southern and Eastern Europe, particularly over Hungary it exceeds 3 °C in summer (however, in winter it remains under 0.5 °C). In the case of ECHAM-driven simulation, the seasonal 2-m temperature differences between the model results and observations are small indicating rather perfect annual simulation with less than 1 °C systematic error over the large part of the domain. Some exceptions are found over the Alps and Southeast from the Carpathian Basin: while in the elevated points the model underestimates the mean temperature, over the southeastern area it predicts too warm climate for the 1961–1990 reference period. Nevertheless, the errors do not reach 3 °C even at these "critical" regions and over Hungary the departure remains under 0.5 °C in every season. It has to be remarked here, that the abovementioned mean temperature errors are calculated by comparison of the raw simulation data and the CRU dataset interpolated to the 0.22-degree resolution model grid. If an altitude correction is applied for the raw temperature values to eliminate the differences between the orography of the model and reference dataset, the large over- and underestimations become deeper exceeding 3 °C over both regions. This is due to the structure of the differences between the two topographies: the northern and eastern slopes (i.e., the lee side) of the high mountains are more highly elevated, while the western and southern parts are lower in REMO than in the case of observations. This pattern implies that there is a shift between the orography fields. Nevertheless, same correction method is not utilized for the results over Hungary, because the target region is a relatively flat area without high orographic features.

The different performance of the two simulations (which differ from each other only in their lateral boundary conditions) can be also demonstrated in the relation between the driven regional and the driving global fields. It can be seen in Fig. 1, that the RCM introduces a systematic heating into the global results. This warming is almost constant along the reference period, however, its magnitude is lower in the ECHAM-driven case than in the ERA40-driven experiment (annually 0.9 and 2.2 °C for Hungary, respectively; Table 1). The higher warming effect in the REMO_ERA simulation led to the identified large temperature overestimation over Hungary (since apart from its minor negative bias, the ERA40 dataset reproduces realistically the observed mean temperature values), whereas in the ECHAM-driven case it manifested as an improvement of the too low temperature values provided by ECHAM (Fig. 1). Based on the outcomes of the re-analysis driven simulation, one can conclude that

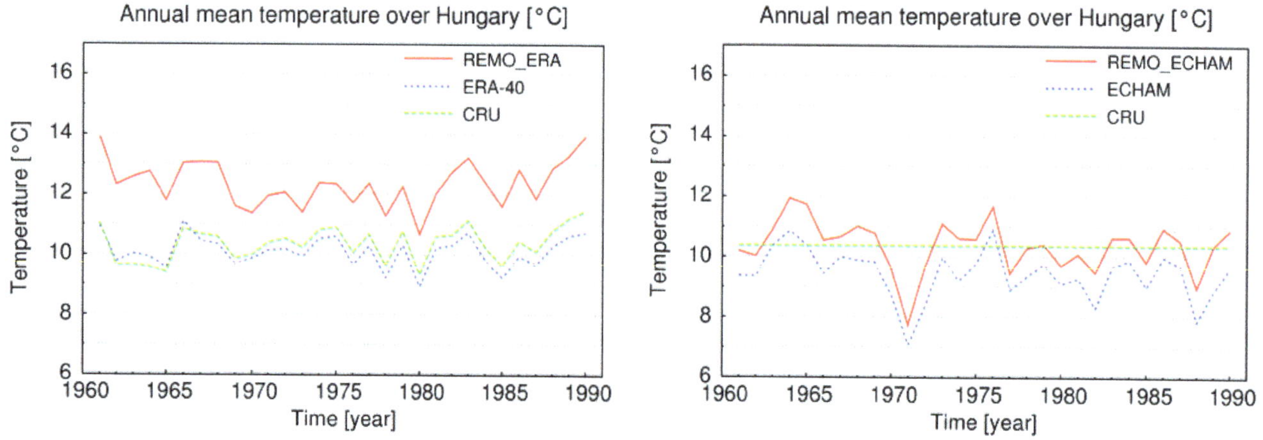


**Figure 1.** Evolution of the annual mean 2-m temperature [°C] in the REMO simulations (red curves: ERA40-driven REMO on the left, ECHAM-driven REMO on the right), the corresponding global fields (blue curves: ERA40 on the left, ECHAM5/MPI-OM on the right) and the CRU1.2 observation (green curve on the left) and its 30-year mean (green line on the right) over Hungary for 1961–1990.

**Table 1.** Mean temperature differences [°C] between the REMO simulations and the corresponding driving fields (ERA40-driven REMO and ERA40, ECHAM-driven REMO and ECHAM5/MPI-OM), the two regional simulations (REMO_ERA and REMO_ECHAM) and the two global fields (ERA40 and ECHAM5/MPI-OM) at 2 m and 850 hPa over Hungary for 1961–1990.

| | REMO_ERA – ERA | | REMO_ECHAM – ECHAM | | REMO_ERA – REMO_ECHAM | | ERA–ECHAM | |
|---|---|---|---|---|---|---|---|---|
| | T2m | T850 | T2m | T850 | T2m | T850 | T2m | T850 |
| Annual | 2.2 | 2.9 | 0.9 | 0.8 | 2.0 | 1.8 | 0.7 | −0.3 |
| Spring | 1.7 | 2.6 | 0.8 | 0.8 | 1.8 | 1.8 | 0.9 | 0 |
| Summer | 3.6 | 3.1 | 1.2 | 0.7 | 3.5 | 2.8 | 1.1 | 0.4 |
| Autumn | 3.1 | 3.1 | 1.1 | 0.9 | 2.6 | 2.3 | 0.6 | 0.1 |
| Winter | 0.5 | 2.7 | 0.6 | 0.9 | 0.1 | 0.5 | 0.2 | −1.3 |

REMO tends to overestimate the temperature over Hungary. Moreover, considering the re-analysis fields as perfect LBCs, this overestimation is coming (almost) purely from the regional climate model. This heating effect with respect to the global forcing exists also in the ECHAM-driven simulation, however, in this case the errors stemming from the RCM and GCM (over- and underestimation, respectively) have the same magnitude. Consequently, they cancel each other resulting in a "perfect" temperature simulation over the area of our interest.

The temperature departure between the regional and the corresponding global results is varying with the altitude, as well (Table 1): compared the 850-hPa differences to the 2-m ones, they mainly enhance in the re-analysis driven case, especially in winter (and excluding summer), while using GCM forcings it is slightly reduced in most seasons (except winter). Parallelly to that, the deviation between the two global datasets is decreasing at higher level: it is lower at 850 hPa than at 2 m, though the ECHAM fields are still colder than the ERA40 ones, apart from winter. Based on these results one can conclude, that the divergence between the two regional temperature simulations is larger for the near-surface, and it is decreasing with the altitude, except winter (Table 1). In winter, contrary to its general "behaviour", ECHAM is much warmer at 850 hPa than ERA40 (the difference exceeds 1 °C). This is valid not only over Hungary, but also over the northern and eastern parts of the domain (Fig. 2), where the two regional simulations look more alike than it is the case for their global counterparts. Nevertheless, over South the abovementioned structure can be concluded again: the ERA40 and REMO_ERA fields have higher 850-hPa temperature values than the ECHAM and REMO_ECHAM ones, respectively.

Looking at the relation between the mean 2-m temperature errors and the simulated temperature values (for Hungary), a further important and interesting feature is found. In Fig. 3 the monthly mean departures from the observations are displayed as a function of the mean values provided by the RCM, averaged for the 30-year reference period in each gridpoint. The simulation errors are strongly enhancing with the

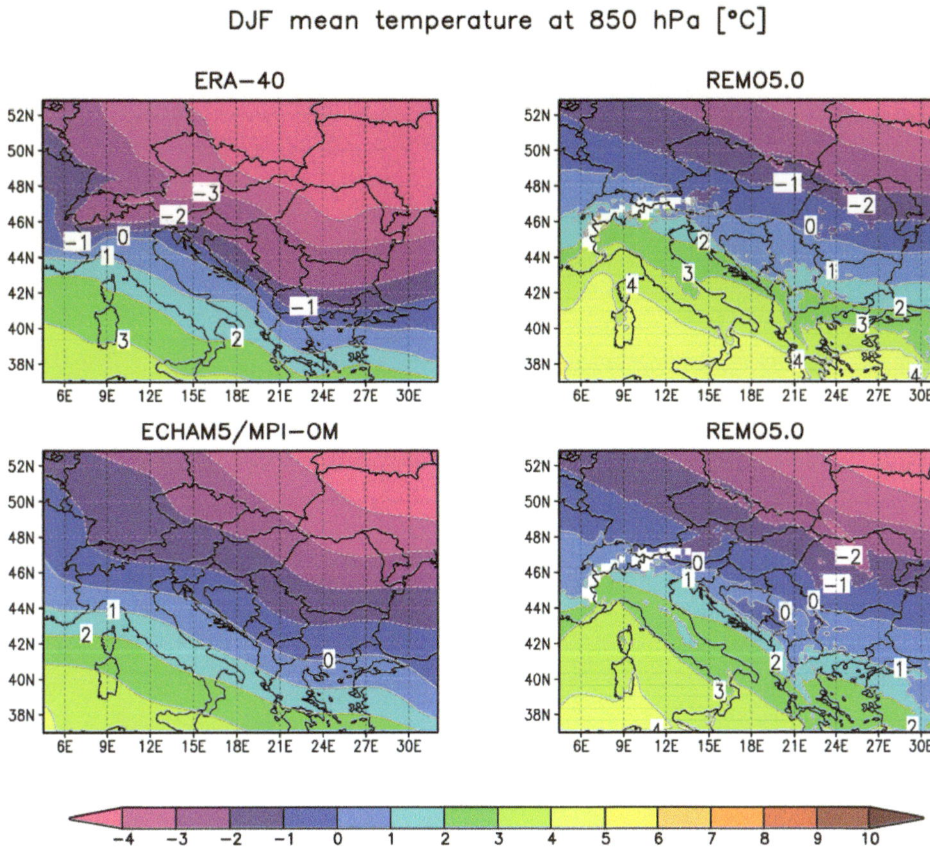


**Figure 2.** Winter mean temperature [°C] at the level of 850 hPa in the REMO simulations (top right: ERA40-driven REMO, bottom right: ECHAM-driven REMO) and the corresponding global fields (top left: ERA40, bottom left: ECHAM5/MPI-OM) for 1961–1990.

increasing simulated value at the re-analysis driven results: the errors are between −1 and +2 °C if the projected monthly mean temperature is between −5 and 5 °C (i.e., in winter), while the errors reach even 5 °C if the range of the simulated temperature covers 20–25 °C (in summer and autumn). Using ECHAM fields as LBCs, the same behaviour cannot be observed and the temperature errors are principally independent of the simulated value: they are basically between −1 and +2 °C for every temperature range. It is proved (but not shown here) that this special feature does not originate from the regional model itself, but the global driving fields already possess these error characteristics. I.e., the re-analysis dataset has the largest inaccuracies in the summer-autumn period, while the performance of the ECHAM5/MPI-OM results is basically not sensitive to the simulated temperatures. In order to find some explanation for this phenomenon it is worthwhile to investigate the link between temperature and precipitation simulations, which is discussed in the next subsection.

## 3.2 Precipitation and relative humidity

Regarding the mean annual precipitation errors (Fig. 4), one can immediately notice that REMO_ECHAM generates more precipitation than REMO_ERA. REMO (using either re-analysis LBCs or GCM ones) exaggerates the precipitation over the highest orographic features of Europe, the Alps and the Carpathians (to a lesser extent). However, the strongest overestimation is not always located over the mountain peaks, but rather over the slopes. The southern parts of the domain are characterized with underestimation (varying between 10 and 50%) even over the mountains, like the Adriatic side of the Alps, the Dinaric Alps or the southern ranges of the Carpathians. The Carpathian Basin is situated in an intermediate zone: North from Hungary the regional simulations have positive bias (precipitation overestimation), while in South rather the underestimation is typical.

As far as the results over Hungary are concerned, even though REMO with ERA40 driving produces realistic seasonal precipitation cycle (not shown), in summer and autumn the results are hampered by some underestimation. Using ECHAM results as LBCs, the regional simulation generates too high precipitation amounts, except in summer, when REMO reproduces the mean precipitation with the greatest

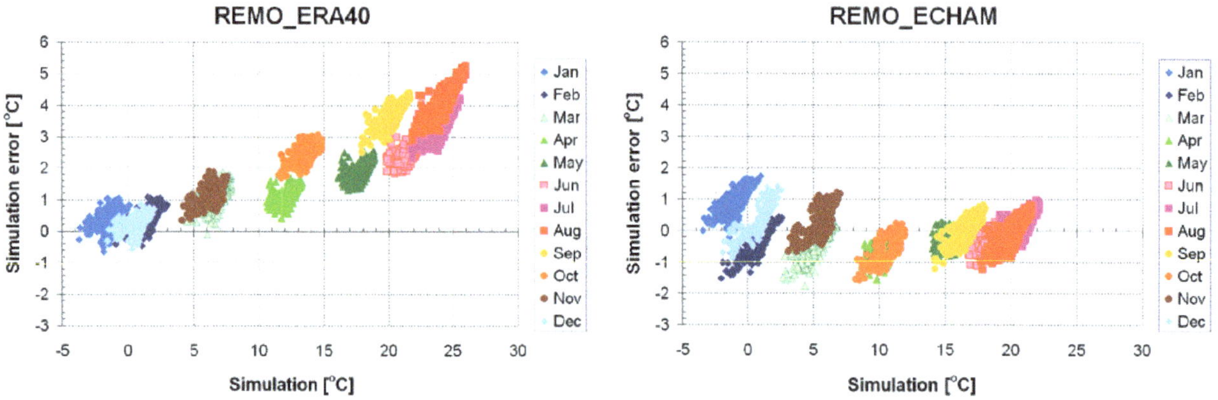


**Figure 3.** ERA40-driven (left panel) and ECHAM-driven (right panel) REMO simulation errors for monthly mean 2-m temperature [°C] with respect to the simulated values [°C] averaged for 1961–1990 in every gridpoint over Hungary. The reference dataset was provided by CRU1.2.

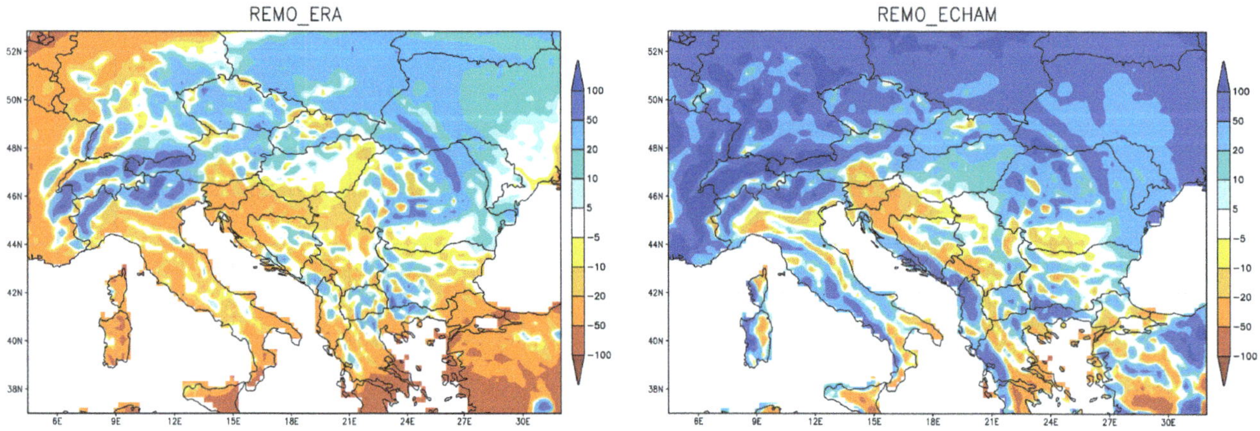


**Figure 4.** Relative annual precipitation difference [%] between the REMO simulations (ERA40-driven REMO on the left, ECHAM-driven REMO on the right) and the CRU1.2 observations for 1961–1990.

skill. (More details about the ECHAM-driven simulation results can be read in Szépszó and Horányi, 2008.) Investigating simultaneously the mean temperature and precipitation simulation errors (Fig. 5), it can be discovered, that in the case of REMO_ERA the higher temperature overestimations (over 2.5 °C) correspond with the deepest precipitation underestimations (exceeding 15%) in the August-October period. The large precipitation deficit in summer and autumn leads to drying the soil, which makes even worse the general positive error in temperature and this process can induce the results seen in Fig. 3. The same mechanism cannot be experienced in the case of REMO_ECHAM: here a precipitation surplus can be rather concluded at the end of summer and beginning of autumn, which is accompanied by only small temperature errors not larger than 0.5 °C.

Comparing the mean precipitation fields in the driving global and the corresponding regional models (not shown), REMO predicts significantly more precipitation than the global fields especially over the highly elevated areas, but also in vicinity of some borders of the integration domain. The excessive precipitation over the mountains in the fine-resolution models compared to the coarse-resolution ones is a well-known property of the numerical models. The influence of the orography is visible over the Alpine region also in the relative humidity at the level of 700 hPa (see Fig. 6 for summer). The spurious features near to the domain borders might come from the numerical treatment of lateral boundaries, when false reflecting waves appear along the edges, which are caused by the "imbalances" between the global and regional fields in the boundary zone. However, it has to be remarked, that in this context precipitation acts as an indicator of this effect, since it is not a coupled, but a diagnostic variable. Therefore, it is more proper to investigate the departures between the regional and global fields for a humidity-related prognostic variable (for instance specific humidity). In our study, the 700-hPa humidity field is considered, however, the noises at higher levels are naturally "smoother" than at the near-surface. At 700 hPa, the abovementioned boundary effect is more emphasized in the case of REMO_ECHAM simulation, since the driving ECHAM fields supply more

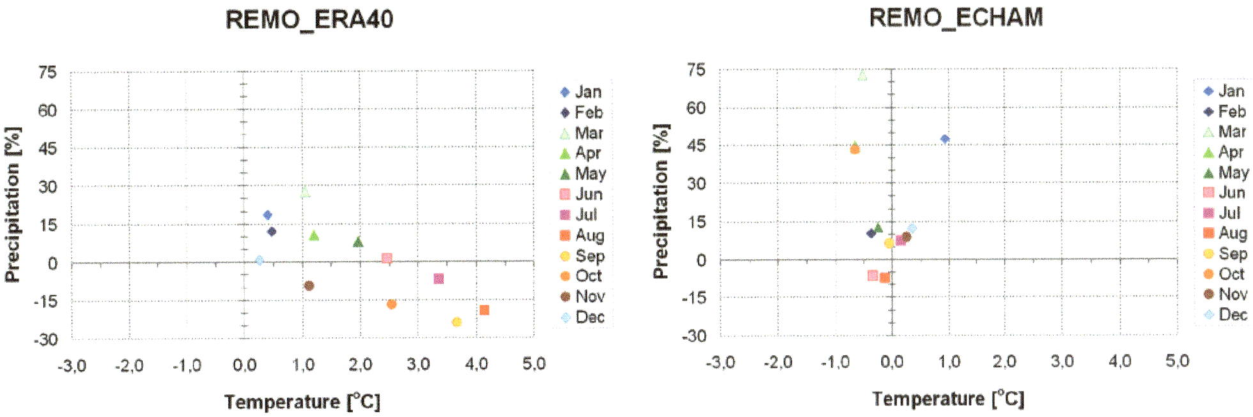


**Figure 5.** Monthly mean precipitation and temperature differences [% and °C, respectively] between the REMO simulations (ERA40-driven REMO on the left, ECHAM-driven REMO on the right) and the CRU1.2 observations over Hungary for 1961–1990.

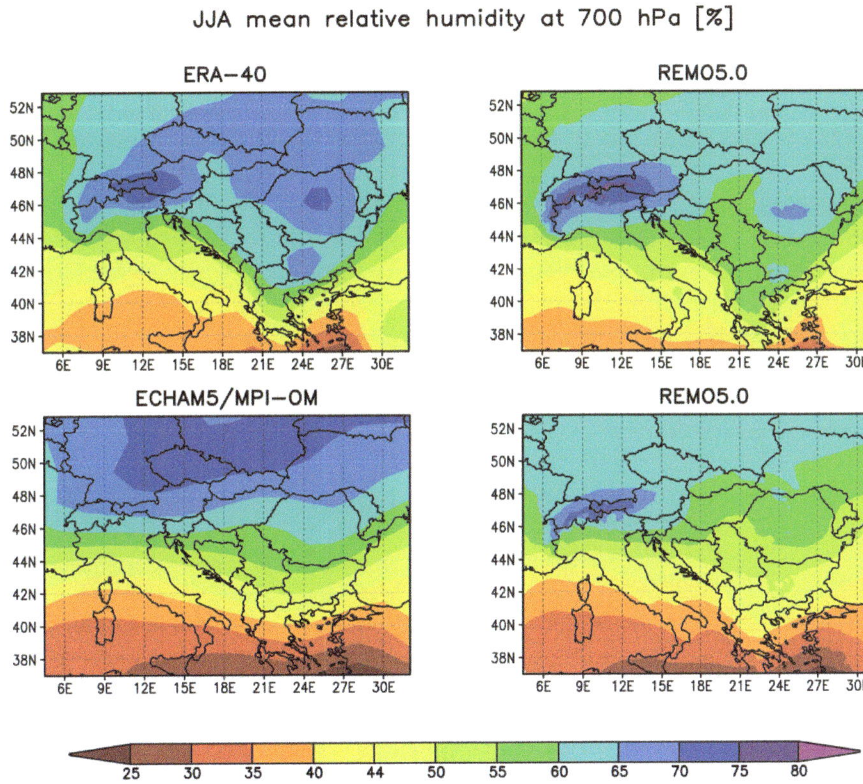


**Figure 6.** Summer mean relative humidity [%] at the level of 700 hPa in the REMO simulations (top right: ERA40-driven REMO, bottom right: ECHAM-driven REMO) and the corresponding global fields (top left: ERA40, bottom left: ECHAM5/MPI-OM) for 1961–1990.

relative humidity, particularly over the northern part of the domain than the re-analyses (Fig. 6). Nevertheless, the regional fields at this level are dryer than the global ones, and linking this fact together with the surface precipitation surplus, it allows to conclude, that the higher-level humidity disappears from the atmosphere in the form of precipitation. (The confirmation of this issue requires the investigation of near-surface humidity and specific humidity.)

## 4 Conclusions and discussion

In the present study the results of two REMO5.0 regional climate model experiments were evaluated: a past simulation driven by the measurement-based (therefore, supposed to be quasi-perfect) ERA40 re-analysis data and a transient climate change simulation where the large-scale forcings were originated from the ECHAM5/MPI-OM AOGCM. The regional model results were analysed for the period of 1961–1990 in order to validate them with respect to the observations and to estimate the impact of the different lateral boundary conditions on the regional outcomes.

According to the validation it can be concluded, that the re-analysis driven experiment provides generally warm and in summer dry past climate over the Carpathian Basin, whereas lower temperature and higher precipitation values are obtained when the lateral boundary information is derived from a global climate model simulation. Over Hungary REMO5.0 brings a heating into the global results, which is larger and increasing with the height in ERA40-driven case. The same but more moderate effect in the GCM-driven simulation improves the originally too low near-surface temperature values provided by ECHAM. The 700-hPa results indicate that the regional model drains the global fields, which is more robust in the case of initially more humid GCM LBCs. Regarding the precipitation results, with the ECHAM LBCs REMO5.0 produces high overestimation over large part of the integration domain. There are some open issues not fully answered by this diagnostic study, hereafter they are briefly listed with some possible explanation:

- The GCM-driven regional temperature simulation performs better compared to the re-analysis driven one. In the latter case, due to the quasi-perfect characteristics of the LBCs, (in theory) validation provides information exclusively about the deficiencies of the regional model. Contrary to that, in the GCM-driven case the errors of the RCM and the GCM cannot be separated, they interact and might cancel each other. However, since also the ERA40 dataset has some known deficiencies, they determine the performance of the RCM results in the same way as using driving fields from a GCM.
- Applying re-analyses as LBCs, the temperature errors are sensitive to the simulated values, at the same time this feature is not detected at GCM-driven results. This different behaviour might be caused by the co-occurrences of high temperature overestimations and large precipitation underestimations at the end of summer and beginning of autumn in the re-analysis driven experiment. This "constellation" can affect the soil wetness towards drying, and the soil circumstances providing surface forcings for the model has great influence on the atmospheric simulation and can make more severe the positive temperature bias. Same coincidences regarding temperature and precipitation cannot be concluded in the case of REMO_ECHAM simulation.
- REMO inevitably provides some additional near-surface heating with respect to the global driving fields, however this warming is smaller in the case of ECHAM5 driving than that of the ERA40 one (its reason is to be explored). The REMO_ECHAM 2-m temperature simulations are rather perfect due to the error cancellation between the global and regional models. This heating effect makes dry the regional fields at higher levels, but for precipitation the opposite is valid. This latter outcome suggests that the humidity, which "disappears" at the higher atmospheric levels from the regional model, falls out near the surface as precipitation. This conclusion raises the necessity of the investigation of specific humidity, which is an absolute measure of the atmospheric moisture and providing more accurate information about the quantity of precipitable water vapour in the atmosphere.

As summary, this diagnostic study raises several questions, which were not completely explained, but are worthwhile to be deeply examined. The ideas gathered in current paper pinpoint the "track" of our future investigations.

**Acknowledgements.** This work was supported by the European Commission's 6th Framework Programme in the framework of CLAVIER project (contract number 037013). The ECHAM5/MPI-OM data were provided by the World Data Center for Climate in Hamburg. The author is grateful to the colleagues at the Max Planck Institute for the Meteorology in Hamburg for their help in adaptation of the REMO model. The useful comments and suggestions of the two anonymous referees are also highly appreciated.

Edited by: A. M. Sempreviva

## References

Denis, B., Laprise, R., Caya, D., and Cote, J.: Downscaling ability of one-way nested regional climate models: the Big-Brother Experiment, Clim. Dynam., 18, 627–646, 2002.

Giorgi, F. and Bates, G.: The climatological skill of a regional model over complex terrain, Mon. Weather Rev., 117, 2325–2347, 1989.

Jacob, D. and Podzun, R.: Sensitivity studies with the regional climate model REMO, Meteorol. Atmos. Phys., 63, 119–129, 1997.

Marsland, S. J., Haak, H., Jungclaus, J. H., Latif, M., and Röske, F.: The Max Planck Institute global ocean/sea-ice model with orthogonal curvilinear coordinates, Ocean Model., 5, 91–127, 2003.

Mitchell, T. D and Jones, P. D: An improved method of constructing a database of monthly climate observations and associated high-resolution grids, Int. J. Climatol., 25, 6, 693–712, 2005.

Nakicenovic, N., Alcamo, J., Davis, G., de Vries, B., Fenhann, J., Gaffin, S., Gregory, K., Grübler, A., Jung, T. Y., Kram, T., La Rovere, E. L., Michaelis, L., Mori, S., Morita, T., Pepper, W., Pitcher, H., Price, L., Raihi, K., Roehrl, A., Rogner, H.-H., Sankovski, A., Schlesinger, M., Shukla, P., Smith, S., Swart, R., van Rooijen, S., Victor, N., and Dadi, Z.: IPCC special report on emissions scenarios, Cambridge University Press, Cambridge, 2000.

Radu, R., Déqué, M., and Somot, S.: Spectral nudging in a spectral regional climate model, Tellus A, 60, 5, 898–910, 2008.

Roeckner, E., Bäuml, G., Bonaventura, L., Brokoff, R., Esch, M., Giorgetta, M., Hagemann, S., Kirchner, I., Kornblueh, L., Manzini, E., Rhodin, A., Schlese, U., Schulzweida, U., and Tompkins, A.: The atmospheric general circulation model ECHAM5. Part I: model description, Report No. 349, Max Planck Institute for Meteorology, Hamburg, Germany, 2003.

Szépszó, G. and Horányi, A.: Transient simulation of the REMO regional climate model and its evaluation over Hungary, Időjárás, 112, 203–232, 2008.

Uppala, S. M., Kållberg, P. W., Simmons, A. J., Andrae, U., da Costa Bechtold, V., Fiorino, M., Gibson, J. K., Haseler, J., Hernandez, A., Kelly, G. A., Li, X., Onogi, K., Saarinen, S., Sokka, N., Allan, R. P., Andersson, E., Arpe, K., Balmaseda, M. A., Beljaars, A. C. M., van de Berg, L., Bidlot, J., Bormann, N., Caires, S., Chevallier, F., Dethof, A., Dragosavac, M., Fisher, M., Fuentes, M., Hagemann, S., Hólm, E., Hoskins, B. J., Isaksen, L., Janssen, P. A. E. M., Jenne, R., McNally, A. P., Mahfouf, J.-F., Morcrette, J.-J., Rayner, N. A., Saunders, R. W., Simon, P., Sterl, A., Trenberth, K. E., Untch, A., Vasiljevic, D., Viterbo, P., and Woollen, J.: The ERA-40 re-analysis, Q. J. Roy. Meteorol. Soc., 131, 2961–3012, 2005.

# Observations and modelling of $1/f$-noise in weather and climate

**R. Blender, X. Zhu, and K. Fraedrich**
University of Hamburg, KlimaCampus, Hamburg, Germany

**Abstract.** Data with power spectra close to $S(f) \sim 1/f$ is denoted as $1/f$ or flicker noise. High resolution measurements during TOGA/COARE for temperature, humidity, and wind speed (1 min resolution) reveal $1/f$ spectra while precipitation shows no power-law scaling during the same period. However, a binary time series indicating the precipitation events (1 for precipitation, 0 for no precipitation) shows a clear $1/f$ spectrum in line with the remaining boundary layer data. For extreme events in time series with $1/f$ spectra the return time distribution is well approximated by a Weibull-distribution for short and long return times. The daily discharge of the Yangtze river shows high volatility which is linked to the intra-annual $1/f$ spectrum. The discharge fluctuations detected in different time windows are represented by a single function (a so-called data collapse) similar to the universal behavior found for turbulence and various physical systems at criticality. The collapse is well described by the Gumbel distribution.

## 1 Introduction

Increasing variability at lower frequencies is ubiquitous in meteorology and hydrology (Jiang et al., 2005; Yano et al., 2004; Fraedrich and Blender, 2003). If changes of measurement devices and external impacts can be excluded this volatility is attributed to the internal variability of weather and climate. The low frequency part of the power spectra of these time series can typically be approximated by a power law, $S(f) \sim f^{-\beta}$, with $\beta > 0$, and is denoted as long term memory since the correlation function does not decay in a finite time.

The so-called $1/f$ or flicker noise denotes spectra close to $1/f$. An important property of $1/f$ noise is that the variance of running mean time series does not decay, thus statistical analyses and skillful statistical predictions are hampered. Since exponents in the wide range $\beta = 0.5 \ldots 1.5$ are allowed (see for example Voss and Clarke, 1975) these spectra are sometimes called "$1/f$ like". This definition should not be confused with the mathematically sharp definition $\beta = 1$ used in statistics, which is the upper bound for stationarity. In finite observational datasets $1/f$ spectra are limited at low frequencies, hence the process can be considered to be stationary (see Stoisiek and Wolf (1976) for a discussion of stationarity of $1/f$ processes).

Data with $1/f$ power spectra cover wide ranges of frequencies from minutes to millennia, for example, tropical boundary layer data with 1 min resolution (TOGA/COARE, Yano et al., 2004) and reconstructed temperatures up to the Milankovitch cycle (Huybers and Curry, 2006). Updating the overview by Fraedrich et al. (2009) we present new results on (a) the variability of binary precipitation events in high resolution TOGA/COARE measurements and (b) on the universality of the fluctuations observed in the Yangtze river discharge. Models and the impact of $1/f$ spectra on extreme event distributions are considered briefly.

## 2 Observing $1/f$ variability

Examples for $1/f$ spectra are found in wide ranges from minutes to millennia: (i) tropical boundary layer observations (TOGA/COARE, Yano et al., 2004), (ii) the discharge of the Yangtze river in the intra-annual frequency range (Wang et al., 2008), (iii) the sea surface temperature in a region of the North Atlantic and in the southern ocean (reproduced by models, Fraedrich and Blender, 2003), (iv) reconstructed near surface temperatures up to the Milankovitch cycle (Huybers and Curry, 2006).

Tropical rainfall at the TOGA/COARE research vessel Kexue shows no clear power-law scaling in contrast to temperature, humidity, and wind speed which reveal clear $1/f$ spectra. In order to detect a similar behavior in the precipitation time series (Fig. 1a) data transformations are assessed

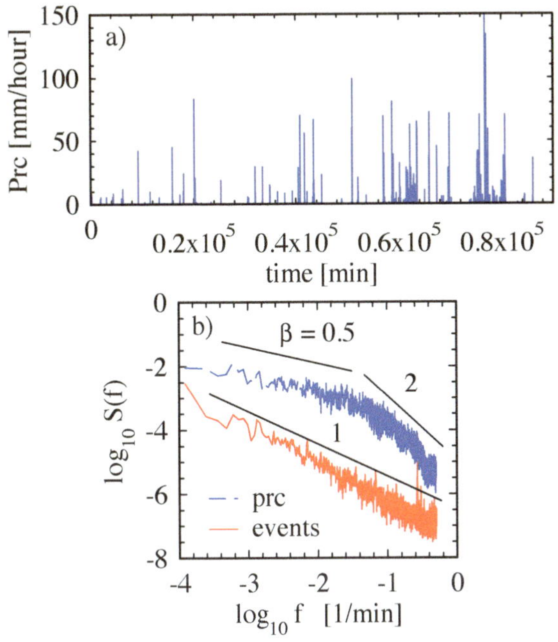


**Figure 1.** **(a)** Precipitation at the station Kexue (TOGA/COARE), resolution 1 min, **(b)** power spectra (10 window averages) for precipitation and the binary event time series. The slopes are guides for the eye.

providing the following result: precipitation *events*, i.e. the binary sequence defined by 1 for nonvanishing precipitation and 0 otherwise, show a clear $1/f$ spectrum (Fig. 1b). A possible explanation for this feature is that precipitation events are coupled to boundary layer processes (e.g. instability), while the rainfall amount at single stations is determined by other processes.

## 3 Modeling $1/f$ variability

Whereas red noise is modeled by autoregressive processes (AR), for example with an AR-1 where the coefficient $a$ determines the threshold frequency, there is no simple or generic model for $1/f$ spectra. An increase of the AR-1 coefficient $a$ towards $a \to 1$ increases memory and variance, but does not alter the low frequency plateau of the red noise spectrum. Long term memory and $1/f$ can be mimicked by autoregressive processes in two ways: (i) superposition of AR-1 processes with a continuous band of threshold frequencies and (ii) extension to an infinite number of coefficients defining a fractional AR process (FAR).

A physical model for $1/f$ behavior is given by diffusion in two adjacent compartments with two distinctly different diffusion coefficients (Fraedrich et al., 2004). An upper mixed layer is driven by an uncorrelated atmospheric heat flux and coupled to a deep abyssal ocean with lower diffusivity. This model enables the simulation of the observed spectrum of sea surface temperatures in the North Atlantic and the southern ocean. Based on this model the abyssal diffusivity can be estimated if long term observations of the surface variability are available.

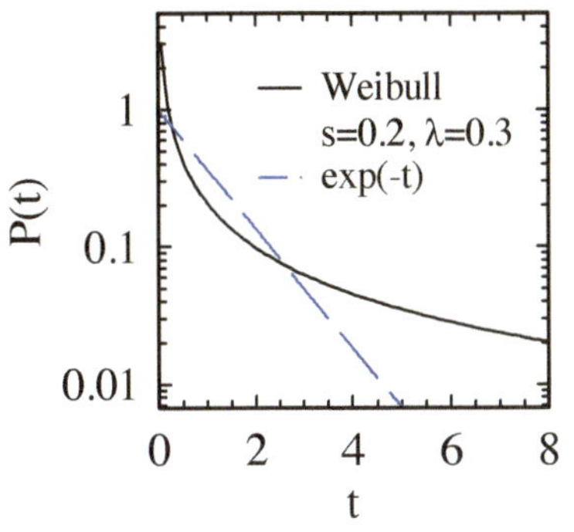


**Figure 2.** Return time distribution for extremes in $1/f$ (solid, Weibull) compared to the exponential distribution for uncorrelated data (dashed).

## 4 Extremes and $1/f$ variability

For long term memory processes the statistical properties of extreme event return times are altered. While the return times are exponentially distributed for uncorrelated data, Bunde et al. (2005) suggested a so-called stretched exponential, $\exp(-(t/\lambda)^k)$, for large return times. Blender et al. (2008) observed that the return time distribution for $1/f$ spectra is well described by a Weibull distribution, $f(t) = (k/\lambda)(t/\lambda)^{k-1}\exp(-(t/\lambda)^k)$, which includes a power-law for small return times and an asymptotic stretched exponential for large times (occurring as first and second factors). A main finding is the long term predictability of successive return times.

Employing a surrogate $1/f$ data (simulated by a fractional autoregressive process, FAR) the parameters of the Weibull distribution are estimated (parameters shape $s = 0.2$ and scale $\lambda = 0.3$) leading to a distinctly different behavior of the return times compared to the exponential (Poisson) distribution (Fig. 2). The responsible mechanism is, given a positive threshold, the presence of long term positive (negative) anomalies, which favor long periods of enhanced (reduced) threshold crossings and hence short (long) return times.

## 5 Universality and $1/f$ variability

The daily Yangtze river discharge measured at the station Yichang (since 1892, located downstream the Three Gorges Project) reveals high intra-annual and inter-annual variability (Fig. 3a). For time scales between two weeks and 1 yr the spectrum follows a $1/f$ power-law (Fig. 3b), which hampers the calculation of a mean annual cycle. Due to this extreme variability the hydrological conditions along the Yangtze

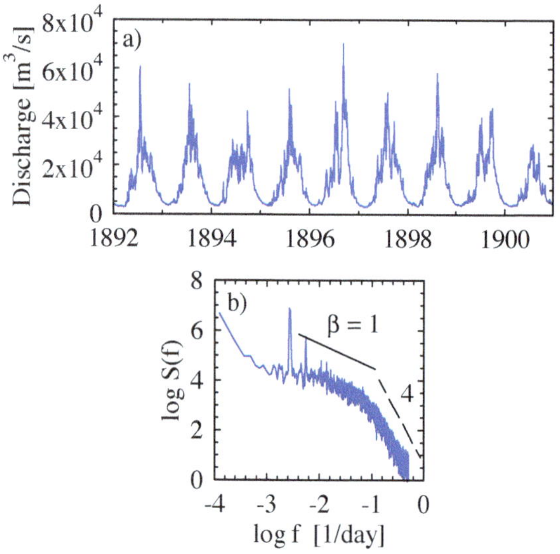


**Figure 3.** Yangtze river discharge at the station Yichang, **(a)** snapshot of 9 yr, **(b)** power spectrum (1892–2004, 6 segment average), slopes indicate the power-law exponent $\beta$.

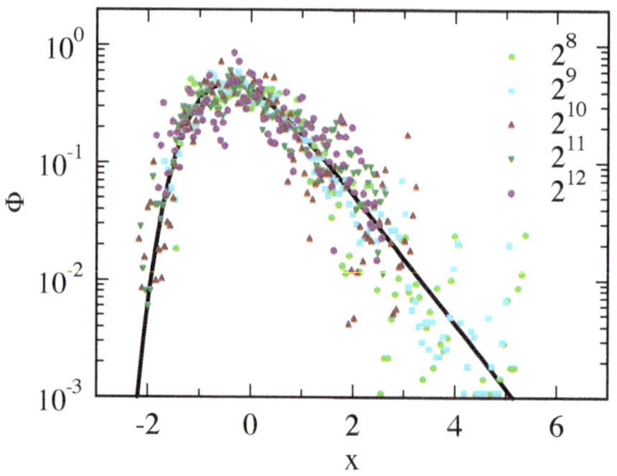


**Figure 4.** Rescaled frequency distribution of variances in the Yangtze river discharge at Yichang for different segment lengths (in days, see also Fig. 3), the solid line represents the Gumbel distribution (for parameters see text).

river were characterized by extreme floods before the construction of the dam in the year 2006. Precipitation can be excluded as a source for this type of variability since it is less correlated. On interannual time scales the discharge reveals weaker long term memory ($\beta \approx 0.3$, see Wang et al., 2008).

Data with a $1/f$ spectrum are associated with an extreme value distribution (Antal et al., 2001). For a time series $h(t)$, fluctuations $w_T = [h(t) - \overline{h}(t)]^2$ for different segments $T$ (time windows) are determined; $\mu$ is the mean and $\sigma$ the standard deviation of the fluctuations $w_T$. The distribution $P(w_T)$ of the fluctuations is transformed to a rescaled function $\Phi(x) = \sigma P(w_T)$ depending on the standardized fluctuation $x = (w_T - \mu)/\sigma$. Employing this transformation (Bramwell et al., 1998) leads to a remarkable overlapping of the data (a so-called data collapse) for a large number of physical systems including fluid turbulence. The data for $\Phi(x)$ obtained for the daily Yangtze river discharge versus the standardized fluctuations $x$ collapses for different time segments (Fig. 4). As found by Antal et al. (2001) the result is well represented by the extreme value (Gumbel) distribution $\Phi(x) = a\ \exp(-(ax+\gamma) - e^{-(ax+\gamma)})$, with $a = \pi/\sqrt{6} \approx 1.282$, and the constant $\gamma \approx 0.5772$, capturing the skewness of the non-Gaussian distribution (note that there is no fit applied).

## 6 Conclusions

Power spectra close to $1/f$ variability are ubiquitous in meteorological and hydrological time series. We present new results obtained for precipitation data with 1 min resolution and for the daily Yangtze river discharge. The frequency range of $1/f$ spectra is limited: at low frequencies either by the duration of the time series or by a regime with less memory, and at high frequencies by the Nyquist frequency or a short term memory regime (for example $S(f) \sim f^{-2}$ or steeper). Since $1/f$ noise is close to nonstationarity, statistical analyses have to be applied carefully because the mean cannot be estimated using averages in finite intervals.

High resolution precipitation data from TOGA/COARE shows no power-law scaling while temperature, humidity and wind speed follow a $1/f$ spectrum for the same measurement period. A binary time series obtained for precipitation events, however, shows a clear $1/f$ power-law scaling. This new finding hints to a coupling of the precipitation events with boundary layer processes, while the rainfall amount is controlled by an independent process.

The exponential distribution of extreme event return times is altered for data with $1/f$ spectra since the low frequency variability can lead to long periods below the threshold (depressions) associated with large return time intervals. Short return times are favored during periods, whose longer term mean is high. We have demonstrated how the Weibull distribution captures both effects.

The daily Yangtze river discharge shows a $1/f$ spectrum on intra-annual time scales. After an adequate rescaling, the discharge fluctuations for different time scales are represented by a single function which is well described by the Gumbel distribution. Thus the fluctuations of the Yangtze river discharge follow a universal distribution known from turbulence and critical phenomena. The physical mechanisms relevant for $1/f$ behavior are still unclear. The ubiquity hints to a general ("universal") or generic property of complex systems, and the absence of characteristic time scales is explained by the scale invariance due to many degrees of freedom. The result for the Yangtze river discharge might be associated with the multitude of time scales involved.

**Acknowledgements.** This work has been supported by the KlimaCampus Hamburg; K. F. acknowledges support by the Max Planck Society. The helpful comments of the reviewers are acknowledged.

Edited by: M. Brunet-India

## References

Antal, T., Droz, M., Györgyi, G., and Rácz, Z.: $1/f$ Noise and extreme value statistics, Phys. Rev. Lett., 87, 240601, doi:10.1103/PhysRevLett.87.240601, 2001.

Blender, R., Fraedrich, K., and Sienz, F.: Extreme event return times in long-term memory processes near $1/f$, Nonlin. Processes Geophys., 15, 557–565, doi:10.5194/npg-15-557-2008, 2008.

Bunde, A., Eichner, J. F., Kantelhardt, J. W., and Havlin S., Long-term memory: A natural mechanism for the clustering of extreme events and anomalous residual times in climate records, Phys. Rev. Lett., 94, 048701, doi:10.1103/PhysRevLett.94.048701, 2005.

Bramwell, S. T., Holdsworth, P. C. W., and Pinton, J.-F.: Universality of rare fluctuations in turbulence and critical phenomena, Nature, 396, 552–554, 1998.

Fraedrich, K. and Blender, R.: Scaling of atmosphere and ocean temperature correlations in observations and climate models, Phys. Rev. Lett., 90, 108501, doi:10.1103/PhysRevLett.90.108501, 2003.

Fraedrich, K., Luksch, U., and Blender, R.: $1/f$-model for long time memory of the ocean surface temperature, Phys. Rev. E, 70, 037301, doi:10.1103/PhysRevE.70.037301, 2004.

Fraedrich, K., Blender, R., and Zhu, X.: Continuum climate variability: long-term memory, extremes, and predictability, Int. J. Mod. Phys. B, 23 (28 & 29), 5403–5416, 2009.

Huybers, P. and Curry, W.: Links between annual, Milankovitch and continuum temperature variability, Nature, 441, 329–332, 2006.

Jiang, T., Zhang, Q., Blender, R., and Fraedrich, K.: Yangtze delta floods and droughts of the last millennium: Abrupt changes and long term memory, Theor. Appl. Climatol., 82, 131–141, doi:10.1007/s00704-005-0125-4, 2005.

Stoisiek, M. and Wolf, D.: Recent investigations on the stationarity of $1/f$ noise, J. Appl Phys., 47, 362–364, 1976.

Voss, R. F. and Clarke, J.: "$1/f$ noise" in music and speech, Nature, 258, 317–318, 1975.

Wang, G., Jiang, T., Blender, R., and Fraedrich, K.: Yangtze $1/f$ discharge variability and the interacting river-lake system, J. Hydrol., 351, 230–237, 2008.

Yano, J.-I., Blender, R., Zhang, C., and Fraedrich, K.: $1/f$-Noise and pulse-like events in the tropical atmospheric surface variabilities: Convective self-criticality and impacts on ENSO, Q. J. Roy. Meteor. Soc., 130, 1697–1721, 2004.

# The historical pathway towards more accurate homogenisation

**P. Domonkos[1], V. Venema[2], I. Auer[3], O. Mestre[4], and M. Brunetti[5]**

[1]Center for Climate Change, Univ. Rovira i Virgili, Campus Terres de l'Ebre, Tortosa, Spain
[2]Meteorological institute of the University of Bonn, Germany
[3]Zentralanstalt für Meteorologie und Geodynamik, Wien, Austria
[4]Meteo France, Ecole Nationale de la Meteorologie, Toulouse, France
[5]Institute of Atmospheric Sciences and Climate (ISAC-CNR), Bologna, Italy

*Correspondence to:* P. Domonkos (peter.domonkos@urv.cat)

**Abstract.** In recent years increasing effort has been devoted to objectively evaluate the efficiency of homogenisation methods for climate data; an important effort was the blind benchmarking performed in the COST Action HOME (ES0601). The statistical characteristics of the examined series have significant impact on the measured efficiencies, thus it is difficult to obtain an unambiguous picture of the efficiencies, relying only on numerical tests. In this study the historical methodological development with focus on the homogenisation of surface temperature observations is presented in order to view the progress from the side of the development of statistical tools. The main stages of this methodological progress, such as for instance the fitting optimal step-functions when the number of change-points is known (1972), cutting algorithm (1995), Caussinus – Lyazrhi criterion (1997), are recalled and their effects on the quality-improvement of homogenisation is briefly discussed. This analysis of the theoretical properties together with the recently published numerical results jointly indicate that, MASH, PRODIGE, ACMANT and USHCN are the best statistical tools for homogenising climatic time series, since they provide the reconstruction and preservation of true climatic variability in observational time series with the highest reliability. On the other hand, skilled homogenizers may achieve outstanding reliability also with the combination of simple statistical methods such as the Craddock-test and visual expert decisions. A few efficiency results of the COST HOME experiments are presented to demonstrate the performance of the best homogenisation methods.

## 1 Introduction

Recently the COST Action HOME (ES0601) (hereafter: HOME) has evaluated the methodological progress in homogenisation of climatic time series. The project, which terminated in October 2011, fostered the development of new statistical tools for homogenising time series, as well as the objective assessment of the effectiveness of traditional and newly-developed methods. After this fruitful period it is timely to examine the recent methodological advances and new scientific results in a wider historical context. In this study a brief overview of the historical development of homogenisation methods is presented, focusing on those steps that turned out to have lasting influence on the practical solutions of the homogenisation task. Most findings are valid for the homogenisation of annual and monthly surface temperature datasets of high spatial correlations in networks, or for the homogenisation of other climatic variables when some basic statistical properties (length of time series, spatial correlations, etc.) are comparable with those of the surface temperature datasets. The review is supplied with some illustrations about the efficiencies of the best methods relying on recent test results of HOME.

## 2 The birth of statistical homogenisation

The idea that technical-originated biases should be eliminated from climatic time series through spatial comparisons of the data is as old as the existence of professional meteorological observing networks, and documents prove that homogenisation, i.e. correction of local biases for sections of observed time series took place as early as in the 19th century (Kreil, 1854). The earliest methods were the subjective estimation of biases based on visual inspection of the observed data and the use of experimental measurements. For example, surface temperature measurements of Milan (Italy) between 1763 and 1834 were adjusted (Kreil, 1854) due to changes in the daily routine of observations. That time daily 7 observations were done in Milan (at 00:00, 03:00, 06:00, 09:00, 12:00, 18:00 and 21:00 LT), but Kreil and his colleagues wanted to produce more accurate daily mean temperatures than the average of such seven data daily. For this reason, a 16 month long experiment of hourly observation performed in Padua, and the bias of the average of the seven observations from the true daily mean was revealed, then the daily mean temperatures of Milan were corrected.

The first documented application of statistical homogenisation is known from 1925. The Austrian climatologist Victor Conrad searched possible change-points by splitting the series of annual precipitation totals into two parts at each year and assuming that the break is in the year for which the statistic shows the most significant difference between the two parts (Conrad, 1925). In the referred study the ratio of the upper and lower quartiles for two time series were compared, but Conrad also proposed two further statistical methods for detecting change-points. The scale of the applicable statistical tools is wide, since a change-point may cause significant difference in the means, accumulated anomalies, mean rank-order values, etc. between two parts of a time series. If the shift in the change-point is large compared to the noise level, these differences can even easily be identified visually (Fig. 1). When the signal-to-noise ratio is lower, statistical significance examination may help to make the distinction between random fluctuations and true inhomogeneities. Since the appearance of the first statistical homogenisation method, a large number of types and versions of homogenisation methods have been developed. Note that in spite of the great development of statistical tools, subjective decisions have not disappeared completely from the homogenisation of climatic time series.

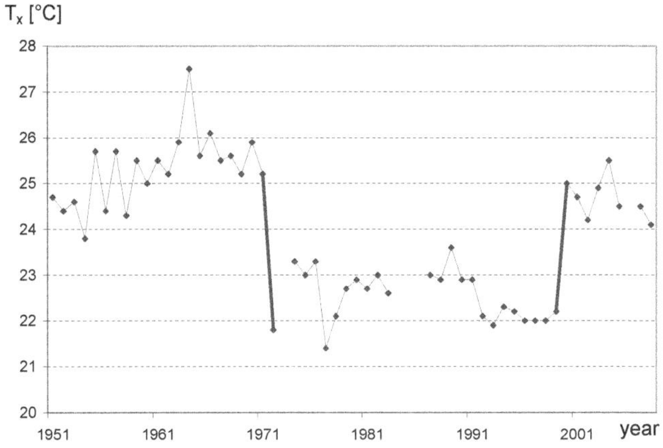


**Figure 1.** Summer daily maximum temperatures ($T_x$) of Tarifa (Spain) between 1951–2008. The two spectacularly large shifts are marked with thick lines.

## 3 Until 1990: single change-point models

In the 20th century, a large number of studies dealt with the problem of time series homogenisation, but before 1990 only the single change-point problem was analysed intensively. Around and after the mid 20th century the double mass analysis became a popular method (Kohler, 1949). In this old method the accumulated anomalies (i.e. progressive sums of anomalies from the beginning of the examined period) are visually compared between the tested series (candidate series) and another series (reference series). Among the later developed methods, the Craddock-test (Craddock, 1979) and the Buishand tests (Buishand, 1982) are also based on the examination of accumulated anomalies. Buishand (1982) provided empirically-developed significance thresholds to his methods, eliminating the need of subjective decisions about the significance of possible change-points. By contrast, the Craddock-test is essentially subjective.

Beyond the examination of accumulated anomalies, several other kinds of tests were developed for solving the single change-point problem. A few of the most used methods are mentioned here. Wilcoxon Rank Sum Test (Wilcoxon, 1945, also known as Mann-Whitney-Wilcoxon test) is a non-parametric method. It is based on the calculation of rank sum statistics before and after the potential change-point. The Maronna-Yohai test (Maronna and Yohai, 1978) is based on maximum likelihood estimations. In Standard Normal Homogeneity Test (SNHT, Alexandersson, 1986) the section-means before and after the potential change-point are compared. Solow (1987) searched for the change-points by fitting two-phase linear regression to the data points. The mentioned four kinds of approach (i.e. non-parametric method, maximum likelihood estimation, comparison of section-means, fitting linear regressions) have several other representatives. Among the contemporary methods the maximum likelihood methods were found to be the best theoretically and they show the highest performance in efficiency test examinations (Domonkos, 2008). However, none of the contemporary statistical methods treats the complex interactions of multiple change-points on the examined test-statistics or their effects on the calculation of correction-terms.

A seeming change-point in a climatic time series may have a macro-climatic or a local, technical origin, and it was known from the beginning that one must distinguish these two during the homogenisation (Conrad and Pollak, 1950). Therefore homogenisation is applied on differences or ratios of two series (relative homogenisation) instead of on one time series (absolute homogenisation), with very few exceptions. In relative homogenisation one of the crucial problems is that the detected change-points often could be originated from more than one inhomogeneous time series and it is not easy to find the true "culprit". An important step forward was the creation of reference series from composite series of the nearby stations (WMO, 1966).

The best-known and most widely applied homogenisation method is SNHT. Alexandersson (1986) did not construct better method than his contemporaries, but his merits are that he set a good example of practical application from the selecting of time series until the final interpretation of detection results and provided a user-friendly description of his statistical method.

There was one additional line of methodological development at that time. Hawkins (1972) constructed the method of optimal segmentation of step-functions for the case of known number of steps. The presented method was not only optimal in statistical sense, but also economical in the use of computer-time, therefore this method is known also as dynamic programming algorithm. With that step research came close to the solution of the multiple inhomogeneities problem. However, at that time the importance of the multiple breaks problem in climatic time series was not widely recognized. Instead of following this line of research, in the following years Hawkins turned back to the examination of single change-point model (Hawkins, 1977).

## 4 Fast development from the 90's

Around 1990 global climate change started to be seen as a potential serious problem and consequently the homogeneity of the observed climatic data became more important. In 1994 the first seminar on data quality control and homogenisation was held, and since then this series of seminars, which is supported by the World Meteorological Organisation (WMO), is held regularly in Budapest. New research lines started in the framework of these seminars, but also independently from them. In the first homogenisation seminar appeared the initial versions of some excellent statistical tools. That time the first attempts to detect multiple change-points as a coherent structure (joint detection of multiple change-points) was published (Szentimrey, 1996; Caussinus and Mestre, 1996, as well as the stepwise comparison of time series, Caussinus and Mestre, 1996) and the use of multiple reference series (Szentimrey, 1996). In the same year as the first homogenization seminar, Peterson and Easterling (1994) published a new method to create reference series from the weighted averages of values in neighbouring stations, where the weights are the squared correlations of the increment series. Since then the latter three methods can be considered recommended ways for spatial comparisons of time series.

Easterling and Peterson (1995) published a hierarchic method for identifying multiple change-points. Later this method became popular, and it is often referred as cutting algorithm. Modern efficiency examinations show (e.g. Domonkos, 2011a) that apart from the joint detection of multiple change-points the cutting algorithm is the best available tool. A complete new homogenisation method was also published in the same study (the combination of two phase regression method with multiple permutation procedure), but that has no better performance than the other contemporary methods.

In 1996 the first attempt to detect both change-points and trend-like inhomogeneities was published, Lanzante (1996) applied the modified version of Wilcoxon Rank Sum Test. However, we do not know of any later applications of that test. Later studies that also considered trend-like inhomogeneities, always applied the late version of SNHT (Alexandersson and Moberg, 1997) or Multiple Linear Regression (MLR, Vincent, 1998). However, efficiency-tests show that these detection methods have often poorer performance than the best of the other methods (Domonkos, 2008, 2011a). The performance of the newer version of SNHT is poorer than that of the earlier version (Alexandersson, 1986). On the other hand, Moberg and Alexandersson (1997) set a good example how to practically apply SNHT or other homogenisation methods that do not include joint detection of multiple change-points. They applied the reference creation by Peterson and Easterling (1994) and also the cutting algorithm. They developed a seemingly appealing semi-hierarchic algorithm for detecting multiple change-points, but its effect is neutral on the efficiency of detection results (Domonkos, 2011a).

In the second part of the 90's the development of methods with joint detection of multiple change-points was the most important research-line. Multiple Analysis of Series for Homogenisation (MASH, Szentimrey, 1999) counts all the possible combination of change-point positions and selects the most likely one based on hypothesis tests. The method uses multiple reference series to reduce the impact of inhomogeneities in reference series on the detection of inhomogeneities in the candidate series, and it has a specific philosophy for keeping the false alarm rate low, i.e. the lower limits of confidence intervals are used as adjustment-factors. MASH is rather complicated and it is more time consuming than any other homogenisation method, but at the end of the 90's it was likely the most effective method. Considering the reliability and preciseness of change-point detection, the Caussinus-Mestre method (its modern version is known as PRODIGE, Caussinus and Mestre, 1996; Mestre, 1999), could be its only competitor, but at that time the Caussinus-Mestre method had not reached its final form yet. PRODIGE applies multiple stepwise comparisons instead of reference series. Its detection part uses the maximum likelihood principle with the Caussinus-Lyazrhi criterion (Caussinus and Lyazrhi, 1997) for finding the optimal number of change-points in best fitting step-functions. This criterion combined with the dynamic programming algorithm to optimize the break positions (Hawkins, 1972) makes PRODIGE a powerful tool. A few years later the detection algorithm of

PRODIGE was refined and the ANOVA correction method was introduced (Caussinus and Mestre, 2004). Since that development the efficiency of PRODIGE is similar to the efficiency of MASH.

## 5 The 21th century: intercomparisons

Around 2000 the first review studies about homogenisation methods appeared (Peterson et al., 1998; Aguilar et al., 2003; Auer et al., 2005, etc.) and there are some studies about the efficiency of change-point detection by different methods (Easterling and Peterson, 1995; Lanzante, 1996; Ducré-Robitaille et al., 2003, etc.). However, the first experiments for characterising objectively the efficiencies of homogenisation methods suffered from substantial shortcomings: (i) simple models were used to generate test-datasets, thus the resemblance between test datasets and true observational datasets was generally low; (ii) detection skill (i.e. the skill in finding the positions of breaks in time series, see e.g. Menne and Williams Jr., 2005) was calculated only, and this measure does not always characterise well the skill in reconstructing and preserving the true characteristics of climatic variability in homogenised time series, (iii) small number of arbitrarily selected methods were tested only.

From 2004 our general knowledge about the statistical properties of inhomogeneities in observational datasets widened. In experiments with a Hungarian observed temperature dataset, statistical characteristics of inhomogeneity-detection results were compared for true and simulated datasets. The empirical results showed that high similarity between the characteristics for the observed and artificial datasets can be obtained only when large number of short-term, platform-like inhomogeneities are included in simulated time series and most inhomogeneities have small magnitude. These characteristics could remain hidden in direct examination of observed time series, because short-term inhomogeneities and small-size inhomogeneities often cannot be detected at all because of the noise. The newly discovered feature of observed time series is important, because small-size inhomogeneities have impact on the performance in detecting and correcting large-size inhomogeneities (Domonkos, 2004, 2011a).

Menne and Williams Jr. (2005) examined the detection skills of homogenisation methods using a simulated dataset that included change-point sizes determined by a normal distribution. Domonkos (2008) examined the efficiency of detection parts of fifteen homogenisation methods with a test dataset whose properties were similar to the observed temperature dataset of Hungary. Beyond detection skill, other efficiency-measures, such as RMSE and the accuracy of linear trend-slopes in homogenised time series were calculated. The influence of parameter-choices on the detection results was analysed in that study as well.

## 6 The HOME period

Between 2007 and 2011 HOME provided favourable conditions for the further developments. A benchmark dataset was created (Venema et al., 2010), which contains surrogate climate networks mimicking the statistical properties of monthly temperature and precipitation time series in European observational networks well. In the benchmark the frequency of inhomogeneities is diverse, and the size-distribution of inhomogeneities approaches well the true characteristics (much more small inhomogeneities than large ones). In the benchmark the resemblance of simulated networks to the real world is demonstrably high which was not common in earlier validation studies of homogenisation methods. HOME and the benchmark provided an excellent opportunity to evaluate the performances of a large number of whole homogenisation procedures with the participation of homogenisers from different countries (Venema et al., 2012). During the benchmark homogenization the true positions of the breaks remained unknown for the homogenizers. The blind test results were evaluated calculating RMSE of monthly and annual values, RMSE of trend-slope estimatons, detection power, false alarm rate, and some other efficiency measures. These tests confirmed that PRODIGE and MASH are indeed among the best homogenisation tools, but several other results may have been less expected. One main finding of the experiments was that the difference between the efficiency of PRODIGE and MASH on the one hand, and the other known methods such as SNHT, Penalised Maximum t-test (PMT, Wang et al., 2007), etc. is substantially larger than the differences between detection performances according to Domonkos (2008, 2011a). The found larger differences for performances of whole procedures are likely due to the over-simplified correction algorithms of several methods.

During HOME a new homogenisation method, Applied Caussinus-Mestre Detection Algorithm for homogenising Networks of Temperature series (ACMANT, Domonkos, 2011b) was developed for homogenising monthly temperature series. The ACMANT includes the step-function fitting and ANOVA correction segments of PRODIGE, but applies a bivariate-test for detecting change-points. Two annual variables are used in the detection, one is the annual mean, and the other is the summer-winter difference. Due to radiation-connected biases, joint inhomogeneities of these variables are frequent in mid-latitude temperatures (Drougue et al., 2005; Domonkos and Štěpánek, 2009; Brunet et al., 2011, etc.), thus the detection model of ACMANT is a powerful tool. The benchmark results show that ACMANT is the most effective homogenisation method for temperature, where it should be noted that the presented contribution by ACMANT was submitted after the deadline and the test was thus not blind. ACMANT applies the method of Peterson and Easterling (1994) to build the reference series and with its inclusion the procedure could be made fully automatic. ACMANT is able to homogenise time series with data-gaps and networks of different-length time series in an automatic way. It applies sophisticated tools for treating the connections between annual- and monthly-scale examinations and corrections. Note that the use of bivariate search for change-point positions is not limited to ACMANT; it has recently been applied in another study to homogenise solar radiation and sunshine duration (Guijarro, 2011).

The methodological development has also been continued overseas. A new automatic homogenisation method, USHCN (Menne and Williams Jr., 2009), was published for homogenising huge temperature datasets, as found in the United States. The detection part of USHCN includes the early version of SNHT, cutting algorithm, Bayesian-based decisions about the form of inhomogeneities, i.e. trend-like inhomogeneities can also be detected, and a special purpose significance test. Important novelties of USHCN are that it applies pairwise comparisons in automated way and automatically uses metadata. USHCN applies homogeneity-adjustments only when the individual estimates from pairwise comparisons concordantly indicate the need for same-sign adjustment. USHCN was tested in the HOME experiment. Its general efficiency turned out to be slightly poorer than that of PRODIGE, MASH and ACMANT, but USHCN only performed annual correction and has the lowest false alarm rate. This latter positive feature of USHCN might become crucially important when datasets with low frequency of large-size inhomogeneities are examined.

Another important step in the methodological development is the introduction of perturbed parameter experiments to the test-process of automatic homogenisation methods (McCarthy et al., 2008; Domonkos, 2008; Titchner et al., 2009; Williams et al., 2012). In these examinations some parameters of the homogenisation methods are varied randomly (in ensemble tests) or systematically (in sensitivity tests) and from the synthesis of the results a more complete picture can be obtained about the connections between the dataset-properties and the performance of the applied homogenisation method.

Although the main line of achieving more precise homogenisation methods is the development of more powerful statistical tools, an old and very simple method, namely the Craddock-test was proven to be a very effective homogenisation method in the blind test experiments of HOME. In case of the Craddock-test, the main protagonist is the homogeniser. Skilled homogenisers may assess the timing and size of change-points well by examining the time series subjectively with the help of some simple statistical characteristics such as the series of accumulated anomalies for the differences of the compared time series. The relation between a skilled homogeniser and an automatic homogenisation method is similar to that between a chess-master and a chess-automat. In the HOME experiments, the results of Gregor Vertacnik (Slovenia) and Michele Brunetti (Italy) were as good as that of the best objective and semi-objective homogenisation methods. Note that the Craddock-

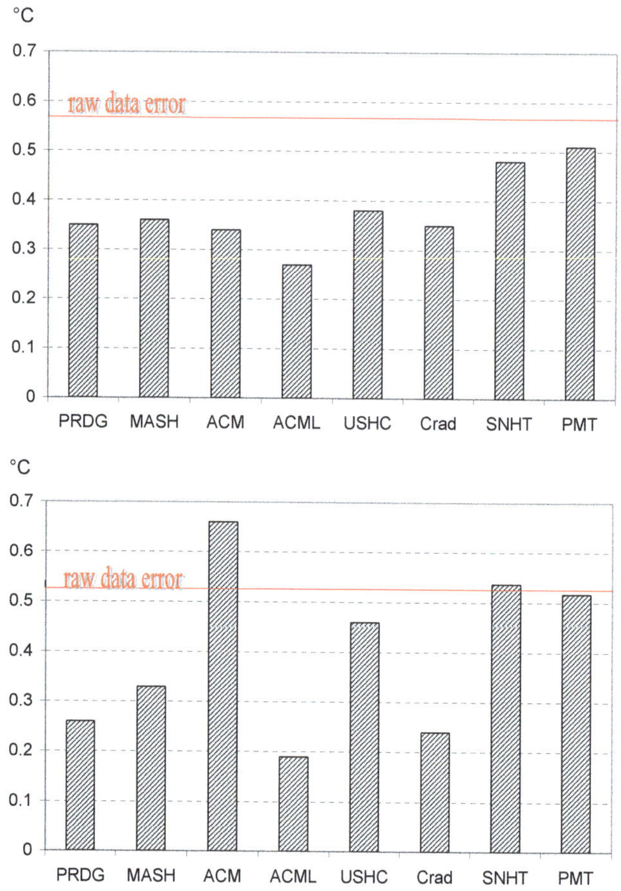


**Figure 2.** Efficiencies of monthly temperature homogenisation for selected homogenisation methods. PRDG = PRODIGE, ACM = early ACMANT, ACML = late ACMANT, USHC = USHCN, Crad = Craddock-test. **(a)** (upper) RMSE of monthly values. **(b)** (bottom) RMSE of biases in network-mean linear trends.

homogenisers used some selected parts of the benchmark only, thus their results are not fully comparable with the results of complete experiments.

The HOME experiments show that the best homogenisation methods are PRODIGE, MASH, ACMANT, USHCN and the Craddock-test. Note that these results refer primarily to the homogenisation of surface temperature time series of relatively dense European or North American observing networks. In Fig. 2 some efficiency results are illustrated for the seven best methods participated in the Benchmark homogenisation. In the construction of these figures full contributions (that used all the 15 surrogated temperature networks; Venema et al., 2012) were taken into account, except for the Craddock-test as Vertacnik's partial contribution contains only 7 networks. When authors produced more than one full experiment with different versions of their methods, the average error of these versions is shown. Most of the efficiencies shown in Fig. 2 are identical to the equivalent results of Venema et al. (2012), except the results for ACMANT late which method was finalised later than the HOME experiments. One can see from the figures that the RMSE error of homogenised time series is lower than that of the raw data for all the methods shown. On the other hand, the efficiency can be negative in such an important climatic characteristic as the accuracy of the network-wide mean linear trend. For this characteristic, significant positive efficiencies are produced only by PRODIGE, MASH, Craddock-test and ACMANT late. Note that after the HOME experiments ACMANT late was subjected to further tests, both to blind tests and perturbed parameter tests, and that results (Domonkos, 2012) confirmed the good performance of ACMANT late (Fig. 2).

Based these theoretical considerations and the experiments with the benchmark, the HOME team developed a software-package for homogenising temperature and precipitation time series (www.homogenisation.org). This package incorporates segments from the best homogenisation methods examined, thus its efficiency is likely similar or better than the other best methods.

Finally, it must be noted that progress in homogenisation methods is never limited to the development and application of statistical tools. A recent example is the Spanish screen bias experiment in which the inhomogeneities caused by the change of the thermometer-screens in the early 20th century have been quantitatively assessed relying on experimental parallel measurements from the recent years (Brunet et al., 2011).

## 7 Concluding remarks

These conclusions refer particularly to the homogenisation of annual and monthly surface temperature time series of sufficiently dense observing networks, although some are more generally valid.

According to our present knowledge six homogenisation methods can be recommended. They are PRODIGE, MASH, ACMANT, USHCN, the Craddock-test and the HOME-software.

The appropriateness of the six methods listed above is often markedly different in solving particular tasks. For instance ACMANT is a highly efficient tool for homogenising temperature datasets of mid-latitudes, but is not tailored to other variables. For homogenising huge datasets USHCN or ACMANT are recommendable, because these methods are fully automatic. The HOME-software, PRODIGE and MASH are usable in wide range of tasks, but certain expertise is needed for their use. The Craddock-test is subjective and is inappropriate for homogenising large datasets.

Further tests are needed to understand the performance of homogenisation methods better. The characteristics of climatic time series are diverse, thus a large number of experiments with varied dataset properties is needed.

The general advantages of fully automatic methods are that they easily can be tested in multiple experiments; their test results are objective and can be reconstructed at any time.

The use of automatic homogenisation methods still needs some expertise in the time series preparation and in the interpretation of the homogenisation results.

**Acknowledgements.** The research was supported by the European projects COST ES0601 and EURO4M FP7-SPACE-2009-1/242093, by the Spanish project "Tourism and Environment" ECO 2010-18158 and by the German Science Foundation project "Daily Stew" VE366/5-1.

Edited by: M. Brunet-India

## References

Aguilar, E., Auer, I., Brunet, M., Peterson, T. C., and Wieringa, J.: WMO Guidelines on climate metadata and homogenization, WCDMP-No. 53, WMO-TD No 1186, WMO, Geneva, 2003.

Alexandersson, H.: A homogeneity test applied to precipitation data, J. Climatol., 6, 661–675, doi:10.1002/joc.3370060607, 1986.

Alexandersson, H. and Moberg, A.: Homogenization of Swedish temperature data. Part I: Homogeneity test for linear trends, Int. J. Climatol., 17, 25–34, 1997.

Auer, I., Böhm, R., Jurković, A., Orlik, A., Potzmann, R., Schöner, W., Ungersböck, M., Brunetti, M., Nanni, T., Maugeri, M., Briffa, K., Jones, P., Efthymiadis, D., Mestre, O., Moisselin, J.-M., Begert, M., Brazdil, R., Bochnicek, O., Cegnar, T., Gajić-Čapka, M., Zaninović, K., Majstorović, Ž., Szalai, S., Szentimrey, T., and Mercalli, L.: A new instrumental precipitation dataset for the greater Alpine region for the period 1800–2002, Int. J. Climatol., 25, 139–166, doi:10.1002/joc.1135, 2005.

Brunet, M., Asin, J., Sigró, J., Bañon, M., García, F., Aguilar, E., Palenzuela, J. E., Peterson, T. C., and Jones, P.: The minimization of the screen bias from ancient Western Mediterranean air temperature records: an exploratory statistical analysis, Int. J. Climatol., 31, 1879–1895, doi:10.1002/joc.2192, 2011.

Buishand, T. A.: Some methods for testing the homogeneity of rainfall records, J. Hydrol., 58, 11–27, 1982.

Caussinus, H. and Lyazrhi, F.: Choosing a linear model with a random number of change-points and outliers, Ann. Inst. Statist. Math., 49, 761–775, doi:10.1023/A:1003230713770, 1997.

Caussinus, H. and Mestre, O.: New mathematical tools and methodologies for relative homogeneity testing, in: Proceedings of the First Seminar for Homogenization of Surface Climatological Data, edited by: Szalai, S., Hungarian Meteorlogical Service, Budapest, Hungary, 63–82, 1996.

Caussinus, H. and Mestre, O.: Detection and correction of artificial shifts in climate series, J. Roy. Stat. Soc., Series C53, 405–425, doi:10.1111/j.1467-9876.2004.05155.x, 2004.

Conrad, V.: Homogenitätsbestimmung meteorologischer Beobachtungsreihen, Meteorol. Z., 482–485, 1925.

Conrad, V. and Pollak, C.: Methods in Climatology, Harvard University Press, Cambridge, MA, p. 459, 1950.

Craddock, J. M.: Methods of comparing annual rainfall records for climatic purposes, Weather, 34, 332–346, 1979.

Domonkos, P.: Testing of homogenisation methods, in: 5th European Conference of Applied Climatology, 26–30 September, Nice, France, EMS04-A-00098, 2004.

Domonkos, P.: Testing of homogenisation methods: purposes, tools and problems of implementation, Proceedings of the 5th Seminar and Quality Control in Climatological Databases, WCDMP-No. 71, WMO-TD 1493, WMO, Geneva, 126–145, 2008.

Domonkos, P.: Efficiency evaluation for detecting inhomogeneities by objective homogenisation methods, Theor. Appl. Climatol., 105, 455–467, doi:10.1007/s00704-011-0399-7, 2011a.

Domonkos, P.: Adapted Caussinus-Mestre Algorithm for Networks of Temperature series (ACMANT), Int. J. Geosci., 2, 293–309, doi:10.4236/ijg.2011.23032, 2011b.

Domonkos, P.: ACMANT: Why is it efficient? Proceedings of the 7th Seminar and Quality Control in Climatological Databases, WMO-HMS, www.c3.urv.cat/publicacions/publicacions2012.html, in press, 2012.

Domonkos, P. and Štěpánek, P.: Statistical characteristics of detectable inhomogeneities in observed meteorological time series, Studia Geoph. et Geod., 53, 239–260, doi:10.1007/s11200-009-0015-9, 2009.

Drogue, G., Mestre, O., Hoffmann, L., Iffly, J.-F., and Pfister, L.: Recent warming in a Small region with semi-oceanic climate, 1949–1998: what is the ground truth?, Theor. Appl. Climatol., 81, 1–10, doi:10.1007/s00704-004-0088-x, 2005.

Ducré-Robitaille, J.-F., Vincent, L. A., and Boulet, G.: Comparison of techniques for detection of discontinuities in temperature series, Int. J. Climatol., 23, 1087–1101, doi:10.1002/joc.924, 2003.

Easterling, D. R. and Peterson, T. C.: A new method for detecting undocumented discontinuities in climatological time series, Int. J. Climatol., 15, 369–377, 1995.

Guijarro, J. A.: Joint homogenization of sunshine duration and global solar radiation Spanish series. 10th European Conference on Applied Meteorology, 12–16 September, Berlin, Germany, EMS2011-679, 2011.

Hawkins, D. M.: On the choice of segments in piecewise approximation, J. Inst. Math. Appl., 9, 250–256, doi:10.1093/imamat/9.2.250, 1972.

Hawkins, D. M.: Testing a sequence of observations for a shift in location, J. Amer. Stat. Assoc., 72, 180–186, 1977.

Kohler, M. A.: Double-mass analysis for testing the consistency of records and for making adjustments, B. Am. Meteorol. Soc., 30, 188–189, 1949.

Kreil, K.: Mehrjährige Beobachtungen in Wien vom Jahre 1775 bis 1850, Jahrbücher der k.k. Central-Anstalt für Meteorologie und Erdmagnetismus, I. Band – Jg. 1848 und 1849, 35–74, 1854.

Lanzante, J. R.: Resistant, robust and non-parametric techniques for the analysis of climate data: theory and examples, including applications to historical radiosonde station data, Int. J. Climatol., 16, 1197–1226, 1996.

Maronna, R. and Yohai, V. J.: A bivariate test for the detection of a systematic change in mean, J. Amer. Stat. Assoc., 73, 640–645, 1978.

McCarthy, M. P., Titchner, H. A., Thorne, P. W., Tett, S. F. B., Haimberger, L., and Parker, D. E.: Assessing bias and uncertainty in the HadAT adjusted radiosonde climate record, J. Climate, 21, 817–832, 2008.

Menne, M. J. and Williams Jr., C. N.: Detection of undocumented changepoints using multiple test statistics and composite reference series, J. Climate, 18, 4271–4286, doi:10.1175/JCLI3524.1, 2005.

Menne, M. J. and Williams Jr., C. N.: Homogenization of temperature series via pairwise comparisons, J. Climate, 22, 1700–1717, doi:10.1175/2008JCLI2263.1, 2009.

Mestre, O.: Step-by-step procedures for choosing a model with change-points, Second Seminar for Homogenization of Surface Climatological Data, edited by: Szalai, S., Szentimrey, T., and Szinell, Cs., WCDMP 41, WMO-TD 962, WMO, Geneva, 15–25, 1999.

Moberg, A. and Alexandersson, H.: Homogenization of Swedish temperature data. Part II: Homogenized gridded air temperature compared with a subset of global gridded air temperature since 1861, Int. J. Climatol., 17, 35–54, 1997.

Peterson, T. C. and Easterling, D. R.: Creation of homogeneous composite climatological reference series, Int. J. Climatol., 14, 671–679, doi:10.1002/joc.3370140606, 1994.

Peterson, T. C., Easterling, D. R., Karl, T. R., Groisman, P., Nicholls, N., Plummer, N., Torok, S., Auer, I., Böhm, R., Gullett, D., Vincent, L., Heino, R., Tuomenvirta, H., Mestre, O., Szentimrey, T., Salinger, J., Førland, E. J., Hanssen-Bauer, I., Alexandersson, H., Jones, P., and Parker, D.: Homogeneity adjustments of in situ atmospheric climate data: a review, Int. J. Climatol., 18, 1493–1517, 1998.

Solow, A. R.: Testing for climate change: an application of the two-phase regression model, J. Clim. Appl. Meteor., 26, 1401–1405, 1987.

Szentimrey, T.: Statistical procedure for joint homogenization of climatic time series, in: Proceedings of the First Seminar for Homogenization of Surface Climatological Data, edited by: Szalai, S., Hungarian Meteorlogical Service, Budapest, Hungary, 47–62, 1996.

Szentimrey, T.: Multiple Analysis of Series for Homogenization (MASH), in: Second Seminar for Homogenization of Surface Climatological Data, edited by: Szalai, S., Szentimrey, T., and Szinell, Cs., WCDMP 41, WMO-TD 962, WMO, Geneva, 27–46, 1999.

Titchner, H. A., Thorne, P. W., McCarthy, M. P., Tett, S. F. B., Haimberger, L., and Parker, D. E.: Critically reassessing tropospheric temperature trends from radiosondes using realistic validation experiments, J. Climate, 22, 465–485, doi:10.1175/2008JCLI2419.1, 2009.

Venema, V., Mestre, O., and the COST HOME Team: Benchmark database, EGU General Assembly, Vienna, Austria, 3–7 May 2010, EGU CL4.6-13357, 2010.

Venema, V. K. C., Mestre, O., Aguilar, E., Auer, I., Guijarro, J. A., Domonkos, P., Vertacnik, G., Szentimrey, T., Stepanek, P., Zahradnicek, P., Viarre, J., Mller-Westermeier, G., Lakatos, M., Williams, C. N., Menne, M. J., Lindau, R., Rasol, D., Rustemeier, E., Kolokythas, K., Marinova, T., Andresen, L., Acquaotta, F., Fratianni, S., Cheval, S., Klancar, M., Brunetti, M., Gruber, C., Prohom Duran, M., Likso, T., Esteban, P., and Brandsma, T.: Benchmarking homogenization algorithms for monthly data, Clim. Past, 8, 89–115, doi:10.5194/cp-8-89-2012, 2012.

Vincent, L. A.: A technique for the identification of inhomogeneities in Canadian temperature series, J. Climate, 11, 1094–1104, 1998.

Wang, X. L., Wen, Q. H., and Wu, Y.: Penalized maximal t test for detecting undocumented mean change in climate data series, J. Appl. Meteor. Climatol., 46, 916–931, doi:10.1175/JAM2504.1, 2007.

Wilcoxon, F.: Individual comparisons by ranking methods, Biometrics Bull., 1, 80–83, 1945.

Williams, C. N., Menne, M. J., and Thorne, P. W.: Benchmarking the performance of pairwise homogenization of surface temperatures in the United States, J. Geophys. Res.-Atmos., 117, D05116, doi:10.1029/2011JD016761, 2012.

WMO: Climatic Change, Report of a working group of the Commission for Climatology, Technical Note 79, WMO – No. 195. TP. 100, pp. 79, 1966.

# 36

# National Climate Observing System of Switzerland (GCOS Switzerland)

**G. Seiz and N. Foppa**

Swiss GCOS Office, Federal Office of Meteorology and Climatology MeteoSwiss, Switzerland

**Abstract.** In recent decades, the global observation of climate and climate change has become increasingly important. The Global Climate Observing System (GCOS) established in 1992 addresses the entire climate system including physical, chemical and biological properties of atmosphere, ocean and land surface. This paper describes the GCOS implementation in Switzerland and highlights some major achievements over the last few years. The Swiss GCOS Office was established at the Federal Office of Meteorology and Climatology MeteoSwiss in February 2006, to coordinate all climate-relevant measurements in Switzerland. The first-ever inventory of the country's long-term climatological data series and international data centres, including an assessment of their future prospects, was compiled in 2007. The National Climate Observing System of Switzerland (GCOS Switzerland) includes long-term climatological data series in the atmosphere and terrestrial domains, international data and calibration centres, satellite-based products and support of climate observations in developing countries. A major milestone in the surface-based atmospheric observations was the definition of the Swiss National Basic Climatological Network (NBCN), consisting of 29 stations of greatest climatological importance. The NBCN was further densified for precipitation with 46 additional daily precipitation stations (NBCN-P). Analysis of the homogenized timeseries of the average temperature in Switzerland shows a total warming of +1.6 °C from 1864 to 2010. In the terrestrial domain, substantial advances were made in all three subdomains hydrosphere, cryosphere and biosphere. As example for the use of satellite data within GCOS Switzerland, the 10-yr MODIS monthly mean cloud fraction climatology over Switzerland from March 2000 to February 2010 is presented, which illustrates the differences in cloud cover between mountainous regions and flatland regions in winter, as well as the north-south gradient in cloud cover over Switzerland in summer.

## 1 Introduction

In recent decades – especially following the adoption of the Climate Convention in 1992 – the demand for observations of climate and climate change has steadily increased. For scientific conclusions on climate change, the attribution of anthropogenic influences and future climate scenarios, long-term, high-quality data series are essential. The recently published Fourth Assessment Report of the Intergovernmental Panel on Climate Change (IPCC, 2007) summarizes the current state of knowledge on climate change and its global impacts. For Switzerland, the regional impacts have been analyzed in a recent report by the Swiss Advisory Body on Climate Change (OcCC, 2007). According to the OcCC report, Switzerland will face autumn, winter and spring temperatures to increase by around 2 °C and summer temperatures by almost 3 °C by 2050 compared to 1990. Precipitation levels are projected to rise by about 10% in winter and to fall by about 20% in summer. In addition, the frequency of extreme precipitation events is expected to increase, especially in winter. Climate change scenarios for Switzerland, however, are quantitatively still very uncertain. Early detection of long-term climate variations is therefore essential for the planning and management of climate change adaptation strategies.

This paper gives an overview about the National Climate Observing System in Switzerland (GCOS Switzerland), covering the atmospheric and terrestrial domains, and including international data and calibration centres in Switzerland, the role of satellite data within GCOS Switzerland and the support of climate observations abroad. Finally, an outlook is given on the future strategy of GCOS Switzerland.

## 2 Global Climate Observing System (GCOS)

### 2.1 GCOS overview

The Global Climate Observing System (GCOS) was established in 1992, parallel to the establishment of the UN Framework Convention on Climate Change (UNFCCC). The aims and requirements of systematic observation are specified in Article 4 and 5 of the UNFCCC and in Article 10 of the subsequent Kyoto Protocol. GCOS is designed to ensure that the observations and information needed to address climate-related issues are obtained and made available to all potential users. GCOS is co-sponsored by the World Meteorological Organization (WMO), the Intergovernmental Oceanographic Commission of UNESCO, the UN Environment Programme (UNEP) and the International Council for Science (ICSU). Building on existing networks and systems, GCOS encompasses the total climate system including observations of physical, chemical and biological properties of the atmosphere, the ocean and the land surface. Within GCOS, a set of so-called Essential Climate Variables (ECV) has been defined, subdivided into atmospheric, oceanic and terrestrial domain (WMO, 2003). The implementation of GCOS is based on the Second Adequacy Report (WMO, 2003) and the Implementation Plan (WMO, 2004). The GCOS Implementation Plan (GIP-04) and its 2010 Update (WMO, 2010a; GIP-10) thereby clearly highlights the role of every country to set up national coordination mechanisms and national plans for the provision of systematic observation of the climate system. Such mechanisms are usually best sustained when a national coordinator (or at least a focal point) is designated and assigned responsibility to coordinate planning and implementation of systematic climate observing systems across the many departments and agencies involved. Regular reporting on climate observations is also an official duty for each Party to the UNFCCC as part of the UNFCCC National Communication (Chapter "Research and Systematic Observation").

### 2.2 National GCOS implementation in Switzerland

Switzerland has a long tradition of climate observation. Temperature and precipitation series of more than 150 yr, the world's longest total ozone series and glacier measurements dating back to the end of the 19th century are only a few highlights of Switzerland's contribution to global and regional climate monitoring. The Fourth IPCC Assessment Report makes reference to various papers concerning Swiss climatological time series, e.g. for precipitation (Schmidli and Frei, 2005), radiation (Philipona et al., 2005; Wild et al., 2005), snow (Scherrer et al., 2004), glaciers (Zemp et al., 2005), permafrost (Vonder Mühll et al., 2004) and phenology (Defila and Clot, 2001).

Following the ratification of the Kyoto Protocol by the Swiss Parliament in summer 2003, the national GCOS coordination was strengthened at the Federal Office of Meteorology and Climatology MeteoSwiss. On 1 February 2006, the Swiss GCOS Office was established, building on the former GCOS Focal Point at MeteoSwiss. The Swiss GCOS Office is responsible for coordinating climatological observations carried out in Switzerland by federal offices, research institutes and universities. Once a year, a so-called "National Swiss GCOS Roundtable" is organized to foster the information exchange between the different partners and to guarantee a coherent strategy. The coordination also includes long-term planning to ensure continuous and representative observations, e.g. by identifying the risk of discontinuity ahead of time and engaging in remedial action. As far as possible, new measurement techniques are also considered in the integrated observation system. In addition, the Swiss GCOS Office identifies resource-related problems affecting the operation of international data and calibration centres in Switzerland and provides financial and technological support for selected observations abroad.

## 3 National Climate Observing System (GCOS Switzerland)

In 2006, the Swiss GCOS Office and ProClim (i.e. the Forum for Climate and Global Change of the Swiss Academy of Sciences SCNAT) performed a survey among Swiss universities, research institutes and federal offices with the aim of documenting climatological series and identifying risks to their continuation. The criteria were defined on the basis of similar studies for the selection of climatological stations (WMO, 1997). The main criteria specified were that a series should: (a) cover a period of more than 50 yr, (b) be longer than comparable series abroad, or (c) concern recently introduced climate variables or observation methods. The secondary criteria included participation in international agreements or data centres, geographical representativeness, data quality and availability of metadata. Respondents were also asked to indicate their requirements for funding in order to ensure continued operation of stations. Based on the survey results, a report about the National Climate Observing System (GCOS Switzerland) has been completed (Seiz and Foppa, 2007). The report is the first inventory of Switzerland's long-term climatological data series and international data centres, including an assessment of their future prospects. The inventory was compiled in cooperation with ProClim and the responsible federal offices, research institutes and universities advised by a steering committee consisting of scientists and policymakers.

Based on the report, the Swiss Federal Council has agreed in June 2008 on a long-term financial contribution to GCOS Switzerland, starting from 2010. This financial contribution will cover the operation costs of several long-term climatological data series of carbon dioxide, freeze dates of lakes, snow water equivalent, glaciers, permafrost and

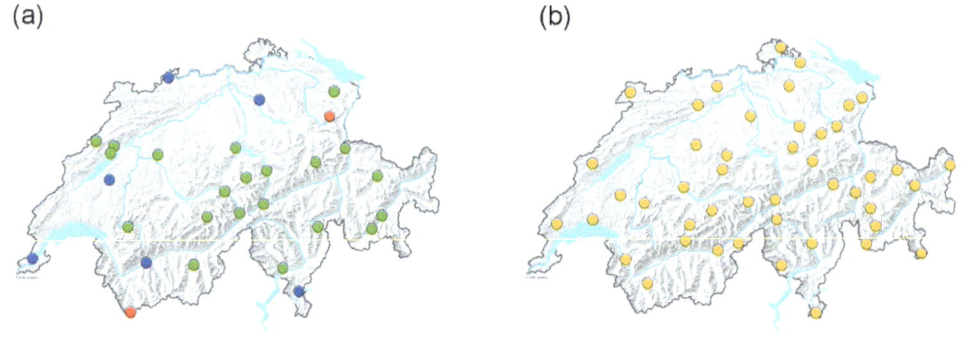


**Figure 1.** Swiss National Basic Climatological Network (NBCN). **(a)** 29 NBCN stations; 2 stations also belong to the GCOS Surface Network GSN (red) and 8 to the Regional Basic Climatological Network RBCN (red + blue). **(b)** additional 46 precipitation stations which complement the NBCN stations to the densified network for precipitation (NBCN-P).

**Table 1.** Essential climate variables of the atmospheric domain, together with additional variables of relevance for Switzerland (in italics), which are part of GCOS Switzerland.

| | |
|---|---|
| Surface | Air temperature, Precipitation, Air pressure, Surface radiation budget, Wind speed and direction, Water vapour |
| Upper Air | Earth radiation budget (including solar irradiance), Upper air temperature, Wind speed and direction, Water vapour, Cloud properties |
| Composition | Carbon dioxide, Methane, Ozone, Other long-lived greenhouse gases, Aerosol properties, *Pollen* |

phenology, and of two international data centres in Switzerland (World Glacier Monitoring Service WGMS, palaeo-historic database Euro-Climhist), whose future were identified to be uncertain in the inventory report.

## 3.1 Atmospheric domain

The most important systematic observations of the atmospheric domain in Switzerland concern the surface, upper air and composition (Table 1). Pollen is not listed as ECV in the GCOS Second Adequacy Report, but its measurement has a long tradition in Switzerland and consequently, its series are an important part of the Swiss climate observing system.

In the following, some examples from the surface, upper air and atmospheric composition observations are given. The systematic measurements of surface-based atmospheric variables extend as far back as December 1863, when Switzerland's first nationwide meteorological observation network came into operation. At many of the sites chosen in 1863, stations are still in operation today. In the NORM90 project (Begert et al., 2003, 2005), for each of Switzerland's twelve major climate regions, a station was selected where measurement data have been collected since at least 1900. These long time series were analyzed for artificial discontinuities and trends caused, for example, by station relocation, change of instrumentation and calibrations, and homogenized (Begert et al., 2005). In order to increase the density of stations, particularly in the Central Alpine region characterized by large differences in altitude, 16 additional stations were selected with time series from at least 1900 (exception: Jungfraujoch only from 1930). These stations of the greatest climatological importance (28 in all) have recently been designated as the Swiss National Basic Climatological Network (NBCN) (Begert et al., 2007). Two NBCN stations are also GCOS Surface Network (GSN) stations – Säntis and Grand St. Bernard. Seven NBCN stations (Säntis, Grand St. Bernard, Geneva, Sion, Basel, Zurich and Lugano) belong to the WMO Regional Basic Climatological Network (RBCN). Additionally, the aerological station Payerne, as GCOS Upper Air Network (GUAN) station, has been added to the NBCN selection, leading to a total of 29 NBCN stations (Fig. 1a). For precipitation, the Swiss NBCN stations are complemented by 46 precipitation stations (with daily measurements) – selected on the basis of a cluster analysis – to densify the Swiss National Basic Climatological Network for Precipitation (NBCN-P) (Begert, 2008) (Fig. 1b). The long time series of precipitation are particularly valuable for assessing the effects of climate change on the water cycle, glacier mass balance and the frequency of flooding events.

Long-term temperature series are crucial for the observation, analysis and quantification of climate change. The deviation of the annual mean temperature in Switzerland 1864–2010 from the multiyear average (norm 1961–1990), based on 12 stations (Begert et al., 2005), offers a striking example of climate change (Fig. 2). The linear trend between 1864 and 2010 is +1.1 °C per 100 yr (+0.1 °C per 10 yr), yielding a total warming of +1.6 °C from 1864 to 2010. Worldwide, the global linear trend for the same time period over land is +0.6 °C per 100 yr, yielding a total warming of +0.9 °C from 1864 to 2010.

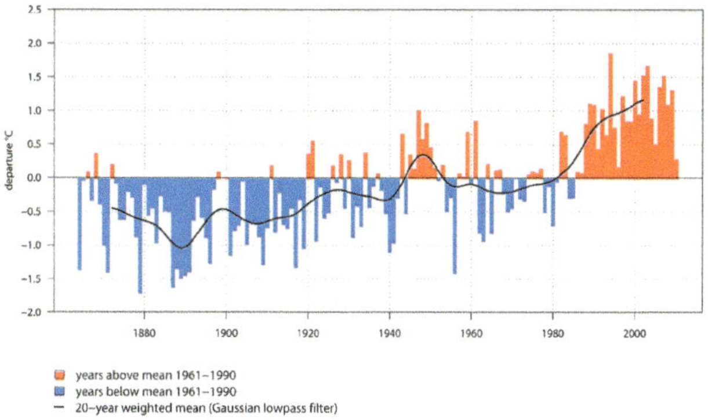


**Figure 2.** Deviation of the annual mean temperature in Switzerland 1864–2010 from the multiyear average (norm 1961–1990).

Since January 2008 the Payerne aerological station is part of the GCOS Upper Air Network (GUAN), meeting higher quality requirements for long-term monitoring of the climate (WMO, 2010b). Moreover, following a recommendation by the meeting on the implementation of the GCOS Reference Upper Air Network (GRUAN) in Lindenberg (Germany) in February 2008, Payerne has been invited to become part of the initial set of 15 sites forming the GRUAN (WMO, 2009).

Switzerland contributes substantially to the WMO Global Atmosphere Watch (GAW) programme, with long-term observations of radiation (Payerne, Jungfraujoch, Davos, Locarno-Monti), ozone (Arosa, Payerne), aerosols (Jungfraujoch), greenhouse gases (Jungfraujoch, Dübendorf) and air pollutants (Payerne, Jungfraujoch, Rigi). Ten years after its launch, an evaluation of the Swiss GAW (GAW-CH) programme by three international experts was organized in 2005. The experts concluded that the transition from the consolidation phase to the permanent phase in 2007 requires at least the same level of funding as successful monitoring needs continuous improvement from research and development. They also recommended to launch a Swiss GAW_Plus programme which adds or extends components going beyond the current scope of GAW-CH. In the meantime, these recommendations have been fully implemented in the Swiss GAW_Plus programme 2010–2013.

## 3.2 Terrestrial domain

Switzerland's most important systematic climate observations in the terrestrial domain concern the hydrosphere, cryosphere and biosphere (Table 2). Isotopes and Phenology are not listed as ECVs in the GCOS Second Adequacy Report, but their measurements have a long tradition in Switzerland and consequently, their series are an important part of the Swiss climate observing system.

In the following, some examples from hydrosphere, cryosphere and biosphere observations are given. The longest time series of freeze and thaw dates for lakes in Switzerland is from Lake St. Moritz; starting in 1832 and extending without interruption up to present. This dataset is unique for central Europe. It is part of a compilation of lake freeze-up and breakup timeseries (Hendricks Franssen and Scherrer, 2008; Magnuson et al., 2000).

**Table 2.** Essential climate variables of the terrestrial domain, together with additional variables of relevance for Switzerland (in italics), which are part of GCOS Switzerland.

| | |
|---|---|
| Hydrosphere | River discharge, Lake levels, Ground water, Water use, *Isotopes* |
| Cryosphere | Snow cover, Glaciers and ice caps, Permafrost and seasonally-frozen ground |
| Biosphere | Albedo, Land cover, Leaf area index, Photosynthetic activity, Biomass, Fire disturbance, *Phenology* |

As a thermal subsurface phenomenon, permafrost reacts sensitively to changes in climate such as the rising air temperatures currently observed. Warming or thawing of alpine permafrost can influence the stability of steep mountain slopes and lead to problems with high-mountain infrastructure or natural hazards. Permafrost observations (i.e. borehole temperatures, velocities of permafrost creep) are coordinated by the network PERMOS (Permafrost Monitoring Switzerland), which was established by a number of university-based partner institutes in the 1990's (Vonder Mühll et al., 2004); the PERMOS coordination office is currently located at the University of Zurich. The PERMOS network currently includes 14 borehole sites and 13 kinematics sites. Funding for PERMOS on a sustained level has been arranged by MeteoSwiss, the Federal Office for the Environment (FOEN) and the Swiss Academy of Sciences (SCNAT). PERMOS will contribute systematically and on a long-term to the global network of permafrost (GTN-P) providing valuable information on alpine permafrost.

A changing climate affects forests by altering the length of the vegetation period – affecting the future distribution limits of individual tree species. Under the federal Long-term Forest Ecosystem Research (LWF) project, intensive and wide-ranging studies have been pursued since 1994 as part of an integrated approach to forest monitoring. At 18 LWF monitoring sites in Switzerland (Fig. 3), (a) external anthropogenic and natural influences (air pollution, climate) are evaluated, (b) changes in important components of the forest ecosystem are assessed, (c) forest health indicators are developed, and (d) comprehensive risk assessments are conducted under various stress scenarios. To this end, numerous site-specific variables are permanently monitored. At the LWF sites, meteorological measurements are carried out automatically according to international standards, with one station located in the stand and a second one in a nearby unstocked area. In addition, stand, vegetation, soil and nutrient data are collected

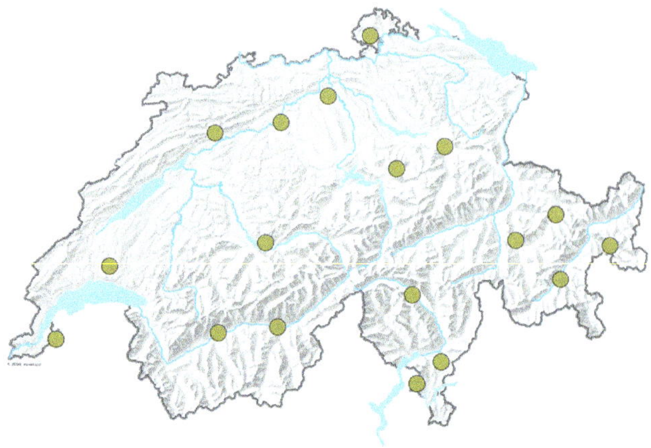

**Figure 3.** The 18 Long-term Forest Ecosystem Research (LWF) sites.

at varying temporal resolution (hourly to yearly). The aims of the LWF project are in agreement with those of the International Co-operative Programme on Assessment and Monitoring of Air Pollution Effects on Forests (ICP Forests), a programme launched in 1985 under the Convention on Long-range Transboundary Air Pollution (CLRTAP) of the United Nations Economic Commission for Europe (UNECE). The LWF project is also part of the International Long-Term Ecological Research Network (ILTER).

### 3.3 International data centres in Switzerland

To determine state and variability of the climate system, measurements need to be globally standardized. The necessary activities of data collection, storage and redistribution are thereby guaranteed by international data and calibration centres around the world. International data centres have been established for many of the essential climate variables networks and systems. In addition, international calibration centres – with their reference instruments and regular calibration activities – make a vital contribution to the quality of global observation programmes.

In Switzerland, the ETH Zurich maintains the Global Earth Balance Archive (GEBA) database at the ETH Zurich, which provides quality controlled records of monthly mean energy fluxes from 1600 stations worldwide to the climate research community. GEBA incorporates various energy balance components, with a total of 19 different variables. These include, for example, global radiation, short- and long-wave radiation and turbulent heat fluxes. Since 1986, the World Glacier Monitoring Service (WGMS) in Zurich has been coordinating the international collection and publication of glacier data, as well as being responsible for the Global Terrestrial Network for Glaciers (GTN-G) within GCOS/GTOS. The WGMS database currently holds more than 34000 length change measurements from 1725 glaciers and 3000 mass balance observations from 200 glaciers worldwide. The basic funding for WGMS at the University of Zurich is assured since January 2010 by long-term funding support through the Swiss GCOS Office.

The Physical Meteorological Observatory (PMOD) was founded at Davos in 1907 to carry out research in the field of solar radiometry and to study the effects of climate and weather conditions on humans, animals and plants. In 1971, the World Radiation Center (WRC) was established at the PMOD on the recommendation of the World Meteorological Organization (WMO). The World Calibration Centre (WCC) for Surface Ozone, Carbon Monoxide and Methane (WCC-Empa) was established at Empa in 1996 at the request of the WMO. The goal of the WCC is to ensure that measurements carried out at different GAW stations are fully traceable to the designated reference. The GAW Quality Assurance/Scientific Activity Centre (QA/SAC Switzerland) was established at Empa in 2000 and is one of four such facilities worldwide.

Euro-Climhist is a database developed at the Institute of History at the University of Bern, including records of early instrumental measurements, daily to seasonal weather reports and observations on lake and river freezing, snow cover, phenology and the impacts of natural disasters, as well as reports on perceptions of weather. The database is a valuable resource for analysis of the climate history of Switzerland (from the late 15th to the early 19th century) and Europe with a focus upon the Medieval Period. Altogether the database currently holds more than 1.2 million records, of which 620000 are available in digital form. The continuation of Euro-Climhist at the University of Bern is assured since July 2010 by long-term funding support through the Swiss GCOS Office.

## 4 The role of satellite data within GCOS Switzerland

Satellite observations are essential to obtain observations of the climate system from a near-global perspective and to compare the behaviour of different parts of the globe. So, a detailed global climate record for the future critically depends upon a major satellite component within GCOS. Therefore, a report about the systematic observation requirements for satellite-based products for climate has been compiled by the WMO GCOS Secretariat in 2006 as a supplement to GCOS Implementation Plan (GIP-SS; WMO, 2006). To emphasize the role of space agencies in the implementation of GCOS, the UNFCCC Subsidiary Body on Science and Technological Advice (SBSTA-25 in 2006) invited the Parties that support space agencies to enable these agencies to implement, to the extent possible, the actions identified in the Committee on Earth Observation Satellites (CEOS) report (CEOS, 2006) and to continue responding in a coordinated manner through CEOS to the efforts to meet these needs.

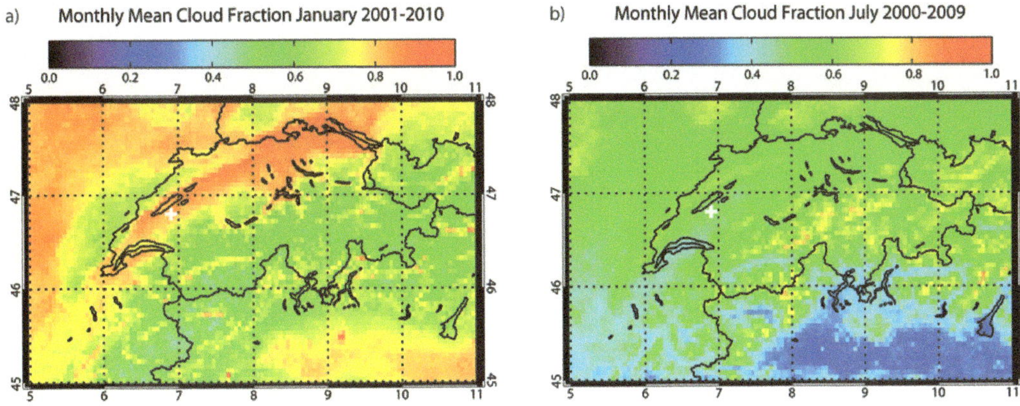


**Figure 4.** 10-yr monthly mean cloud fraction from MODIS (Terra), at 0.05° resolution, derived from MOD35 product. **(a)** January 2001–2010, **(b)** July 2000–2009.

Switzerland is member of the European Space Agency (ESA) and the European Organisation for the Exploitation of Meteorological Satellites (EUMETSAT). ESA, focused on research and development projects, as well as EUMETSAT, more orientated towards pre-operational and operational projects, have activities in the field of climate monitoring. As high-quality surface and in-situ observations remain of central importance for calibration and validation of satellite data, Switzerland with its many unique observation systems and climate records can play an important role in advancing the integration of satellite data into fundamental climate data records. The Swiss GCOS Office is fostering the exploitation of new measurement techniques, in particular from satellites, within GCOS Switzerland (Seiz et al., 2009, 2011).

As an example for the potential of satellite data for climatological analysis over Switzerland, the 10-yr cloud climatology derived from the MODerate resolution Imaging Spectroradiometer (MODIS) data is presented. MODIS onboard Terra was launched in late 1999 with data stream beginning in late February 2000, followed by MODIS onboard Aqua in May 2002. MODIS features spectral and spatial resolution in key atmospheric bands in order to expand the capability to globally retrieve cloud properties. The MODIS atmosphere products are archived into two categories (Platnick et al., 2003): pixel-level retrievals (Level-2 products) and global gridded statistics at a resolution of 1° (Level-3 products). The Level-2 MODIS Cloud Mask product (MOD35) is a daily, global product generated at 1 km spatial resolution. The Level-3 Atmosphere Products (e.g. MOD08) contain statistical datasets from the Level-2 products, summarized over a 1° by 1° global equal-angle grid and temporally aggregated into daily, eight-day and monthly files.

To assess the usability of cloud coverage from satellite sensors for Swiss climatological studies, the Level-3 parameter "Monthly Cloud Fraction" from MODIS onboard Terra (i.e. MOD08 product) was analysed for the area over the GRUAN station Payerne (6°56′ E, 46°48′ N) from March 2000 until December 2008, showing a good agreement in general between the MODIS Terra cloud fraction product (MOD08) and the Synop Payerne data (Seiz et al., 2009). In a follow-up study, the higher resolution MOD35 data were analysed over Switzerland for the 10-yr period March 2000 until February 2010. A daily and monthly cloud fraction was calculated on a $0.05° \times 0.05°$ grid, based on the MOD35 data. For the calculation, the same methodology as for the operational MOD08 product generation was applied, i.e. assigning a cloud fraction of 100% to the MOD35 classes "cloudy" and "uncertain/probably cloudy", and a cloud fraction of 0% to the MOD35 classes "probably clear" and "confident clear". Figure 4 shows the 10-yr monthly mean cloud fraction for the area of Switzerland for January and July. The seasonal variation in cloud cover between winter and summer is clearly visible. Furthermore, the difference between mountainous regions with lower cloud cover and low altitude regions (Swiss middleland, Northern Italy) with higher cloud cover is well apparent for January. The characteristic north-south gradient is particularly visible in July.

## 5 Observations outside Switzerland

A comprehensive global climate observing system relies on an appropriate spatial distribution of climate-relevant observations. However, as a result of the limited technical and financial resources, the continuation of observations in developing countries is often at risk. It is particularly in these regions that the impacts of climate change are one of the main threats to stability, security and socio-economic development. To respond to this problem, a multilateral financing mechanism, the GCOS Cooperation Mechanism, was launched in 2002. The goal of the mechanism is to improve climate observations primarily in developing and emerging countries. It should also strengthen cooperation between the national GCOS offices and the development aid agencies. The observations currently carried out by Swiss institutions abroad cover the following climate variables: ozone (Nairobi, Kenya), trace gases (Assekrem, Algeria; Mt. Kenya, Kenya; Bukit Koto Tabang, Indonesia) and glaciers (30 glaciers worldwide in 9 mountain ranges). Furthermore, the GUAN stations of Dar es Salaam (Tanzania) and Harare (Zimbabwe) have recently been reactivated successfully with the assistance of the Swiss Agency for Development and Cooperation (SDC).

## 6 Conclusions and outlook

Since the establishment of GCOS in 1992, the global climate observing systems have developed and produced promising results. The assessment reports of international climate experts have benefited from the improved information. However, there are still substantial gaps in the measurement networks and many existing observations are threatened, globally as well as in Switzerland. The Swiss GCOS Office at the Federal Office of Meteorology and Climatology MeteoSwiss, established in 2006, is responsible for the national GCOS implementation in Switzerland and seeks to protect the long-term series of essential climate variables and international data and calibration centres in Switzerland that are at risk. Additional aspects deserving special attention within the current and future Swiss GCOS Office activities include metadata, quality assurance and data delivery to the official GCOS international data centres. Furthermore, new measurement technologies on the ground and on satellites will be fostered to help complement the current measurement systems as far as possible. A major challenge for the future will be to ensure continuation of the important historical measurement series as well as to start new measurement series (e.g. soil moisture, soil carbon) for future generations.

**Acknowledgements.** The Swiss GCOS Office would like to acknowledge all contributors from various federal agencies, research institutes, universities and private companies for their valuable support to prepare this paper. The MODIS MOD35 and MOD08 data were obtained from the "Level 1 and Atmospheric Archive and Distribution System" (LAADS Web).

Edited by: S. Rösner

## References

Begert M.: Die Repräsentativität der Stationen im Swiss National Basic Climatological Network (Swiss NBCN), Arbeitsberichte der MeteoSchweiz, 217, 40 pp., 2008.

Begert, M., Seiz, G., Schlegel, T., Musa, M., Baudraz, G., and Moesch, M.: Homogenisierung von Klimamessreihen der Schweiz und Bestimmung der Normwerte 1961–1990, Schlussbericht des Projektes NORM90, Veröffentlichungen der MeteoSchweiz, Nr. 67, 170 pp., 2003.

Begert, M., Schlegel, T., and Kirchhofer, W.: Homogeneous temperature and precipitation series of Switzerland from 1864 to 2000, Int. J. Climatol., 25, 65–80, 2005.

Begert, M., Seiz, G., Foppa, N., Schlegel, T., Appenzeller, C., and Müller, G: Die Überführung der klimatologischen Referenzstationen der Schweiz in das Swiss National Basic Climatological Network (Swiss NBCN), Arbeitsberichte der MeteoSchweiz, 215, 43 pp., 2007.

CEOS: Satellite Observation of the Climate System: The Committee on Earth Observation Satellites (CEOS) Response to the Global Climate Observing System (GCOS) Implementation Plan (IP), available at: http://www.ceos.org (last access: 12 May 2011), 2006.

Defila, C. and Clot, B.: Phytophenological trends in Switzerland, Int. J. Biometeorol., 45, 203–207, 2001.

Hendricks Franssen, H.-J. and Scherrer, S.C.: Freezing of lakes on the Swiss Plateau in the period 1901–2006, Int. J. Climatol., 28, 421–433, doi:10.1002/joc.1553, 2008.

IPCC: Climate Change 2007: The Physical Science Basis, in: Contribution of Working Group I to the Fourth Assessment Report of the Intergovernmental Panel on Climate Change, edited by: Solomon, S., Qin, D., Manning, M., Chen, Z., Marquis, M., Averyt, K. B., Tignor, M., and Miller, H. L., Cambridge University Press, Cambridge, United Kingdom and New York, NY, USA, 996 pp., 2007.

Magnuson, J. J., Robertson, D. M., Benson, B. J., Wynne, R. H., Livingstone, D. M., Arai, T., Assel, R. A., Barry, R. G., Card, V., Kuusisto, E., Granin, N. G., Prowse, T. D., Stewart, K. M., and Vuglinski, V. S.: Historical trends in lake and river ice cover in the Northern Hemisphere, Science, 289, 1743–1746, 2000.

OcCC: Klimaänderung und die Schweiz 2050, Erwartete Auswirkungen auf Umwelt, Gesellschaft und Wirtschaft, OcCC/ProClim, Bern, 168 pp., 2007.

Philipona, R., Dürr, B., Ohmura, A., and Ruckstuhl, C.: Anthropogenic greenhouse forcing and strong water vapor feedback increase temperature in Europe, Geophys.. Res. Lett., 32, L19809, doi:10.1029/2005GL023624, 2005.

Platnick, S., King, M. D., Ackerman, S. A., Menzel, W. P., Baum, B. A., Riédi, J. C., and Frey, R. A.: The MODIS Cloud Products: Algorithms and Examples from Terra, IEEE T. Geosci. Remote, 41, 2, 459–473, 2003.

Scherrer, S. C., Appenzeller, C., and Laternser, M.: Trends in Swiss alpine snow days – the role of local and large scale climate variability, Geophys. Res. Lett., 31, L13215, doi:10.1029/2004GL020255, 2004.

Schmidli, J. and Frei, C.: Trends of heavy precipitation and wet and dry spells in Switzerland during the 20th century, Int. J. Climatol., 25, 753–771, 2005.

Seiz, G. and Foppa, N.: National Climate Observing System (GCOS Switzerland), Publication of MeteoSwiss and ProClim, 92 pp., 2007.

Seiz, G., Foppa, N., and Walterspiel, J.: Use of satellite-based products within the National Climate Observing System (GCOS Switzerland), EUMETSAT Conference Proceedings, p. 55, 2009.

Seiz, G., Foppa, N., Meier, M., and Paul, F.: The Role of Satellite Data Within GCOS Switzerland, Remote Sens., 3(4), 767–780, doi:10.3390/rs3040767, 2011.

Vonder Mühll, D., Nötzli, J., Makowski, K., and Delaloye, R.: Permafrost in Switzerland 2000/2001 and 2001/2002, Glaciological Report (Permafrost) No. 2/3, Glaciological Commission of the Swiss Academy of Sciences, Zurich, 86 pp., 2004.

Wild, M., Gilgen, H., Roesch, A., Ohmura, A., Long, C., Dutton, E., Forgan, B., Kallis, A., Russak, V., and Tsvetkov, A.: From Dimming to Brightening: Decadal Changes in Solar Radiation at Earth's Surface, Science, 308, 847–850, 2005.

WMO: Initial Selection of a GCOS Surface Network, GCOS-34, WMO TD 799, 1997.

WMO: Second Report on the Adequacy of the Global Observing Systems for Climate in Support of the UNFCCC, GCOS-82, WMO TD 1143, 2003.

WMO: Implementation Plan for the Global Observing System for Climate in Support of the UNFCCC, GCOS-92, WMO TD 1219, 2004.

WMO: Systematic Observation Requirements for Satellite-Based Products for Climate, GCOS-107, WMO TD 1338, 2006.

WMO: GRUAN Implementation Plan 2009-2013, GCOS-134, WMO TD 1506, 2009.

WMO: Implementation Plan for the Global Observing System for Climate in Support of the UNFCCC (2010 Update), GCOS-138, WMO TD 1523, 2010a.

WMO: Guide to the GCOS Surface Network (GSN) and GCOS Upper-Air Network (GUAN), GCOS-144, WMO TD 1558, 2010b.

Zemp, M., Frauenfelder, R., Haeberli, W., and Hoelzle, M.: Worldwide glacier mass balance measurements: general trends and fi rst results of the extraordinary year 2003 in Central Europe, in: XIII Glaciological Symposium, Shrinkage of the Glaciosphere: Facts and Analyses, St. Petersburg, Russia edited by: Science, R. A. O., Data of Glaciological Studies [Materialy glyatsiologicheskikh issledovaniy], Moscow, Russia, 3–12, 2005.

# Sunshine duration climate maps of Belgium and Luxembourg based on Meteosat and in-situ observations

**M. Journée, C. Demain, and C. Bertrand**

Royal Meteorological Institute of Belgium, Brussels, Belgium

*Correspondence to:* M. Journée (michel.journee@meteo.be)

**Abstract.** Sunshine duration was historically the traditional parameter to quantify solar radiation. In the present study, we propose to benefit from the measurements made with Campbell-Stokes recorders at 18 sites in Belgium and Luxembourg to generate maps of daily sunshine duration for the period 1995–2005. In order to provide maps with the highest degree of confidence as possible, ancillary data of high spatial resolution derived from Meteosat satellites images are used within the mapping process. Different methods combining in situ sunshine duration measurements and Meteosat-derived data are evaluated. The best performing method is then used to compile a dataset of gridded daily sunshine duration data over Belgium and Luxembourg. The 1995–2005 sunshine duration climate map is finally provided. Various areal mean values are computed and discussed.

## 1 Introduction

Solar radiation was historically monitored by measuring the sunshine duration with Campbell-Stokes recorders. An estimate of the global solar radiation was then obtained through the well-known Ångström–Prescott equation (Ångström, 1924; Prescott, 1940). Although pyranometers are nowadays available to directly measure the global solar radiation, the sunshine duration is still an essential climatological parameter that is still monitored in many meteorological stations in order to extend the historical time series.

The Royal Meteorological Institute of Belgium (RMI) operates since the 50's a network of Campbell-Stokes recorders. This network has been complemented in the 90's with a network of automatic weather stations measuring the global solar radiation and, at a reduced number of sites, the sunshine duration according to the 120 $W m^{-2}$ WMO definition (WMO, 2008).

In this study, we propose to use these time series to generate daily maps of sunshine duration over Belgium and Luxembourg. Such a set of daily maps can be used for several purposes: for instance, to fill the gaps in the measurements time series, to estimate climatological normals, variability and extremes for specific locations as well as for areas (such as climatological or administrative regions, countries, etc.) and to derive climate maps.

Because the network of Campbell-Stokes recorders is relatively sparse (i.e., the network was deployed to a maximum of 18 sites over this area), interpolation errors can be high. Therefore, we aim at improving the confidence of the interpolation process by exploiting observations from the first generation of European geostationary meteorological satellites (i.e., the Meteosat First Generation (MFG) satellites) in addition to the in-situ measurements. In particular, the present analysis will rely on the Meteosat solar radiation climate data records released by the EUMETSAT Satellite Application Facility for Climate Monitoring (CM SAF). This dataset provides MFG-derived estimates of the global and direct solar radiations spatially distributed on the MFG field on view for the period 1983 to 2005 (Posselt et al., 2011). Because of a processing error in this dataset for the time period 1983 to 1994 (Journée et al., 2011), the present study is limited to the period 1995 to 2005, i.e., to 11 yr of data. It is important to note that an improved version of this MFG-based global solar radiation dataset has been recently proposed for the Benelux countries (i.e., Belgium, the Netherlands, and Luxembourg) by merging the satellite estimates with pyranometer measurements (Journée et al., 2012).

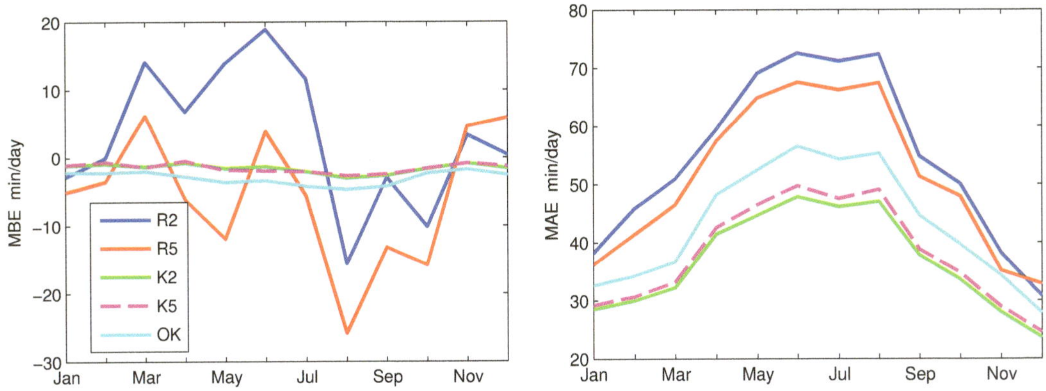


**Figure 1.** Cross-validation MBE (left panel) and MAE (right panel) for selected methods as a function of the month.

The main focus of this paper concerns the identification of the best way to combine the sunshine duration records with the MFG-derived solar radiation data. Several approaches are considered and evaluated in a cross-validation framework. The resulting dataset of daily spatially distributed sunshine duration data is then discussed, and sunshine duration climate maps over Belgium and Luxembourg are finally provided.

## 2 Solar radiation data

Two types of solar radiation data are considered in this study: in-situ daily sunshine duration time series acquired by Campbell-Stokes recorders and gridded daily solar radiation data derived from MFG observations.

Regarding the sunshine duration data, it should be noted that the network of Campbell-Stokes recorders has slightly varied over the period 1995–2005: the number of sites has changed from 15 sites in 1995 to 17 sites in 2005, with a maximum of 18 sites in 2000. Missing data are occasionally reported. A few additional sites are measuring the sunshine duration by means of automatic sensors. These data are not considered here in order to avoid spatial heterogeneities due to different measurement instruments. The location of the 18 Campbell-Stokes recorders is illustrated on Fig. 1.

Gridded solar radiation data were extracted from the Meteosat solar radiation climate data records released by the EUMETSAT Satellite Application Facility for Climate Monitoring (CM SAF), which provides MFG-derived estimates of the global and direct solar radiation (Posselt et al., 2011). This dataset was generated by an adaptation of the well-known Heliosat method (Cano et al., 1986; Rigollier et al., 2004), which determines the effective cloud albedo (or cloud index) from the reflectance measured by the broadband visible channel of MFG. The effective cloud albedo enables the all-sky global and direct irradiances to be estimated through the use of models that simulate the solar irradiances in clear sky conditions. The CM SAF dataset currently covers the period 1983 to 2005 and is provided on a regular $0.03° \times 0.03°$ latitude-longitude grid after triangulation regridding from the original MFG grid, which has a spatial resolution of 2.5 km at sub-satellite point. As previously mentioned, the present study is restricted to the period 1995 to 2005 (i.e., 11 yr of data). Regarding global solar radiation, a composite dataset based on the CM SAF data and pyranometer measurements made in the Benelux countries was shown to be significantly more accurate than the original CM SAF data as well as gridded datasets based on spatially interpolated in-situ measurements (Journée et al., 2012). In this study, both the CM SAF data (providing daily global and direct solar radiations) and the composite dataset (providing the daily global solar radiation) are considered.

## 3 Data merging approaches

Several approaches can be considered to combine two types of solar radiation data in order to generate daily sunshine duration maps with the highest degree of confidence. Two methods will be investigated here.

First, the Ångström–Prescott equation provides a linear relationship between the relative sunshine duration (i.e., the sunshine duration, SD, normalized by the maximum possible sunshine duration, $SD_{max}$) and the clearness index (i.e., the global horizontal solar radiation, $G$, normalized by the extra-atmospheric radiation, $E$), i.e.,

$$\frac{SD}{SD_{max}} = a + b\frac{G}{E}. \tag{1}$$

The site-dependent coefficients $a$ and $b$ can be estimated at each sunshine duration measurement site, and then spatially interpolated at the same spatial resolution as the ancillary global solar radiation dataset. Equation (1) is finally applied to each grid point to generate daily sunshine duration maps. The same approach can be considered for further regression models that estimate the relative sunshine duration based on the beam ratio $B/G$, where $B$ is the beam horizontal solar radiation (Journée and Bertrand, 2011a), e.g.,

$$\frac{\mathrm{SD}}{\mathrm{SD}_{\max}} = a + b\frac{B}{G}. \tag{2}$$

The clearness index and the beam ratio can also be combined within the same regression model,

$$\frac{\mathrm{SD}}{\mathrm{SD}_{\max}} = a + b\frac{G}{E} + c\frac{B}{G}. \tag{3}$$

A second method consists in the use of geostatistical techniques, such as kriging with external drift (KED), that are able to handle densely sampled ancillary variables highly correlated with the parameter of interest (Wackernagel, 1995). In the context of relative sunshine durations, KED can be used with either the clearness index or the beam ratio as a drift, or simultaneously with two drifts.

## 4 Cross-validation analysis

The three variants of the two merging approaches are validated by leave-one-out cross-validation based on the 18 sunshine duration measurement sites and compared against the ordinary kriging (OK) spatial interpolation of the in-situ measurements. Regarding the linear regression models, this means that the coefficients $a$, $b$ and $c$ at one of the sites are successively spatially interpolated by ordinary kriging from the coefficients at the other sites. For the kriging-based methods, the measurement at one of the sites is successively omitted in the merging/interpolation process. In this way, the estimation at a specific site is independent of the measurements made at that site, which can thus be used to evaluate the quality of the estimation. The following statistics are provided in Table 1:

– the cross-validation mean bias error

$$\mathrm{MBE} = \frac{1}{n}\sum_{k=1}^{n}(\bar{\mathrm{SD}}_k - \mathrm{SD}_k),$$

– the cross-validation mean absolute error

$$\mathrm{MAE} = \frac{1}{n}\sum_{k=1}^{n}|\bar{\mathrm{SD}}_k - \mathrm{SD}_k|,$$

where $\mathrm{SD}_k$ is the measured daily sunshine duration and $\bar{\mathrm{SD}}_k$ is the corresponding estimation. The summation is done on all daily instances from 1995 to 2005 and all stations with available in-situ measurements. Regarding ancillary information on global solar radiation, both the CM SAF data (denoted by $G$ in Table 1) and the composite dataset (denoted by $\bar{G}$ in Table 1) are considered. Because of the limited number of measurement sites, the variogram used by OK and KED to model the spatial correlation of the relative sunshine duration is not estimated on a daily basis but kept constant on the entire validation period. Based on Journée and Bertrand (2011b), we used an exponential variogram model with zero nugget and a range fixed to 300 km and 150 km for OK and KED, respectively.

**Table 1.** Cross validation mean bias error (MBE) and mean absolute error (MAE) in minutes per day. $G$ and $\bar{G}$ denote the solar radiation data from the CM SAF data and the composite dataset, respectively.

| Method | | MBE | MAE |
|---|---|---|---|
| R1 | Regression model (1) based on $G/E$ | −1.51 | 59.1 |
| R2 | Regression model (1) based on $\bar{G}/E$ | 3.09 | 54.5 |
| R3 | Regression model (2) based on $B/G$ | 3.41 | 63.3 |
| R4 | Regression model (3) based on $G/E$ and $B/G$ | −5.25 | 55.7 |
| R5 | Regression model (3) based on $\bar{G}/E$ and $B/\bar{G}$ | −5.66 | 51.2 |
| K1 | KED with $G/E$ as drift | −1.50 | 37.6 |
| K2 | KED with $\bar{G}/E$ as drift | −1.59 | 36.7 |
| K3 | KED with $B/G$ as drift | −1.34 | 37.6 |
| K4 | KED with $G/E$ and $B/G$ as drifts | −1.27 | 38.9 |
| K5 | KED with $\bar{G}/E$ and $B/\bar{G}$ as drifts | −1.55 | 37.9 |
| OK | Ordinary kriging (spatial interpolation) | −2.98 | 43.1 |

The results in Table 1 indicate that the best performance is obtained with KED. The sensitivity to the different considered drifts is rather low. The smallest MAE is however reached when the clearness index is used as a drift and information on the global solar radiation is taken from the in-situ/MFG composite gridded dataset. Surprisingly, the simple spatial interpolation of the in-situ sunshine duration measurements outperforms the use of linear regression models. A possible reason for this is that, in contrast to the linear regression models, the kriging methods used with zero-nugget variograms perform an exact interpolation, i.e., estimations at measurement sites equal the measurements. Regarding the linear regression models, the best performance in terms of MAE is achieved when three parameters need to be fitted and when global solar radiation data is taken from the composite dataset.

The seasonal variability of the cross-validation indices is illustrated in Fig. 1. While the kriging-based methods exhibit a small MBE all the year around, the MBE of the linear regression models is very variable from month to month. This has to be related to the regression coefficients that are fixed to constant values over the entire period 1995–2005 and not fitted independently for each month. Regarding the MAE, the highest errors are observed in summer for all methods, i.e., when the daily sunshine duration can reach high values.

In Fig. 2, the cross-validation MAE of the regression model R2 and the kriging approach K2 are computed separately for each station and spatially interpolated by OK. For both methods, the MAE is the highest for isolated stations (e.g., in Luxembourg) and the smallest in densely covered regions (e.g., center area of Belgium). The difference between these two error maps (Fig. 2, right panel) indicates that the

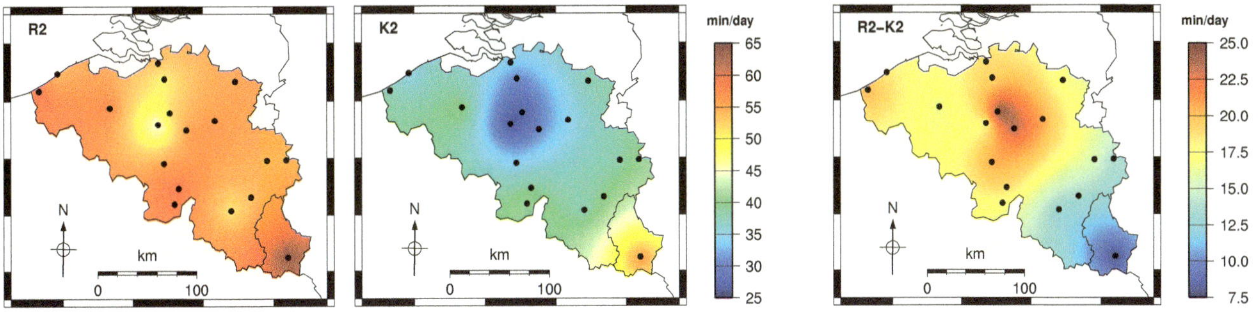


**Figure 2.** Spatial distributions of the cross-validation MAE for the methods R2 (left panel) and K2 (center panel), and spatial distribution of their difference (right panel).

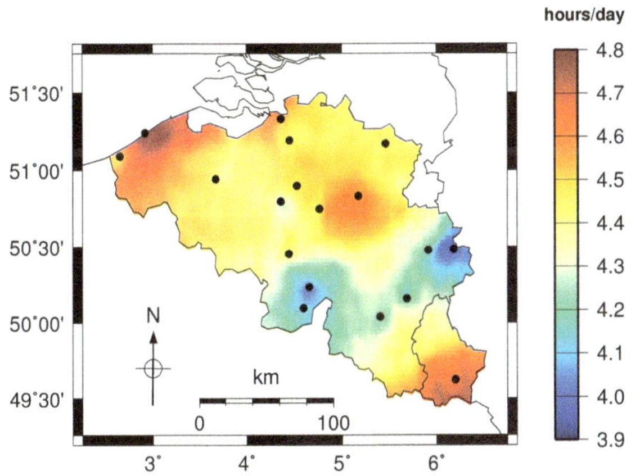


**Figure 3.** 1995–2005 mean map of the daily sunshine duration.

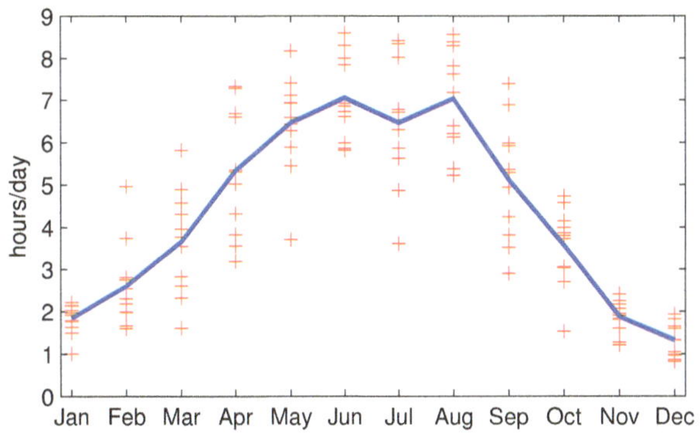


**Figure 4.** 1995–2005 monthly means (red dots) and monthly normals (blue line) of the daily sunshine duration spatially averaged over Belgium and Luxembourg.

benefits of K2 versus R2 is the highest for densely covered areas and the smallest for sparsely covered areas. For both methods, but especially for K2, the access to daily sunshine durations from neighboring countries' stations located in the border area would be highly valuable.

## 5 Sunshine duration climate maps

In view of the results discussed in Sect. 4, the KED method with the clearness index based on the in-situ/MFG composite dataset as a drift (i.e., K2) was used to generate maps of the daily sunshine duration on a regular latitude-longitude grid of resolution $0.03° \times 0.03°$. In order to deal with missing sunshine duration data as well as with the varying number of sunshine duration measurement sites, such maps are computed on a daily basis in order to generate a complete time series over the period 1995–2005.

Several outputs dedicated to climate analysis can be extracted from such a gridded database of daily data. First, the 1995–2005 sunshine duration climate map of Belgium and Luxembourg was computed by taking the average of all the daily maps. As shown in Fig. 3, the largest values are observed over the Belgian coastline and the South of Luxembourg, while the smallest ones are observed around the highest point of Belgium, in the plateau of the Hautes Fagnes. This distribution has many similarities with the distribution of the global solar radiation (Journée et al., 2012), which is largely dominated by clouds. On average over Belgium and Luxembourg, the mean daily sunshine duration amounts to 4 h and 24 min.

The generated database allows also to compute monthly mean values at the scale of a country through the computation of areal means. As shown in Fig. 4, the daily mean sunshine duration is the largest in June and August. Regarding the month of July, the sunshine duration was very variable over the period 1995–2005 (from 03:40 per day in July 2000 to 08:26 per day in July 2003).

Finally, we illustrate in Fig. 5 the time series of the daily sunshine duration spatially averaged over Belgium and Luxembourg. A one-year moving window averaging has been performed in order to analyse the annual variability. On a yearly basis, the smallest and largest sunshine duration values were observed in 1998 and 2003, respectively. No significant trend was detected.

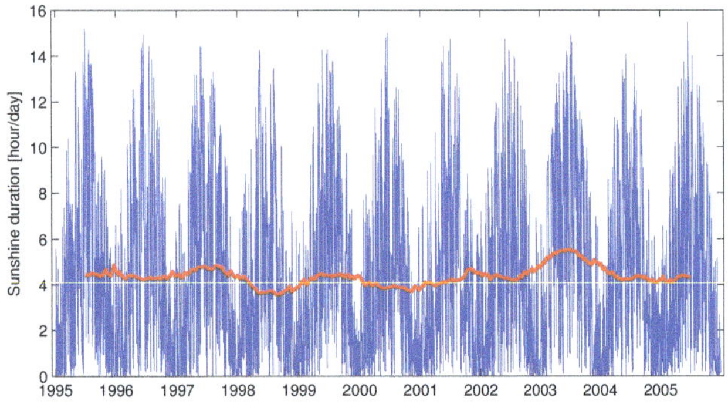


**Figure 5.** Time series of the daily sunshine duration spatially averaged over Belgium and Luxembourg (blue curve) and the corresponding one-year moving window average (red curve).

## 6 Conclusions

In this study, a time series of daily sunshine duration maps was compiled for Belgium and Luxembourg by combining in-situ sunshine duration measurements with high resolution ancillary data derived from Meteosat First Generation (MFG) satellites images. Two classes of methods to combine the two data types were compared: linear regression models, such as the well-known Ångström–Prescott equation, as well as geostatistical interpolation techniques such as kriging with external drift (KED). The cross-validation analysis indicated that the KED method outperforms the linear regression models. KED was tested with various ancillary data as drift: the clearness index, the beam ratio as well as both of them. Although the sensitivity to the type of drift is rather low, the best performance is obtained by using only the clearness index as drift and extracting information on global solar radiation from the in-situ/MFG composite dataset. The 1995–2005 sunshine duration climate map of Belgium and Luxembourg was finally proposed. The sunshine duration is the highest in the coastal area as well as in the South of Luxembourg. Monthly mean values were computed for Belgium and Luxembourg, and the annual variability was analyzed. No significant trend in the daily sunshine duration was found for the period 1995–2005 over Belgium and Luxembourg.

**Acknowledgements.** The Meteosat data was processed by MeteoSwiss within the SAF framework through the Satellite Application Facility on Climate Monitoring (CMSAF), that is partly funded by EUMETSAT. This work was supported by the Belgian Science Policy Office (BELSPO) through the ESA/PRODEX program PRODEX-9 contract No. 4000102777 "Surface Solar Radiation".

Edited by: M. Dolinar

## References

Ångström, A.: Solar and terrestrial radiation, Q. J. Roy. Meteor. Soc., 50, 121–125, 1924.

Cano, D., Monget, J. M., Albuisson, M., Guillard, H., Regas, N., and Wald, L.: A method for the determination of the global solar radiation from meteorological satellite data, Sol. Energy, 37, 31–39, 1986.

Journée, M. and Bertrand, C.: Quality control of solar radiation data within the RMIB solar measurements network, Sol. Energy, 85, 72–86, 2011a.

Journée, M. and Bertrand, C.: Geostatistical merging of ground-based and satellite-derived data of surface solar radiation, Adv. Sci. Res., 6, 1–5, doi:10.5194/asr-6-1-2011, 2011b.

Journée, M., Müller, R., and Bertrand, C.: Towards a climatology of surface incoming solar radiation over the Benelux by merging long time-series of Meteosat-derived estimations and ground-based measurements, Proceedings of the EUMETSAT Meteorological Satellite Conference, Oslo, Norway, 2011.

Journée, M., Müller, R., and Bertrand, C.: Solar resource assessment in the Benelux by merging Meteosat-derived climate data and ground measurements, Sol. Energy, 86, 3561–3574, 2012.

Posselt, R., Müller, R., Stöckli, R., and Trentmann, J.: CM SAF Surface Radiation MVIRI Data Set 1.0 – Monthly Means/Daily Means/Hourly Means. Satellite Application Facility on Climate Monitoring, doi:10.5676/EUM_SAF_CM/RAD_MVIRI/V001, 2011.

Prescott, J. A.: Evaporation from water surface in relation to solar radiation, Trans. R. Soc. Aust., 64, 114–125, 1940.

Rigollier, C., Lefevre, M., and Wald, L.: The method Heliosat-2 for deriving shortwave solar radiation from satellite images, Sol. Energy, 77, 159–169, 2004.

Wackernagel, H.: Multivariate geostatistics: an introduction with applications, Springer-Verlag, Berlin, 1995.

World Meteorological Organization (WMO): Measurement of sunshine duration, Guide to Meteorological Instruments and Methods of Observation (Chapter 8), WMO-No 8, Geneva, Switzerland, 2008.

# 38

# Quality control of 10-min air temperature data at RMI

**C. Bertrand, L. Gonzalez Sotelino, and M. Journée**

Royal Meteorological Institute of Belgium, Brussels, Belgium

*Correspondence to:* C. Bertrand (cedric.bertrand@meteo.be)

**Abstract.** In the '90s, the Royal Meteorological Institute (RMI) of Belgium started to replace its conventional "manual" meteorological network by automated weather stations (AWSs). The meteorological measurement network is now fully automated. RMI counts 18 AWSs that made automated observations centrally available in our headquarters in Uccle, Brussels to internal as well as external users. Due to the large increase in the data amount associated with the automation, quality assurance (QA) procedures are being automated. However, human operators continue to play an essential role in the data validation processes. This contribution describes our newly developed semi-automatic quality control (QC) of 10-min air temperature data. After an existence test, the data are checked for limits consistency, temporal consistency and spatial consistency. At the end of these automated checks, a decision algorithm attributes a flag to each particular data. Each day the QC staff analyzes the preceding day observations in the light of the quality flags assigned by automated QA procedures during the night. It is the human decision whether or not a value is accepted.

## 1 Introduction

The value of any meteorological measurement is dependent on the accuracy and precision with which it represents the physical quantity being measured. Factors such as instrument calibration, long-term field exposure to the elements, and instrument maintenance all play a role in collecting what ultimately becomes a good or bad data. In the past, little attention was paid to data quality control, believed to be less important than the improvement of numerical weather prediction and data assimilation techniques, and considered as a less "glamour" topic. Quite early, though, it was recognized that the insufficiency of the quality control applied to the observations was an obstacle to the quality of the analysis, also crucial for the skill of numerical forecasts. Since the eighties more effort has gone in the study and formalization of quality control procedures. Literature of the last two decades suggests an evolution toward complex quality assurance (QA) and quality control (QC) in practice with meteorological data processing. Complex QA is distinguished from more traditional ("simple") QA by the use of several different tests and a decision tree to weigh all of the evidence before flagging data (Gandin, 1988; Eskridge et al., 1995). The guiding principle is that no decision to flag a datum should be made until all available approaches have been applied toward the assessment of its validity. Among the first complex QA systems to be developed was for radiosonde data (Gandin, 1988). In more recent years, however, complex QA has been extended to daily (e.g., Reek et al., 1992; Kunkel et al., 1998), hourly (e.g., Meek and Hatfield, 1994; De Gaetano, 1997; Graybeal et al., 2004) and real-time (e.g., Shafer et al., 2000; Liljegren et al., 2009) surface meteorological data. A three-pronged framework was developed at the Royal Meteorological Institute (RMI) of Belgium for implementing automated component checks on the 10-min air temperature records and decision-making process weighting flags by hierarchical flag type. The framework is similar to Gandin's concept of complex QA, in that it approaches the question of the validity of a given datum from several different angles and considers errors of different types. However, in the present approach, the automated QA functions are included in a larger QA protocol involving manual inspections. The goal is to provide a high-quality 10-min dataset with a trustworthy assessment of the confidence in each datum to both internal and external users.

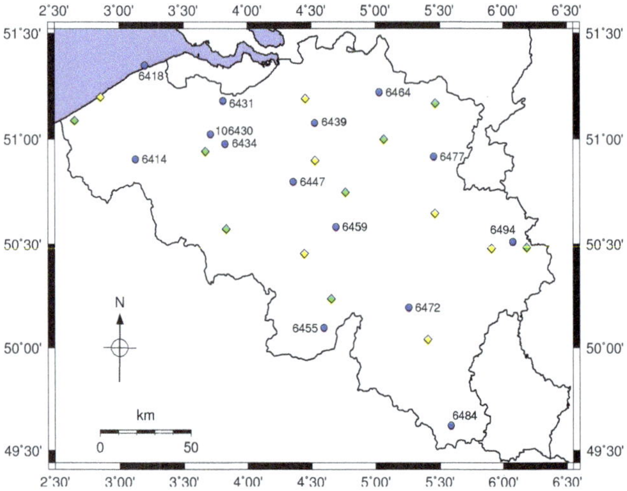


**Figure 1.** Location of the automatic air temperature measurements performed within the Belgian territory. Blue dots are for the RMI's AWSs while yellows and green diamonds are for the Belgo-control and Meteo Wing AWSs. See Table 1 for the RMI's stations names and associated air temperature records.

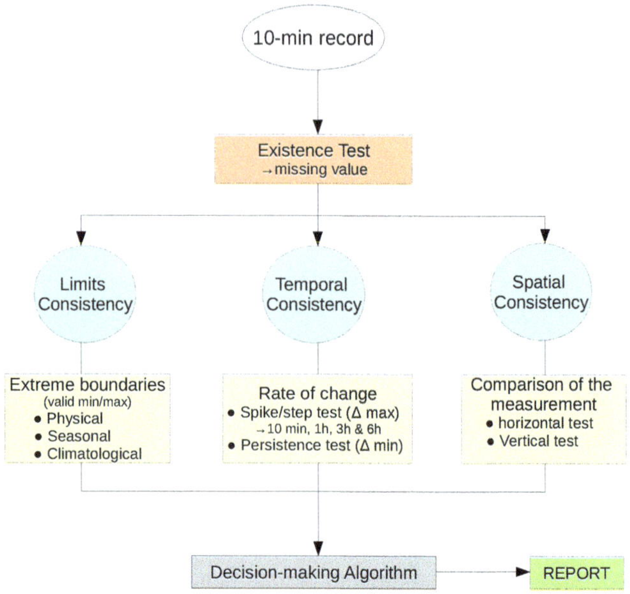


**Figure 2.** Complex QA of 10-min air temperature records and decision tree.

## 2 10-min air temperature records

Automatic weather stations (AWSs) operated by RMI range from basic climatological stations to fully equipped synoptic stations performing a complete set of meteorological observations. Air temperature measurements are performed in 14 of them (blue dots in Fig. 1). Note that in addition to the RMI's measurements, automated air temperature records are also performed in the synoptic AWSs operated by Belgocontrol, the belgian public company in charge of the air traffic safety in the civil airspace (7 stations, yellow diamonds in Fig. 1) as well as by the Meteorological Wing of the Air Component of Defense (8 stations, green diamonds in Fig. 1) but RMI is not in charge of their data QC. Due to the large heterogeneity within the RMI's AWSs, five groups based on the recorded air temperature have been distinguished for the automated data QC (see Table 1).

## 3 Complex QA and the decision tree

RMI's AWSs are built around a programmable data logger that measures the sensors, then processes, stores and transmits the data to the central data base (DB) in Uccle, Brussels. Once converted to digital values a first processing is performed on the raw data at the data logger level allowing calculation of 10-min air temperature values from the 5-s measurements. To ensure that gross errors are trapped before being further transmitted in the central DB a first basic QC is performed on all air temperature values once acquired centrally. Automated procedures monitor the data to make sure they are collected and that system performance is acceptable. After an existence test, a module checks for physical limits and flags the data violating these limits (erroneous when data lie outside physical limits and suspect when lying outside basic long-term climatological extremes that do not take into account the time of year and location). A list of missing and flagged data is automatically produced after each control cycle and transmitted to the AWS network maintenance team for further intervention.

Second, every night automated QA procedures check the 10-min air temperature values for more subtle errors. Implementation of the complex automated QA is diagrammed in Fig. 2. Data are checked for plausibility using adjusted limits to reflect climatic conditions more precisely than in the first real time range test, temporal consistency (step test and persistence test) determined from several years of prior data, and spatial consistency.

### 3.1 Physical limit check or plausible value check

The aim of the test is to verify whether the values are within acceptable range limits depending on the climatic conditions of the measurement site and the season. The check provides information as to whether the values are erroneous or suspect.

### 3.2 Temporal consistency or variability tests

When a sensor fails it will often report a constant value; thus the standard deviation, $\sigma$, will become smaller, and if the sensor is out for an entire reporting period, $\sigma$ will be zero. In other cases, the instrument may work intermittently and produces reasonable values interspierced with zero values, thereby greatly increasing the variability for the period. The

**Table 1.** List of the 14 RMI's Automatic Weather Stations performing at least one air temperature record and the associated measurements.

| AWS | | Air Temperature Measurement(s) | | | | | QC |
|---|---|---|---|---|---|---|---|
| Code | Name | 1.5 m | 1.5 m* | 2 m | 10 m | 30 m | group |
| 6414 | Beitem | X | X | X | X | | 2 |
| 6418 | Zeebrugge | | X | | | | 5 |
| 6431 | Zelzate | X | | | | | 5 |
| 6434 | Melle | X | X | X | X | X | 1 |
| 6439 | Sint Katelijn Waver | X | X | | | | 4 |
| 6447 | Uccle | X | X | X | | | 3 |
| 6455 | Dourbes | X | X | X | X | | 2 |
| 6459 | Ernage | X | X | X | X | | 2 |
| 6464 | Retie | X | X | X | X | | 2 |
| 6472 | Humain | X | X | X | X | X | 1 |
| 6477 | Diepenbeek | X | X | X | X | | 2 |
| 6484 | Buzenol | X | X | X | X | | 2 |
| 6494 | Mont Rigi | X | X | X | X | | 2 |
| 106 430 | Gent Sterre | | X | | | | 5 |

* RH-T sensor

spike or step test ($\Delta$max) checks for a pausible rate of change from a preceding acceptable level. The maximum probable change is based on the 99.9th percentile change for several years of previous data. Values are checked for 10-min, 1 h, 3 h, and 6 h time steps. A value which does not pass the test is flagged as doubtful or suspect. Because in case of extreme meteorological conditions, an unusual variability in the air temperature may occur, data may be flagged as suspect, although correct. To prevent from this, the algorithm does not report a spike/step anomaly if air temperatures at different levels at the same site fails the spike/step test.

In the persistence (or $\Delta$min) test, the measurements that fail to change by more than a minimum amount within a selected time range are flagged as suspect. Similarly to the spike/step test, the minimum change is based on the 99.9th percentile change for several years of previous data. Temperatures are checked for 10-min, 1 h, 3 h, and 6 h time steps. To minimize the possibility of a false positive identification the algorithm does not report a persistence anomaly if two or more air temperatures at the same place fail the persistence test when the ambient temperature is above 0 °C. Below 0 °C a persistence anomaly is reported based on the assumption that freezing conditions can affect multiple sensors.

### 3.3 Spatial consistency tests

Two types of spatial tests are applied to the data where possible: (i) horizontal comparisons at the same height on different stations, and (ii) vertical comparisons at different heights on the same station. The horizontal check works in two steps. First, an outlier detection is performed on both the station data being quality checked and the data of the surrounding stations using the daily 10-min temperature timeseries of each station. If an outlier is detected for the station being

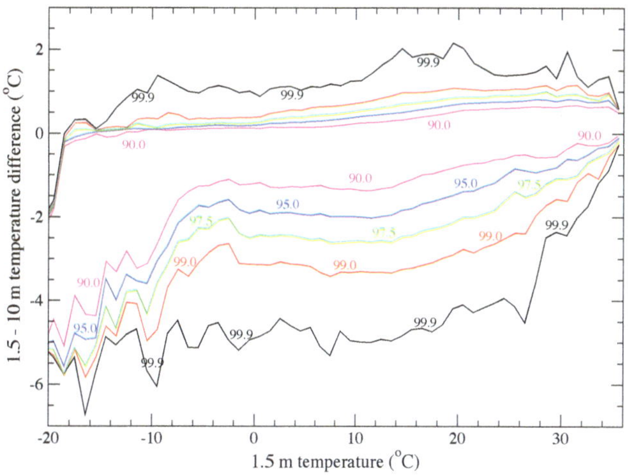


**Figure 3.** Probable vertical air temperature differences (1.5–10 m) as a function of the 1.5-m temperature.

quality checked, then the data fails the horizontal consistency test. Otherwise, we test on a 10-min basis whether the analyzed station value falls inside a confidence interval formed from surrounding stations data that were not classified as outliers. Measurements that fail the test are flagged as suspect or erroneous depending upon the departure of the data from the confidence interval. Because the horizontal consistency test has been found to report numerous false positive identification in the 10-min air temperature records performed at our two remotest and highest stations (i.e., AWSs code 6484 and 6494 in Fig. 1, respectively), the algorithm does not report a horizontal anomaly for these stations. Only the result of the outlier detection is considered by the decision making algorithm.

The vertical check provides a more stringent constraint than simple valid maximum/minimum limit tests by requiring consistency among the vertical temperature profile as well as consistency with historical data. Basically, the differences between air temperature measurements at any two levels as a function of a given air temperature level are compared to probable differences determined from several years of prior data screened to eliminate periods of erroneous values that would affect the results. As an example, the contours in Fig. 3 show which combinations of air temperature at 1.5 m and 10 m fall within a given percentile of the joint probability density. Following a review of the values that fall outside the 99.9 % boundary, the 99.9th percentile was selected as the boundary of acceptability. Combinations that fall outside the 99.9 % boundary are considered as erroneous and suspect when falling between the 99.0 % and 99.9 % boundaries. Because two comparisons are necessary to unambiguously identify which level is problematic, at least two vertical tests (comparing three levels) must be performed. Consequently, the vertical check is only applied at stations perfoming air temperature measurements at a minimum of 3 different levels (AWS QC groups 1 and 2 in Table 1). The following tests are applied where possible:

- $(Tair_{1.5m} - Tair_{2.0m})$ vs. $Tair_{1.5m}$
- $(Tair_{1.5m} - Tair_{10.0m})$ vs. $Tair_{1.5m}$
- $(Tair_{1.5m} - Tair_{30.0m})$ vs. $Tair_{1.5m}$
- $(Tair_{2.0m} - Tair_{10.0m})$ vs. $Tair_{2.0m}$
- $(Tair_{2.0m} - Tair_{30.0m})$ vs. $Tair_{2.0m}$
- $(Tair_{10.0m} - Tair_{30.0m})$ vs. $Tair_{10.0m}$

### 3.4 Decision algorithm

At the end of the checks, when all 10-min daily air temperatures time series recorded in a given station have been analyzed, a decision algorithm interprets the results of the individual checks described above to attribute a flag (i.e., erroneous, suspect or valid) to each particular data given the weight of evidence and produces for each station a report for the QC staff. Similarly to Liljegren et al. (2009) the algorithm proceeds sequentially through each step until a failure mode is identified; if no failure mode is identified the measurement is judged to be valid.

If the measurement is missing or if the quality checks reveal that the measurement is outside the valid range, no interpretation is necessary. Otherwise, if, for example, the measurement fails the range test, which is based on the statistical likelihood of occurence of the value in the past, then the horizontal and vertical test results are used (if available) to determine whether the measurement is likely to be problematic or whether the value is simply outside the statistical range. If the measurement also fails the horizontal or vertical tests, it is judged to have failed the range check. If it passes all of the horizontal and vertical tests, then it is judged to have succeeded the limits consistency check. Similar reasonings hold for the temporal consistency checks as illustrated above in the spike and persistence tests description. If the measurement passes the limits consistency and temporal consistency checks but still fails the horizontal test (i.e., it does not agree with similar measurements at other stations), then it is judged suspect. Finally, if the measurement fails vertical comparisons with two other levels, then it is judged to be erroneous. Two vertical tests must fail to determine which level has the problem.

## 4 Manual QA

Each day, the QC staff analyses the preceding day 10-min air temperature records in the light of the assigned quality flags from the automated QA system. It is the human decision whether or not a value is accepted. As a result of this assessment, "trouble tickets" are issued where needed to the maintenance team so that sensors can be replaced or repaired. In addition, where errors are found corrections are supplied and where values are missing estimates are made if possible. Operators are supported in this task by automated correction and estimation procedures. Any modification or validation introduced by the operator in the 10-min air temperature records is automatically echoed in the central DB and the associated quality flag updated accordingly.

## 5 Conclusions

Automation of the RMI's AWSs data quality control is ongoing. The purpose of this automated data screening is to objectively identify abnormal data values for subsequent review by an experienced data analyst. The review is necessary to determine whether an anomaly results from a problem with instrumentation or whether it accurately reflects unusual meteorological conditions. The automated quality control of air temperature data is nearly completed. A last module comparing the data standard deviation over a longer period that one day (i.e., a few months) to the expected limits (variability test) will be added in the forthcoming months. Because such a test uses aggregate statistics, it cannot discern which observations within the time period being checked are responsible for a possible offense. Consequently, all data values of the parameter failing the test will be flagged over the tested time period for further checking. Finally, we plan to add the 1.5 m/2 m air temperature records from the Belgo-control and Wing Meteo AWSs in our spatial consistency check – horizontal tests – as soon as they will be available at RMI on a 10-min basis. This is expected to improve the performance of the horizontal tests at our two remotest stations (i.e., AWSs code 6484 and 6494, respectively) at least for the 1.5 m/2 m level(s).

Edited by: S. Rösner

## References

De Gaetano, A. T.: A quality control procedure for hourly wind data, J. Atmos. Ocean. Tech., 14, 308–317, 1997.

Eskridge, R. E., Alduchov, O. A., Chernykh, I. V., Panmao, Z., Polansky, A. C., and Doty, S. R.: A comprehensive aerological reference data set (CARDS): Rough and systematic errors, B. Am. Meteorol. Soc., 76, 1759–1775, 1995.

Gandin, L. S.: Complex quality control of meteorological observations, Mon. Weather Rev., 116, 1137–1156, 1988.

Graybeal, D. Y., De Gaetano, A. T., and Eggleston, K. L.: Complex quality assurance of historical hourly surface airways meteorological data, J. Atmos. Ocean. Tech., 21, 1156–1169, 2004.

Kunkel, K. E., Andsager, K., Conner, G., Decker, W. L., Hillaker, H. J., Knox, P. N., Nurnberger, F. V., Rogers, J. C., Scheeringa, K., Wendland, W. M., Zandlo, J., and Angel, J. R.: An expanded digital daily database for climatic resources applications in the midwestern United States, B. Am. Meteorol. Soc., 79, 1357–1366, 1998.

Liljegren, J. C., Tschopp, S., Rogers, K., Wasmer, F., Liljegren, L., and Myirski, M.: Quality Control of Meteorological Data for the Chemical Stockpile Emergency Preparedness Program, J. Atmos. Ocean. Tech., 26, 1510–1526, 2009.

Meek, D. W. and Hatfield, J. L.: Data quality checking for single station meteorological databases, Agr. Forest Meteorol., 69, 85–109, 1994.

Reek, T., Doty, S. R., and Owen, T. W.: A deterministic approach to the validation of historical daily temperature and precipitation data from the Cooperative Observer Network, B. Am. Meteorol. Soc., 73, 753–762, 1992.

Shafer, M. A., Fiebrich, C. A., Arndt, D. S., Fredrickson, S. E., and Hughes, T. W.: Quality Assurance procedures in the Oklahoma Mesonetwork, J. Atmos. Ocean. Tech., 17, 474–494, 2000.

# 39

# INCA-CE: a Central European initiative in nowcasting severe weather and its applications

**A. Kann, G. Pistotnik, and B. Bica**

Central Institute for Meteorology and Geodynamics, Vienna, Austria

*Correspondence to:* A. Kann (alexander.kann@zamg.ac.at)

**Abstract.** The INCA-CE (**I**ntegrated **N**owcasting through **C**omprehensive **A**nalysis – **C**entral **E**urope) project aims at implementing a transnational weather information system as well as applications for different socio-economic sectors to reduce risks of major economic damage and loss of life caused by severe weather. Civil protection and also stakeholders from economic sectors are in a growing need of accurate and reliable short-term weather forecasts. Within INCA-CE, a state-of-the art nowcasting system (INCA) is implemented at weather services throughout the European Union's CE (Central Europe) Programme Area, providing analyses and short term forecasts to the aforementioned end-users. In a coherent approach, the INCA (Integrated Nowcasting through Comprehensive Analysis) system will be adapted for implementation and use in a number of partner countries. Within transregional working groups, the gap between short-term weather information and its downstream activities in hydrological disaster management, civil protection and road management will be bridged and best practice management and measure plans will be produced. A web-based platform for outreach to related socio-economic sectors will initiate and foster a dialogue between weather services and further stakeholders like tourism or the insurance sector, flood authorities for disaster management, and the construction industry for cost-efficient scheduling and planning. Furthermore, the project will produce a compact guideline for policy makers on how to combine structural development aspects with these new features. In the present paper, an outline of the project implementation, a short overview about the INCA system and two case studies on precipitation nowcasts will be given. Moreover, directions for further developments both within the INCA system and the INCA-CE project will be pointed out.

## 1 Introduction

Technological and scientific developments in nowcasting during the last 10 yr have opened up new opportunities in public safety, in risk management and environmental protection, and in cost effectiveness of services provided by the public and private sector. Since weather phenomena do not "stop at borders", the development of a truly integrated nowcasting system (which includes the whole chain from modeling to protective action) is best achieved in a transnational collaboration. Cooperation of several trans- and international institutions is needed to improve the quality of nowcasting for all partners. INCA-CE is a cooperation of 16 partners from 8 Central European countries which bridges the gap between the development of a nowcasting system and the specific demands of the application aspect on a transnational basis. Six meteorological and hydrological institutions work together to enhance the nowcasting system INCA (Integrated Nowcasting Through Comprehensive Analysis; Haiden et al., 2010, 2011) as part of a transnational framework. Together with 10 partners from the scientific and application fields, the INCA-CE community sets up pilot implementations for applications in hydrology, civil protection and road safety. INCA-CE is co-financed by the Central Europe Programme of the European Union (http://www.central2013.eu).

## 2 Project motivation and implementation

Generally, the INCA-CE project aims at "Reducing Risks and Impacts of Natural and Man-made Hazards" (area of intervention as defined by the CE Programme of the European Union). The INCA-CE project specifically has the goal of setting up a web-based trans-national weather information system that uses state-of-the-art nowcasting methods developed by several countries. It enables users in the public and private sector to take into account weather-related risks and hazards in a more timely fashion, more precisely, and in greater geographical detail. More precise information about heavy rainfall and associated rise of water levels (flooding) will help to set up improved procedures and strategies in the management of mitigating measures for the protection of buildings, roads, and other infrastructure. Civil protection will benefit from a more comprehensive assessment of meteorological threats, and a more detailed and timely forecast, leading to more efficient warning protocols and dissemination strategies. Road safety will be enhanced by a more detailed road weather forecast made available both to the road management authorities as well as to the general public.

### 2.1 Expected benefits

The importance of real-time, high-resolution weather information has increased during the past decades. A continuously growing number of human activities critically depends on meteorological conditions. The project addresses the problem of regional differences in the quality and availability of this information, especially in areas which are highly topographically structured, in order to develop a high-resolution real-time meteorological information system. Secondly, the implications and benefits for potential stakeholders e.g. for risk warning, road maintenance and civil protection need to be based on evaluated best practice examples.

This project establishes the scientific, technical and administrative cooperation that is needed for significant progress in nowcasting and weather warnings. Economically, there is an increasing market for reliable, quantitative, and problem-oriented meteorological information. One of the outcomes of this project is that each Central European country will have at its disposal a state-of-the-art system from which it can provide meteorological services and products. A web-based information system for the general public will provide weather information in a much higher quality than in the past, which will increase the ability of citizens to plan their commercial as well as other weather-sensitive activities in better accordance with meteorological conditions and avoid unnecessary hazards.

The necessity of this project also results from the fact that we are facing changing weather conditions as a result of climate change, with serious socio-economic repercussions. Paradoxically, technology has made our society and industry at the same time less dependent on, but more sensitive to, weather and its effects. The better prepared a region enters into this new phase of human-induced climate change, the better it will be able to cope with its day-to-day effects. An efficient nowcasting and warning procedure based on meteorological and hydrological modeling and an optimized dissemination strategy can help communities to implement mitigating action and reduce damage.

### 2.2 Project implementation

The INCA-CE project is divided into 6 work packages (WPs), two of which are committed to project management and communication/knowledge management, respectively. This strategy ensures a strong coordination and control over the project activities as well as dedicated mechanisms for internal and external communication. Both work packages are depicted in light blue boxes of Fig. 1 which indicate the framework of the project. Four work packages (light orange in Fig. 1) are dedicated to the actual development and setup of an integrated warning system for hydrological, civil protection and road safety purposes. The major points to be tackled by these work packages are described in the text boxes. The project's core outputs are the establishment of a permanent cooperation, the implementation and joint use of a common tool, and the increase in public safety and environmental conditions as a result of an improved, transnational strategy. Corresponding to the three transnational Working Groups "Operational Hydrology", "Civil Protection", and "Road Safety" in WP3, there will be three core outputs in the form of transregional strategies, formulated in a strategy paper, for optimal use of nowcasting and weather warnings in the respective application. A coordinated evaluation of the actual benefit for targeted users is carried out during the pilot studies (WP5). This cooperation includes both providers of weather information and warnings as well as user groups, and ensures the implementation of application-oriented strategies and tested recommendations. At least three different types of hazards including flood, storm and snowfall events will be evaluated in the pilot studies.

The common nowcasting tool (INCA) will be developed and refined, respectively, in WP4 (Adaptation and refinement of nowcasting for users) in a joint effort by the provider-level partners. The core output of WP4 is a set of warning system modules incorporating the hydro-meteorological developments specifically carried out towards improved application in Operational Hydrology, Civil Protection, and Road Safety. The user level-partners will implement the methods and feedback the experiences to provider level as a basis for the joint definition, by providers and users, of recommendations and guidelines (WP6). Another core output is a web portal which will be accessible for stakeholders and the general public on the www.

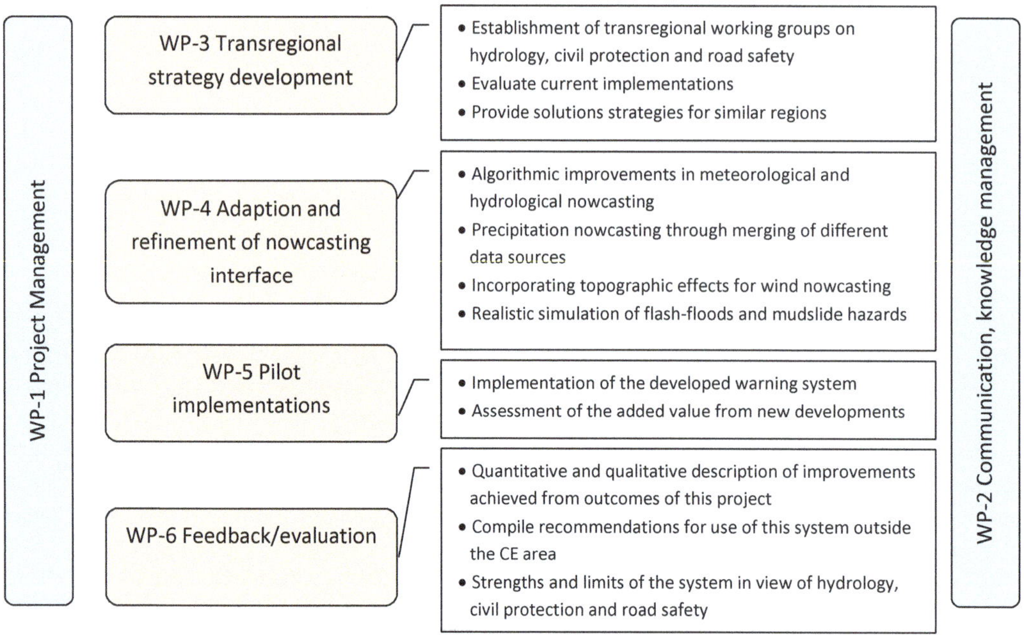


**Figure 1.** High level description of the INCA-CE work packages (WPs).

## 3 The nowcasting system INCA

### 3.1 Generalities

Most existing observation-based forecasting systems focus on the prediction of precipitation and convective activity (Browning and Collier, 1989; Li et al., 1995; Hand, 1996; Golding, 1998; Pierce et al., 2000; Seed, 2003). During the World Weather Research Program (WWRP) Forecast Demonstration Project of the 2000 Sydney Olympics several of these methods were tested and compared (Pierce et al., 2004). In the same project, a wind analysis and nowcasting system was evaluated (Crook and Sun, 2004). However, research has generally focused not so much on forecasting the wind field as such but its effect on the initiation and development of deep convection (Wilson and Schreiber, 1986; Wilson et al., 2004). Real-time flood-warning systems are implemented based on hydrological models that require meteorological input at small scales and short lead times. Weather services face the challenge of issuing weather warnings at a high update frequency and with more precise geographical specification. In order to satisfy these requirements the INCA system is developed.

In the case of the 3-D INCA analyses of temperature, humidity, and wind, NWP forecasts provide the first guess on which corrections based on observations are superimposed. For this purpose the output of the limited area model Aire Limitée Adaptation Dynamique Développement International (ALADIN) running operationally at ZAMG is used. The NWP fields are 1-hourly, at a resolution of 9.6 km, with 60 levels in the vertical (Wang et al., 2006). Recently, the operational ALADIN version has been updated to a resolution of 5 km (ALARO5). The most important data sources for the INCA system are surface stations. In Austria, ZAMG operates a network of about 250 automated stations which provide measurements in one minute intervals. Data are transferred every 5–10 min. Additionally, a large number of third party observations are taken into account, in particular the hydro-meteorological stations network which provides temperature and precipitation data. The Austrian radar network is operated by the civil aviation administration (Austro Control). It consists of five radar stations located at Vienna airport, near the city of Salzburg, near the city of Innsbruck (on Patscherkofel mountain), in southern Austria (on Zirbitzkogel mountain) and in western Austria (on Valluga mountain). ZAMG operationally obtains 2-D radar data synthesized from these 5 locations, containing column maximum values in 14 intensity categories, at a time resolution of 5 min. Ground clutter has already been removed from the data. The Meteosat 2nd Generation (MSG) satellite products used in INCA are "Cloud Type" (Derrien and Le Gléau, 2005), which consists of 17 categories, and the visible brightness image. The cloud types distinguish between different degrees of opaqueness and whether clouds are more likely to be convective or stratiform. The 1-km topography used in INCA is obtained through bilinear interpolation from the global 30″ elevation dataset provided by the US Geological Survey.

A full description of the INCA analysis and nowcasting scheme is given in Haiden et al. (2010) and (2011). In the present paper, a brief overview describing the methods used is presented.

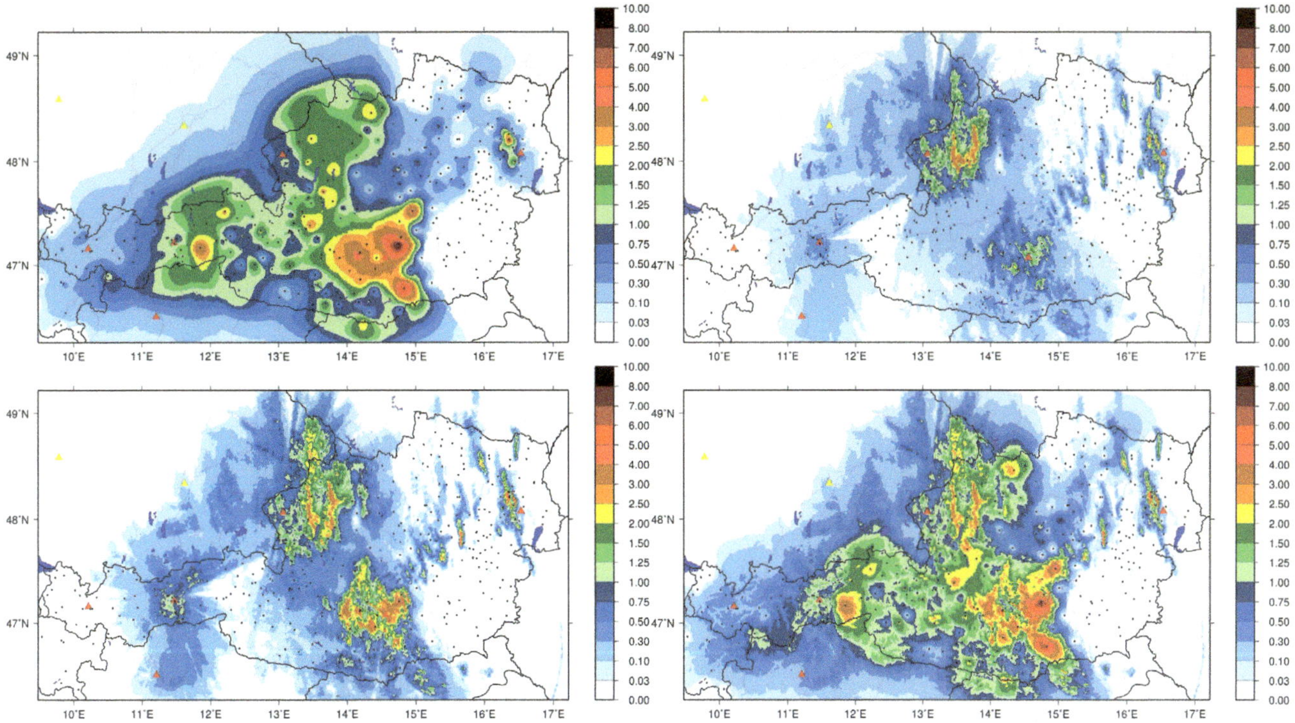


**Figure 2.** Example of a 15-min INCA precipitation analysis [mm/15 min] (issued for 18 July 2009, 07:30 UTC) based on the combination of station and radar data. Upper left panel: pure station interpolation, upper right panel: uncorrected radar field, bottom left panel: corrected and scaled radar field, bottom right panel: final INCA precipitation analysis.

Those fields which are analyzed three-dimensionally (temperature, humidity, wind), are generated as follows. The NWP (ALADIN) fields are interpolated tri-linearly to the INCA grid (horizontal resolution 1 km, vertical resolution 200 m for temperature and humidity, 125 m for wind). Within valleys that are not represented by the NWP model, a downward shifting of the first guess to a so-called valley-floor surface is performed. Differences between station observations and first guess are interpolated three-dimensionally. The interpolation uses a distance-squared weighting method with geometrical distance weighting in the horizontal and distance weighting in potential temperature space in the vertical. Finally, the resulting difference field is added to the first guess. For the wind analysis, a relaxation algorithm is additionally applied to ensure mass-consistency of the wind field with respect to the INCA topography. In the nowcasting mode, the trend of the NWP forecast is superimposed on the most recent INCA analysis. At lead times beyond +6 h, the INCA nowcast is blended into the bias-corrected NWP forecast. The effect of errors in the NWP cloud forecast is also taken into account in the temperature nowcast.

Turning to the precipitation module, surface station data are algorithmically combined with radar data. In this way the higher quantitative accuracy of the station data, and the better spatial coverage of the radar data, is utilized. The resulting analysis reproduces the observed values at the station locations. In-between it contains the spatial structure given by the remote sensing data. An intensity-dependent parameterization of elevation effects is used (Haiden and Pistotnik, 2009) in addition to the station-radar combination. Figure 2 illustrates the individual steps of the precipitation analysis procedure. In the nowcasting mode, motion vectors are computed using a correlation method from consecutive analyses. The resulting vectors are filtered statistically by setting a threshold for the correlation, and meteorologically by comparing it with the NWP wind at the 500 and 700 hPa levels. Using these vectors, the nowcast of precipitation is computed. Between +2 and +6 h, the nowcast is merged with the NWP forecast with a linearly decreasing weight, so that from +6 h onwards the pure NWP forecast is used. The most important application of the INCA precipitation forecast is operational flood prediction (Komma et al., 2007). For improved nowcasting of convective cells (Steinheimer and Haiden, 2007) a number of additional 2-D fields are computed, such as Convective Available Potential Energy (CAPE), Convective Inhibition (CIN), Moisture Flux Convergence and others.

Cross validation of the temperature analysis typically shows a mean absolute error (MAE) near 1 K, and a root mean square error (RMSE) near 1.5 K (not shown, see Haiden et al., 2011). The main reason for larger analysis errors, especially in winter, is insufficient information about inversion heights and about patterns of Foehn-induced mixing in mountain areas. Averaged over all stations and seasons,

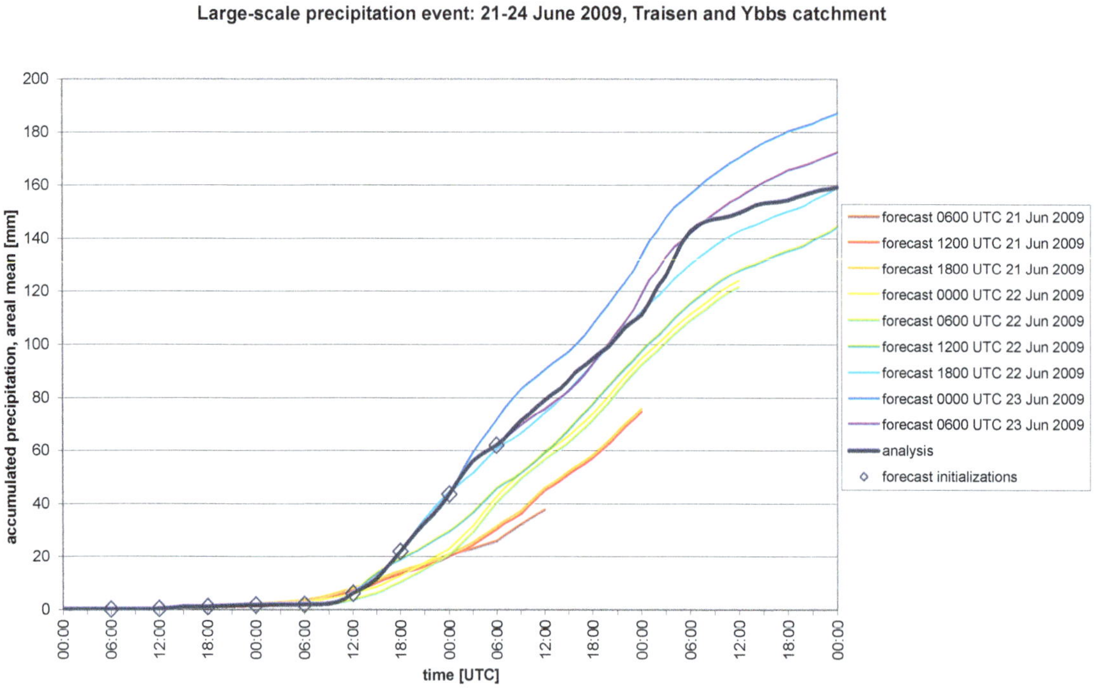


**Figure 3.** Hydrographs of analyzed precipitation (thick black curve) and forecasted precipitation (colored curves) with various initialization times (open diamonds) during the 21–24 June 2009 precipitation event, depicted as areal mean values over the Traisen and Ybbs catchments in Lower Austria.

the nowcast of temperature is significantly better than that of the NWP model during the first 6 h of the forecast (not shown). Beyond +6 h there is a small but non-negligible benefit from the downscaling procedure. Verification of the precipitation nowcast shows a similar result, with significant improvements relative to the NWP forecast in the nowcasting range. However, as it can be expected, the benefit of the nowcasting vanishes earlier (at forecast ranges of 2–3 h) in the case of precipitation (Haiden et al., 2011).

## 3.2 Case study: nowcasting of precipitation events

### 3.2.1 Large-scale precipitation event: 21–24 June 2009

In late June 2009, many parts of Austria were affected by a heavy precipitation event with severe flooding that lasted for several days. Following the passage of a cold front from the Northwest that advected a cool maritime air mass into Central Europe, a secondary low developed over Italy on 20 June 2009 and moved eastward to the Balkan Peninsula. At its Northern periphery, the return of wrapped-around warm and humid air from the Northeast began on 21 June 2009. The strong warm air advection caused extensive rainfalls, which were further enhanced by orographic lifting along the Northern Alpine rim. Most of the precipitation occurred in form of steady and stratiform rainfalls from 22 to 24 June 2009, followed by widespread (though less abundant) convective precipitation only later when the warm air mass finally established itself also at lower levels. This weather pattern is well-known for bearing a potential of excessive rainfalls in the Alpine region; however, this time its degree of strength and its persistence over several days were extraordinary.

One of the areas that were particularly affected was the Northern Alpine region of Lower Austria. The Traisen, the Ybbs and several smaller rivers drain this moderate-sized physical region (5222 square kilometers) towards the North and discharge into the Danube River. The area is very prone to "Stau" (upslope flow) enhancement of precipitation when the steering level air flow is from Northerly directions. Consequently, the areal mean of the 3-day precipitation sum in the Traisen and Ybbs catchment from 22 to 24 June 2009 amounted to approximately 160 mm, with highest peaks reaching up to 230 mm. The resulting floods of these two rivers and their smaller neighbors had statistical return periods of 10 to 50 yr.

Figure 3 shows forecasted and analyzed hydrographs for this precipitation event in the Traisen and Ybbs catchment. The thick black line is the areal mean of the accumulated precipitation according to hourly INCA precipitation analyses. The colored lines show the areal mean of accumulated INCA precipitation forecasts (with the open diamonds denoting the various initialization times), starting at 06:00 UTC 21 June 2009 and following with an update interval of six hours in order to illustrate the temporal development of forecast quality. It should be noted that the actual update interval of INCA precipitation forecasts is 15 min; however, most of the updated information affects only the nowcasting range, i.e. the first 6 forecast hours. New NWP information is available only four times a day, namely new ALADIN runs usually around 06:00 and 18:00 UTC, and new ECMWF runs usually around 08:00 and 20:00 UTC. For the reconstruction of this case study, an availability of both NWP model data at 09:00 and 21:00 UTC has been assumed, which corresponds to the rare "worst case scenario" in case of a delay due to technical problems. A thinning of the update interval to six hours was considered sufficient to address the issue of the forecast quality development for this large-scale precipitation event.

It can be seen in Fig. 3 that the initial forecast, started at 06:00 UTC 21 June 2009 (dark red curve) – approximately 36 h prior to the start of the heavy rain – considerably underestimated the precipitation sums. Stepwise improvement of forecast quality can be seen for the 12:00 UTC initialization of the same day and the 00:00 and 12:00 UTC initializations of 22 June 2009, indicating that each new NWP model runs provided more valuable information. On the other hand, the 18:00 and 06:00 UTC initializations in-between, which could only rely on new nowcasting information, could not further improve the performance, due to the fact that the event was still out of the nowcasting range.

The first INCA forecast that showed a very good quality was the 18:00 UTC 22 June 2009 initialization, shortly after the start of the main precipitation event. Since precipitation was already present in the latest analyses, also the nowcasting part could begin to provide reliable estimations for the following 6 h and contribute to the further rise of forecast quality. The subsequent INCA forecasts kept a fairly good performance, with forecast errors less than 20 % of the actual precipitation sums.

It also corresponds to our experience that the totals of large-scale precipitation events can usually be well predicted at the latest when the event actually starts. This is due to the good predictability of these events by NWP models, as well as to the "reliable" behavior of these precipitation fields in terms of nowcasting.

### 3.2.2 Convective precipitation event: 23 July 2009

Convective precipitation events form the other wing of the spectrum of precipitation. They are characterized by higher intensities and shorter durations than their stratiform counterparts, plus an often high spatial and temporal variability that reduces their predictability for NWP models as well as for nowcasting methods.

The chosen day for a case of convective precipitation is 23 July 2009, an outstanding thunderstorm day in Austria and other parts of Central Europe. Ahead of a cold front that was approaching from the Northwest, extremely hot and energetic air from Northwestern Africa and the Iberian Peninsula was advected to Central Europe. In the early evening, a pre-frontal convergence line triggered a first round of thunderstorms that travelled eastward along the Northern Alps, followed by a second and more violent surge of storms a few hours later when the cold front arrived. Severe wind gusts and large hail caused enormous and widespread damage; the high precipitation intensities also led to localized flooding, although the fast translation speed of the thunderstorms kept this threat rather low and subordinate.

Figure 4 shows hourly INCA precipitation analyses at 17:00, 18:00, 19:00, and 20:00 UTC on this evening. Again, we focus on the area of the Ybbs and Traisen catchment, which is sketched by a red rectangle in the 17:00 UTC analysis (upper left). The first thunderstorms, which were associated with the passage of the convergence line, can be seen in the 17:00 (upper left) and 18:00 UTC analyses (upper right). They were moving from West to East and affected only the Northern part of the area of interest, before the second and more severe round of thunderstorms arrived shortly before 20:00 UTC (lower right).

Figure 5 illustrates the temporal evolution of the INCA precipitation forecasts for this event. Again, the thick black line is the hydrograph of the areal mean of the hourly precipitation analyses in the Ybbs and Traisen catchment, and the colored lines show the succession of INCA precipitation forecasts with selected initialization times depicted by the open diamonds.

The initial 06:00 UTC forecast (dark red curve) considerably underestimated the forthcoming precipitation event, with an areal mean of only 2.4 mm compared to almost 10 mm according to the verifying analyses. Even the 12:00 UTC initialization (bright red curve), though it had new NWP information to its disposal, could only insignificantly raise the expected areal precipitation mean to 3.1 mm. Since the next NWP update would not arrive before evening, no new information could be expected from the NWP part for this event any more.

The 15:00 UTC (orange curve) and 16:00 UTC (yellow curve) initializations further reduced the expected precipitation amounts, because still no signals were present in the precipitation analyses. Therefore, the nowcasting part could not react adequately and wrongly began to dampen the NWP data, which showed first weak precipitation signals starting from 18:00 UTC. It was not before 17:00 UTC (bright green curve) that the nowcasting picked up the signals of the first thunderstorms and correctly moved them over the Northern part of the catchment. Finally at 18:00 UTC (dark green curve), the first signal for the second surge of thunderstorms appeared in the INCA forecast, though still two hours too late between 21:00 and 22:00 UTC. In reality, the second line of thunderstorms continuously accelerated and arrived much earlier than anticipated. The time lag of the precipitation nowcasts was reduced in the 19:00 UTC initialization (bright blue curve) but did not completely vanish before 20:00 UTC (dark blue curve), when the severe thunderstorm event was already in full swing.

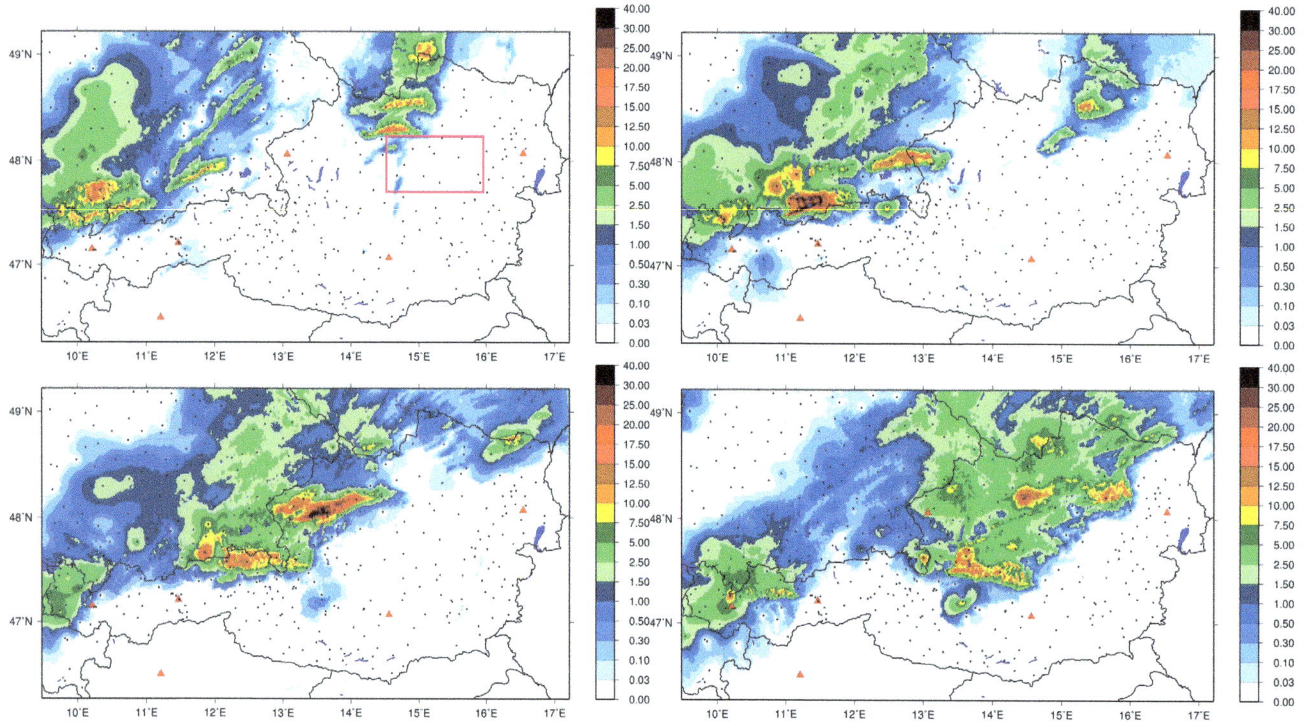

**Figure 4.** INCA precipitation analyses [mm $h^{-1}$] at 17:00 UTC (upper left), 18:00 UTC (upper right), 19:00 UTC (lower left) and 20:00 UTC (lower right) on 23 July 2009. The area of interest for Fig. 5 is sketched by the red rectangle.

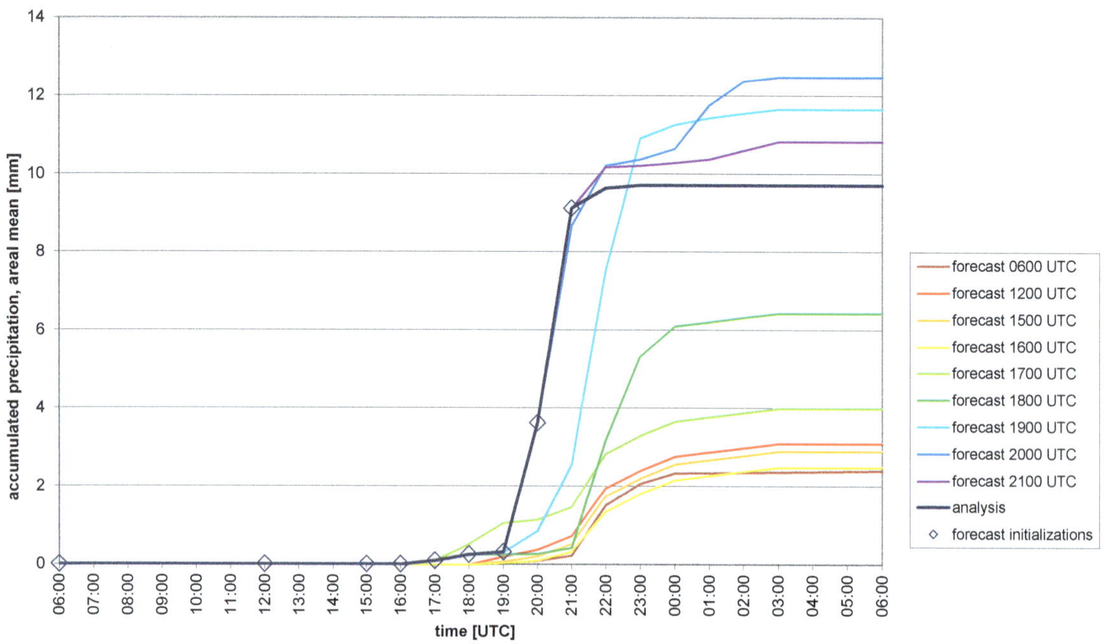


**Figure 5.** As in Fig. 3, but for the 23 July 2009 precipitation event. Note the different scaling of both axes.

The decline of quality of INCA forecasts, as compared to a large-scale precipitation event, is evident. Indeed, experience shows that many convective precipitation events are still poorly predictable even when they have actually started. It is the high spatial and temporal variability, plus the repeated cycle of formation, growth and decay of individual convective cells, that keeps the predictability low. This applies to NWP model forecasts and to nowcasting procedures likewise.

## 4 New developments in INCA-CE

The case studies presented above reveal both the requirements and challenges that need to be addressed when introducing new or improving existing modules of the INCA system. Within the project duration, a refinement of algorithms in INCA is currently carried out. Special focus in the developments is given to parameters which are important for applications in flood warning, civil protection (mostly storm forecasts) and road safety.

For hydrological applications, the prediction of convective precipitation needs to be improved. Currently, only a translation of convective cells is used for forecasts whereas an approach including the life cycle of convective cells is needed. Based on diagnostic fields already present in INCA, such as CAPE, CIN, moisture flux convergence and others, the susceptibility of certain areas to initiation, intensification, weakening and dissipation of convective cells could be assessed. The most critical factor for storm evolution is the existence of boundary layer convergence zones. Hence, convective nowcasting algorithms rely heavily on the quality of the near-surface wind field forecasted by the NWP models. Developments in wind nowcasting are focusing on improvements in the way topographic effects (downslope windstorms, channeling) are treated, as well as in a better analysis of the downburst potential of thunderstorm cells. For Road Safety, the nowcasting of surface temperature is refined, especially in the temperature range close to the freezing point. However, all developments with respect to precipitation, wind and temperature are contributing to improved road safety applications.

## 5 Conclusions

An overview of the INCA-CE project is presented. The core module, i.e. the INCA nowcasting system, which forms the basis for downstream applications in hydrology, civil protection and road safety, is introduced briefly. The INCA-CE project provides the framework to jointly develop a high-level nowcasting system with respect to the specific needs of applications in hydrology, civil protection and road safety. The user-oriented development is guaranteed by the feedback mechanisms implemented in the project structure which allows for a permanent revision and improvement of the nowcasting tools and methods. Trainings including practical examples will be provided to the end-users in order to ensure a proper use and to avoid misinterpretations of nowcasting products, especially for convective events. Thus, the project can be regarded as a proto-type of demand-side scientific research and development, bridging the gap between science and application.

**Acknowledgements.** INCA-CE is a project involving 16 partners from 8 different Central European countries (see http://www.inca-ce.eu for details on the partnership). We would like to express our thanks to each partner of this project. The INCA-CE project is implemented through the Central Europe Programme co-financed by the European Regional Development Fund. The authors are thankful to two anonymous reviewers for their suggestions and comments that led to an improvement of the manuscript.

Edited by: B. Reichert

## References

Browning, K. A. and Collier, C. G.: Nowcasting of precipitation systems, Rev. Geophys., 27, 345–370, 1989.

Crook, N. A. and Sun, J.: Analysis and forecasting of the low-level wind during the Sydney 2000 Forecast Demonstration Project, Weather Forecast., 19, 151–167, 2004.

Derrien, M. and Le Gléau, H.: MSG/SEVIRI cloud mask and type from SAFNWC, Int. J. Remote Sens., 26, 4707–4732, 2005.

Golding, B. W.: Nimrod: A system for generating automated very short range forecasts, Meteorol. Appl., 5, 1–16, 1998.

Haiden, T. and Pistotnik, G.: Intensity-dependent parameterization of elevation effects in precipitation analysis, Adv. Geosci., 20, 33–38, doi:10.5194/adgeo-20-33-2009, 2009.

Haiden, T., Kann, A., Pistotnik, G., Stadlbacher, K., and Wittmann, C.: Integrated Nowcasting through Comprehensive Analysis (INCA) – System description, ZAMG, Vienna, 60 pp., http://www.zamg.ac.at/fix/INCA_system.pdf, 2010.

Haiden, T., Kann, A., Wittmann, C., Pistotnik, G., Bica, B., and Gruber, C.: The Integrated Nowcasting through Comprehensive Analysis (INCA) system and its validation over the Eastern Alpine region, Weather Forecast., 26, 166–183, doi:10.1175/2010WAF2222451.1, 2011.

Hand, W. H.: An object-oriented technique for nowcasting heavy showers and thunderstorms, Meteorol. Appl., 3, 31–41, 1996.

Komma, J., Reszler, C., Blöschl, G., and Haiden, T.: Ensemble prediction of floods – catchment non-linearity and forecast probabilities, Nat. Hazards Earth Syst. Sci., 7, 431–444, doi:10.5194/nhess-7-431-2007, 2007.

Li, L., Schmid, W., and Joss, J.: Nowcasting of motion and growth of precipitation with radar over a complex topography, J. Appl. Meteorol., 34, 1286–1300, 1995.

Pierce, C. E., Hardaker, P. J., Collier, C. G., and Haggett, C. M.: GANDOLF: a system for generating automated nowcasts of convective precipitation, Meteorol. Appl., 7, 341–360, 2000.

Pierce, C. E., Ebert, E., Seed, A. W., Sleigh, M., Collier, C. G., Fox, N. I., Donaldson, N., Wilson, J. W., Roberts, R., and Mueller, C. K.: The nowcasting of precipitation during Sydney 2000: An appraisal of the QPF algorithms, Weather Forecast., 19, 7–21, 2004.

Seed, A. W.: A dynamic and spatial scaling approach to advection forecasting, J. Appl. Meteorol., 42, 381–388, 2003.

Steinheimer, M. and Haiden, T.: Improved nowcasting of precipitation based on convective analysis fields, Adv. Geosci., 10, 125–131, doi:10.5194/adgeo-10-125-2007, 2007.

Wang, Y., Haiden, T., and Kann, A.: The operational Limited Area Modelling system at ZAMG: ALADIN-AUSTRIA, Österr. Beiträge Meteorol. Geophys., 37, ISSN 1016-6254, 2006.

Wilson, J. W. and Schreiber, W. E.: Initiation of convective storms by radar-observed boundary layer convergence lines, Mon. Weather Rev., 114, 2516–2536, 1986.

Wilson, J. W., Ebert, E. E., Saxen, T. R., Roberts, R. D., Mueller, C. K., Sleigh, M., Pierce, C. E., and Seed, A.: Sydney 2000 Forecast Demonstration Project: Convective storm nowcasting, Weather Forecast., 19, 131–150, 2004.

# 40

# On the predictability of outliers in ensemble forecasts

**S. Siegert[1], J. Bröcker[1,*], and H. Kantz[1]**

[1]Max Planck Institute for the Physics of Complex Systems, Dresden, Germany
[*]visiting research fellow: Centre for the Analysis of Time Series, London School of Economics, London, UK

*Correspondence to:* S. Siegert (siegert@pks.mpg.de)

**Abstract.** In numerical weather prediction, ensembles are used to retrieve probabilistic forecasts of future weather conditions. We consider events where the verification is smaller than the smallest, or larger than the largest ensemble member of a scalar ensemble forecast. These events are called outliers. In a statistically consistent $K$-member ensemble, outliers should occur with a base rate of $2/(K+1)$. In operational ensembles this base rate tends to be higher. We study the predictability of outlier events in terms of the Brier Skill Score and find that forecast probabilities can be calculated which are more skillful than the unconditional base rate. This is shown analytically for statistically consistent ensembles. Using logistic regression, forecast probabilities for outlier events in an operational ensemble are calculated. These probabilities exhibit positive skill which is quantitatively similar to the analytical results. Possible causes of these results as well as their consequences for ensemble interpretation are discussed.

## 1 Motivation

An *ensemble* is a collection of $K$ individual forecasts for the same target. Here, we consider only forecasts of scalar variables, such as the temperature at a certain location. The individual ensemble members may differ in their initial conditions, in the model parameters, or they can even be produced by entirely different models. The objective of an ensemble forecast is to obtain an estimate of the flow-dependent forecast uncertainty, expressed, for example, in terms of a probability distribution function. The heterogeneities of the ensemble members reflect the lack of confidence in the initial state and the relevant physical processes. Under the special working hypothesis that all ensemble members are equally likely model scenarios, the final measurement – the *verification* – should behave like just another ensemble member in a statistical sense. An ensemble that satisfies this condition is called *statistically consistent* (Anderson, 1997).

*Outliers* in ensemble forecasts are events where the verification is smaller than the smallest or larger than the largest ensemble member (see Fig. 1). In this context, the term "outlier" was coined by Buizza and Palmer (1998). An ensemble of $K$-members defines $K+1$ intervals into which the verification can fall. If the ensemble is statistically consistent, none of these intervals should be preferred by the verification and each should occur with the same *base rate* (average relative frequency) of $1/(K+1)$. Two of the intervals correspond to outliers, hence the outlier base rate should be neither larger nor smaller than $2/(K+1)$ if the ensemble is consistent.

In operational forecast ensembles, however, outliers typically occur at a higher base rate. In Fig. 2 the outlier base rate in the ECMWF[1] temperature ensemble that was operational between 2001 and 2006 is shown. The location is Dresden, Germany (WMO10488). Throughout this study, we use ensemble data at the nearest grid point to this station and do not include the control run in the ensemble. For short-range forecasts outliers occur around 35 % of the time, for long lead times the outlier base rate saturates at around 10 %. If this 50-member ensemble were consistent, the outlier base rate should not exceed $2/51 \approx 3.9\,\%$. An increased frequency of outliers can have several causes such as conditional bias, lack of ensemble dispersion, model and discretization error, as well as observation noise (Hamill, 2001; Saetra et al., 2004).

Outliers are of interest because they can be interpreted as "unexpected" events if the range of possible scenarios is defined by the ensemble range. The outlier at lead time 1 day in Fig. 1 represents a particularly serious example of this interpretation. All of the 50 ensemble members predict

[1]European Centre for Medium-Range Weather Forecasts

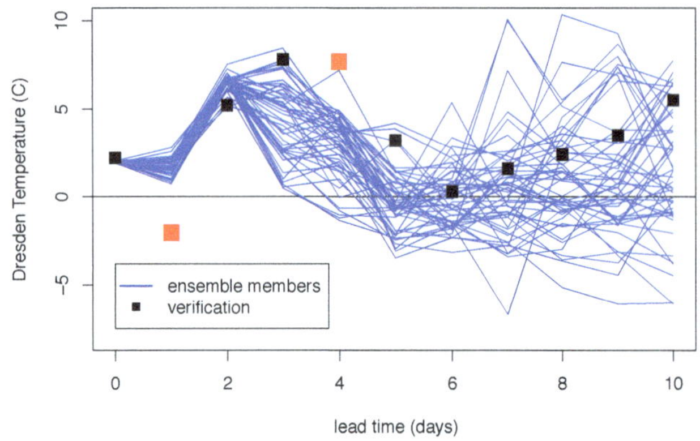


**Figure 1.** An ensemble temperature forecast issued by the ECMWF. Initialization date (lead time 0) is 22 December 2004, the location is WMO10488 (Dresden, Germany). The outliers at lead times 1 and 4 days are marked red.

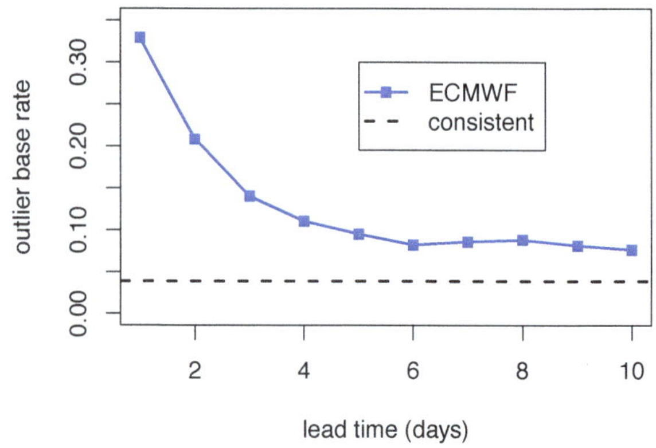


**Figure 2.** The markers indicate outlier base rate over lead time for the ECMWF temperature ensemble. The time period is 2001–2006 and the location is Dresden, Germany. The dashed line indicates the outlier base rate 2/51 of a consistent 50-member ensemble.

temperature above zero degrees but the verification drops below zero. A forecast user might incur damage from this outlier not due to the extremeness of the event, but due to not being prepared for the ensuing weather conditions, such as freezing rain. Outliers (such as the ones in Fig. 1) are not necessarily extreme with respect to climatology, but they are extreme events with respect to the forecast distribution. Note that, in the present study, we do not consider the distance of the verification from the ensemble, but only whether it is inside or outside the ensemble range.

In summary, the increased frequency of occurrence of outliers on the one hand, and their interpretation as unexpected events on the other hand motivate us to consider their predictability. In particular we consider the following question: can probabilistic predictions be issued for outliers that are more skillful than their unconditional base rate of occurrence?

In Sect. 2 we review the Brier Skill Score which we use to assess the quality of probabilistic predictions. In Sect. 3 we provide theoretical arguments as to the predictability of outliers in a statistically consistent ensemble, thereby providing a benchmark result. This result is contrasted to prediction skill of outliers in an operational ensemble in Sect. 4. Section 5 concludes.

A similar study has recently been published in Siegert et al. (2011). The present study is different in its discussion of outlier predictability in the context of Brier Score decomposition, in the statistical prediction algorithm employed, a more careful discussion of the base rate effect, and in the details of the ensemble prediction system.

## 2 Evaluating probabilistic predictions by means of the Brier Skill Score

The *Brier Score* (abbr. BS, Brier, 1950) and *Brier Skill Score* (abbr. BSS, e.g. Wilks, 2006) provide means to evaluate and compare, respectively, the skill of probabilistic forecasts. For a single probabilistic prediction $p$ of a binary event $y$ the BS is given by

$$\mathrm{BS}(p,y) = (y-p)^2, \tag{1}$$

where the occurrence or non-occurrence of the event is coded by $y=1$ or $y=0$, respectively. The BS is negatively oriented and vanishes only for a perfect forecast. In the present study our goal is to compare the prediction skill of a forecast $p$ to that of a base rate forecast $q=\bar{y}$. The BSS of $p$ with respect to $q$ is given by

$$\mathrm{BSS}(p,q) = 1 - \frac{\overline{\mathrm{BS}(p,y)}}{\overline{\mathrm{BS}(q,y)}}, \tag{2}$$

where the overbar indicates either the mathematical expectation value or an empirical average. In the present context, the numerical value of the BSS can be interpreted as the percentage of improvement of the forecast $p$ over the base rate forecast.

## 3 Theoretical prediction skill in a consistent forecast ensemble

In this section a benchmark result as to the predictability of outliers in consistent forecast ensembles is derived. Two assumptions are made: the ensemble $\boldsymbol{e}=(e_1,\cdots,e_K)$ is statistically consistent and the cumulative distribution function $F(\cdot)$, from which ensemble members and the verification $\eta$ are independently drawn, is known to the forecaster. If $F$ and $\boldsymbol{e}$ are known, the "true" outlier probability $\sigma$ can be calculated by

$$\sigma := \mathbb{P}(\eta \notin [e_{[1]}, e_{[K]}] \mid F, \boldsymbol{e}) = F(e_{[1]}) + 1 - F(e_{[K]}), \tag{3}$$

where $e_{[i]}$ denotes the $i$-th order statistic, that is, the value of the $i$-th largest ensemble member. Obviously, $\sigma$ is not a constant but a fluctuating random quantity. In Siegert et al. (2011) we show that the probability distribution function (pdf) of $\sigma$, evaluated at the value $x$, is given by

$$p_\sigma(x) = K(K-1)(1-x)^{K-2}x, \tag{4}$$

and thus independent of $F$. The universality of the pdf of $\sigma$ is a result of the fact that $F(x)$ is uniformly distributed on the unit interval if $x$ is a random sample drawn from $F$ (Mood et al., 1974). Note that under the distribution given by Eq. (4), the expectation value of $\sigma$ (the base rate) is correctly given by $2/(K+1)$.

For the calculation of the BSS the expectation value of the BS of $\sigma$, i.e. $\mathbb{E}(y-\sigma)^2$, as well as the expectation value of the BS of the base rate forecast, i.e. $\mathbb{E}(y-\frac{2}{K+1})^2$ are required. The calculation of the BS of $\sigma$ involves conditional expectations (Mood et al., 1974) and the equality $\mathbb{P}(y=1\mid\sigma)=\sigma$. One can show that $\mathbb{E}(y-\sigma)^2 = \mathbb{E}\sigma(1-\sigma)$, which can then be explicitly calculated using Eq. (4). Combining the two resulting Brier Scores using Eq. (2) yields for the BSS of outlier prediction

$$\mathrm{BSS}\left(\sigma, \frac{2}{K+1}\right) = \frac{1}{K+2}, \tag{5}$$

under the assumptions of a consistent ensemble and a known forecast distribution. We conclude that under this idealized scenario, outliers are indeed better predictable than merely by their unconditional base rate. The result is universal since it is independent of the forecast distribution. The improvement over the base rate forecast is the higher, the smaller the ensemble is and vanishes as $K\to\infty$. The positive skill is obtained by predicting the true (fluctuating) outlier probability instead of its mean. In the 50-member ECMWF-ensemble, for example, we would get a BSS of close to 2 % improvement over the base rate forecast under the idealized scenario.

As it was shown in Murphy (1973), the Brier Score can be decomposed into a sum of three terms, namely an *uncertainty* term, a property only of the process to be predicted, a *reliability* term which quantifies the deviance of the forecast from reliability, and a *resolution* term which quantifies how much the individual predicted probabilities differ from each other. In fact, this decomposition is possible for every proper score (Bröcker, 2009). In the outlier prediction problem, the $\sigma$-forecast obtains its improvement in skill from a non-vanishing resolution term – the uncertainty terms of the $\sigma$-forecast and the base rate forecast are equal and their reliability terms vanish. The pdf of $\sigma$ given by Eq. (4) is a Beta distribution whose variance decreases as the ensemble size $K$ increases. In fact, for any reliable forecast (such as $\sigma$ here), the resolution term of the BS is given by the variance of the forecast probabilities. Thus, the vanishing skill of $\sigma$ resulting from a decreasing resolution can equivalently be explained by a decreasing variance of $\sigma$ which leads to the $\sigma$-forecast becoming more and more similar to the base rate forecast. This example provides an intuitive illustration of the concept of resolution in probabilistic forecasting.

## 4 Outlier prediction skill in an operational ensemble

In an operational setting the assumptions that lead to the universal result given by Eq. (5) are not satisfied. Neither is the ensemble consistent (see Fig. 2), nor is the true distribution from which the ensemble is drawn known to the forecaster. Notwithstanding that, outlier probabilities that are conditional on the present state of the ensemble can still be estimated by means of statistical learning.

We trained a *logistic regression* model with Lasso regularization (Tibshirani, 1996) that takes as its input vector $\boldsymbol{x}_t$ at time instance $t$ a constant intercept, both the smallest and the largest ensemble member and also seasonal varying inputs with frequency $\phi = 2\pi/(365\text{ days})$:

$$\boldsymbol{x}_t = (1, \sin(\phi t), \cos(\phi t), e^t_{[1]}, e^t_{[K]}). \tag{6}$$

Logistic regression fits a linear superposition of the inputs to the log-odds ratio of the outlier probability $p_t$, that is

$$\boldsymbol{x}_t\boldsymbol{\beta} = \log\left(\frac{p_t}{1-p_t}\right) \tag{7}$$

where $\boldsymbol{\beta}$ is the vector of coefficients. The coefficients are obtained using cyclical coordinate descent to maximize the log-likelihood of the data given the coefficients (Friedman et al., 2010). The Lasso penalty leads to shrinkage of unimportant inputs, thus avoiding overfitting. We apply 10-fold cross-validation to estimate out-of-sample performance. Our data set comprises 6 yr of daily temperature ensemble forecasts ($K=50$) for Dresden (WMO10488) issued by the ECMWF ensemble between 2001 and 2006.

At each instance we issue the Lasso-estimated outlier probability $p_t$. Simultaneously we issue the constant base rate forecast, where we use as the base rate the outlier frequency known from the training data. The Brier Scores of these predictions are compared using the BSS.

The BSS over lead time is shown in Fig. 3. At all lead times, BSS is significantly greater than zero, indicating significant predictive skill of the Lasso-estimated outlier probabilities. At short lead times (for which a lot of outliers occur), the BSS is about 10 % improvement over the base rate forecast. This is a large improvement compared to the analytical result of about 2 %. The ensemble at short lead times has an increased outlier base rate and thus, qualitatively, behaves like an ensemble with fewer than 50 members. Since, according to Eq. (5), the BSS is larger in smaller ensembles, the observed skill in the short lead time ensembles is larger than the analytical BSS using $K=50$. At higher lead times, the BSS saturates at a small, but significantly positive, value which is close to the analytical result for the consistent ensemble, despite the operational ensemble not being consistent. This observation does not change if we apply statistical downscaling of the ensemble using the four neighboring grid points, instead of using ensemble data only at the nearest grid point. Our downscaling corrects for seasonal bias, which reduces the outlier base rate shown in Fig. 2 by around 3 % at

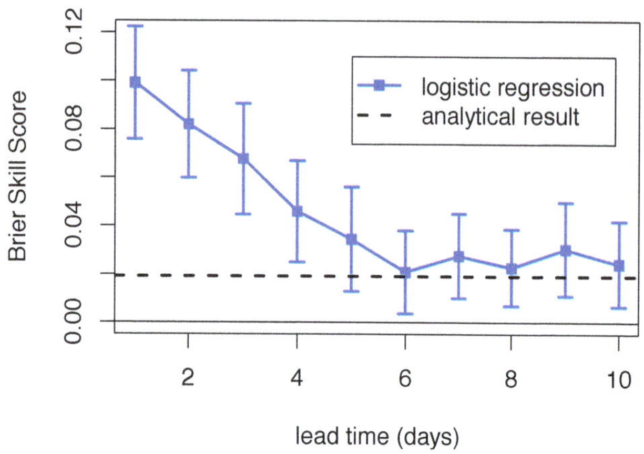


**Figure 3.** Brier Skill Score over lead time of estimated outlier probability compared to base rate forecast. Error bars indicate 95 % confidence intervals. For reference, the analytical result obtained in Sect. 3 using $K = 50$ is shown.

all lead times. But the behavior of Brier Skill Score shown in Fig. 3 remains the same. Further, the BSS is invariant if we use ensemble and verification data at a different station (Heligoland, WMO10015) and if we include the control as an additional ensemble member. Note, however, that the results presented here might not be representative of the ECMWF ensemble which is currently operational.

We have tested further potential predictors for the logistic regression estimates. Examples include the full ensemble $e^t$, higher frequencies of the seasonal inputs, ensembles at neighboring stations, ensembles at smaller and larger lead times and ensembles of different variables at the same station. None of these could significantly improve the predictive skill over that of the predictors given in Eq. (6).

## 5 Summary and conclusions

We have considered the predictability of outliers in ensemble forecasts. The main result is that outliers are predictable better than by their unconditional base rate. This was shown in terms of the BSS both for a hypothetical ensemble which is perfectly statistically consistent, and for an operational temperature forecast ensemble. For both ensembles, the predictive skill is obtained from the fact that the outlier probability is conditional on the current state of the ensemble. In the operational ensemble, additional skill is obtained from seasonality.

A more detailed analysis shows that outliers become more likely if the range of the forecast ensemble is low. In fact, using the range as a predictor instead of $e^t_{[1]}$ and $e^t_{[K]}$ does not alter the BSS significantly. The fact that a small range leads to a higher outlier probability seems trivial but at the same time it contradicts to some extent the common view that a very narrow ensemble is an indicator of low forecast uncertainty. Sampling effects can lead to an increase or decrease of the outlier probability compared to the base rate.

The empirical results of Sect. 4 suggest that the members of the operational ensemble should be interpreted as order statistics drawn from a distribution and not as quantiles of a distribution. If ensemble members were quantiles, the mass of probability concentrated between the ensemble members, and outside the ensemble, would be a constant. The outlier probability conditional on the ensemble would then be equal to the unconditional base rate. However, by allowing this probability to change as a function of certain characteristics of the ensemble we obtain significant forecast skill.

A combination of an increased base rate, seasonality, and the performance of an empirical statistical learning algorithm leads to approximately the same forecast skill in the operational ensemble that we expect if the ensemble were consistent and if we knew the distribution from which the members were drawn. Neither of the latter assumptions is satisfied in operational ensembles. Based on these differences, we should not overinterpret the apparent similarity in predictive skill. The main result is that predictive skill of outlier events in the operational ensemble is small, yet significant. However, the analytical benchmark skill score made the observed skill more interpretable, which motivates us to further address predictability problems in that way.

**Acknowledgements.** We are grateful to Renate Hagedorn and the ECMWF for providing ensemble and station data for Dresden and Heligoland. We thank Christopher Ferro and an anonymous referee for helpful comments on an earlier draft of this article.

Edited by: H. Böttger

## References

Anderson, J.: The impact of dynamical constraints on the selection of initial conditions for ensemble predictions: Low-order perfect model results, Mon. Weather Rev., 125, 2969–2983, 1997.

Brier, G.: Verification of forecasts expressed in terms of probability, Mon. Weather Rev., 78, 1–3, 1950.

Bröcker, J.: Reliability, sufficiency, and the decomposition of proper scores, Q. J. Roy. Meteorol. Soc., 135, 1512–1519, 2009.

Buizza, R. and Palmer, T. N.: Impact of ensemble size on ensemble prediction, Mon. Weather Rev., 126, 2503–2518, 1998.

Friedman J., Hastie, T., and Tibshirani, R.: Regularization Paths for Generalized Linear Models via Coordinate Descent, J. Stat. Softw., 33, 1–22, 2010.

Hamill, T. M.: Interpretation of rank histograms for verifying ensemble forecasts, Mon. Weather Rev., 129, 550–560, 2001.

Mood, A., Graybill, F., and Boes, D.: Introduction to the theory of statistics (3rd Edn.), McGraw-Hill, New York, NY, 1974.

Murphy, A. H.: A new vector partition of the probability score, J. Appl. Meteorol., 12, 595–600, 1973.

Saetra, Ø., Hersbach, H., Bidlot, J. R., and Richardson, D. S.: Effects of observation errors on the statistics for ensemble spread and reliability, Mon. Weather Rev., 132, 1487–1501, 2004.

Siegert, S., Bröcker, J., and Kantz, H.: Predicting outliers in ensemble forecasts, Q. J. Roy. Meteorol. Soc., 137, 1887–1897, 2011.

Tibshirani, R.: Regression shrinkage and selection via the Lasso, J. Roy. Stat. Soc. B, 58, 267–288, 1996.

Wilks, D. S.: Statistical methods in the atmospheric sciences (2nd Edn.), Academic Press, New York, NY, 2006.

# Permissions

 Every chapter published in this book has been scrutinized by our experts. Their significance has been extensively debated. The topics covered herein carry significant findings which will fuel the growth of the discipline. They may even be implemented as practical applications or may be referred to as a beginning point for another development.

The contributors of this book come from diverse backgrounds, making this book a truly international effort. This book will bring forth new frontiers with its revolutionizing research information and detailed analysis of the nascent developments around the world.

We would like to thank all the contributing authors for lending their expertise to make the book truly unique. They have played a crucial role in the development of this book. Without their invaluable contributions this book wouldn't have been possible. They have made vital efforts to compile up to date information on the varied aspects of this subject to make this book a valuable addition to the collection of many professionals and students.

This book was conceptualized with the vision of imparting up-to-date information and advanced data in this field. To ensure the same, a matchless editorial board was set up. Every individual on the board went through rigorous rounds of assessment to prove their worth. After which they invested a large part of their time researching and compiling the most relevant data for our readers.

The editorial board has been involved in producing this book since its inception. They have spent rigorous hours researching and exploring the diverse topics which have resulted in the successful publishing of this book. They have passed on their knowledge of decades through this book. To expedite this challenging task, the publisher supported the team at every step. A small team of assistant editors was also appointed to further simplify the editing procedure and attain best results for the readers.

Apart from the editorial board, the designing team has also invested a significant amount of their time in understanding the subject and creating the most relevant covers. They scrutinized every image to scout for the most suitable representation of the subject and create an appropriate cover for the book.

The publishing team has been an ardent support to the editorial, designing and production team. Their endless efforts to recruit the best for this project, has resulted in the accomplishment of this book. They are a veteran in the field of academics and their pool of knowledge is as vast as their experience in printing. Their expertise and guidance has proved useful at every step. Their uncompromising quality standards have made this book an exceptional effort. Their encouragement from time to time has been an inspiration for everyone.

The publisher and the editorial board hope that this book will prove to be a valuable piece of knowledge for researchers, students, practitioners and scholars across the globe.

# List of Contributors


**A. Baklanov**
Danish Meteorological Institute, Lyngbyvej 100, 2100 Copenhagen, Denmark

**J. Hiebl**
Central Institute for Meteorology and Geodynamics, Vienna, Austria

**S. Reisenhofer**
Central Institute for Meteorology and Geodynamics, Vienna, Austria

**I. Auer**
Central Institute for Meteorology and Geodynamics, Vienna, Austria

**R. Böhm**
Central Institute for Meteorology and Geodynamics, Vienna, Austria

**W. Schöner**
Central Institute for Meteorology and Geodynamics, Vienna, Austria

**J. Lémond**
Direction de la Climatologie, Météo-France, 42 av. G. Coriolis, 31057 Toulouse Cedex, France
CNRM-GAME, URA1357, Météo-France, CNRS, 42 av. G. Coriolis, 31057 Toulouse Cedex, France

**Ph. Dandin**
Direction de la Climatologie, Météo-France, 42 av. G. Coriolis, 31057 Toulouse Cedex, France

**S. Planton**
CNRM-GAME, URA1357, Météo-France, CNRS, 42 av. G. Coriolis, 31057 Toulouse Cedex, France

**R. Vautard**
LSCE, UMR8212, IPSL, FR636, laboratoire CEA-CNRS-Université Versailles Saint Quentin, Orme des Merisiers, 91191 Gif sur Yvette, France

**C. Pagé**
Scinces de l'Univers au CERFACS, URA1875, 42 av. G. Coriolis, 31057 Toulouse Cedex 01, France

**M. Déqué**
CNRM-GAME, URA1357, Météo-France, CNRS, 42 av. G. Coriolis, 31057 Toulouse Cedex, France

**L. Franchistéguy**
Direction de la Climatologie, Météo-France, 42 av. G. Coriolis, 31057 Toulouse Cedex, France

**S. Geindre**
Direction de la Climatologie, Météo-France, 42 av. G. Coriolis, 31057 Toulouse Cedex, France

**M. Kerdoncuff**
Direction de la Climatologie, Météo-France, 42 av. G. Coriolis, 31057 Toulouse Cedex, France

**L. Li**
LMD, UMR8539, IPSL, FR636, CNRS-Université Pierre et Marie Curie, 4 place Jussieu, 75252 Paris Cedex 05, France

**J. M. Moisselin**
Direction de la Climatologie, Météo-France, 42 av. G. Coriolis, 31057 Toulouse Cedex, France

**T. Noël**
LSCE, UMR8212, IPSL, FR636, laboratoire CEA-CNRS-Université Versailles Saint Quentin, Orme des Merisiers, 91191 Gif sur Yvette, France

**Y. M. Tourre**
Direction de la Climatologie, Météo-France, 42 av. G. Coriolis, 31057 Toulouse Cedex, France

**M. B. Gavrilov**
Chair of Physical Geography, Faculty of Sciences, University of Novi Sad, Trg D. Obradovića 3, 21000 Novi Sad, Serbia

**G. R. Jovanović**
Hydrometeorological Service of Serbia, Kneza Višeslava 66, 11000 Belgrade, Serbia

**Z. Janjić**
National Centres for Environmental Predictions, Washington, USA

**K. Wapler**
Deutscher Wetterdienst, Offenbach, Germany
Hans-Ertel-Centre for Weather Research, Atmospheric Dynamics and Predictability Branch, Germany

**M. Goeber**
Deutscher Wetterdienst, Offenbach, Germany
Hans-Ertel-Centre for Weather Research, Optimal Application of Weather Forecasts Branch, Germany

**S. Trepte**
Deutscher Wetterdienst, Offenbach, Germany

**D. Mastrangelo**
Institute of Atmospheric Sciences and Climate, National Research Council, Bologna, Italy

**P. Malguzzi**
Institute of Atmospheric Sciences and Climate, National Research Council, Bologna, Italy

**C. Rendina**
Institute of Atmospheric Sciences and Climate, National Research Council, Bologna, Italy

**O. Drofa**
Institute of Atmospheric Sciences and Climate, National Research Council, Bologna, Italy

**A. Buzzi**
Institute of Atmospheric Sciences and Climate, National Research Council, Bologna, Italy

**J. Rapp**
Deutscher Wetterdienst, Regional Climate Centre Node on Climate Monitoring (WMO RA VI), Frankfurter Str. 135, 63067 Offenbach am Main, Germany

**H. Nitsche**
Deutscher Wetterdienst, Regional Climate Centre Node on Climate Monitoring (WMO RA VI), Frankfurter Str. 135, 63067 Offenbach am Main, Germany

**P. Bissolli**
Deutscher Wetterdienst, Regional Climate Centre Node on Climate Monitoring (WMO RA VI), Frankfurter Str. 135, 63067 Offenbach am Main, Germany

**T. P. Legg**
National Climate Information Centre, Met Office, Exeter, UK

**R. Hernández**
Basque Meteorology Agency (EUSKALMET), Vitoria-Gasteiz, Spain TECNALIA, Meteo Unit. Vitoria-Gasteiz, Spain

**M. Maruri**
Basque Meteorology Agency (EUSKALMET), Vitoria-Gasteiz, Spain TECNALIA, Meteo Unit. Vitoria-Gasteiz, Spain

**K. Otxoa de Alda**
Basque Meteorology Agency (EUSKALMET), Vitoria-Gasteiz, Spain TECNALIA, Meteo Unit. Vitoria-Gasteiz, Spain

**J. Egaña**
Basque Meteorology Agency (EUSKALMET), Vitoria-Gasteiz, Spain TECNALIA, Meteo Unit. Vitoria-Gasteiz, Spain

**S. Gaztelumendi**
Basque Meteorology Agency (EUSKALMET), Vitoria-Gasteiz, Spain TECNALIA, Meteo Unit. Vitoria-Gasteiz, Spain

**A. Kann**
Central Institute for Meteorology and Geodynamics, Vienna, Austria

**F. Desiato**
Istituto Superiore per la Protezione e la Ricerca Ambientale (ISPRA), Rome, Italy

**G. Fioravanti**
Istituto Superiore per la Protezione e la Ricerca Ambientale (ISPRA), Rome, Italy

**P. Fraschetti**
Istituto Superiore per la Protezione e la Ricerca Ambientale (ISPRA), Rome, Italy

**W. Perconti**
Istituto Superiore per la Protezione e la Ricerca Ambientale (ISPRA), Rome, Italy

**A. Toreti**
Istituto Superiore per la Protezione e la Ricerca Ambientale (ISPRA), Rome, Italy
Oeschger Centre for Climate Change Research (OCCR) and Institute of Geography, Climatology and Meteorology, University of Bern, Bern, Switzerland

**F. Imbery**
German Weather Service, Offenbach, Germany

**S. Plagemann**
German Weather Service, Offenbach, Germany

**J. Namyslo**
German Weather Service, Offenbach, Germany

**L. Feudale**
OSMER - Osservatorio Meteorologico Regionale dell'ARPA FVG, Visco (UD), Italy

**A. Manzato**
OSMER - Osservatorio Meteorologico Regionale dell'ARPA FVG, Visco (UD), Italy

**S. Micheletti**
OSMER - Osservatorio Meteorologico Regionale dell'ARPA FVG, Visco (UD), Italy

**C. Lussana**
ARPA Lombardia, Milano, Italy

**H. Mächel**
Deutscher Wetterdienst, Klima und Umwelt, Frankfurter Str. 135, 63067 Offenbach, Germany

**A. Kapala**
Meteorological Institute, University Bonn, Auf dem Hügel 20, 53121 Bonn, Germany

**C. Wittmann**
Central Institute for Meteorology and Geodynamics, Vienna, Austria

**T. Haiden**
Central Institute for Meteorology and Geodynamics, Vienna, Austria

**A. Kann**
Central Institute for Meteorology and Geodynamics, Vienna, Austria

**R. Juvanon du Vachat**
Société Météorologique de France, 1, Quai Branly 75007, Paris, France

**I. Krüzselyi**
Hungarian Meteorological Service, Budapest, Hungary

**J. Bartholy**
Department of Meteorology, Eötvös Loránd University, Budapest, Hungary

**A. Horányi**
Hungarian Meteorological Service, Budapest, Hungary

**I. Pieczka**
Department of Meteorology, Eötvös Loránd University, Budapest, Hungary

**R. Pongrácz**
Department of Meteorology, Eötvös Loránd University, Budapest, Hungary

**P. Szabó**
Hungarian Meteorological Service, Budapest, Hungary

**G. Szépszó**
Hungarian Meteorological Service, Budapest, Hungary

**Cs. Torma**
Department of Meteorology, Eötvös Loránd University, Budapest, Hungary

**M. Schwarb**
Meteodat GmbH, Zurich, Switzerland

**D. Acuña**
SENAMHI, Lima, Peru

**Th. Konzelmann**
Federal Office of Meteorology and Climatology MeteoSwiss, Zurich, Switzerland

**M. Rohrer**
Meteodat GmbH, Zurich, Switzerland

**N. Salzmann**
Department of Geography, University of Zurich, Switzerland

**B. Serpa Lopez**
SENAMHI, Lima, Peru

**E. Silvestre**
SENAMHI, Lima, Peru

**C. Gallego**
Wind Energy Forecasting Group, CIEMAT. Avd. Complutense 22, 28040 Madrid, Spain

**A. Costa**
Wind Energy Forecasting Group, CIEMAT. Avd. Complutense 22, 28040 Madrid, Spain

**A. Cuerva**
IDR/UPM, E.T.S.I. Aeronáuticos, Universidad Politécnica de Madrid, Pza. de Cardenal Cisneros 3, 28040 Madrid, Spain

**E. V. van der Plas**
Royal Netherlands Meteorological Institute, De Bilt, The Netherlands

**B.Wichers Schreur**
Royal Netherlands Meteorological Institute, De Bilt, The Netherlands

**K. Kok**
Royal Netherlands Meteorological Institute, De Bilt, The Netherlands

**I. Zalachori**
IRSTEA, Hydrology Group, UR HBAN, 1 rue Pierre-Gilles de Gennes CS 10030, 92761 Antony, France

**M.-H. Ramos**
IRSTEA, Hydrology Group, UR HBAN, 1 rue Pierre-Gilles de Gennes CS 10030, 92761 Antony, France

**R. Garçon**
EDF-DTG, Electricité de France, Direction Technique de Grenoble, France

**T. Mathevet**
EDF-DTG, Electricité de France, Direction Technique de Grenoble, France

**J. Gailhard**
EDF-DTG, Electricité de France, Direction Technique de Grenoble, France

**C. Román-Cascón**
Dpto. Geofísica y Meteorología, Universidad Complutense de Madrid, Spain

**C. Yagüe**
Dpto. Geofísica y Meteorología, Universidad Complutense de Madrid, Spain

**M. Sastre**
Dpto. Geofísica y Meteorología, Universidad Complutense de Madrid, Spain

**G. Maqueda**
Dpto. Astrofísica y Ciencias de la Atmósfera, Universidad Complutense de Madrid, Spain

**F. Salamanca**
Lawrence Berkeley National Laboratory (LBNL), Berkeley, CA, USA

**S. Viana**
AEMET, Barcelona, Spain

**D. Bellafiore**
ISMAR-CNR, Institute of Marine Science - National Research Council, Venice, Italy

**E. Bucchignani**
CIRA Italian Aerospace Research Centre, Capua, Italy
CMCC Centro Euro-Mediterraneo per i Cambiamenti Climatici, Bologna, Italy

**S. Gualdi**
CMCC Centro Euro-Mediterraneo per i Cambiamenti Climatici, Bologna, Italy

**S. Carniel**
ISMAR-CNR, Institute of Marine Science - National Research Council, Venice, Italy

**V. Djurdjeviæ**
SEE/VCCC - Institute of Meteorology, Faculty of Physics, University of Belgrade, Serbia

**G. Umgiesser**
ISMAR-CNR, Institute of Marine Science - National Research Council, Venice, Italy
Coastal Research and Planning Institute (CORPI), Klaipeda University, Klaipeda, Lithuania

**C. Agnese**
Università degli Studi di Palermo, Dip. SAGA, Palermo, Italy

**G. Baiamonte**
Università degli Studi di Palermo, Dip. SAGA, Palermo, Italy

**C. Cammalleri**
Università degli Studi di Palermo, DICAM, Palermo, Italy

**D. Cat Berro**
Società Meteorologica Italiana, Castello Borello, Bussoleno (TO), Italy

**S. Ferraris**
Università degli Studi di Torino, DIST, Torino, Italy

**L. Mercalli**
Società Meteorologica Italiana, Castello Borello, Bussoleno (TO), Italy

**S. Alessandrini**
RSE, Ricerca Sistema Energetico, Milano, Italy

**S. Sperati**
RSE, Ricerca Sistema Energetico, Milano, Italy

**P. Pinson**
Technical University of Denmark (DTU), DTU Informatics, Lyngby, Denmark

**S. Federico**
ISAC-CNR, UOS of Lamezia Terme, Lamezia Terme (CZ), Italy
CRATI, Rende (CS), Italy

**E. Avolio**
ISAC-CNR, UOS of Lamezia Terme, Lamezia Terme (CZ), Italy

**F. Fusto**
ARPACal, Catanzaro, Italy

**R. Niccoli**
ARPACal, Catanzaro, Italy

**C. Bellecci**
CRATI, Rende (CS), Italy
Engeenering Department, University of "Tor Vergata", Rome, Italy

**C. Bertrand**
Royal Meteorological Institute of Belgium, Brussels, Belgium

**L. González Sotelino**
Royal Meteorological Institute of Belgium, Brussels, Belgium

**M. Journée**
Royal Meteorological Institute of Belgium, Brussels, Belgium

**A. Obregón**
Deutscher Wetterdienst, WMO RA VI Regional Climate Centre Network Offenbach Node on Climate Monitoring, Frankfurter Str. 135, 63067 Offenbach, Germany

**H. Nitsche**
Deutscher Wetterdienst, WMO RA VI Regional Climate Centre Network Offenbach Node on Climate Monitoring, Frankfurter Str. 135, 63067 Offenbach, Germany

**M. Körber**
Deutscher Wetterdienst, WMO RA VI Regional Climate Centre Network Offenbach Node on Climate Monitoring, Frankfurter Str. 135, 63067 Offenbach, Germany

**A. Kreis**
Deutscher Wetterdienst, WMO RA VI Regional Climate Centre Network Offenbach Node on Climate Monitoring, Frankfurter Str. 135, 63067 Offenbach, Germany

**P. Bissolli**
Deutscher Wetterdienst, WMO RA VI Regional Climate Centre Network Offenbach Node on Climate Monitoring, Frankfurter Str. 135, 63067 Offenbach, Germany

**K. Friedrich**
Deutscher Wetterdienst, WMO RA VI Regional Climate Centre Network Offenbach Node on Climate Monitoring, Frankfurter Str. 135, 63067 Offenbach, Germany

**S. Rösner**
Deutscher Wetterdienst, WMO RA VI Regional Climate Centre Network Offenbach Node on Climate Monitoring, Frankfurter Str. 135, 63067 Offenbach, Germany

**I. Antolović**
Faculty of Electronic Engineering, Niš, Serbia

**V. Mihajlović**
Faculty of Electronic Engineering, Niš, Serbia

**D. Rančić**
Faculty of Electronic Engineering, Niš, Serbia

**D. Mihić**
Hydrometeorological Service of Serbia, Belgrade, Serbia

**V. Djurdjević**
Hydrometeorological Service of Serbia, Belgrade, Serbia

**M. A. Wlasak**
Met Office, FitzRoy Road, Exeter, Devon, EX1 3PB, UK

**M. J. P. Cullen**
Met Office, FitzRoy Road, Exeter, Devon, EX1 3PB, UK

**T. Kabas**
Wegener Center for Climate and Global Change (WEGC), and Institute for Geophysics, Astrophysics and Meteorology/Institute of Physics (IGAM/IP), University of Graz, Graz, Austria

**A. Leuprecht**
Wegener Center for Climate and Global Change (WEGC), and Institute for Geophysics, Astrophysics and Meteorology/Institute of Physics (IGAM/IP), University of Graz, Graz, Austria

**C. Bichler**
Wegener Center for Climate and Global Change (WEGC), and Institute for Geophysics, Astrophysics and Meteorology/Institute of Physics (IGAM/IP), University of Graz, Graz, Austria

**G. Kirchengast**
Wegener Center for Climate and Global Change (WEGC), and Institute for Geophysics, Astrophysics and Meteorology/Institute of Physics (IGAM/IP), University of Graz, Graz, Austria

**G. Szépszó**
Hungarian Meteorological Service, P.O. Box 38, 1525 Budapest, Hungary

**R. Blender**
University of Hamburg, KlimaCampus, Hamburg, Germany

**X. Zhu**
University of Hamburg, KlimaCampus, Hamburg, Germany

**K. Fraedrich**
University of Hamburg, KlimaCampus, Hamburg, Germany

**P. Domonkos**
Center for Climate Change, Univ. Rovira i Virgili, Campus Terres de l'Ebre, Tortosa, Spain

**V. Venema**
Meteorological institute of the University of Bonn, Germany

**I. Auer**
Zentralanstalt für Meteorologie und Geodynamik, Wien, Austria

**O. Mestre**
Meteo France, Ecole Nationale de la Meteorologie, Toulouse, France

**M. Brunetti**
Institute of Atmospheric Sciences and Climate (ISAC-CNR), Bologna, Italy

**G. Seiz**
Swiss GCOS Office, Federal Office of Meteorology and Climatology MeteoSwiss, Switzerland

**N. Foppa**
Swiss GCOS Office, Federal Office of Meteorology and Climatology MeteoSwiss, Switzerland

**M. Journée**
Royal Meteorological Institute of Belgium, Brussels, Belgium

**C. Demain**
Royal Meteorological Institute of Belgium, Brussels, Belgium

**C. Bertrand**
Royal Meteorological Institute of Belgium, Brussels, Belgium

**C. Bertrand**
Royal Meteorological Institute of Belgium, Brussels, Belgium

**L. Gonzalez Sotelino**
Royal Meteorological Institute of Belgium, Brussels, Belgium

**M. Journée**
Royal Meteorological Institute of Belgium, Brussels, Belgium

**A. Kann**
Central Institute for Meteorology and Geodynamics, Vienna, Austria

**G. Pistotnik**
Central Institute for Meteorology and Geodynamics, Vienna, Austria

**B. Bica**
Central Institute for Meteorology and Geodynamics, Vienna, Austria

**S. Siegert**
Max Planck Institute for the Physics of Complex Systems, Dresden, Germany

**J. Bröcker**
Max Planck Institute for the Physics of Complex Systems, Dresden, Germany
visiting research fellow: Centre for the Analysis of Time Series, London School of Economics, London, UK

**H. Kantz**
Max Planck Institute for the Physics of Complex Systems, Dresden, Germany